Growth *and* Mineral Nutrition *of* Field Crops

BOOKS IN SOILS, PLANTS, AND THE ENVIRONMENT

Soil Biochemistry, Volume 1, edited by A. D. McLaren and G. H. Peterson
Soil Biochemistry, Volume 2, edited by A. D. McLaren and J. Skujiņš
Soil Biochemistry, Volume 3, edited by E. A. Paul and A. D. McLaren
Soil Biochemistry, Volume 4, edited by E. A. Paul and A. D. McLaren
Soil Biochemistry, Volume 5, edited by E. A. Paul and J. N. Ladd
Soil Biochemistry, Volume 6, edited by Jean-Marc Bollag and G. Stotzky
Soil Biochemistry, Volume 7, edited by G. Stotzky and Jean-Marc Bollag
Soil Biochemistry, Volume 8, edited by Jean-Marc Bollag and G. Stotzky
Soil Biochemistry, Volume 9, edited by G. Stotzky and Jean-Marc Bollag

Organic Chemicals in the Soil Environment, Volumes 1 and 2, edited by C. A. I. Goring and J. W. Hamaker
Humic Substances in the Environment, M. Schnitzer and S. U. Khan
Microbial Life in the Soil: An Introduction, T. Hattori
Principles of Soil Chemistry, Kim H. Tan
Soil Analysis: Instrumental Techniques and Related Procedures, edited by Keith A. Smith
Soil Reclamation Processes: Microbiological Analyses and Applications, edited by Robert L. Tate III and Donald A. Klein
Symbiotic Nitrogen Fixation Technology, edited by Gerald H. Elkan
SoilWater Interactions: Mechanisms and Applications, Shingo Iwata and Toshio Tabuchi with Benno P. Warkentin
Soil Analysis: Modern Instrumental Techniques, Second Edition, edited by Keith A. Smith
Soil Analysis: Physical Methods, edited by Keith A. Smith and Chris E. Mullins
Growth and Mineral Nutrition of Field Crops, N. K. Fageria, V. C. Baligar, and Charles Allan Jones
Semiarid Lands and Deserts: Soil Resource and Reclamation, edited by J. Skujiņš
Plant Roots: The Hidden Half, edited by Yoav Waisel, Amram Eshel, and Uzi Kafkafi
Plant Biochemical Regulators, edited by Harold W. Gausman
Maximizing Crop Yields, N. K. Fageria

Transgenic Plants: Fundamentals and Applications, edited by Andrew Hiatt
Soil Microbial Ecology: Applications in Agricultural and Environmental Management, edited by F. Blaine Metting, Jr.
Principles of Soil Chemistry: Second Edition, Kim H. Tan
Water Flow in Soils, edited by Tsuyoshi Miyazaki
Handbook of Plant and Crop Stress, edited by Mohammad Pessarakli
Genetic Improvement of Field Crops, edited by Gustavo A. Slafer
Agricultural Field Experiments: Design and Analysis, Roger G. Petersen
Environmental Soil Science, Kim H. Tan
Mechanisms of Plant Growth and Improved Productivity: Modern Approaches, edited by Amarjit S. Basra
Selenium in the Environment, edited by W. T. Frankenberger, Jr., and Sally Benson
Plant–Environment Interactions, edited by Robert E. Wilkinson
Handbook of Plant and Crop Physiology, edited by Mohammad Pessarakli
Handbook of Phytoalexin Metabolism and Action, edited by M. Daniel and R. P. Purkayastha
Soil–Water Interactions: Mechanisms and Applications, Second Edition, Revised and Expanded, Shingo Iwata, Toshio Tabuchi, and Benno P. Warkentin
Stored-Grain Ecosystems, edited by Digvir S. Jayas, Noel D. G. White, and William E. Muir
Agrochemicals from Natural Products, edited by C. R. A. Godfrey
Seed Development and Germination, edited by Jaime Kigel and Gad Galili
Nitrogen Fertilization in the Environment, edited by Peter Edward Bacon
Phytohormones in Soils: Microbial Production and Function, W. T. Frankenberger, Jr., and Muhammad Arshad
Handbook of Weed Management Systems, edited by Albert E. Smith
Soil Sampling, Preparation, and Analysis, Kim H. Tan
Soil Erosion, Conservation, and Rehabilitation, edited by Menachem Agassi
Plant Roots: The Hidden Half, Second Edition, Revised and Expanded, edited by Yoav Waisel, Amram Eshel, and Uzi Kafkafi
Photoassimilate Distribution in Plants and Crops: Source–Sink Relationships, edited by Eli Zamski and Arthur A. Schaffer
Mass Spectrometry of Soils, edited by Thomas W. Boutton and Shinichi Yamasaki
Handbook of Photosynthesis, edited by Mohammad Pessarakli
Chemical and Isotopic Groundwater Hydrology: The Applied Approach, Second Edition, Revised and Expanded, Emanuel Mazor
Fauna in Soil Ecosystems: Recycling Processes, Nutrient Fluxes, and Agricultural Production, edited by Gero Benckiser
Soil and Plant Analysis in Sustainable Agriculture and Environment, edited by Teresa Hood and J. Benton Jones, Jr.
Seeds Handbook: Biology, Production, Processing, and Storage: B. B. Desai, P. M. Kotecha, and D. K. Salunkhe

Modern Soil Microbiology, edited by J. D. van Elsas, J. T. Trevors, and E. M. H. Wellington

Growth and Mineral Nutrition of Field Crops: Second Edition, N. K. Fageria, V. C. Baligar, and Charles Allan Jones

Fungal Pathogenesis in Plants and Crops: Molecular Biology and Host Defense Mechansims, P. Vidhyasekaran

Plant Pathogen Detection and Disease Diagnosis, P. Narayanasamy

Additional Volumes in Preparation

Agricultural Systems Modeling and Simulation, edited by Robert M. Peart and R. Bruce Curry

Agricultural Biotechnology, edited by Arie Altman

Plant–Microbe Interactions and Biological Control, edited by Gregory J. Boland and L. David Kuykendall

Sulfur in the Environment, edited by Doug Maynard

Handbook of Soil Conditioners, edited by Arthur Wallace and Richard E. Terry

SECOND EDITION,
REVISED AND EXPANDED

Growth *and* Mineral Nutrition *of* Field Crops

N. K. FAGERIA

National Rice and Bean Research Center
Empresa Brasileira de Pesquisa Agropecuária (EMBRAPA)
Goiânia-Goias, Brazil

V. C. BALIGAR

Appalachian Soil and Water Conservation Research Laboratory
Agricultural Researh Service
United States Department of Agriculture
Beckley, West Virginia

CHARLES ALLAN JONES

Texas Agricultural Experiment Station
Texas A&M University System
College Station, Texas

MARCEL DEKKER, INC. NEW YORK • BASEL • HONG KONG

Library of Congress Cataloging-in-Publication Data

Fageria, N. K.
Growth and mineral nutrition of field crops / N. K. Fageria, V. C. Baligar, Charles Allan Jones. — 2nd ed., rev. and expanded.
p. cm. — (Books in soils, plants, and the environment ; v. 57)
Includes bibliographical references and index.
ISBN: 0-8247-0089-9 (hardcover : alk. paper)
1. Field crops—Growth. 2. Field crops—Nutrition. I. Baligar, V. C. II. Jones, C. Allan. III. Title. IV. Series.
SB185.5.F34 1997
631. 8—dc21 97-13975
CIP

The publisher offers discounts on this book when ordered in bulk quantities. For more information, write to Special Sales/Professional Marketing at the address below.

This book is printed on acid-free paper.

MARCEL DEKKER, INC.
270 Madison Avenue, New York, New York 10016
http:/www.dekker.com

Current printing (last digit)
10 9 8 7 7 6 5 4 3 2 1

PRINTED IN THE UNITED STATES OF AMERICA

Preface to the Second Edition

The main aim of this revised edition is to cover soil-plant-climatic relationships that govern the growth and mineral nutrition of most of the important crops of temperate and tropical regions. Crop growth and mineral nutrition in agricultural science are dynamic processes that need to be updated regularly to take into account the new information generated through scientific experimentation. The structure of this book has not been altered; however, the sequence of chapters has been changed and a new chapter relating to nutrient management in degraded soils has been added. The contents of each chapter have been revised, and about 90 new figures have been added. Further, about 20% of the references are new to reflect recent research advances. We have tried to cite references from different regions of the world to provide a broader perspective of the various subjects covered. Mostly the discussion is supported by experimental results to make the book as practical as possible. It is our opinion that this edition is complete in all respects, and it should prove to be a valuable reference tool to assist researchers, professors, extension personnel, and students who are involved in the areas of soil fertility, plant nutrition, crop breeding, crop physiology, and crop production.

Preparing a book of this nature involves the assistance and cooperation of many people, to whom we are especially grateful. Our thanks to Iris Lilly for her excellent assistance in typing the manuscript and Susan Boyer for her help in drafting the figures.

Many scientific societies and publishers have granted permission to reproduce figures and texts from copyrighted material, and we sincerely thank all of them. Dr. N. K. Fageria expresses his sincere thanks to the National Rice and Bean Research Center of EMBRAPA, Brazil, for granting special leave during the preparation of the second edition.

Last, but not least, our heartfelt thanks to our families, who not only have been patient with us, but have encouraged us to complete the task of writing this book while we spent many, many hours at home on weekends and during weekday evenings preparing this book.

N. K. Fageria
V. C. Baligar
Charles Allan Jones

Preface to the First Edition

Although dramatic production gains have been made in the last 20 years, particularly from cereals, average yields of most field crops are still much below their potential. Potential yields can only be achieved with ideal crop cultivars under ideal management in an optimal physical and chemical environment. Supplying optimal quantities of mineral nutrients to growing crop plants is one way to improve crop yields. This book covers most of the important temperate and tropical crops as well as soil–plant relationships that govern crop growth and mineral nutrrition. Special emphasis has been given to soil as the substrate for plant growth. New concepts and principles developed by soil scientists, agrononomists, and physiologists are discussed. A special feature of this book is the interpretation of these principles in terms of modern agricultural practices. Most of the discussions are supported by experimental evidence to make the book as practical as possible. Important cereal, legume, and pasture crops are covered in a single book to provide the basic and applied information needed by agricultural specialists to improve the management of field crops. We sincerely hope that this book will serve as a textbook for agronomy students, a guide for extension, and a resource for research scientists involved in the areas of soil fertility, plant nutrition, breeding, and crop production.

Preparing a book of this nature involves the assistance and cooperation of many people, to whom we are especially grateful. Outstanding scientists selected to review the first draft of this book include Dr. R. J. Wright, Soil Scientist, USDA/ARS, Beckley, West Virginia; Dr. W. R. Kussow, Professor of Soil Science, University of Wisconsin, Madison; Dr. G. E. Wilcox, Professor of Horticulture, Purdue University, West Lafayette, Indiana; Dr. Thomas E. Staley, Microbioloigst, USDA/ARS, Beckley, West Virginia; Dr. Dale Richey, Soil Scientist, USDA/ARS, Mayaguez, Puerto Rico; Dr. J. R. Kiniry, Crop Physiologist, USDA/ARS, Temple, Texas; Dr. T. J. Gerik, Crop Physiologist, Texas Agricultural Experiment Station, Texas A&M University, Temple, Texas; and Dr. J. R. Williams, Hydrologic Engineer, USDA/ARS, Temple, Texas. These reviewers made valuable suggestions for improving the quality of the book and the authors gratefully acknowledge their contributions.

Our thanks to Antoinette Erickson, Shirley Athey and Pam Vansa for their excelent assistance in typing the text, to Susan Boyer and Norman Eskin for their help in drafting the figures, and to Sheba Wheeler for assistance in reading the proofs.

Many scientific societies and publishers have granted permission to reproduce figures and tables from copyright material, and we sincerely thank all of them. Among these are American Society of Agronomy, Soil Science Society of America, Crop Science Society of America, Brazilian Society of Soil Science, EMBRAPA (Brazil), International Potash Institute (Switzerland), CIAT (Columbia), IRRI (Phillipines), ICRASAT (India), Association of Applied Biologists (England), The Cotton Foundation of USA, Academic Press, John Wiley & Sons, Springer-Verlag, Elsevier Science Publishers, McGraw-Hill Book Company, The Iowa State University Press, The Kluwer Academic Publishers, *Plant & Soil*, Praeger Publishers, *Canadian Journal of Soil Science*, and *Euphytica*. We also appreciate the permission given by several authors to reproduce their figures and tables.

The authors wish to thank Dr. R. Paul Murrman, Director of USDA/ARS, Appalachian Soil and Water Conservation Research Laboratory; Dr. Clarence Richardson, USDA/ARS, Grassland, Soil and Water Research Laboratory, and Texas Agricultural Experiment Station, Blackland Research Center, for providing the necessary facilities to write this book. Dr. N. K. Fageria expresses his sincere thanks to the National Rice and Bean Research Center of EMBRAPA, Brazil, for study leave and to the Brazilian Scientific and Technological Research Council (CNPq) for providing a fellowship.

Last, but not least, our heartfelt thanks to our families, who not only have been patient with us, but have encouraged us to complete the task of writing this book.

N. K. Fageria
V. C. Baligar
Charles Allan Jones

Contents

1

Field Crops and Mineral Nutrition

I. INTRODUCTION

Economic development of modern society depends on field crops as materials necessary, directly or indirectly, for human consumption. Supplying adequate amounts of mineral nutrients to crops is one of the most important factors in achieving higher productivity. This book is a broad treatise on the mineral nutrition of field crops.

II. FIELD CROPS

Field crops, often referred to as agronomic crops, are crops grown on a large scale for human consumption, livestock feeding, or raw materials for industrial products. Crops are grouped or classified on the basis of their botanical characteristics or of their utility, or both. It is by these criteria that field crops are commonly grouped into cereals, pulses or grain legumes, fiber crops, oil crops, root crops, and rubber crops (Donald and Hamblin, 1983). The most important food crops are cereals and feed grains, such as wheat, rice, corn, barley, oats, and rye, and an assortment of legumes. Feed crops include several of the same species of plants used for food, as well as forage and pasture crops and in many instances food and fiber crop residues from some type of processing. Forages, when preserved by anaerobic fermentation, are termed silage crops; when dried, they are called hay crops.

The cereals such as wheat, rice, and corn dominate as food crops. Their total global production is much higher than that of other food crops. These crops are dominant primarily because more breeding work has been done for them and secondarily because they are grown mostly on better and usually irrigated soils. Also, these crops occupy a large area in major parts of the world. The dominance of cereals is increasing because they tend to replace lower-yielding and less management-responsive crops. Nevertheless, excessive displacement of pulses by cereals can be undesirable, in terms of not only creating imbalance in human diets but also reducing soil fertility, increasing the incidence of pests and diseases, and disrupting the stability of farming systems.

World average yields for all the pulse crops are only about half of those for cereals and have shown a slower rate of increase in recent years. Record yields are also much lower. However, the lower biomass yields of pulses are offset in part by the higher calorie contents of proteins and lipids in the oilseed crops such as soybean, rape, and sunflower. From comparisons of the known energy requirements of the various metabolic pathways, nutritionally equivalent seed yields are 60 for rape seed, 70 for soybean, 90 for peas, and 100 for wheat and corn (Evans, 1980).

The higher protein content of pulse seeds appears to be related to early mobilization of protein out of leaves, leading to a decline in their capacity for further photosynthesis and low biomass yield. The growth habit of legumes, particularly their progressive flowering and seed setting compared with the synchronous flowering of cereal crops, may also reduce their yield potential (Evans, 1980). Finally, the growth cycle of legumes is generally shorter than that of cereals. This shorter vegetative growth period allows less time for nutrient uptake, which is why pulses typically require more fertile soils for high production than do cereals.

Among food crops, root crops such as potato, sweet potato, and cassava play an important role in supplying calories for the world population. It is generally thought that food produced by root crops is of inferior quality. The root crops are especially thought to contain less protein than grain crops. It is true that cereal grains generally contain 7–11% protein, whereas root crops contain only 0.4–2.8% (Vries et al., 1967). But for the determination of food quality, parameters other than protein must also be measured (Table 1.1). Rice, the most heavily consumed food crop in the world, does not compare favorably with root crops from a broad nutritional perspective (Table 1.1). It actually contains a little more protein than most root crops, but in other parameters of food quality the root crops are superior to rice. This is also

Table 1.1 Composition per 100 Calories Edible Portion of Some Important Grain and Root Crops

Crop	Protein (g)	Ca (mg)	Fe (mg)	Vitamin A (IU)	Thiamine (mg)	Riboflavin (mg)	Nicotinamide (mg)	Ascorbic acid (mg)
Rice	2.0	1.4	0.28	±0	0.02	0.01	0.28	0
Wheat	3.2	5.8	0.73	±0	0.09	0.02	0.58	0
Corn	2.8	3.3	0.69	30–170	0.09	0.04	0.55	0
Sorghum	2.9	9.0	1.26	±0	0.14	0.03	0.98	0
Sweet potato	0.35–2.5	21.9	0.90	0–3500	0.09	0.04	0.61	26
Cassava	0.46	16.3	0.65	±0	0.46	0.02	0.46	20
Yam	1.9	9.6	1.1	0–190	0.09	0.03	0.38	9

Source: Compiled from Platt, 1965.

true for the other grain crops. Therefore, it may be concluded that the grain crops are not superior to root crops in terms of food quality.

Root crops deserve more attention in breeding and selection programs insofar as they have far higher production of edible calories per day of vegetative growth than do cereals.

III. MINERAL NUTRITION

Mineral nutrition includes the supply, absorption, and utilization of essential nutrients for growth and yield of crop plants. No one knows with certainty when humans first incorporated organic substances, manures, or wood ashes as fertilizer in soil to stimulate plant growth. However, it is documented in writings as early as 2500 B.C. that humans recognized the richness and fertility of alluvial soils in valleys of the Tigris and Euphrates rivers (Tisdale et al., 1985). Forty-two centuries later, scientists were still trying to determine whether plant nutrients were derived from water, air, or soil ingested by plant roots. Early progress in the development of understanding of soil fertility and plant nutrition concepts was slow, although the Greeks and Romans made significant contributions in the years 800 to 200 B.C. (Westerman and Tucker, 1987). It was mainly to the credit of Justus von Liebig (1803–1873) that the scattered information concerning the importance of mineral elements for plant growth was collected and summarized and that mineral nutrition of plants was established as a scientific discipline (Marschner, 1983).

In 1840 Liebig published results from his studies on the chemical analysis of plants and the mineral contribution of soils. These studies initiated modern research on plant nutrition and highlighted the importance of individual minerals in stimulating plant growth. From these studies evolved the concept that individual minerals were limiting factors on the growth potential of plants (Sinclair and Park, 1993). These findings led to a rapid increase in the use of chemical fertilizers. By the end of the 19th century, large amounts of potash, superphosphate, and, later, inorganic nitrogen were used in agriculture and horticulture to improve plant growth, especially in Europe (Marschner, 1995). Notwithstanding these, it was not until the 20th century that the list of 16 essential elements was completed and the fundamental concepts of plant nutrition were developed. The quest for an understanding of plant nutrition is not yet complete, however (Glass, 1989).

Plants contain small amounts of 90 or more elements, but only 16 elements are known to be essential (Epstein, 1972; Fageria, 1984). Essential nutrients are divided into two groups on the basis of the quantity required by plants. Those required in large quantities are classified as macronutrients and those required in small amounts as micronutrients. Carbon, hydrogen, oxygen, nitrogen, phosphorus, potassium, calcium, magnesium, and sulfur are macronutrients. Micronutrients include iron, manganese, boron, zinc, copper, molybdenum, and chlorine. Sodium, silicon, and cobalt are beneficial for some plants but have not been established as essential elements for all higher plants (Mengel and Kirkby, 1982).

The macro- and micronutrient classification is simply based on the amount required. All nutrients are equally important for plant growth. If deficiency of any nutrient occurs in the growth medium, plant growth is adversely affected. Soil and plant analyses are the common practices to identify nutritional deficiencies in crop production. The best criterion, however, for diagnosing nutritional deficiencies in annual crops is through evaluation of crop responses to applied nutrients. If a given crop responds to an applied nutrient in a given soil, this means that the nutrient is deficient for that crop. The relative decrease in yield in the absence of a nutrient as compared to an adequate soil fertility level can give an idea of the magnitude of nutrient deficiency. A study was conducted at the National Rice and Bean Research Center of EMBRAPA, Goiania, Brazil, to provide evidence of which nutrient is most yield-limiting for annual crop production in an Oxisol (Figure 1.1).

The magnitude of yield reduction of five annual crops without application of 12 essential plant nutrients in an Oxisol varied from crop to crop and nutrient to nutrient. Phosphorus and calcium were the most yield-limiting nutrients among five crops tested except the upland rice, which was not sensitive to Ca deficiency. Among micronutrients, B followed by Zn were the most yield-limiting nutrients for common bean. Molybdenum and zinc were the most yield-limiting for upland rice, Zn, Cu, and Mn for soybean, Zn, B, and Mo for corn, and Mn was the most yield-limiting for the wheat crop. Among the

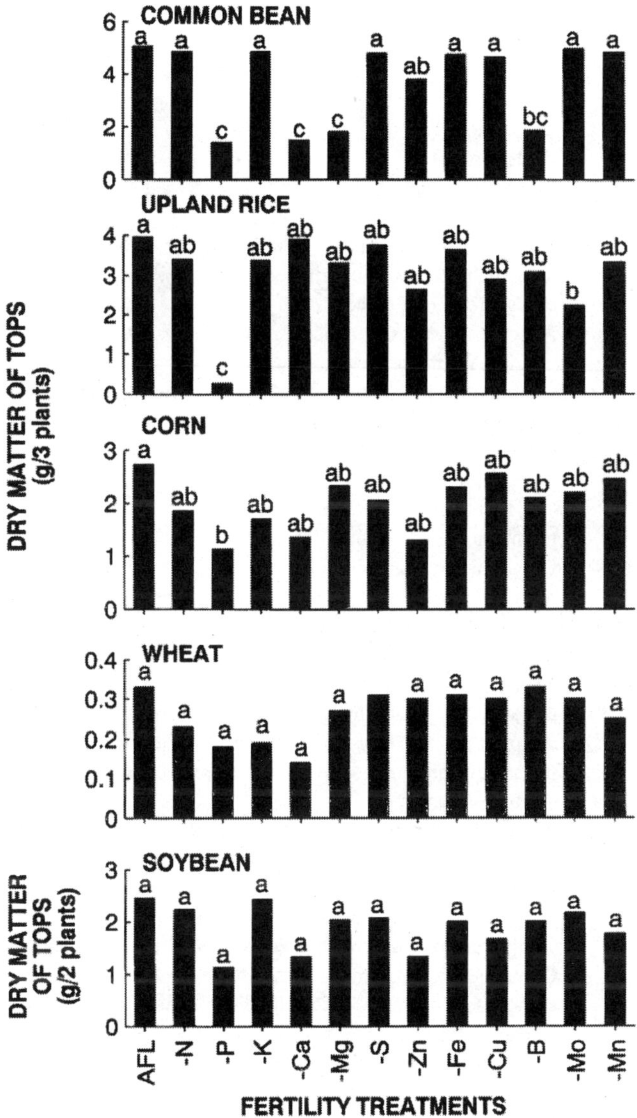

Figure 1.1 Response of five annual crop species to different fertility levels on an Oxisol. AFL = adequate fertility level, and minus (–) sign against each nutrient means without application of that nutrient. Different letters above each bar indicate significant differences between treatments by Tukey's test at 5% probability level.

Table 1.2 Functions of Essential Nutrients in Plants

Nutrient	Function
Carbon	Basic molecular component of carbohydrates, proteins, lipids, and nucleic acids.
Oxygen	Oxygen is somewhat like carbon in that it occurs in virtually all organic compounds of living organisms.
Hydrogen	Hydrogen plays a central role in plant metabolism. Important in ionic balance and as main reducing agent and plays a key role in energy relations of cells.
Nitrogen	Nitrogen is a component of many important organic compounds ranging from proteins to nucleic acids.
Phosphorus	Central role in plants is in energy transfer and protein metabolism.
Potassium	Helps in osmotic and ionic regulation. Potassium functions as a cofactor or activator for many enzymes of carbohydrate and protein metabolism.
Calcium	Calcium is involved in cell division and plays a major role in the maintenance of membrane integrity.
Magnesium	Component of chlorophyll and a cofactor for many enzymatic reactions.
Sulfur	Sulfur is somewhat like phosphorus in that it is involved in plant cell energetics.
Iron	An essential component of many heme and nonheme Fe enzymes and carriers, including the cytochromes (respiratory electron carriers) and the ferredoxins. The latter are involved in key metabolic functions such as N fixation, photosynthesis, and electron transfer.
Zinc	Essential component of several dehydrogenases, proteinases, and peptidases, including carbonic anhydrase, alcohol dehydrogenase, glutamic dehydrogenase, and malic dehydrogenase, among others.
Manganese	Involved in the O_2-evolving system of photosynthesis and is a component of the enzymes arginase and phosphotransferase.
Copper	Constituent of a number of important enzymes, including cytochrome oxidase, ascorbic acid oxidase, and laccase.
Boron	The specific biochemical function of B is unknown but it may be involved in carbohydrate metabolism and synthesis of cell wall components.
Molybdenum	Required for the normal assimilation of N in plants. An essential component of nitrate reductase as well as nitrogenase (N_2 fixation enzyme).
Chlorine	Essential for photosynthesis and as an activator of enzymes involved in splitting water. It also functions in osmoregulation of plants growing on saline soils.

Source: Compiled from Oertli, 1979; Ting, 1982; and Stevenson, 1986.

five crops tested, the susceptibility to P deficiency based on top dry matter yield was in the order of upland rice > common bean > soybean > corn > wheat. The order of susceptibility for Ca deficiency was common bean > wheat > corn > soybean > upland rice. This means upland rice was most tolerant to soil acidity, and common bean had the least tolerance among the crops tested.

Macronutrients are needed in concentrations of 1000 $\mu g\ g^{-1}$ of dry matter or more, whereas micronutrients are needed in tissue concentrations equal to or less than 100 $\mu g\ g^{-1}$ of dry matter (Oertli, 1979). The low requirement of plants for trace or micronutrients can be accounted for by the participation of these elements in enzymatic reactions and as constituents of growth hormones rather than as components of major plant products such as structural and protoplasmic tissue (Stevenson, 1986). Macronutrients play a major role in plant structure, whereas micronutrients are principally involved in enzymatic processes (Table 1.2).

In the literature, the term mineral nutrition is very common and is often used to refer to essential plant nutrients. This is a slight misnomer in that these elements, or plant nutrients, are not mineral. The term comes from the fact that most essential elements were combined with other elements in the forms of minerals, which eventually broke down into their component parts. Mineral nutrients include all essential plant nutrients other than carbon, hydrogen, and oxygen, which are derived from CO_2 and H_2O, and nitrogen which originally came from atmospheric N_2 (Bennett, 1993).

Essential plant nutrients can also be classified as metals or nonmetals. The group metals includes K, Ca, Mg, Fe, Zn, Mn, Cu, and Mo. The nonmetals include N, P, S, B, and Cl (Bennett, 1993).

According to Mengel and Kirkby (1982), classification of plant nutrients based on their biochemical behavior and their physiological functions seems more appropriate. Based on such a physiological approach, plant nutrients may be divided into the following four groups:

Group 1: C, H, O, N, and S. These nutrients are major constituents of organic material, involved in enzymic processes and oxidation-reduction reactions.

Group 2: P and B. These elements are involved in energy transfer reactions and esterification with native alcohol groups in plants.

Group 3: K, Ca, Mg, Mn, and Cl. This group plays osmotic and ion balance roles, plus more specific functions in enzyme conformation and catalysis.

Group 4: Fe, Cu, Zn, and Mo. Present as structural chelates or metalloproteins, these elements enable electron transport by valence change.

Even though some of these elements have been known since ancient times, their essentiality has been established only within the last century (Tamhane

Table 1.3 Essential Nutrients for Plant Growth, Their Principal Forms for Uptake, and Discovery

Nutrient	Chemical symbol	Principal forms for uptake	Year	Discovered essential to plants by
Carbon	C	CO_2	1882	J. Sachs
Hydrogen	H	H_2O	1882	J. Sachs
Oxygen	O	H_2O, O_2	1804	T. De Saussure
Nitrogen	N	NH_4^+, NO_3^-	1872	G. K. Rutherford
Phosphorus	P	$H_2PO_4^-$, HPO_4^{2-}	1903	Posternak
Potassium	K	K^+	1890	A. F. Z. Schimper
Calcium	Ca	Ca^{2+}	1856	F. Salm-Horstmar
Magnesium	Mg	Mg^{2+}	1906	Willstatter
Sulfur	S	SO_4^{2-}, SO_2	1911	Peterson
Iron	Fe	Fe^{2+}, Fe^{3+}	1860	J. Sachs
Manganese	Mn	Mn^{2+}	1922	J. S. McHargue
Boron	B	H_3BO_3	1923	K. Warington
Zinc	Zn	Zn^{2+}	1926	A. L. Sommer and C. B. Lipman
Copper	Cu	Cu^{2+}	1931	C. B. Lipman and G. MacKinney
Molybdenum	Mo	MoO_4^{2-}	1938	D. I. Arnon and P. R. Stout
Chlorine	Cl	Cl^-	1954	T. C. Broyer et al.

et al., 1966; Marschner, 1995). The discovery of essential nutrients, their chemical symbols, and their principal forms for uptake are presented in Table 1.3.

Now the question arises, "What are the criteria of essentiality of nutrients for plant growth?" Arnon and Stout (1939) and Arnon (1950) long ago proposed certain criteria of essentiality of mineral nutrients, and these criteria are still valid. According to these researchers, the essentiality of a nutrient is based on the following criteria:

1. Omission of the element in question results in abnormal growth, failure to complete the life cycle (i.e., from viable seeds), or premature death of the plant.
2. The element forms part of a molecule or constituent of the plant that is itself essential in the plant. Examples are N in protein and Mg in chlorophyll.
3. The element exerts its effect directly on growth or metabolism and not by some indirect effect such as antagonism of another element present at a toxic level.

IV. SUMMARY

Crops are plants grown on large scale for human consumption, livestock feeding, and raw materials for industrial products. On the basis of area planted and world production, the important field crops are wheat, rice, corn, barley, sorghum, potato, soybeans, sugarcane, sweet potatoes, cassava, sugar beet, rye, groundnuts, and cotton. This does not mean that other crops are not important. Supply of adequate amounts of essential nutrients is one of the most important factors in obtaining good yields of these crops. Sixteen nutrients are essential for plant growth. These are C, H, O, N, P, K, Ca, Mg, S, Zn, Cu, Fe, Mn, B, Mo, and Cl. The first three (C, H, O) make up about 95% of plant dry weight and are supplied to plants by air and water. Chlorine is also supplied through atmospheric air. Humans have to furnish the rest (12 nutrients) in adequate amounts to obtain good yields of crops if there are inadequate amounts in the growth medium. Requirements of mineral nutrients for important field crops are discussed in the chapters devoted to those crops.

REFERENCES

Arnon, D. I. 1950. Criteria of essentiality of inorganic micronutrients. *In*: W. D. McElroy and B. Glass (eds.). Trace elements in plant physiology. Chronica Botanica, Waltham, Massachusetts.

Arnon, D. I. and P. R. Stout. 1939. The essentiality of certain elements in minute quantity for plants with special reference to copper. Plant Physiol. 145: 371–375.

Bennett, W. F. 1993. Plant nutrient utilization and diagnostic plant symptoms, pp. 1–7. *In*: W. F. Bennett (ed.) Nutrient deficiencies and toxicities in crop plants. APS Press, The American Phytopathological Society, St. Paul, Minnesota.

Donald, C. M. and J. Hamblin. 1983. The convergent evolution of annual seed crops in agriculture. Adv. Agron. 36: 97–143.

Epstein, E. 1972. Mineral nutrition of plants: Principles and perspectives. Wiley, New York.

Evans, L. T. 1980. The natural history of crop yield. Am. Sci. 68: 388–397.

Fageria, N. K. 1984. Fertilization and mineral nutrition of rice. EMBRAPA-CNPAF/Editora Campus, Rio de Janeiro.

Glass, A. D. M. 1989. Plant nutrition: an introduction to current concepts. Jones and Bartlett Publishers, Boston.

Marschner, H. 1983. General introduction to the mineral nutrition of plants, pp. 5–60. *In*: A. Lauchli and R. L. Bieleski (eds.). Encyclopedia of plant physiology, New Ser., Vol. 15A. Springer-Verlag, New York.

Marschner, H. 1995. Mineral nutrition of higher plants. 2nd ed. Academic Press, New York.

Mengel, K. and E. A. Kirkby. 1982. Principles of plant nutrition. 3rd ed. International Potash Institute, Bern.

Oertli, J. J. 1979. Plant nutrients, pp. 382–385. *In*: R. W. Fairbridge and C. W. Finkl, Jr. (eds.). The encyclopedia of soil science, Part 1. Dowden, Hutchinson and Ross, Stroudsburg, Pennsylvania.

Platt, B. S. 1965. Tables of representative values of foods commonly used in tropical countries. Med. Res. Coun. Spec. Rep. Ser., London, No. 302, 46 pp.

Sinclair, T. R. and W. I. Park. 1993. Inadequacy of the Liebig limiting-factor paradigm for explaining varying crop yields. Agron. J. 85: 742–746.

Stevenson, F. J. 1986. The micronutrient cycle, pp. 321–367. *In*: Cycles of soil. Wiley, New York.

Tamhane, R. V., D. P. Motiramani, Y. P. Bali, and R. L. Donahue. 1966. Soils: Their chemistry and fertility in tropical Asia. Prentice-Hall of India, New Delhi.

Ting, I. P. 1982. Plant mineral nutrition and ion uptake, pp. 331–363. *In*: Plant physiology. Addison-Wesley, Reading, Massachusetts.

Tisdale, S. L., W. L. Nelson, and J. D. Beaton. 1985. Soil fertility—past and present, pp. 5–18. *In*: Soil fertility and fertilizers, 4th ed. Macmillan, New York.

Vries, C. A., J. D. Ferwerda, and M. Flach. 1967. Choice of food crops in relation to actual and potential production in the tropics. Neth. J. Agric. Sci. 15: 241–248.

Westerman, R. L. and T. C. Tucker. 1987. Soil fertility concepts: Past, present and future, pp. 171–179. *In*: Future developments in soil science research. Soil Sci. Soc. Am., Madison, Wisconsin.

2

Factors Affecting Production of Field Crops

I. INTRODUCTION

Many factors are recognized as limiting to production of field crops. Some factors, such as water, cultivars, nutrients, insects, and diseases, are subject to a measure of control, and most crop management practices are directed at balancing the levels of control to obtain maximum economic return. When such controls are successful and these factors are not limiting to the growth of plants suitably adapted to the prevailing climate, maximum productivity depends principally on rates of light interception and carbon dioxide assimilation by the crop surface (Loomis and Williams, 1963).

Before discussing factors affecting production of field crops, it is important to define the term crop productivity. Crop productivity or yield is the measurable produce of a crop (Zadoks and Schein, 1979). According to Westlake (1963), the weight of new organic matter created by photosynthesis over a period, expressed as a rate, becomes productivity. Crop productivity or yield is a function of environmental, plant, management, and socioeconomical factors and their interaction. Mathematically, crop yield can be expressed by the following equation:

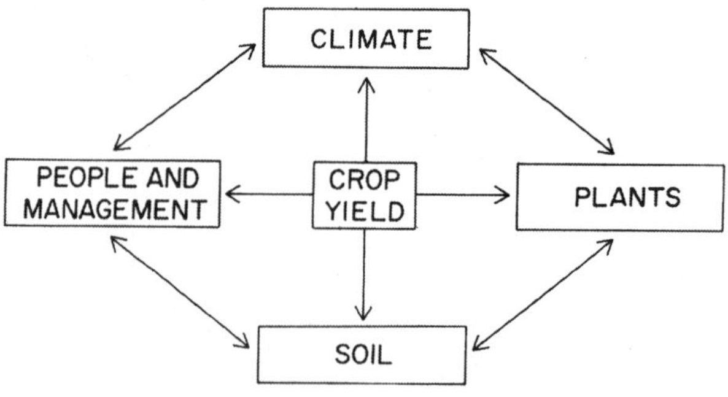

Figure 2.1 Factors affecting crop yield.

$$Y = f(E,P,M,S)$$

where Y = yield, E = environment, P = plant, M = management, and S = socioeconomic factors. Figure 2.1 shows factors affecting crop yield. Maximum yield of a crop in a given environment is possible only when all these factors are at optimum levels. If any one of these factors is limiting, crop yield will decrease.

Crop productivity is a very complex phenomenon and it is not an easy task to improve and/or stabilize it. In the past decade yields of important field crops have been improved through the use of improved cultivars, fertilizers, irrigation, fungicides, insecticides, and herbicides and improved cultural practices. All these can be classified as technological factors.

II. ENVIRONMENTAL FACTORS

The environment of a plant may be defined as the sum of all external forces and substances affecting the growth, structure, and reproduction of that plant (Billings, 1952). Crop environment is composed of climatic and soil factors which exert a great influence on plant growth and, consequently, yield. Climatic factors such as temperature, solar radiation, and moisture supply play an important role in crop production. Similarly, soil physical, chemical, and biological properties are directly related to crop productivity.

A. Climatic Factors

Climate is one of the most important factors determining the crops to be grown, nature of the natural vegetation, characteristics of the soils, and type of farming that can be practiced in a given agroecological region. From an agricultural point of view, there are two main types of climates: tropical and temperate. It is generally assumed that temperate climate means cold weather and tropical climate, hot weather. However, in temperate regions, temperature is not always lower than in tropical regions. On the basis of mean annual temperature, the boundaries between tropical and temperate climates, reported by various workers, varies from 18°C (64.4°F) to 23°C (73.4°F). Above this temperature range climate may be considered tropical and below this range may be considered temperate. One important characteristic of the tropical climate is temperature stability in most parts of the year. In the tropics, the mean monthly temperature variation is 5°C (41°F) or less between the average of the three warmest and the average of the three coldest months (Sanchez, 1976).

Temperature

Soil and air temperatures are important and often critical environmental factors for plant growth and productivity. In temperate regions, the temperature determines the length of the growing season, whereas in tropical regions temperature is more important for respiration than for photosynthesis. Under a tropical climate, temperature plays an important role in the supply of water and nutrients. Root temperatures are the critical temperatures as far as plant survival is concerned, and root temperatures are usually lower than air temperature during the growth period. The variation in the temperature of the root zone is less than that of the ambient air to which the tops of plants are subjected (Nielsen, 1974). The optimum temperature for maximum production of root material for several species ranges from 20 to 30°C (Voorhees et al., 1981). Figure 2.2 shows that rice yield had a strong negative relationship between grain yield and increasing canopy temperature from 30 to 38.5°C.

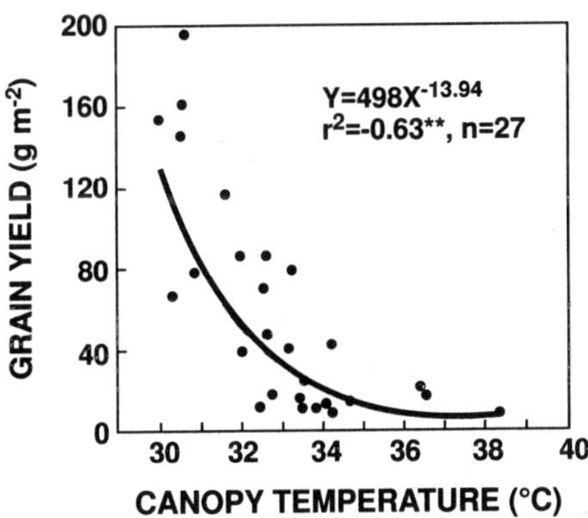

Figure 2.2 Relationship between rice grain yield and canopy temperature at 50% flowering. (From Garrity and O'Toole, 1995.)

Table 2.1 Optimum Soil Temperature for Maximum Yield of Important Field Crops

Crop	Optimal temperature (°C)	Reference
Barley (*Hordeum vulgare* L.)	18	Power et al., 1970
Oats (*Avena sativa* L.)	15–20	Case et al., 1964
Wheat (*Triticum aestivum* L.)	20	Whitfield and Smika, 1971
Corn (*Zea mays* L.)	25–30	Dormaar and Ketcheson, 1960
Cotton (*Gossypium hirsutum* L.)	28–30	Pearson et al., 1970
Potato (*Solanum tuberosum* L.)	20–23	Epstein, 1966
Rice (*Oryza sativa* L.)	25–30	Owen, 1971
Bean (*Phaseolus vulgaris* L.)	28	Mack et al., 1964
Soybean (*Glycine max* L. Merr.)	30	Voorhees et al., 1981
Sugar beet (*Beta vulgaris* L.)	24	Radke and Bauer, 1969
Sugarcane (*Saccharum officinarum* L.)	25–30	Hartt, 1965
Alfalfa (*Medicago sativa* L.)	28	Heinrichs and Nielsen, 1966

Optimum soil temperature for maximum yield of important field crops is presented in Table 2.1. Root temperature is influenced by intensity, quality, and duration of solar radiation; air temperature; surface vegetation; and color and thermal conductivity of the soil.

Reaction rates in plant processes are restricted at low temperatures and increased at higher temperatures. From a practical management standpoint, it appears that crop yield can be expected to benefit most from a soil temperature regime that encourages fast germination and good early growth. A common measure of growth rate in relation to temperature is Q_{10}. The temperature coefficient (Q_{10}) can be defined as the amount by which the process rate increases with a 10°C rise in temperature. Mathematically, it can be expressed as (Voorhees et al., 1981):

$$Q_{10} = \left(\frac{V_2}{V_1}\right)^{10/(T_2 - T_1)}$$

where V_2 and V_1 are the reaction rate at temperatures T_2 and T_1, respectively. Q_{10} may range from 1 for photochemical reactions unaffected by temperature to values as great as 6 for higher-temperature denaturation of enzymes (Voorhees et al., 1981).

Most food crops are suited to a particular type of climate such as tropical or temperate, cannot tolerate extremes in temperature, and have a narrow

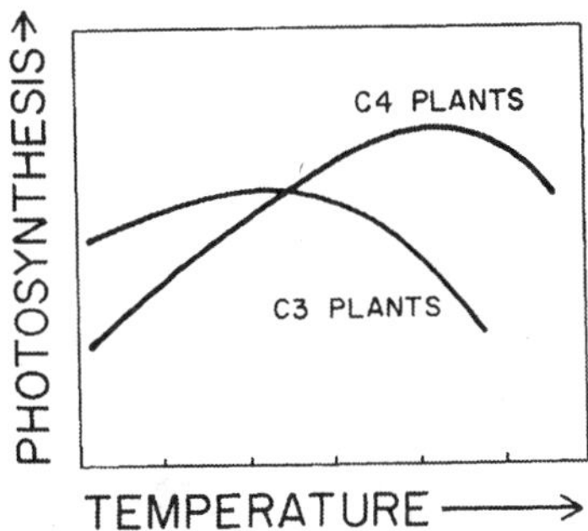

Figure 2.3 Relationship between temperature and photosynthesis in C3 and C4 plants. C3 plants have optimum temperature for photosynthesis around 25°C and C4 plants around 35°C.

temperature range for maximum growth. Although there are exceptions, C_4 plants are generally more tolerant of high temperatures and more sensitive to low temperatures than C_3 plants (Edwards et al., 1983). Figure 2.3 shows the relationship between temperature and photosynthesis in C_3 and C_4 plants. Many C_4 plants, like sugarcane and corn, are better able to grow under high temperatures than are C_3 plants such as wheat and barley. The tolerance of C_3 plants to low temperatures and thus their generally better photosynthetic performance under these conditions, compared to C_4 plants, may be due in part to differences in the levels of certain photosynthetic enzymes, the cold lability of some key enzymes of the C_4 pathway, and differences in the phase transition of lipids in membranes (Edwards et al., 1983).

Moisture Supply

Moisture availability is one of the most important factors determining crop production. The distribution of vegetation in an agroecological region is controlled more by the availability of water than by any other single factor. Water is required by plants for the translocation of mineral elements, for the manufacture of carbohydrates, and for the maintenance of hydration of protoplasm. Crop yield can be reduced at both very low and very high levels of moisture. Excess moisture reduces soil aeration and thus the supply of O_2 available to roots. With poor aeration, activities of beneficial microorganisms and water and nutrient uptake by plants are seriously inhibited. There may be exceptions with aquatic plants such as flooded rice. Severe drought can cause stomata in the leaf to close, reducing photosynthesis. Moisture stress causes reductions in both cell division and cell elongation and, hence, in growth. Figure 2.4 shows the relationship between sunflower head diameter and water supply.

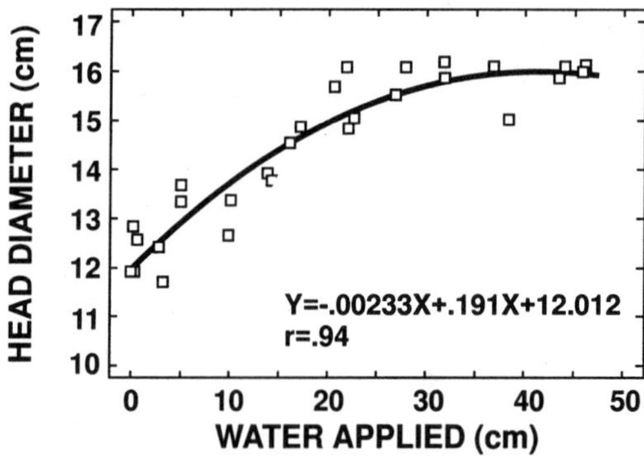

Figure 2.4 Relationship of sunflower head size to various amounts of water applied. (From Hang and Evans, 1985.)

Head size increased with the increase in irrigation water up to about 35 cm of water applied. It then leveled off as the amount of water applied increased up to 46 cm.

The supply of water (*W*), as expressed by the hydrological budget, is equal to precipitation (*P*) plus irrigation (*I*) and the change in storage (*S*), less runoff (*R*) and drainage (*D*) (Loomis, 1983):

$$W = P + I + S - R - D$$

The storage terms relate only to the portion of soil moisture available to the plant. Field capacity and wilting coefficient or permanent wilting point are the upper and lower limits of water availability for crops. The upper limit of water availability to plants is based on water contents after saturated soil has freely drained for 2–3 days or wetted soils have been subjected to pressures in the range from 5 to 30 kPa (kilopascal) or 0.05–0.3 bar (Unger et al., 1981). The lower value is generally applicable to light-textured soils and the higher value to heavy-textured soils.

Plants vary widely in efficiency of water use. The ratio of dry matter production to the amount of water transpired by a crop is known as water use efficiency. Generally, C_4 plants are about twice as efficient as C_3 plants in utilizing water. Efficient use of available water by these plants is due to a slightly higher stomatal resistance to gas exchange through the C_4 pathway of photosynthesis (Edwards et al., 1983).

Water stress is normally most severe in semiarid and arid regions because of long rainless periods and limited soil water storage. Plants under stress however, may also occur in subhumid and humid regions due to short-term drought. Distribution pattern as well as quantity of rainfall is important.

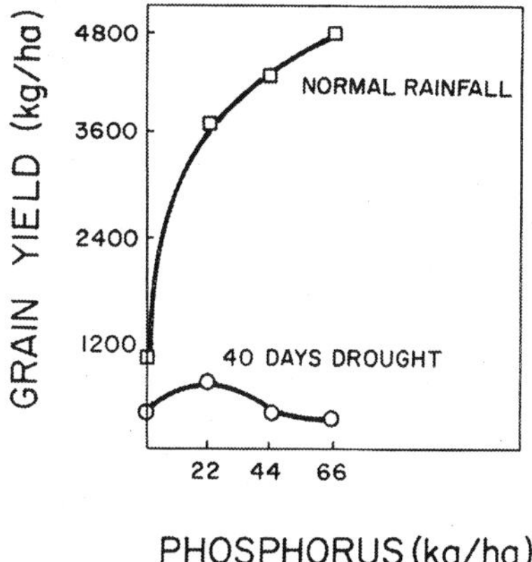

Figure 2.5 Upland rice yield influenced by phosphorus and rainfall in Oxisol of central Brazil. (From Fageria, 1980.)

Sometimes a region has plenty of rainfall but, due to the erratic distribution of the rainfall, crop failure is not uncommon. In the central part of Brazil, the average rainfall is about 1500 mm per year, sufficient to produce at least two crops if evenly distributed. But 2–3 weeks of drought is very common during the rainy season and sometimes results in complete crop failure (Fageria, 1980). Figure 2.5 shows the yield of upland rice in the region with increasing levels of P. It is clear from the figure that during normal rainfall distribution, a yield of about 5 t ha^{-1} was obtained at 66 kg P ha^{-1}. But for crops affected with drought for 40 days from the initiation of primordium floral to flowering, yield was reduced by about 90%. Most field crops are more sensitive to water stress during reproduction and grain-filling stages than the vegetative stage (Fageria, 1980). In cereals, the most sensitive growth stage for water deficiency is around flowering (Yoshida, 1972).

Solar Radiation

Solar radiation is an important climatic factor in plant growth and development. The duration of the solar radiation varies with the latitude and season. At the equator, day and night are of equal length all year. The greatest annual inputs of solar radiation occur in subtropical regions at 20–30° latitude under climates with little cloud cover and correspondingly low rainfall. In tropical humid regions there is comparatively little seasonal variation in energy input, and a steady value of 400–500 cal cm^{-2} per day is often experienced (Cooper, 1970). In the tropics and subtropics the net assimilation rate and relative growth rate can greatly exceed those recorded for cool regions, and these

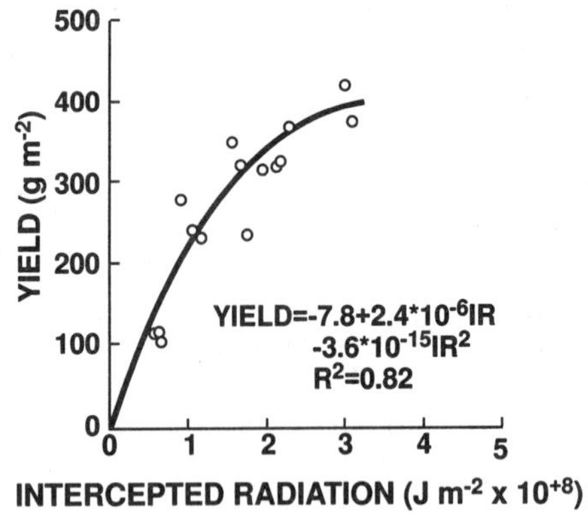

Figure 2.6 Yield of barley as a function of intercepted radiation at the grain-filling stage. (From Whitman et al., 1985.)

differences are attributed to the higher insolation and temperature (Blackman and Black, 1959).

Solar radiation affects photosynthesis and consequently crop productivity. Figure 2.6 shows solar radiation interception and grain yield of barley during grain filling. Grain yield increases in a quadratic fashion with increasing intercepted radiation. This figure also shows that grain yield in barley was found to be sink-limited rather than source-limited because the yields did not linearly increase with intercepted radiation. As radiation passes through the atmosphere to the earth's surface, much energy is lost by absorption and scattering caused by water vapor, dust, CO_2, and ozone. At the upper boundary of the atmosphere, and at the earth's mean distance from the sun, the total irradiance is 1360 J M^{-2} s^{-1}, which includes ultraviolet and infrared wavelengths (Salisbury and Ross, 1985). Approximately 900 J m^{-2} s^{-1} reaches plants, depending on latitude, time of day, elevation, and other factors (Salisbury and Ross, 1985). About half of the radiation is in the infrared region of the light spectrum, roughly 5% is in the ultraviolet, and the rest, approximately 400 J M^{-2} s^{-1}, is at wavelengths between 400 and 700 nm, which are capable of causing photosynthesis. This is called photosynthetically active radiation (McCree, 1981).

Dry matter accumulation can be considered as the product of the amount of photosynthetically active radiation absorbed (PAR_A) and the efficiency with which it is used (Monteith, 1977). The amount of radiation absorbed depends on canopy size (leaf area index) and the radiation extinction coefficient (k). PAR_A is related to leaf area index (LAI) and k via Beer's law (Goyne et al., 1993):

$$PAR_A + PAR_0\,[1 - \exp(-kLAI)]$$

where PAR_0 is the above-canopy or incident radiation.

Dry matter and PAR_A are linearly related when soil moisture and nutrients are not limiting (Gallagher and Biscoe, 1978). The slope of this relationship is referred to as radiation use efficiency (*RUE*) (Stockle and Kiniry, 1990). In reviewing factors affecting cereal yield, Green (1984) concluded that *RUE* was a relatively stable parameter and that most variation in biomass production was due to changes in the amount of radiation intercepted. Table 2.2 shows the growth rate and percent utilization of photosynthetically active radiation (*PAR*) by different crops. It is clear that crop growth rate is directly related to total radiation input and only about 4–10% of the *PAR* is utilized in the process of photosynthesis. Table 2.3 shows how climate affects input of solar radiation and dry matter production. Solar radiation is higher in tropical and subtropical climates than in temperate climates. Similarly, dry matter production is higher in tropical and subtropical climates than in temperate ones.

In temperate as well as tropical plant species, there are differences in light saturation level and hence in maximum photosynthesis. Appreciable variation in maximum photosynthetic rate occurs between varieties and even between individual genotypes and is often associated with differences in the leaf mesophyll structure or in the activity of the carboxylating enzymes. In the subtropical plant species such as corn and sugarcane, photosynthetic rates continue to increase in response to light intensities up to more than 60,000 lux, with maximum values of over 70 mg CO_2 dm^{-2} h^{-1}, equivalent to conversion rates of 5–6% of these high light intensities (Hesketh and Moss, 1963). For most tropical grasses, the production of 1 g of dry matter corresponds to the fixation of about 4130–5020 cal of chemically bound energy, and for most temperate grasses it corresponds to 4250 cal (Butterworth, 1964; Golley, 1961).

Radiation-use efficiency (*RUE*; grams of biomass accumulated, divided by total solar radiation intercepted) has proven to be a useful variable for quantifying biomass accumulation by crops (Sinclair et al., 1992).

Monteith (1977) presented an analysis of the relationship between the accumulation of crop dry matter and intercepted solar radiation and concluded that radiation-use efficiency is ≈ 1.4 g MJ^{-1} for many crops. Ensuing experiments confirmed a reasonable stability in *RUE* within crop species, although important differences among species were found. Kiniry et al. (1989) summarized a number of field experiments, as well as additional data of their own, and reported mean values of *RUE* of 1.6 g MJ^{-1} for maize, 1.3 g MJ^{-1} for sorghum [*Sorghum bicolor* (L.) Moench] and sunflower (*Helianthus annuus* L.), and 1.0 for rice (*Oryza sativa* L.) and wheat (*Triticum aestivum* L.). From a theoretical derivation of *RUE*, Sinclair and Horie (1989) showed that the

Table 2.2 Maximum Growth Rate and Solar Energy Utilization by Some Crops

Crop	Location	Growth rate ($g\ m^{-2}\ day^{-1}$)	Total radiation input ($cal\ cm^{-2}\ day^{-1}$)	Percent utilization of *PAR* [a] ($cal\ cm^{-2}\ day^{-1}$)	Reference
Barley	UK	23	484	4.3	Blackman and Black, 1959
Barley	Netherlands	17.7	450	3.7	Sibma, 1968
Wheat	Netherlands	17.5	450	3.7	Sibma, 1968
Corn	Netherlands	17.1	350	4.6	Sibma, 1968
Corn	Ithaca, NY	52	500	9.8	Wright and Lemon, 1966
Corn	Davis, CA	52	736	6.4	Williams et al., 1968
Corn	New Zealand	29.2	450	6.1	Brougham, 1960
Sorghum	Davis, CA	51	690	6.7	Loomis and Williams, 1963
Potato	Netherlands	23	400	5.4	Sibma, 1968
Pearl millet	Australia	54	510	9.5	Begg, 1965

[a] *PAR* = Photosynthetically active radiation, 400 to 700 nm.

Table 2.3 Estimated Annual Energy Input and Potential Conversion in Different Climates

Climate	Location	Input of total radiation (kcal cm^{-2} $year^{-1}$)	Dry matter from 3% conversion of incoming light[a] (t ha^{-1} $year^{-1}$)
Tropical	Puerto Rico	160	51
	Singapore	155	49
	Lagos, Nigeria	130	41
	Townsville, Australia	180	57
	Hawaii	155	49
Subtropical	Algiers	165	52
	Davis, CA	160	51
	Buenos Aires, Argentina	145	46
	Brisbane, Australia	170	54
Temperate	Wageningen, Netherlands	90	9
	Wellington, New Zealand	115	37
	Wisconsin	120	38

[a]Assuming 1 g DM = 4250 cal.
Source: Compiled from Cooper, 1970.

differences in *RUE* among species should be expected. In addition, they showed that *RUE* will vary within a species, depending on the light-saturated rate of leaf photosynthesis.

B. Soil

Soil is the unconsolidated mineral material on the immediate surface of the earth that serves as a natural medium for plant growth. Better yield of crops can be expected only when conditions are optimal or favorable in the growth medium. Optimal conditions for plant growth are difficult to define because they vary with plant species, type of soil, and agroclimatic region. Some conditions or factors which affect plant growth can be related to soil physical, chemical, and biological properties. These factors, directly or indirectly, affect plant root growth, absorption of water and nutrients, and, consequently, plant growth and yields. How these factors affect plant growth is discussed in this section.

Physical Properties

Physical properties are the characteristics, processes, or reactions of a soil which are caused by physical forces and which can be described by or expressed in physical or chemical equations. Soil physical properties play an important role in the growth and development of plants. The important soil physical properties which affect plant growth are texture, structure, consistency, pore space and density, soil tilth, and soil color. These physical properties are interrelated, and a change in one may cause a change in others that may be favorable or adverse. The creation of favorable soil physical conditions for plant growth is a very complex phenomenon.

Texture. Soil texture refers to the relative proportions of various soil separates such as sand, silt, and clay in a soil. Most soils are mixtures of particles of various sizes. To facilitate the description of these mixtures, classes have been defined according to relative proportions of sand, silt, and clay. The most common classification used in agriculture is that given by the U.S. Department of Agriculture (USDA) (Soil Survey Staff, 1951). According to this classification, soil material that contains 85% or more of sand and a percentage of silt plus 1.5 times clay that does not exceed 15 is known as sand. Silt is the soil material that contains 85% or more silt and less than 12% clay. Clay is the soil material that contains 40% or more clay, less than 45% sand, and less than 40% silt. Sand separates have diameters of 0.05–2 mm, silt 0.05–0.002 mm, and clay less than 0.002 mm. Particles greater than 2 mm in diameter are known as coarse fragments.

Soil texture affects productivity of crops in several ways. It affects the water-holding capacity of soil, aeration, temperature, cation exchange capacity, nutrient-supplying power, and hence growth and production. The general tendency is for productivity to be better on medium-textured soils than on soils that are either light or heavy. Soil texture is relatively permanent soil property and is little influenced by tillage or other manipulation unless the modification is drastic. It can be altered by soil loss through erosion or by deposition of new materials from wind or water.

Structure. The binding of soil particles into aggregates results in structure. Soil structure, in combination with texture, governs the porosity of the soil and thus affects aeration, water infiltration, root penetration, and microbiological activities of soil flora and fauna. Primary factors in the development of soil structure are shrink and swell phenomena during wetting and drying. Pressure is also exerted by plant roots. Soil separates are bonded together by clays, iron, aluminum compounds, and organic substances such as humus, polyuronides, polysaccharides, and proteins (Baver et al., 1972). Soil structure plays an important role in plant growth and consequently in crop production.

Soil must have favorable structure for high productivity. A good soil structure provides adequate aeration and drainage, sufficient water storage capacity, good root growth, and access to nutrients (Russell, 1973). No one structure is completely ideal, however, because requirements differ among crop plants. Moreover, crops will generally grow satisfactorily over a range of structural conditions (Low, 1979). Soil structure can be modified much more readily through cultural practices than can texture.

Soil structure can be extremely important to root growth in fine-textured soils, but soil strength usually is more important than soil structure in sandy soils. Soil strength is defined as the ability or capacity of a particular soil in a particular condition to resist or endure an applied force (Gill and Vandenberg, 1967).

Consistency. Soil consistency is the degree of cohesion or adhesion of the soil mass. Cohesion and adhesion, which are surface phenomena, are largely a function of the clay or organic matter contents in soil and the structural state of soil. The importance of soil consistency in agriculture is related to the stability of soil structure, the suitability of soil for plowing, and its susceptibility to erosion.

Pore Space and Density. Pore space is the total space of soil not occupied by soil particles, whereas density is the mass per unit volume including pore space. Soil density, which includes pore space, is known as bulk density. If the mass of a soil, as determined by weighing, is divided by the measured volume of the solids making up the soil, a value expressing the density of the solids or the particle density is obtained (Hausenbuiller, 1972). Soil structure, to a large extent, determines the bulk density of a soil. As a rule, the higher the bulk density, the more compact the soil, the more poorly defined the structure, and the smaller the amount of pore space. Bulk density is really a measure of pore space in the soil. The higher the bulk density for a given textural class, the smaller the amount of pore space present.

The particle density varies widely, but the types most prevalent in the soil have density values in the range of 2.6–2.7 g cm^{-3}, with an average of about 2.65 g cm^{-3}.

Information on particle density is needed for estimates of porosity, air filled voids, settling rates of particles in fluids, and transport of particles by wind or water. If the bulk density and particle density are known, the porosity of a soil can be calculated from the formula

$$Porosity(\%) = \left(1 - \frac{Bd}{Pd}\right) \times 100$$

where *Bd* stands for bulk density and *Pd* for particle density.

Soil porosity and bulk density are inversely related. Therefore, any practice that affects one also affects the other. Differences in optimum porosity exist, but values reported in literature are 6–10% for sudangrass, 10–15% for wheat and oats, and 15–20% for barley and sugar beet (Grable, 1966).

Tilth. Tilth describes the physical condition of soil as related to its ease of tillage, fitness as a seedbed, and impedance to seedling emergence and root penetration. A good tilth means adequate supply of water and air to plants. Soil structure, particularly the degree of aggregation of primary particles, contributes to tilth. Good tilth is more critical in fine-textured than in coarse-textured soils and is influenced by tillage. The specific objectives of tillage in crop production are: 1) to physically loosen soils and break hard pans to facilitate the infiltration of water and air; 2) to improve germination and root development; 3) to reduce large aggregates to a desirable size range; 4) to incorporate crop residues, lime, and fertilizers; and 5) to level the soil to facilitate irrigation, weed control, and planting and harvesting operations.

Strategies to Improve Soil Physical Properties

Soil physical properties such as structure, porosity, bulk density, and tilth can be improved through appropriate soil management practices.

Maintenance of Organic Matter. The term soil organic matter refers to all materials of vegetable and animal origin formed in or added to soils, regardless of stage of decomposition (Finkl, 1979). Soil organic matter thus includes the highly decomposed and colloidal soil fraction known as humus, as well as organic residues that have not lost their anatomic structure (Brady, 1974). It is the humus fraction that contributes most to soil properties that are of great importance in crop production. Many important soil properties, including absorption and retention of water, reserves of exchangeable cations, the capacity to supply N, P, and S to growing plants, stability of soil structure, and adequacy of aeration, are dependent to some degree on the quantity of organic matter present (Broadbent, 1965).

Figure 2.7 shows a relationship between organic matter (OM) and cation exchange capacity of soil. The marked effect of organic matter on soil cation exchange capacity (CEC) can be explained by the high CEC of organic matter. The CEC-OM relationship (Figure 2.7) shows that an incremental 1% increase in OM on a dry-weight basis (starting near zero) resulted in a corresponding increase of 1.7 cmol (p^+) kg^{-1} in soil CEC (Kapland and Estes, 1985). Similarly, Figures 2.8 and 2.9 show a relationship between organic matter and soil extractable Zn and Mn. As the soil organic matter content in this soil was increased, there was a linear increase in extractable Zn and Mn. Further, organic matter is the main source of energy for soil microorganisms and interacts with fertilizers, pesticides, and herbicides added to soil.

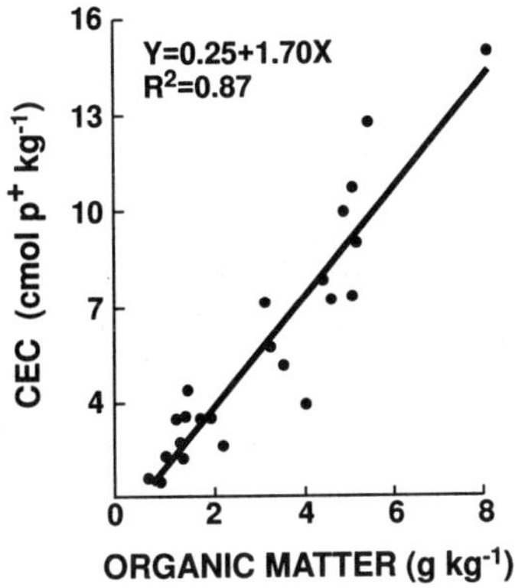

Figure 2.7 Relationship between cation exchange capacity and organic matter. (From Kapland and Estes, 1985.)

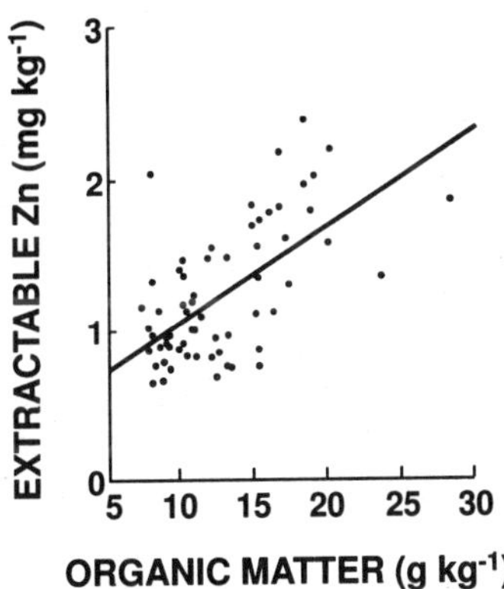

Figure 2.8 Relationship between Mehlich 1 extractable Zn and organic matter. (From Kapland and Estes, 1985.)

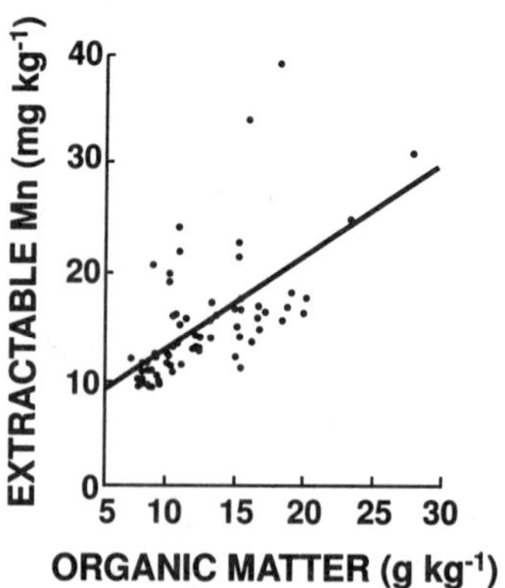

Figure 2.9 Relationship between Mehlich 1 extractable Mn and organic matter. (From Kapland and Estes, 1985.)

The organic matter in a soil sample is measured indirectly by determining the organic carbon and using the factor 1.724 to convert organic C to organic matter. This factor is based on an assumed C content of 58% for the organic matter (Stevenson, 1986). The numerous combinations under which the soil-forming factors operate account for the great variability in the organic matter contents of soils even in a very localized area. In uncultivated soils, the amount of organic matter present is governed by the soil-forming factors in the following order of importance: climate > vegetation > topography = parent material > age (Jenny, 1930).

The absolute amounts of C vary considerably from one soil to another, and can be as low as 1% or less in coarse-textured soils (sands) and as much as 3.5% in prairie grassland soils (Stevenson, 1986). Poorly drained soils (aquepts) often have C contents approaching 10%. Tropical soils are generally low in organic matter content. The low organic matter content in tropical soils under cultivation is attributed to the higher activities of microorganisms at the higher temperature as shown in Figure 2.10. Organic matter content usually decreases rapidly when soils are brought under cultivation. This may be related to improved aeration, moisture supply, and soil reaction which increases microbial activity and decomposition of organic compounds.

Organic matter in the soil can be maintained by incorporating crop residues, adding organic manures (animal as well as green plants), keeping the land under pasture and forests, and including a sod crop in the rotation. Figure 2.11 shows a relationship between plant residue addition and organic C content (0–5 cm) of the soil surface. There was a positive quadratic relationship between the amount of plant residues added and the surface soil organic concentration.

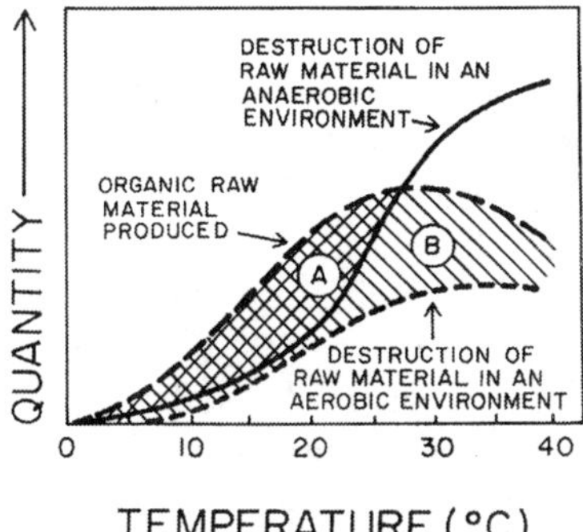

Figure 2.10 Influence of temperature on organic raw material production through photosynthesis and organic matter distribution by microorganisms. Accumulation of humus under aerobic soil conditions are confined to zone A; accumulation under anaerobic conditions occur over the entire temperature range (A–B). (From Senstius, 1958, and Stevenson, 1986: John Wiley, New York, reproduced with permission.)

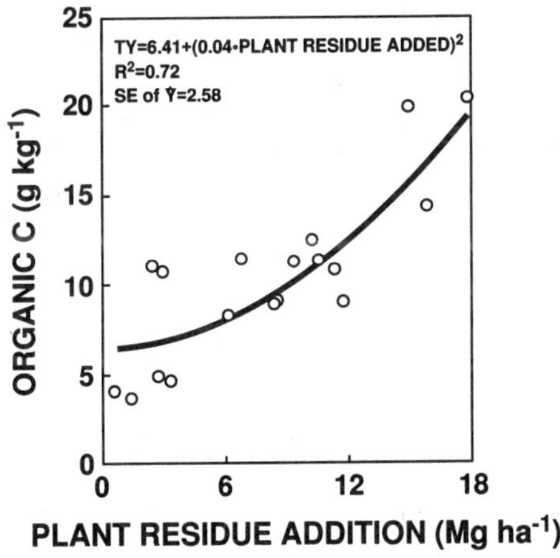

Figure 2.11 Relationship between soil organic C concentration in the top 0–5 cm soil layer and plant residue addition. (From Wood et al., 1990.)

Conservation Tillage. Tillage is defined as the mechanical manipulation of soil for crop production, and conservation tillage is any tillage sequence which reduces loss of soil by water relative to conventional tillage (Papendick and Elliott, 1983). More recently, the Conservation Technology Information Center (CTIC) (1992) defined conservation tillage as "any tillage and planting systems that maintain at least 30% of the soil surface covered by residue after planting to reduce water erosion; or where soil erosion by wind is a primary concern, maintain at least 455 kg of flat, small grain residue equivalent on the surface during the critical wind erosion period."

Soil tillage is perhaps as old as settled agriculture, yet its impacts on soil degradation, soil resilience, ecological stability, environmental quality, and agricultural sustainability are more important now than ever before. Properly used, tillage can be an important restorative tool that can alleviate soil-related constraints in achieving potential productivity or utility. When improperly used, tillage can set in motion a wide range of degradative processes such as deterioration in soil structure, accelerated erosion, depletion of soil organic matter and fertility, and disruption in cycles of H_2O, C, and essential plant nutrients (Lal, 1993).

The main features of conservation tillage are increased soil surface roughness and maintenance of crop residues on the soil surface. This system of tillage controls wind and water erosion, reduces energy use, and conserves soil and water (Unger and McCalla, 1980). The success with reduced tillage in crop production depends on soil type, drainage, climate, and management practices (Papendick and Elliott, 1983). Development of crop cultivars specially adapted to conservation tillage environments may be one way to increase and/or stabilize crop production (Kaspar et al., 1987).

The agronomic and economic performance of conservation-effective tillage is extremely location-specific. Problems important in semi-arid regions may not be significant in humid tropical areas. Therefore, to be successful, a conservation-effective tillage program needs to be flexible enough to be adapted to a variety of economic, geographic, and land use and related variables (Benites and Ofori, 1993).

Organic Farming. Organic farming means minimizing the use of chemical fertilizers or pesticides in crop production. In this type of farming, farmers maintain the top soil in good physical condition through regular use of soil-building crops and animal manures. Weeds, insect pests, and plant diseases are kept in check through use of crop rotation, timely tillage, and biological controls. Several thousand farmers in the United States are operating commercial farms profitably with organic farming (Papendick and Elliott, 1983).

Chemical Properties

Soil chemical properties such as nutrient deficiencies and toxicities, pH, cation exchange capacity, oxidation-reduction, and salinity are the important properties determining growth and production of crops. These soil properties can be modified through management practices for higher crop production. At present, sufficient know-how is available almost all over the world to improve these chemical properties if unfavorable conditions exist for crop production. But sometimes it is not possible to apply the know-how in a particular situation for economic reasons. Any technology in agricultural science is adopted by farmers if it satisfies their needs. Farmers must make a profit. If they are sure that a given technology will increase profit, they will apply that technology if they have the resources to do so. The belief that farmers are hesitant to adopt new technology is outdated. This has been proved by the adoption of high-yielding wheat and rice cultivars in Asia, where the majority of the farms are small and the farmers are uneducated.

Nutrient Deficiencies. Inadequate nutrient supply restricts plant growth and yield. Nutrient deficiencies vary among soils and areas (Table 2.4), but nitrogen and phosphorus are the most deficient nutrients in temperate as well as tropical soils all around the world. Among the essential nutrients, potassium is the nutrient which is absorbed in maximum amounts by modern improved crop cultivars. The nutrient deficiencies in a particular soil are related to parent material, weathering, cultivation, and erosion.

The nutrient-supplying power of a soil is normally evaluated through soil and plant analysis and visual symptoms in plants. But crop response to applied nutrients is the best indicator of the nutritional status of a soil. In such evaluations, all other growth factors should be at optimum levels. Nutrient deficiencies can be alleviated through application of fertilizers and amendments and use of efficient cultivars.

Nutrient/Elemental Toxicities. The toxicities most commonly found in food crops are those of aluminum, manganese, and iron. Aluminum and manganese toxicities are most common in acid soils, whereas iron toxicity occurs in flooded rice under reduced soil conditions. Table 2.4 presents nutrient or elemental toxicities in important soil groups. Crop yield reduction varies with the intensity of toxicity, and the intensity of toxicity varies with plant species, soil, and climatic conditions. Figure 2.12 shows that relative grain yield of Al-susceptible common bean species decreased quadratically with increasing Al saturation while Al-tolerant rice species yield increased with increasing Al saturation in an Inceptisol.

Metal toxicity can be expressed in two ways. Direct toxicity occurs when the excess of the element is absorbed and becomes lethal to the plant cell.

Table 2.4 Element Deficiencies and Toxicities Associated with Major Soil Groups

		Element	
Soil order	Soil group	Deficiency	Toxicity
Andisols (Andepts)	Andosol	P, Ca, Mg, B, Mo	Al
Ultisols	Acrisol	N, P, Ca, and most other	Al, Mn, Fe
Ultisols/Alfisols	Nitosol	P	Mn
Spodosols (Podsols)	Podzol	N, P, K, Ca, micronutrients	Al
Oxisols	Ferralsol	P, Ca, Mg, Mo	Al, Mn, Fe
Histosols	Histosol	Si, Cu	
Entisols (Psamments)	Arenosol	K, Zn, Fe, Cu, Mn	
Entisols (Fluvents)	Fluvisol		Al, Mn, Fe
Mollisols (Aqu), Inceptisols, Entisols, etc. (poorly drained)	Gleysol	Mn	Fe, Mo
Mollisols (Borolls)	Chernozem	Zn, Mn, Fe	
Mollisols (Ustolls)	Kastanozem	K, P, Mn, Cu, Zn	Na
Mollisols (Aridis) (Udolls)	Phaeozem		Mo
Mollisols (Rendolls) (shallow)	Rendzina	P, Zn, Fe, Mn	
Vertisols	Vertisol	N, P, Fe	S
Aridisols	Xerosol	Mg, K, P, Fe, Zn	Na
Aridisols/arid Entisols	Yermosol	Mg, K, P, Fe, Zn, Co, I	Na, Se
Alfisols/Ultisols (Albic) (poorly drained)	Planosol	Most nutrients	Al
Alfisols/Aridisols/Mollisols (Natric) (high alkali)	Solonetz	K, N, P, Zn, Cu, Mn, Fe	Na
Aridisols (high salt)	Solonchak		B, Na Cl

Source: Modified from Dudal (1976), Clark (1982), and personal communications, S. W. Buol and H. Eswaran.

Indirect toxicity can be related to nutritional imbalance. When excess Al, Mn, and Fe are present in the growth medium, they may inhibit the uptake, transport, and utilization of many other nutrients and induce nutritional deficiency. Table 2.5 shows the Al inhibition of nutrient uptake by rice plants grown in nutrient solution. Increased Al concentrations in the solution exerted an inhibiting effect on the concentrations of N, P, K, Ca, Mg, Zn, Fe, Mn, and Cu in the plant tops. Decreases in uptake of these elements were mostly

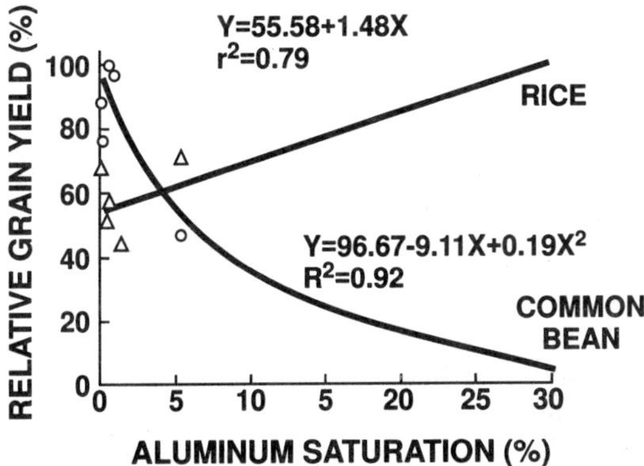

Figure 2.12 Relationship between aluminum saturation and relative grain yield of rice and common bean in an Inceptisol. (From Fageria and Santos, 1996.)

Table 2.5 Influence of Al on Uptake of Nutrients by Rice Plants at 21 Days Growth in Nutrient Solution[a]

	Al concentration (µm)				
Nutrient	0	371	742	1484	2226
N	50.0	44.9	47.6	45.0	41.5
P	6.2	5.1	4.9	3.6	3.3
K	44.4	41.8	39.9	32.8	26.5
Ca	2.0	1.9	1.8	1.5	1.4
Mg	3.8	2.7	2.6	2.6	2.4
Zn	31	26	23	18	14
Fe	299	297	237	172	176
Mn	695	505	449	276	146
Cu	21	24	22	18	18

[a]Values are means for six cultivars. Concentrations of macronutrients are in g kg^{-1} and of micronutrients in mg kg^{-1}.
Source: Compiled from Fageria and Carvalho, 1982.

related to morphological, physiological, and biochemical effects of Al (Fageria and Carvalho, 1982; Fageria et al., 1988). These effects can be explained by the following hypotheses:

1. Aluminum inhibits root growth, thereby causing the uptake of these nutrients to be reduced (Fageria, 1982).
2. Aluminum reduces cellular respiration in plants, inhibiting the uptake of all ions (Aimi and Murakami, 1964).
3. Aluminum increases the viscosity of protoplasm in plant root cells and decreases overall permeability to salts (Aimi and Murakami, 1964; McLean and Gilbert, 1927).
4. Aluminum blocks, neutralizes, or reverses the negative charges on the pores of the free space and thereby reduces the abilities of such pores to bind Ca (Clarkson, 1971).
5. Aluminum may compete for common binding sites at or near the root surface and thereby reduce uptake of K, Ca, Mg, and Cu (Harward et al., 1955; Hiatt et al., 1963).
6. Aluminum reduces Ca uptake by completely inactivating part of the Ca accumulation mechanism (Johnson and Jackson, 1964).
7. In general, aluminum interferes with cell division in plant roots, decreases root respiration, interferes with certain enzymes governing the deposition of polysaccharides in cell walls, increases cell wall rigidity, and interferes with the uptake, transport, and use of several elements such as K, Ca, and Mg (Foy, 1974).
8. Aluminum injures plant roots and reduces Ca uptake (Lance and Pearson, 1969).
9. Aluminum decreases the sugar content, increases the ratio of nonprotein to protein N, and decreases the P contents of leaves from several plants grown on acid soils (Foy, 1974).

Similarly, high Fe concentrations in the growth medium reduced uptake of nutrients (Table 2.6). Among macronutrients, uptake of P was highly affected, followed by K and N. Among micronutrients, absorption of Mn and Zn was most affected. These results suggest that when there is a higher concentration of iron in lowland or flooded rice, P, K, and Zn deficiencies will be first to appear if concentrations of these nutrients in the soil are not sufficiently high. This means that one way to solve the iron toxicity problem in lowland rice is to increase P, K, and Zn supplies through fertilization.

pH. Soil pH is one of the most important soil chemical properties. It indicates the need for lime, the likelihood of excess phytotoxic ions, the activity of microorganisms, and the relative availability of most inorganic nutrients. The pH of soil indicates whether it is acid, neutral, or alkaline. Neutrality

Table 2.6 Uptake of Nutrients in the Roots and Shoots of Rice Cultivars[a]

	Iron concentration (mM)					
	0.09		0.89		1.73	
Nutrients	Concentration, g kg^{-1} or mg kg^{-1}	Content, mg or μg per 4 plants	Concentration, g kg^{-1} or mg kg^{-1}	Content, mg or μg per 4 plants	Concentration, g kg^{-1} or mg kg^{-1}	Content, mg or μg per 4 plants
Roots						
N	28.2	23	27.6	11	25.3	6
P	3.3	2.66	2.7	1	3.9	0.96
K	29.5	25	17.3	6	14.6	4
Ca	0.8	0.65	1.0	0.38	1.1	0.26
Mg	1.2	1	1.1	0.42	1.1	0.26
Zn	44	37	26	10	38	10
Cu	19	15	22	8	23	6
Mn	22	18	27	10	38	9
Fe	2258	1806	12,717	4658	37,458	9202
Shoots						
N	40.9	186	33.8	51	41.8	50
P	4.8	21	1.8	3	2.6	3
K	29.5	133	19.4	26	21.7	25
Ca	1.7	8	2.4	3	2.2	3
Mg	4.3	19	3.9	5	2.2	5
Zn	24	109	18	24	21	25
Cu	14	62	16	22	17	20
Mn	199	874	139	183	152	184
Fe	350	1578	2008	2627	4233	4988

[a]Values are mean for 12 cultivars. Concentrations of macronutrients are in g kg^{-1} and of micronutrients in mg kg^{-1}. Similarly, macronutrient contents are in mg and micronutrient contents in μg.
Source: Fageria, 1988.

Table 2.7 Critical Soil pH for Important Crops and Their Classification to Soil Acidity

Crop	Critical soil pH	Classification
Alfalfa (*Medicago sativa* L.)	6.0–6.5	Susceptible
Red clover (*Trifolium pratense* L.)	6.0–6.5	Susceptible
Sugar beet (*Beta vulgaris* L.)	6.0–6.5	Susceptible
Barley (*Hordeum vulgare* L.)	5.5–6.0	Moderately tolerant
Cotton (*Gossypium hirsutum* L.)	5.5–6.0	Moderately tolerant
Sorghum (*Sorghum vulgare* Pers.)	5.5–6.0	Moderately tolerant
Soybean (*Glycine max* L. Merr.)	5.5–6.0	Moderately tolerant
Wheat (*Triticum aestivum* L.)	5.5–6.0	Moderately tolerant
Common bean (*Phaseolus vulgaris* L.)	5.5–6.0	Moderately tolerant
Corn (*Zea mays* L.)	5.0–5.5	Tolerant
Oats (*Avena sativa* L.)	5.0–5.5	Tolerant
Peanuts (*Arachis hypogaea* L.)	5.0–5.5	Tolerant
Potato (*Solanum tuberosum* L.)	5.0–5.5	Tolerant
Rice (*Oryza sativa* L.)	5.0–5.5	Tolerant

Source: Compiled from Adams, 1981.

occurs at a pH of 7.0. Acidity is identified with any pH value less than 7.0 and alkalinity with any value above 7.0. At pH 7.0 the concentration of H^+ ions and OH^- ions is the same. In alkaline systems OH^- concentration exceeds that of H^+, whereas in acid systems the reverse is true. The pH values of most agricultural soils are in the range of 4–9. The most useful soil pH is the minimum pH above which liming will not increase crop yield. This is conveniently called the "critical" pH (Adams, 1981). Critical soil pH values for various crops are given in Table 2.7. These values should be used with caution because critical pH will vary with soil type and among cultivars of the same species.

The amount of lime required to raise soil pH to a specific value is determined by the soil's pH buffer capacity and is directly proportional to the cation exchange capacity of highly weathered soils (Adams, 1981). At low base saturation and low pH, soils are highly buffered because of the hydrolytic reactions of Al^{3+}, after it has exchanged for Ca^{2+} or Mg^{2+}, according to the reaction (Adams, 1981)

$$Al^{3+} + 3H_2O = Al(OH)_3 + 3H^+$$

Figure 2.13 shows the relative dry matter yield of tops of corn, soybean, and wheat in a Brazilian Inceptisol. Yields of wheat and soybean were significantly ($P < 0.05$) increased in a quadratic response with increasing soil

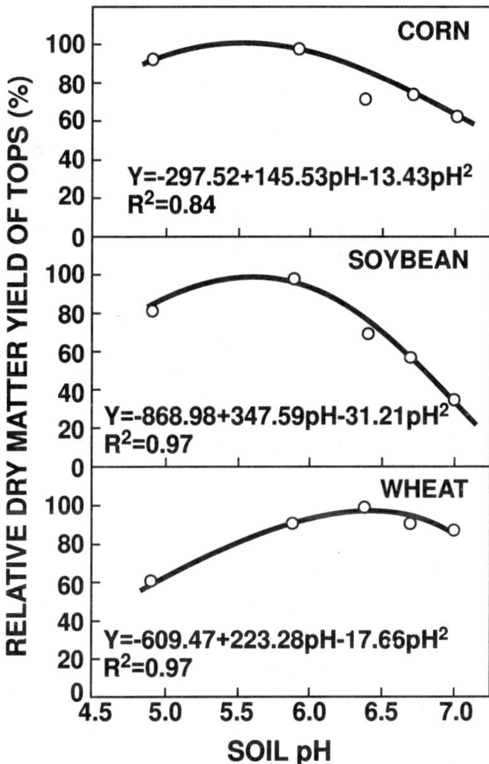

Figure 2.13 Relationship between soil pH and relative dry matter yield of tops of corn, soybean, and wheat in an Inceptisol. (From Fageria and Zimmermann, 1996.)

pH from 4.7 to 7.0. The maximum relative dry weight of tops of wheat was achieved at pH 6.3, corn at pH 5.4, and soybean at pH 5.6, as calculated by quadratic equation. For soils low in organic matter, the amount of lime required to change soil pH by one unit is least in the pH range of about 5–6, where exchangeable Al^{3+} is nil and the HCO_3^- system is insignificant.

Low soil pH stress is a major growth-limiting factor for crop production in many regions of the world. Figure 2.14 shows the distribution of acid soils in the world. Soil acidity may be due to the parent material being acidic and naturally low in the basic cations such as Ca^{2+}, Mg^{2+}, K^+, and Na^+ or to leaching of these elements down the soil profile by excess rains (Kamprath and Foy, 1971). Soil acidity may also be produced by use of ammonium fertilizers for a long time, removal of cations in the harvested portion of crops, and decomposition of plant residues or organic wastes into organic acids.

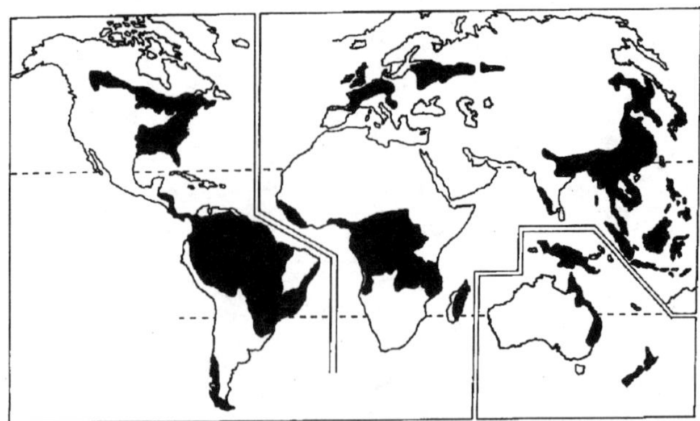

Figure 2.14 Distribution of acid soils in climates warmer than cryic. (From Wambeke, 1976.)

Cation Exchange Capacity. Cation exchange capacity (CEC) is defined as the sum of the exchangeable cations retained by soil. It is expressed in milliequivalents per 100 g or in cmol of cations per kg of soil. Cation exchange in soil is a reversible chemical reaction and corresponds to the negative charge of the soil. The cation exchange capacity of soils is highly variable. The principal factors which determine CEC are amount and type of clay present, organic matter content, and soil pH. Representative CEC values of common exchange materials in soils at pH 7.0 are organic matter 200–400, vermiculite 100–150, montmorillonite 60–100, illite 20–40, kaolinite 2–16, and sesquioxides 0 meq/100 g (Hausenbuiller, 1972). The negative charge, and hence the CEC, increases as pH rises. The capacity of the soil to adsorb exchangeable cations cannot therefore be defined unless a standard pH for its measurement is agreed on (Bache, 1979). Cation exchange capacity is most commonly determined as the quantity of cations absorbed from salt solution buffered at pH 7 with NH_4OAc or at pH 8.2 with $BaCl_2$–triethanolamine. For soils with field pH values of 7–8.2, these measurements adequately reflect the CEC. They are also adequate if the soil has little or no variable charge (Sanchez, 1976). But this method is generally not suitable for tropical soils which exhibit a significant amount of variable charge or for temperate soils with significant organic matter contents. The measurement that reflects more accurately the total charge at the actual soil pH involves leaching with a neutral, unbuffered salt such as KCl or $CaCl_2$, determined at the pH of the soil (Sanchez, 1976). Coleman and Thomas (1967) called this the effective CEC. Cation exchange capacities obtained by this method are generally lower values than those obtained by other methods. A minimum value of 4 cmol

kg^{-1} is needed to retain most cations susceptible to leaching. Cation exchange capacity can be increased through addition of organic matter and liming of acid soils.

Base Saturation. Base saturation is an important soil chemical property in acid soils, affecting plant growth and nutrient uptake. It can be calculated with the help of the following formula:

$$\text{Base saturation} = \frac{\text{exchangeable Ca, Mg, Na, K}}{\text{CEC at pH 7 or 8.2}} \times 100$$

The base saturation philosophy promotes that maximum yields can be achieved by creating an ideal ratio of calcium, magnesium, and potassium in the soil (Eckert, 1987). From the standpoint of soil chemical properties and reactions, however, the base saturation is more correctly an acidity index or liming index (Bohn et al., 1979). Figure 2.15 shows the relationship between relative dry matter yields of tops of rice, wheat, corn, commonbean, and cowpea in an Oxisol of central Brazil. Upland rice maximum yield was achieved at about 25%, wheat about 60%, commonbean about 67%, and cowpea about 53%, as calculated by quadratic regression equations. Corn dry matter increase was linear between 26 and 80% base saturation in the soil under investigation.

Oxidation-Reduction. Oxidation-reduction is a chemical reaction in which electrons are transferred from a donor to an acceptor (Ponnamperuma, 1972). The donor loses electrons and increases its oxidation number (it is oxidized); the acceptor gains electrons and decreases its oxidation number (it is reduced). The main source of electrons for biological reductions is organic matter. In well-drained soils, oxidation-reduction potentials range from +0.4 to +0.6 V, whereas in waterlogged and submerged soils in the presence of organic substances, reduction processes can decrease the potential to +0.2 V and below (Orlov, 1979). The principal factors controlling the potential level are aeration and the operation of biological processes.

Oxidation-reduction changes the concentrations of the nutrients and hence their availability to plants. Under reduced conditions, Fe^{3+} changes to Fe^{2+} and Mn^{4+} to Mn^{2+}, thus affecting the availability of these nutrients to flooded rice. Sometimes concentrations of these elements become so high that toxicity occurs.

Salinity and Alkalinity. A soil containing sufficient quantities of soluble salts and exchangeable sodium to interfere with the growth of most crop plants is known as a saline-sodic soil. Figure 2.16 shows the relative yield of two canola cultivars as a function of increasing soil salinity. The traditional classification of salt-affected soils has been based on the soluble salt (EC) concentrations in extracted soil solution and on the exchangeable sodium

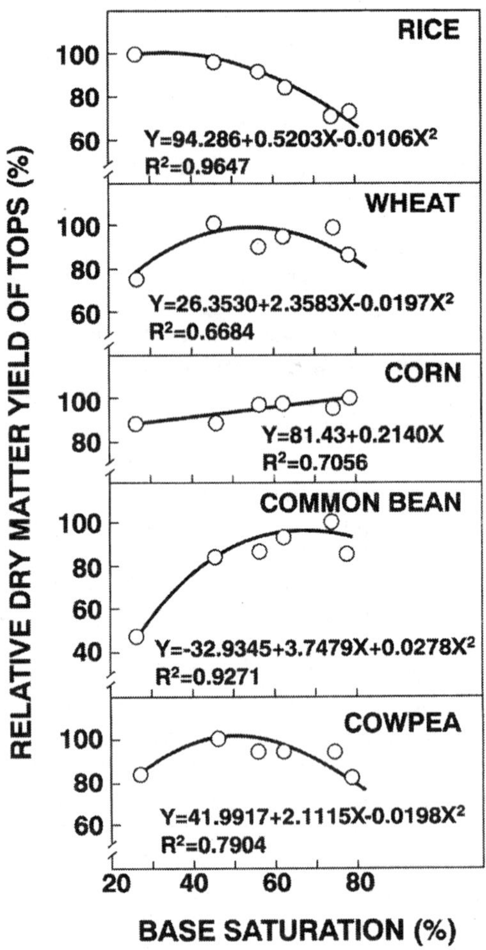

Figure 2.15 Relationship between base saturation and relative dry matter yield of tops of rice, wheat, corn, common bean, and cowpea in an Oxisol.

percentage of the associated soil. The dividing line between saline and non-saline soils was established at 4 dS m^{-1} for water extracts from saturated soil pastes. Salt-sensitive plants, however, can be affected in soils whose saturation extracts are only 2–4 dS m^{-1}. The terminology committee of the Soil Science Society of America lowered the boundary between saline and nonsaline soils to 2 dS m^{-1} in the saturation extract (Bohn et al., 1979). In an alkali soil, more than 15% of the cation exchange capacity is saturated with alkali ions,

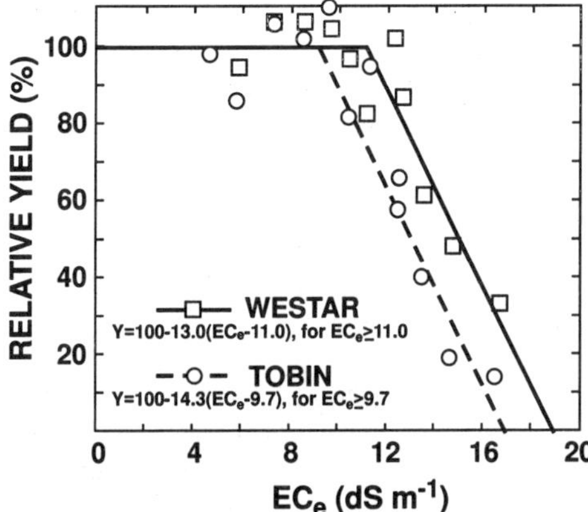

Figure 2.16 Relative seed yield of two canola cultivars as a function of increasing soil salinity. (From Francois, 1994.)

most often sodium. In salt-affected soils, the yield reduction may result from osmotic stress caused by the total soluble salt concentration, from toxicities or nutrient imbalances created when specific salts become excessive, or from reduction of water penetration of soil structure (Hoffman, 1981).

Management practices to correct salinity and alkalinity are related to leaching excess salts below the root zone and use of tolerant species and cultivars within species. Special care is required in saline-alkali soils, however, because removal of the soluble salts without reducing the exchangeable sodium percentage leads to highly undesirable soil structure. The sodium saturation percentage can be lowered before leaching through addition of calcium salts (gypsum) or through the use of acid-forming substances such as sulfur, H_2SO_4, iron sulfate, and organic matter (Oertli, 1979).

Biotic Factors

Biotic factors which affect crop production are related to soil microorganisms such as bacteria, actinomycetes, fungi, and nematodes. These microorganisms carry out a range of activities in the plant rhizosphere which are harmful as well as beneficial for plant growth. Table 2.8 lists some of the various impacts that microorganisms around the root have on plant growth. The more common activities are breakdown of organic matter, nitrogen fixation, secretion of growth substances, and increasing the availability of mineral nutrients. They also cause plant disease or protect the plant from pathogens.

From the point of view of their relationships with plants, microorganisms can be classified into three groups (Barea and Aguilar, 1983): 1) *saprophytes*,

Table 2.8 Examples of Some Interactions between Microorganisms and Plants That are Either Detrimental (Negative) or Beneficial (Positive) to Plant Growth

Negative	Positive
Root pathogens	Nitrogen fixation (*Rhizobium*, *Frankia*, associated N fixation
Subclinical root pathogens	Mycorrhiza
Detrimental rhizobacteria	
Cyanide production	Biocontrol of detrimental microorganisms
Denitrification	Hormone/growth factor production
	Plant-growth-promoting rhizobacteria
	Phosphate solubilization
Nutrient unavailability	Nutrient availability

Source: Bowen and Rovira, 1991.

usually opportunists, but benefactors in some situations; 2) *parasitic symbionts* or *pathogens*, potentially harmful to the plant; and 3) *mutualistic symbionts*, usually called symbionts in the literature, which develop activities beneficial to plant growth.

The most beneficial contribution of soil microorganism to plant development involves the supply of nutrients. Among these microorganisms, those concerned with N fixation and the enhancement of nutrients supplied by diffusion are especially relevant (Barea and Aguilar, 1983). A brief discussion of these two types of microorganism is given in this section.

Symbiotic Nitrogen Fixation. Nitrogen fertilizer is quite expensive due to the high price of petroleum. Under these circumstances, the ability of legumes in symbiosis with rhizobia to obtain atmospheric nitrogen is important in crop production. Biological nitrogen fixation reduces the cost of production and helps reduce pollution. More than 60% of the N input to the natural plant community has a biological origin (Postgate and Hills, 1979). Biological nitrogen fixation in Brazil contributes an estimated 215 million tons of N, representing an economy of 1.8 billion U.S. dollars per year (Dobereiner et al., 1995). The N-fixing bacteria in symbiotic association with plants either convert N into bacterial proteins or make it directly available to plants as NH_3.

Nitrogen fixation in legumes is termed obligatory because both the host plant and the organism are required for fixation to occur (Bezdicek, 1979). A classical example is the association between dicotyledonous plants of the

Table 2.9 Legume Crops Nodulated by Rhizobium Bacteria

Crop	*Rhizobium* species
Alfalfa	*Rhizobium meliloti*
Sweet clover	*Rhizobium meliloti*
Bean	*Rhizobium phaseoli*
Clover	*Rhizobium trifoli*
Cowpea	*Rhizobium japonicum*
Soybean	*Rhizobium japonicum*
Peanut	*Rhizobium japonicum*
Pea	*Rhizobium leguminosarum*

family Leguminosae and members of the bacterial genus *Rhizobium*. Plants in this category include legumes such as soybean, peas, alfalfa, and beans. Certain species of *Rhizobium* can produce nodules only on certain legumes. For example, bean is nodulated only by bacteria called *R. phaseoli*. Table 2.9 shows *Rhizobium*-legume associations.

The quantity of nitrogen fixed by a plant depends on the plant species, soil environment, and management practices. Burns and Hardy (1975) averaged a great many published estimates to arrive at an average figure of 140 kg N fixed per year per hectare of arable land under legumes. Shortly thereafter, that figure was considered by a group of scientists attending a conference on nitrogen-fixing microbes. They concluded that a more realistic figure would be half the Burns-Hardy value (Larue and Patterson, 1981). That is, on average 70 kg N per year per hectare is fixed by legumes. This is a quite impressive figure from a practical agricultural point of view. However, there is no concrete evidence that any legume crop satisfies all its N requirements by fixation, especially at higher yield levels. This means inorganic N fertilizers have to be applied if higher productivity is the goal.

Further, over the last 20 years, many new species of N_2–fixing bacteria have been discovered in association with grasses, cereals, and other non-nodulating crops. Virtually all of these bacteria are microaerophilic, fixing N_2 only in the presence of low partial pressures of oxygen. Until a few years ago, much attention was focused on members of the genus *Azospirillum*, and it was assumed that N_2 fixation was restricted to the rhizosphere or rhizoplane of the host plants. Through the use of N balance and ^{15}N techniques, it has been shown that in the case of lowland rice, several tropical pasture grasses, and especially sugar cane, the contributions of biological N_2 fixation (BNF) are of agronomic significance (Boddey and Dobereiner, 1995).

More detailed study of the N_2-fixing bacteria associated with sugar cane (*Acetobacter diazotrophicus* and *Herbaspirillum* spp.) has shown that they occur in high numbers, not only in roots of this crop but also in the stems, leaves, and trash, but are rarely found in the soil. Some of these endophytic diazotrophs have now also been found in forage grasses, cereals, sweet potato, and cassava, although evidence of significant BNF contributions is still lacking (Boddey and Dobereiner, 1995).

Mycorrhizae. Mycorrhizae have been shown to improve the nutrition of the host plants for nutrients that are diffusion limited, such as P, Zn, Cu, and Fe (Tinker, 1982). Mycorrhizae accomplish this primarily by extension of the root geometry. In this symbiotic association, the fungus utilizes carbohydrates produced by the plant, while the plant benefits by increased uptake of nutrients. The beneficial effect of mycorrhiza is of special importance for plants that have a coarse and poorly branched root system, since the external hyphae can extend as much as 8 cm away from the roots (Mosse, 1981), absorbing nutrients from a much larger soil volume than the absorption zone surrounding a nonmycorrhizal root (Howeler et al., 1987).

Vesicular-arbuscular mycorrhizal (VAM) fungi are present in nearly all natural soils, and these fungi infect the great majority of plants including the major food crops. Many tropical crops and pastures are grown on soils that are very P deficient, and particularly in those soils an efficient mycorrhizal association is of great importance in increasing P uptake and crop yields (Howeler et al., 1987).

A great deal of work has been done on mycorrhizae, but still much has to be learned, especially about their practical application. One problem with mycorrhizae is that the symbiosis appears to form a huge C drain on the host plant. Some ectomycorrhizae consume up to 50% of the C fixed by host photosynthesis (Lauchli, 1987). The ectomycorrhizae are those that primarily colonize forest trees. Estimates of the C drain for VAM, those that colonize root systems for many perennial and annual crop species, are lower, ranging from 6 to 20% of the C fixed by the host plant (Lauchli, 1987). Looking into this aspect, it is clear that much research has to be done to modify the fungus or environment so that the fungus becomes less dependent on the C supply generated by the host plant.

Plant

Crop productivity results from the complex interaction of many characteristics of the plant with each other and with the environment. Plant factors which affect crop productivity are genetic variability, C_3 or C_4 metabolism, photosynthetic efficiency, plant architecture, harvest index, and plant density.

Genetic Variability. Genetic variability in relation to yield is defined as the inheritable characteristics of a plant species or cultivar that make it differ in yield potential in favorable or unfavorable environments as compared to other species or cultivars within species. Genetic variability of crop species or cultivars within species makes a difference in crop or cultivar productivity. Useful genetic variation apparently exists in all major field crops and cultivars within crops for tolerance to mineral stress (Duvik et al., 1981; Lafever, 1981; Clark, 1982; Epstein and Rains, 1987), diseases (Day, 1973; Buddenhagen, 1983), insects (Beck, 1965), and drought (Unger et al., 1981; Bruckner and Frohberg, 1987). Because of this genetic variability, some plant species or cultivars adapt better and produce more under particular environmental conditions than other species or cultivars. According to Mahon (1983), putting the genetic variability of agricultural crops into practice is justified if four criteria are satisfied: 1) the variability must be of sufficient magnitude to justify the cost of assessing the character; 2) if selection for improved expression of a characteristic is to be useful over a range of spatial and temporal environments, some stability of genotype performance is necessary; 3) the variability must be related to agronomic benefit, either in increasing yield or quality or reducing cost of production; and 4) the variability must be measurable in large-scale trials.

Crop genetic resources are principally the *product* of a complex interaction over time between the abiotic and biotic environments, and farmers' handling and selection of the material (Harlan, 1992). This interaction involves introgressions from wild and weedy relatives, hybridization with other cultivars, mutations, and natural and human selection pressure. The results of this evolutionary process are materials or "landraces" which are well adapted to the local abiotic and biotic environmental variation (Weltzien and Fischbeck, 1990). Landraces are often genetically heterogeneous populations; the genetic variation within landraces is supposed to be a consequence of the variation in environmental conditions under which the material evolved (Almekinders et al., 1995).

Because genetic variation has the potential to adapt to environmental variation, it also may be considered a *tool* in agricultural production. Genetic variation within and between crops often favors production stability in time and space through a suppression of pests, diseases, and weeds (Altieri and Liebman, 1986).

Stability may be defined as the variability of a genotype across environments. The coefficient of variation (v) of yields in different environments then is an appropriate stability measure (Lin et al., 1986; Hühn, 1987). Since v may be approximately

$$\sigma = (\Sigma_i c_i)^{1/2}$$

c assesses the contribution of the i-th component to the instability of yield. In breeding programs, the c_i values can help identify key components responsible for yield instability. In order to stabilize yield, it may be a promising strategy to improve the stability of these key components of yield (Piepho, 1995).

Plant species or cultivars which produce better under unfavorable conditions do not necessarily produce better under favorable conditions and vice versa. Therefore, in selecting a species or cultivar under unfavorable conditions, the objective should be to stabilize productivity. Yield stability is a measure of variation between potential and actual yield of a genotype across changing environments (Blum, 1980). Yield stability can result from genetic heterogeneity, yield component compensation, stress tolerance, capacity to recover rapidly from stress, or a combination of these factors (Heinrich et al., 1983).

Under favorable conditions, the objective should be to increase productivity. Under these conditions inputs are higher, and higher productivity is essential to compensate for the cost of production.

Although ample variation exists for many characteristics, their ease of manipulation in a breeding program and eventual incorporation into improved cultivars depend on the type of genetic control, the amount of genetic variation available, and its heritability (Cooper, 1973). Genetic correlations between desirable and undesirable characters may also be important limitations in breeding programs. Finally, success is this regard depends on the collaboration and joint efforts of soil scientists, plant pathologists, plant physiologists, agronomists, and breeders to achieve these objectives.

C_3 and C_4 Plants. Plant species have been classified into C_3 and C_4 groups according to their pathway of carbon dioxide fixation. Plants whose first carbon compound in photosynthesis consists of a three-carbon-atom chain are called C_3 plants, and plants whose first compound in photosynthesis consists of a four-carbon-atom chain are called C_4 plants. Plants in the C_4 group have high photosynthetic efficiency as compared with plants in the C_3 group. Characteristics which distinguish the two groups of higher plants are presented in Table 2.10. According to the characteristics presented in this table, it is clear that the C_4 plants represent an adaption of habitats with high temperature, high irradiance, and limited water supply (Bjorkman, 1971; Black, 1971; Pearcy and Ehleringer, 1984). Important field crops in these groups are presented in Table 2.11. Detailed discussions of C_3 and C_4 plants are by Downton (1971, 1975) and Black (1971).

Photosynthetic Efficiency. Photosynthesis is the basis of all crop yield. It provides 90–95% of plant dry weight. Thus, net photosynthesis of the entire plant canopy integrated over a growing season should mainly determine total

Table 2.10 Characteristics of C_3 and C_4 Plants

Characteristics	C_3	C_4
Photosynthetic efficiency	Low	High
Photorespiration	High	Low
Water utilization efficiency	Low	High
Optimum temperature for photosynthesis	10–25°C	30–45°C
Response to light intensity	Low	High
Response to CO_2 concentration	Low	High
Response to O_2 concentration	Low	High
Major pathway of photosynthetic CO_2 fixation	Reductive pentose phosphate cycle	C_4–dicarboxylic acid and reductive pentose phosphate cycle
Transpiration ratios	High	Low
Leaf chlorophyll *a* to *b* ratio	Low	High

plant dry weight and thereby indirectly determine economic yield (Kueneman et al., 1979).

In the process of photosynthesis, light energy is converted into chemical potential energy. In the photochemical reaction, carbohydrate is produced and O_2 and water are released according to the following equation:

$$6CO_2 + 12H_2O \xrightarrow[\text{energy}]{\text{sunlight}} C_6H_{12}O_6 + 6O_2 + 6H_2O$$

Table 2.11 C_3 and C_4 Field Crops

C_3 crop	C_4 crop
Rice (*Oryza sativa* L.)	Sugarcane (*Saccharum officinarum* L.)
Wheat (*Triticum aestivum* L.)	Sorghum (*Sorghum vulgare* Pers.)
Oat (*Avena sativa* L.)	Corn (*Zea mays* L.)
Barley (*Hordeum vulgare* L.)	Pearl millet (*Pennisetum americanum* L.)
Peanut (*Arachis hypogaea* L.)	
Sugar beet (*Beta vulgaris* L.)	
Soybean (*Glycine max* L. Merr.)	
Cotton (*Gossypium hirsutum* L.)	
Common bean (*Phaseolus vulgaris* L.)	
Cowpea (*Vigna unguiculata* L. Walp.)	

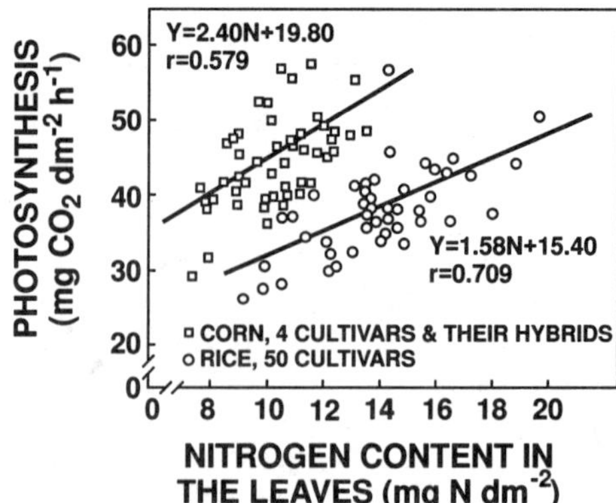

Figure 2.17 Relationship between nitrogen content in the leaves and photosynthesis rate in corn and rice plants. (From Akita, 1995.)

Chloroplasts are important components of green plant cells insofar as they are the site where photosynthesis takes place. Chloroplasts contain chlorophyll, a substance that is capable of absorbing solar energy and, through a set of chemical reactions, converting this light energy into food materials.

The maximum possible efficiency of total solar energy conversion by crops is between 6 and 8% (Loomis and Williams, 1963; Monteith, 1978), but for many crops growing in a wide range of environments the ceiling biological yield is achieved with a radiant energy conversion of approximately 1–2% (Holliday, 1966; Gibbon et al., 1970). Energy conversion efficiency in crop plants can be calculated from the formula

$$\text{Efficiency} = \frac{\text{energy content of plant dry matter}}{\text{total solar energy available}}$$

The low energy conversion efficiency is related to nutrient and water deficiency, inadequate temperature and pest control, and poor management practices.

The photosynthetic rate of leaf tissue varies markedly among plant species. Corn, sorghum, and certain other species have maximum photosynthetic rates of 50–60 mg CO_2 dm^{-2} h^{-1}, whereas 20–30 mg dm^{-2} h^{-1} is the maximum rate in small grains, temperate grasses, and many other plants (Hesketh and Moss, 1963; Murata and Iyama, 1963; Menz et al., 1969). Further, the photosynthetic rate of plant species is also determined by N concentration in the leaves (Figure 2.17).

Considerable genetic variability for photosynthetic capacity exists, and it seems possible that photosynthetic rates can be increased through selection and breeding (Loomis and Williams, 1963). Leaves of most crop plants appear to be reasonably efficient at low light intensities, and the need is to extend these efficiencies to higher light intensities. This means increasing the point of high saturation or reducing excess light absorption above the saturation point. To do this would seem to require a low chlorophyll/enzyme ratio in the photosynthetic unit and the capability of absorbing CO_2 more rapidly (Gaffron, 1960).

Plant Architecture. Plant architecture influences photosynthesis, growth, lodging, and yield. Plant ideotype has been suggested to improve crop yield. An ideotype is a biological model which is expected to perform or behave in a predictable manner within a defined environment and to yield a greater quantity or quality of grain, oil, or other useful product when developed as a cultivar (Donald, 1968; Donald and Hamblin, 1983). Donald and Hamblin (1983) proposed that the principal characteristics of the ideotype for all annual seed crops are as follows:

1. strictly annual habit
2. erect growth form
3. dwarf stature
4. strong stems
5. unbranched or nontillered habit
6. reduced foliage (smaller, shorter, narrower, or fewer leaves)
7. erect leaf disposition
8. determinate habit
9. high harvest index
10. nonphotoperiodic for most but not all situations
11. early flowering for most but not all situations
12. high population density
13. narrow rows or square planted
14. response to high nutrient levels
15. wide climatic adaptation

Harvest Index. Crop production can be measured as total biomass or economically useful parts of the plant. The total yield of plant material is known as biological yield, and the ratio of the yield of grain to the biological yield is harvest index (Donald and Hamblin, 1976). Efficiency of grain production in crop plants is frequently expressed as harvest index. Harvest index, by definition, is a factor less than 1, but some workers prefer to express it as a percentage. Grain yield is directly proportional to harvest index, but biological yield and harvest index are unrelated (Donald and Hamblin, 1976). Increases in harvest index have contributed significantly to increasing yields of rice and wheat, and a value of more than 50% has been achieved in these cereals, but there is clearly a limit to how much further the harvest index can

Table 2.12 Harvest Index (%) of Five Cereal Crops

Crop	Minimum	Maximum	Average
Millet	16	40	26
Sorghum	25	56	27
Corn	25	56	42
Rice	34	55	44
Wheat	35	49	41

Source: Duivenbooden et al., 1996.

be improved. It seems unlikely that it will be able to rise much above 60% in cereals, although it may go further in root and tuber crops (Evans, 1980). Further increases in yield potential will then depend on improvements in the rates of photosynthesis and growth. Harvest indices of important crop species are given in Table 2.12.

Plant Density. Plant density is an important plant factor affecting crop yields. Biological yield increases with density to a maximum value determined by some factor of the environment and at higher densities tends to remain constant, provided there are no interfering factors such as lodging. Grain yield increases to a maximum value but declines as density is further increased. Figure 2.18 shows hypothetical relationships between plant density and biological and grain yield. Optimum plant density should be determined for each crop under each agroecosystem to obtain maximum yield. This parameter has special importance in crop production because it normally costs very little for farmers to adopt appropriate plant densities.

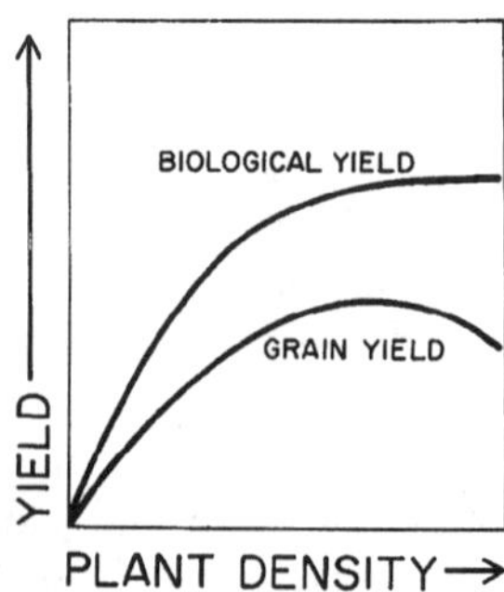

Figure 2.18 Relationship between plant density and yield.

Socioeconomic Factors

Besides climatic, soil, and plant factors, socioeconomic factors play an important role in crop production. Our civilization is being threatened by locally excessive soil erosion and other degradation of natural resources, currently aggravated by rural economic problems. However, recent famines in some parts of the world appear to be more closely related to socioeconomic factors than to any other factors. The quantity and quality of global soil resources seem capable of supporting present civilization at least well into the twenty-first century, but local areas of hunger will continue because of local and regional economic conditions and various forms of political disruptions and repressions (McCracken, 1987).

Marketing. Marketing is an essential part of agricultural production. Farmers have to have a market in which to sell their excess products at a reasonable price. It is to the advantage of farmers to have a market for their products as near as possible to the site of production.

Price. The price of agricultural commodities governs their production for the marketplace. The crop which fetches the best price in the market will occupy more area in an agroecological region. A good example of this is an area under soybean cultivation in the southern and central parts of Brazil. Soybean is an important export commodity and provides better prices to farmers than rice and corn. Farmers also apply more fertilizer and pesticide inputs on soybean fields. This means higher production per unit area. The relationship between input costs and market value of the crop produced strongly influences the farmers' use of high-technology inputs. Government price supports and subsidies often determine whether farmers can stay in business. Price is almost never controlled by supply and demand alone.

Extension Service. The transfer of technology from research institutes to farmers is carried out by extension workers and agribusiness technical representatives. It cannot be carried out effectively if these workers are poorly trained and poorly paid. Good training and higher salaries are an important aspect of successful extension of new technology to farmers. Extension officers need to have credibility with the farmers and must have technical competence. Developing credibility with farmers is claimed to be the single most important influence on the success of advisory services and individual extension personnel. When this is achieved, extension personnel are able to transfer technology and secure adoption at a considerably higher level. Table 2.13 identifies some of these criteria for credibility as perceived by farmers in New South Wales.

Several cogent approaches to improving existing extension programs in developing countries have been suggested (Hubbell, 1995), including:

Table 2.13 Farmers' Perceptions of Attributes that Make Extension Personnel Credible

1. Maintain a practical approach to problem solving
2. Make recommendations that are feasible in economic, technical, and social context
3. Make recommendations visible to the farmer
4. Have experience in the application of new practices on farms
5. Be well informed on the latest developments in agriculture
6. Have an overall knowledge of agriculture
7. Know the trends within industries in agriculture
8. Be accessible to the farmer
9. Be unbiased, honest, trustworthy, and reliable
10. Maintain confidentiality
11. Empathize with farmers and their needs
12. Understand and work within the social rules of the farming community

Source: Guerin and Guerin, 1994.

1. Relieve extension agents of nonproduction responsibilities such as regulatory enforcement.
2. Strengthen training and technical back-up of extension workers.
3. Focus research and extension on improved technologies for priority crops and animals.
4. Increase dialogue between extension workers and researchers, including participation in on-farm research and feedback to the researchers from the farmers.
5. Coordinate public sector extension with mass media approaches and private sector activities.
6. Increase the use of female extension workers and direct attention to the needs of women and other low-resource farmers.

Again, these are issues, considerations, responsibilities, etc. which can and should be effectively addressed by national governments. External scientific, technical, and financial support should, by invitation, be made available to these countries from the international community (Hubbell, 1995).

Availability of Credit. For crop productivity research imperatives to be effectively implemented, it is obligatory for policy makers and administrators to provide appropriate financial support to the production systems. Availability of credit to farmers at the right time and in the right amount to buy fertilizers, good seed, and pesticides is an important aspect of this support. In most of the developing countries, farmers do not receive agricultural credits in a timely manner because of government bureaucracy. Before credit is approved, papers have to pass through several authorities, and sometimes the right planting time is already passed before the credit is approved. The application of modern technology requires capital and governments need to help farmers by providing appropriate credit facilities.

III. SUMMARY

Crop yield is the product of interactions involving climate, soil, plants, and people. Quantitative evaluation of these factors and their interactions should become increasingly important in evaluating systems that optimize yield. The best evidence that these factors are not limiting crop growth is provided when high crop yields are produced. Most of the crop production factors can be modified in favor of higher yields by modifying the environment or the physiological characteristics of the crop and by developing cultural practices and cultivars to exploit a specific agricultural environment.

The dramatic increase in yields of most important food crops in the last three decades is beginning to level off. The favorable mix of genetics and technology that has characterized this area must be improved on to achieve even higher yields in the future.

It is obvious that much research and development activity must be performed to remove the biological, physical, and socioeconomic constraints to increased crop production. Yields obtained under experimental conditions are much higher than those under farm conditions. It is important to identify accurately the constraints limiting the farmers' yields.

The potential for yield increases is greatest in South America, Africa, and Asia, where both land area and yields per unit area can be increased. Some important strategies for improving yields in these regions are: 1) evaluating environmental constraints such as climate and soils and developing economically viable methods for improving these constraints; 2) increasing intensities of cropping to permit more complete utilization of soil and water resources; 3) developing cultivars resistant to diseases, insects, and mineral stress; and 4) whenever possible, increasing irrigation facilities which will permit higher cropping intensity and the use of new technology.

REFERENCES

Adams, F. 1981. Alleviating chemical toxicities: Liming acid soils, pp. 269–301. *In*: G. F. Arkin and H. M. Taylor (eds.). Modifying the root environment to reduce crop stress. Monogr. No. 4, Am. Soc. Agric. Engr., St. Joseph, Michigan.

Aimi, R. and T. Murakami. 1964. Cell physiological studies on the effect of aluminum on the growth of crop plants. Natl. Inst. Agric. Sci. Bull. Ser. D 11: 331–396.

Akita, S. 1995. Economic aspects relation to increase commercial and biological production potential of rice, pp. 57–76. *In*: B. S. Peneiro and E. P. Guimares (eds.) Rice in Latin America: Perspective to increase production and production potential. EMBRAPA-CNPAF, Document No. 60, Goiania, Brazil.

Almekinders, C. J. M., L. O. Fresco, and P. C. Struik. 1995. The need to study and manage variation in agro-ecosystems. Netherland S. J. Agric. Sci. 43: 127–142.

Altieri, M. A. and M. Liebman. 1986. Insect, weed and plant disease management in multiple cropping systems, pp. 183–218. *In*: C. A. Francis (ed.) Multiple cropping systems. Macmillan, New York.

Bache, B. W. 1979. Base saturation, pp. 38–41. *In*: R. W. Fairbridge and C. W. Finkl, Jr. (eds.). The encyclopedia of soil science, Part 1. Dowden, Hutchinson and Ross, Stroudsburg, Pennsylvania.

Barea, J. M. and C. A. Aguilar. 1983. Mycorrhizas and their significance in nodulating nitrogen-fixing plants. Adv. Agron. 36: 1–54.

Baver, L. D., W. H. Gardner, and W. R. Gardner. 1972. Soil physics. Wiley Interscience, New York.

Beck, S. D. 1965. Resistance of plants to insects. Annu. Rev. Entomol.10: 207–232.

Begg, J. E. 1965. High photosynthetic efficiency in a low-latitude environment. Nature (London) 205: 1025–1026.

Benites, J. R. and C. S. Ofori. 1993. Crop production through conservation-effective tillage in the tropics. Soil Tillage Res. 27: 9–33.

Bezdicek, D. F. 1979. Nitrogen fixation, pp. 325–332. *In*: R. W Fairbridge and C. W. Finkl, Jr. (eds.). The encyclopedia of soil science, Part 1. Dowden, Hutchinson and Ross, Stroudsburg, Pennsylvania.

Billings, W. D. 1952. The environmental complex in relation to plant growth and distribution. Q. Rev. Biol. 27: 251–265.

Bjorkman, O. 1971. Comparative photosynthetic CO_2 exchange in higher plants, pp. 18–32. *In*: M. D. Hatch, C. B. Osmond, and R. O. Slayter (eds.). Photosynthesis and photorespiration. Wiley Interscience, New York.

Black, C. C. 1971. Ecological implications of dividing plants into groups with distinct photosynthetic production capacities. Adv. Ecol. Res. 7: 87–114.

Blackman, G. E. and J. N. Black. 1959. Physiological and ecological studies in the analysis of plant environment. XIII. The role of the light factors limiting growth. Ann. Bot. 23: 131–145.

Blum, A. 1980. Genetic improvement of drought adaption, pp. 450–452. *In*: N. C. Turner and P. J. Kramer (eds.). Adaption of plants to water and high temperature stress. Wiley, New York.

Boddey, R. M. and J. Dobereiner. 1995. Nitrogen fixation associated with grasses and cereals: Recent progress and perspectives for the future. Fert. Res. 42: 241–250.

Bohn, H. L., B. L. McNeal, and G. A. O'Conner. 1979. Soil Chemistry. Wiley, New York.

Bowen, G. D. and A. D. Rovira. 1991. The rhizosphere: The hidden half of the hidden half, pp. 641–669. *In*: Y. Waisel, A. Eshel, and U. Kafkafi (eds.) Plant roots: The hidden half. Dekker, New York.

Brady, N. C. 1974. The nature and properties of soils. Macmillan, New York.

Broadbent, F. E. 1965. Organic matter, pp. 1397–1400. *In*: C. A. Black (ed.). Methods of soil analysis. Monogr. 9, Am. Soc. Agron., Madison, Wisconsin.

Brougham, R. W. 1960. The relationship between the critical leaf area index, total chlorophyll content, and maximum growth rate of some pasture and crop plants. Ann. Bot. 24: 463–474.

Bruckner, P. L. and R. C. Frohberg. 1987. Stress tolerance and adaptation in spring wheat. Crop Sci. 27: 31–36.

Buddenhagen, I. W. 1983. Breeding strategies for stress and disease resistance in developing countries. Annu. Rev. Phytopathol. 21: 385–409.

Burns, R. C. and R. W. F. Hardy. 1975. Nitrogen fixation in bacteria and higher plants. Springer-Verlag, New York.

Butterworth, M. H. 1964. The digestible energy content of some tropical forages. J. Agric. Sci. 3: 319–321.

Case, V. W., N. C. Brady, and D. J. Lathwell. 1964. The influence of soil temperature and phosphorus fertilizers of different water solubilities on the yield and phosphorus content of oats. Soil Sci. Soc. Am. Proc. 28: 409–412.

Clark, R. B. 1982. Plant response to mineral element toxicity and deficiency, pp. 71–142. *In*: M. N. Christiansen and C. F. Levis (eds.). Breeding plants for less favorable environments. Wiley, New York.

Clarkson, D. T. 1971. Inhibition of the uptake and long distance transport of calcium by aluminum and other polyvalent cations. J. Exp. Bot. 23: 837–851.

Coleman, N. T. and G. W. Thomas. 1967. The basic chemistry of soil acidity, pp. 1–41. *In*: R. W. Pearson and F. Adams (eds.). Soil acidity and liming. Monograph 12, Am. Soc. Agron., Madison, Wisconsin.

Conservation Technology Information Center (CTIC). 1992. National Survey of Conservation Tillage Practices. CTIC, West Lafayette, Indiana.

Cooper, J. P. 1970. Potential production and energy conversion in temperate and tropical grasses. Herbage Abstr. 40: 1–15.

Cooper, J. P. 1973. Genetic variation in herbage constituents, pp. 379–417. *In*: G. W. Butler and R. W. Bailey (eds.). Chemistry and biochemistry of herbage, Vol 2. Academic Press, New York.

Day, P. R. 1973. Genetic variability of crops. Annu. Rev. Phytopathol. 11: 293–312.

Dobereiner, J., S. Urquaiga, and R. M. Boddey. 1995. Alternative for nitrogen nutrition of crop in tropical agriculture. Fert. Res. 42: 339–346.

Donald, C. M. 1968. The breeding of crop ideotypes. Euphytica 17: 385–403.

Donald, C. M. and J. Hamblin. 1976. The biological yield and harvest index of cereals as agronomic and plant breeding criteria. Adv. Agron. 28: 361–405.

Dormaar, J. F. and J. W. Ketcheson. 1960. The effect of nitrogen form and soil temperature on the growth and phosphorus uptake of corn grown in the greenhouse. Can. J. Soil Sci. 40: 177–184.

Downton, W. J. S. 1971. Adaptive and evolutionary aspects of C4 photosynthesis, pp. 3–17. *In*: M. D. Hatch, C. B. Osmond, and R. O. Slayter (eds.). Photosynthesis and photorespiration. Wiley Interscience, New York.

Downton, W. J. S. 1975. The occurrence of C4 photosynthesis among plants. Photosynthetica 9: 96–109.

Dudal, R. 1976. Inventory of the major soils of the world with special reference to mineral stress hazards, pp. 3–13. *In*: M. J. Wright (ed.). Plant adaptation to mineral stress in problem soils. Proceedings of a Workshop, National Agricultural Library, Beltsville, Maryland, Special Publication of Cornell University Press, Ithaca, New York.

Duivenbooden, N. Van, C. T. de Wit, and H. Van Keulen. 1996. Nitrogen, phosphorus and potassium relations in five major cereals reviewed in respect to fertilizer recommendations using simulation modelling. Fert. Res. 44: 37–49.

Duvik, D. N., R. A. Kleese, and N. M. Frey. 1981. Breeding for tolerance of nutrient imbalances and constraints to growth in acid, alkaline and saline soils. J. Plant Nutr. 4: 111–129.

Eckert, J. 1987. Soil test interpretations: Basic cation saturation ratios and sufficiency levels, pp. 53–54. *In*: J. R. Brown (ed.) Soil testing. Sampling, correlation, calibration and interpretation. SSSA Spec. Publ. No. 21, Madison, Wisconsin.

Edwards, G. E., S. B. Ku, and J. G. Foster. 1983. Physiological constraints to maximum yield potential, pp. 105–119. *In*: T. Kommedahl and P. H. Williams (eds.). Challenging problems in plant health. Am. Phytopathol. Soc., St. Paul, Minnesota.

Edwards, J. H., C. W. Wood, D. L. Thurlow, and M. E. Ruf. 1992. Tillage and crop rotation effects on fertility status of a Hapludult soil. Soil Sci. Soc. Am. J. 58: 1577–1582.

Epstein, E. 1966. Effect of soil temperature at different growth stages on growth and development of potato plants. Agron. J. 58: 169–171.

Epstein, E. and D. W. Rains. 1987. Advances in salt tolerance. Plant Soil. 99: 113–125.

Evans, L. T. 1980. The natural history of crop yield. Am. Sci. 68: 388–397.

Fageria, N. K. 1980. Rice in Cerrado soils with water deficiency and its response to phosphorus. Pesq. Agropec. Bras., Brasilia 15: 259–265.

Fageria, N. K. 1982. Differential aluminum tolerance of rice cultivars in nutrient solution. Pesq. Agropec. Bras., Brasilia 17: 1–9.

Fageria, N. K. 1988. Influence of iron on a nutrient uptake by rice. IRRI Newsletter 13: 20–21,

Fageria, N. K. and J. R. P. Carvalho. 1982. Influence of aluminum in nutrient solutions on chemical composition in upland rice cultivars. Plant Soil 69: 31–44.

Fageria, N. K. and A. B. Santos. 1996. Rice and common bean growth and nutrient uptake as influenced by aluminum on a Varzea soil. Paper presented at IV Int. Symp. on Plant-Soil Interactions at Low pH. March 17–24, 1996, Belo Horizonte, Brazil.

Fageria, N. K. and F. J. P. Zimmermann. 1996. Influence of pH on growth and nutrient uptake by crop species in an Oxisol. Paper presented at IV Int. Symp. on Plant-Soil Interactions at Low pH. March 17–24, 1996, Belo Horizonte, Brazil.

Fageria, N. K., V. C. Baligar, and R. J. Wright. 1988. Aluminum toxicity in crop plants. J. Plant Nutr. 11: 303–319.

Finkl, C. W., Jr. 1979. Organic matter, pp. 348–349. *In*: R. W. Fairbridge and C. W. Finkl (eds.). The encyclopedia of soil science, Part 1. Dowden, Hutchinson and Ross, Stroudsburg, Pennsylvania.

Foy, C. D. 1974. Effects of aluminum in plant growth, pp. 601–642. *In*: E. W. Carson (ed.). The plant root and its environment. University Press of Virginia, Charlottesville.

Francois, L. E. 1994. Growth, seed yield, and oil content of canola grown under saline conditions. Agron. J. 86: 233–237.

Gaffron, H. H. 1960. Energy storage: Photosynthesis. *In*: F. C. Steward (ed.). Plant Physiol., Vol. 1B. Academic Press, New York.

Gallagher, J. N. and P. V. Biscoe. 1978. Radiation absorption, growth and yield of cereals. J. Agric. Sci. (Cambridge) 91: 47–60.

Garrity, D. P. and J. C. O'Toole. 1995. Selection for reproductive stage drought avoidance in rice, using infrared thermometry. Agron. J. 87: 773–779.

Gibbon, D., R. Holliday, F. Mattei, and G. Luppi. 1970. Crop production potential and energy conversion efficiency in different environments. Exp. Agric. 6: 197–204.

Gill, W. R. and G. E. VandenBerg. 1967. Soil dynamics in tillage and traction. USDA Handbook, No. 316.

Golley, F. B. 1961. Energy values of ecological materials. Ecology 42: 581–584.

Goyne, P. J., S. P. Milroy, J. M. Lilley, and J. M. Hare. 1993. Radiation interception, radiation use efficiency and growth of barley cultivars. Aust. J. Agric. Res. 44: 1351–1366.

Grable, A. R. 1966. Soil aeration and plant growth. Adv. Agron. 18: 57–106.

Green, C. F. 1984. Discrimants of productivity in small grain cereals: A review. J. National Inst. Agric. Bot. 16: 453–463.

Guerin, L. J. and T. F. Guerin. 1994. Constraints to the adoption of innovations in agricultural research and environmental management: A review. Aust. J. Exp. Agric. 34: 549–571.

Hang, A. N. and D. W. Evans. 1985. Deficit sprinkler irrigation of sunflower and safflower. Agron. J. 77: 588–592.

Hartt, C. E. 1965. The effects of temperature upon translocation of C^{14} in sugarcane. Plant Physiol. 40: 74–81.

Harward, M. E., W. A. Jackson, W. L. Lott, and D. D. Mason. 1955. Effects of Al, Fe and Mn upon the growth and composition of lettuce. Proc. Am. Soc. Hortic. Sci. 66: 261–266.

Hausenbuiller, R. L. 1972. Soil science principles and practices. W. M. C. Brown Company, Iowa.

Heinrich, G. M., C. A. Francis, and J. D. Eastin. 1983. Stability of grain sorghum yield components across diverse environments. Crop Sci. 23: 209–212.

Heinrichs, D. H. and K. F. Nielsen. 1966. Growth response of alfalfa varieties of diverse genetic origin to different root zone temperatures. Can. J. Plant Sci. 46: 291–298.

Hesketh, J. D. And D. N. Moss. 1963. Variation in the response of photosynthesis to light. Crop Sci. 3: 107–110.

Hiatt, A. J., D. F. Amos, and H. F. Massey. 1963. Effects of aluminum on copper sorption by wheat. Agron. J. 55: 284–287.

Hoffman, G. J. 1981. Alleviating salinity stress, pp. 305–343. *In*: G. F. Arkin and H. M. Taylor (eds.). Modifying the root environment to reduce crop stress. Monogr. 4, Am. Soc. Agric. Eng., St. Joseph, Michigan.

Holliday, R. 1966. Solar energy consumption in relation to crop yield. Agric. Prog. 41: 24–34.

Howeler, R. H., E. Sieverding, and S. Saif. 1987. Practical aspects of mycorrhizal technology in some tropical crops and pastures. Plant Soil 100: 249–283.

Hubble, D. H. 1995. Extension of symbiotic biological nitrogen fixation technology in developing countries. Fert. Res. 42: 231–239.

Hühn, M. 1987. Stability analysis of winter-rape by using plant density and mean yield per plant. J. Agron. Crop Sci. 159: 73–81.

Jenny, H. 1930. A study on the influence of climate upon the nitrogen and organic matter content of the soil. Missouri Agric. Exp. Stn. Res. Bull. 152: 1–66.

Johnson, R. E. and W. A. Jackson. 1964. Calcium uptake and transport by wheat seedlings as affected by aluminum. Soil Sci. Soc. Am. Proc. 28: 381–386.

Kamprath, E. J. and C. D. Foy. 1971. Lime-fertilizer plant interactions in acid soils, pp. 105–151. *In*: R. W. Olsen, T. J. Army, J. J. Hanway, and V. J. Kilmer (eds.). Fertilizer technology and use., 3rd ed. Soil Sci. Soc. Am., Madison, Wisconsin.

Kapland, D. I. and G. O. Estes. 1985. Organic matter relationship to soil nutrient status and aluminum toxicity in alfalfa. Agron. J. 77: 735–738.

Kaspar, T. C., T. M. Crosbie, R. M. Cruse, D. C. Erbach, D. R. Timmons, and K. N. Potter. 1987. Growth and productivity of four corn hybrids as affected by tillage. Agron. J. 79: 477–481.

Kiniry, J. R., C. A. Jones, J. C. O'Toole, R. Blanchet, M. Cabelguenne, and D. A. Spanel. 1989. Radiation-use efficiency in biomass accumulation prior to grain-filling for five grain crop species. Field Crops Res. 20: 51–64.

Kueneman, E. A., D. H. Wallace, and P. M. Ludford. 1979. Photosynthetic measurements of field grown dry beans and their relation to selection for yield. J. Am. Soc. Hortic. Sci. 104: 480–482.

Lafever, H. N. 1981. Genetic differences in plant response to soil nutrient status. J. Plant Nutr. 4: 89–109.

Lal, R. 1993. Tillage effects on soil degradation, soil resilience, soil quality, and sustainability. Soil Tillage Res. 27: 1–8.

Lance, J. C. and R. W. Pearson. 1969. Effect of low concentrations of aluminum on growth and water and nutrient uptake by cotton roots. Soil Sci. Soc. Am. Proc. 33: 95–98.

Larue, T. A. and T. G. Patterson. 1981. How much nitrogen do legumes fix? Adv. Agron. 34: 15–38.

Lauchli, A. 1987. Soil science in the next twenty-five years: Does biotechnology play a role? Soil Sci. Soc. Am. J. 51: 1405–1408.

Lin, C. S., M. R. Binns, and L. P. Leukovitch. 1986. Stability analysis: Where do we stand? Crop Sci. 26: 894–900.

Loomis, R. S. 1983. Crop manipulation for efficient use of water: An overview, pp. 345–374. *In*: H. M. Taylor, W. R. Jordan, and T. R. Sinclair (eds.). Limitations to efficient water use in crop production. Am. Soc. Agron., Madison, Wisconsin.

Loomis, R. S. and W. A. Williams. 1963. Maximum crop productivity: An estimate. Crop Sci. 3: 67–72.

Low, A. J. 1979. Soil Structure, p. 508–514. *In*: R. W. Fairbridge and C. W. Finkl, Jr. (eds.). The encyclopedia of soil science, Part 1. Dowden, Hutchinson and Ross, Stroudsburg, Pennsylvania.

Mack, H. J., S. C. Fang, and S. B. Butts. 1964. Effects of soil temperature and phosphorus fertilization on snap beans and peas. Proc. Am. Soc. Hortic. Sci. 84: 332–338.

Mahon, J. D. 1983. Limitations to the use of physiological variability in plant breeding. Can. J. Plant Sci. 63: 11–21.

McCracken, R. J. 1987. Soils, soil scientists, and civilization. Soil Sci. Soc. Am. J. 51: 1395–1400.

McCree, K. J. 1981. Photosynthetically active radiation, pp. 41–55. *In*: O. L. Lange, P. S. Nobel, C. B. Osmond, and H. Ziegler (eds.). Encyclopedia of plant physiology, New Ser., Vol 12A. Springer-Verlag, New York.

McLean, E. O. and B. E. Gilbert. 1927. The relative aluminum tolerance of crop plants. Soil Sci. 24: 163–175.

Menz, K. M., D. N. Moss, R. Q. Cannell, and W. A. Brun. 1969. Screening for photosynthetic efficiency. Crop Sci. 9: 692–694.

Monteith, J. L. 1977. Climate and efficiency of crop production in Britain. Philosophical Transactions of the Royal Society of London B281: 277–294.

Monteith, J. L. 1978. Reassessment of maximum growth rates for C_3 and C_4 crops. Exp. Agric. 14: 1–5.

Mosse, B. 1981. Vesicular-arbuscular mycorrhiza research for tropical agriculture. Res. Bull. 194, Hawaii Institute of Tropical Agriculture and Human Resources, University of Hawaii.

Murata, Y. and J. Iyama. 1963. Studies on the photosynthesis of forage crops. II. Influence of air-temperature upon the photosynthesis of some forage and grain crops. Proc. Crop Sci. Soc. Jpn. 31: 315–322.

Nielsen, K. F. 1974. Roots and root temperatures, pp. 293–333. *In*: E. W. Carson (ed.). The plant root and its environment. University Press of Virginia, Charlottesville.

Oertli, J. J. 1979. Soil fertility, pp. 453–462. *In*: R. W. Fairbridge and C. W. Finkl, Jr. (eds.). The encyclopedia of soil science, Part 1. Dowden, Hutchinson and Ross, Stroudsburg, Pennsylvania.

Orlov, D. S. 1979. Physical chemistry, pp. 377–382. *In*: R. W. Fairbridge and C. W. Finkl, Jr. (eds.). The encyclopedia of soil science, Part 1. Dowden, Hutchinson and Ross, Stroudsburg, Pennsylvania.

Owen, P. C. 1971. The effects of temperature on the growth and development of rice. Field Crop Abstr. 24: 1–8.

Papendick, R. I. and L. F. Elliott. 1983. Soil physical factors that affect plant health, pp. 168–180. *In*: T. Kommedahl and P. H. Williams (eds.). Challenging problems in plant health. Am. Soc. Phytopathol., St. Paul, Minnesota.

Pearcy, R. W. and J. Ehleringer. 1984. Comparative ecophysiology of C_3 and C_4 plants. Plant Cell Environ. 7: 1–13.

Pearson, R. W., L. F. Ratliff, and H. M. Taylor. 1970. Effect of soil temperature, strength, and pH on cotton seedling root elongation. Agron. J. 62: 243–246.

Piepho, H. P. 1995. A simple procedure for yield components analysis. Euphytica 84: 43–48.

Ponnamperuma, F. N. 1972. The chemistry of submerged soils. Adv. Agron. 24: 29–96.

Postgate, J. R. and S. Hills. 1979. Microbial ecology. Blackwell, Oxford.

Power, J. F., D. L. Grunes, G. A. Reichman, and W. O. Willis. 1970. Effect of soil temperature on rate of barley development and nutrition. Agron. J. 62: 567–571.

Radke, J. F. and R. E. Bauer. 1969. Growth of sugarbeets as affected by root temperatures. Part I. Greenhouse studies. Agron. J. 61: 860–863.

Russell, E. W. 1973. Soil conditions and plant growth, 10th ed. Longman, London.

Salisbury, F. B. and C. W. Ross. 1985. Photosynthesis: Environmental and agricultural aspects, pp. 216–228. Plant physiology, 3rd ed., Wadsworth, Belmont, California.

Sanchez, P. A. 1976. Properties and management of soils in the tropics. Wiley, New York.

Senstius, M. W. 1958. Climax forms of rock weathering. Am. Sci. 46: 355–367.

Sibma, L. 1968. Growth of closed green crop surfaces in the Netherlands. Neth. J. Agric. Sci. 16: 211–216.

Sinclair, T. R. and T. Horie. 1989. Leaf nitrogen, photosynthesis, and crop radiation use efficiency: A review. Crop Sci. 29: 90–98.

Sinclair, T. R., T. Shiraiwa, and G. L. Hammer. 1992. Variation in crop radiation use efficiency with increased diffusion radiation. Crop Sci. 31: 1281–1284.

Soil Survey Staff. 1951. Soil survey manual. USDA Handbook, No. 18.

Stevenson, F. J. 1986. Cycles of soil. Wiley Interscience, New York.

Stockle, C. O. and J. R. Kiniry. 1990. Variability of crop radiation-use efficiency associated with vapour-pressure deficit. Field Crops Res. 25: 171–181.

Tinker, P. B. 1982. Mycorrhizas: the present position. Trans. Int. Congr. Soil Sci., 12th ed. 5: 150–166.

Unger, P. W. and T. M. McCalla. 1980. Conservation tillage systems. Adv. Agron. 33: 1–58.

Unger, P. W., H. V. Eck, and J. T. Musick. 1981. Alleviating plant water stress, pp. 61–96. *In*: G. F. Arkin and H. M. Taylor (eds.). Modifying the root environment to reduce crop stress. Monogr. 4, Am. Soc. Agric. Eng., St. Joseph, Michigan.

Voorhees, W. B., R. R. Allmaras, and C. E. Johnson. 1981. Alleviating temperature stress, pp. 217–266. *In*: G. F. Arkin and H. Taylor (eds.). Modifying the root environment to reduce crop stress. Monogr. 4, Am. Soc. Agric. Eng., St. Joseph, Michigan.

Wambeke, A. Van. 1976. Formation, distribution and consequences of acid soils in agricultural development, pp.15–24. *In*: M. J. Wright (ed.). Plant adaptation to mineral stress in problem soils. Proceedings of a Workshop, National Agricultural Library, Beltsville, Maryland, Special Publication of Cornell University Press, Ithaca, New York.

Weltzien, E. and G. Fischbeck. 1990. Performance and variability of local barley landraces in Near-Eastern environments. Plant Breeding 104: 58–67. *In*: W. S. de Boef, K. Amanor, K. Wellard, and A. Beppington (eds.) Cultivating knowledge. Genetic diversity, farmer experimentation and crop research. IT Publication Ltd., London.

Westlake, D. F. 1963. Comparisons of plant productivity. Biol. Rev. 38: 385–425.

Whitman, C. E., J. L. Hatfield, and R. J. Reginato. 1985. Effect of slope position on the microclimate, growth, and yield of barley. Agron. J. 77: 663–669.

Whitfield, C. J. and D. E. Smika. 1971. Soil temperature and residue effects on growth components and nutrient uptake of four wheat varieties. Agron. J. 63: 297–300.

Williams, W. A., R. S. Loomis, W. G. Duncan, A. Dovrat, and F. Nunez. 1968. Canopy architecture at various population densities and the growth and grain yield of corn. Crop Sci. 8: 303–308.

Wood, C. W., D. G. Westfall, G. A. Peterson, and I. C. Burke. 1990. Impacts of cropping intensity on carbon and nitrogen mineralization under no-till dryland agroecosystems. Agron. J. 82: 1115–1120.

Wright, J. L. and E. R. Lemon. 1966. Photosynthesis under field conditions. IX. Vertical distribution of photosynthesis within a corn crop. Agron. J. 58: 265–269.

Yoshida, S. 1972. Physiological aspects of grain yield. Annu. Rev. Plant Physiol. 23: 437–464.

Zadoks, J. C. and R. D. Schein. 1979. Epidemiology and plant disease management. Oxford University Press, New York.

3

Nutrient Flux in Soil-Plant System

I. INTRODUCTION

Nutrient flux in the soil-plant system is predicted by complex interactions between plant roots, soil microorganisms, chemical reactions, and pathways for loss. The concentration dependence of most of the processes that take place in soil implies that when the immediate supply exceeds the ability of the plant to take up a nutrient, various processes will act to reduce its concentration (Shaviv and Mikkelsen, 1993). Such processes include transformations induced by microbes such as nitrification, denitrification, and immobilization. Further included are chemical processes such as exchange, fixation, precipitation, and hydrolysis as well as physical ones such as leaching runoff and volatilization. The extent by which nutrients are removed from solution by the processes competing with plant uptake can thus affect both nutrient use efficiency and the environment (Shaviv and Mikkelsen, 1993).

Nutrient uptake by plants depends on ion concentrations at the root surfaces, root absorption capacity, and plant demand. It is a dynamic series of processes in which nutrients must be continuously replenished in soil solution from the soil solid phase and transported to roots as uptake proceeds. After uptake by roots, nutrients are translocated to the various plant organs for utilization in different metabolic processes. In this way, nutrient uptake by plants involves several interconnected processes such as nutrient release from the soil solid phase to solution, transport to roots for absorption, and plant translocation and utilization. The consequence is that nutrient uptake by plants is simultaneously influenced by soil, climatic, and plant factors.

Nutrient transport to roots, absorption by roots, and translocation in the shoot all occur simultaneously. This means that a rate change of one process will ultimately influence all other processes involved in uptake. In other words, if one process slows down, it may become the limiting factor in the uptake process. Only inorganic nutrients can be absorbed by plants. If nutrients are present in organic forms, they have to undergo mineralization before uptake by plants can occur.

II. NUTRIENT SUPPLY TO PLANT ROOTS

In the process of nutrient absorption by plants, transport of the nutrients to the vicinity of roots is the first step. Nutrient supply to the roots is governed by nutrient concentrations in the soil solution, nature of the nutrients, soil moisture status, and plant's absorption capacity. At any instance in time, the concentration of a nutrient in the solution immediately adjacent to a root appears to be the best measure of its availability for absorption, though many factors within the plant and the concentrations of other ions in the solution phase may influence the actual rate at which it is absorbed (Russell, 1977).

Plant and soil properties interact in the transfer of nutrients from the soil into the plants (Jungk and Claassen, 1989). The major processes and factors that contribute and interact in the transfer of nutrients from soil to plants are summarized in Table 3.1. In the soil system, nutrients move to plant roots by mass flow, diffusion, and root interception.

A. Mass Flow

Mass flow is the passive transport of nutrients to the root in soil water as it is absorbed by plants. The amounts of nutrients reaching roots by this process depend on the concentrations of nutrients in the soil solution and the rate of water transport to and into the roots. The nutrient supply by mass flow is affected by soil properties, climatic conditions, solubility of the nutrient, and plant species. The level of a particular nutrient in the soil solution near the root may increase, stay the same, or decrease depending on the balance between the rate of supply to the root by mass flow and the rate of absorption into the root (Barber, 1995).

Table 3.1 Processes and Factors Involved in Nutrient Transfer from Soil to Plant

Process	Factors
Root development	Root length Root distribution Root morphology, diameter, and hairs
Nutrient uptake	Concentration at root surface Kinetics of uptake
Transport from soil to root (Mass flow, diffusion)	Transpiration Concentration of gradient
Mobilization by roots (Desorption, dissolution, hydrolysis of organic compounds)	Depletion of soil solution Root exudates (H^+, HCO_3^-, reducing agents, chelates, organic anions) Chemical soil composition pH of soil solution Enzymes (e.g., phosphatases)
Mobilization by associated organisms	Mycorrhizal infection Bacteria

Source: Jungk (1991).

The contribution of mass flow in the process of nutrient supply to the roots can be calculated from the product of the soil solution concentration and the volume of water transpired by the plant. This, however, does not take into account mass flow of the soil solution which is not due to the water uptake by the root, such as gravity-induced movement of water down the soil profile. Mass flow rate can be calculated with the help of the following equation:

$$MF = C \times WU$$

where MF is the contribution to ion uptake by mass flow, C is the solution concentration of any given ion, and WU is the total water uptake, which is the water content in the plant plus the water transpired. Transpiration is not a constant process. It varies with plant species, climate, soil conditions, location of water source in soil, age of plants, and time. At night, mass flow is greatly restricted, which causes diurnal fluctuations at the root surface. Furthermore, nutrient concentrations in plants vary with development stages. Young plants usually have higher concentrations of nutrients; therefore, mass flow would then contribute a smaller fraction of the demand (Jungk, 1991).

Table 3.2 shows the estimates of nutrients supplied to corn roots by three processes. Mass flow can meet the crop's nutrient requirements for all nutri-

Table 3.2 Estimated Amounts of Nutrients Supplied by Mass Flow, Diffusion, and Root Interception to Corn Roots in a Fertile Alfisol

Nutrient	Approximate amount (% of total uptake)		
	Mass flow	Diffusion	Root interception
Nitrogen	79	20	1
Phosphorus	5	93	2
Potassium	18	80	2
Calcium	375	0	150
Magnesium	222	0	33
Sulfur	295	0	5
Iron	66	21	13
Zinc	230	0	43
Manganese	22	35	43
Copper	219	0	6
Boron	1000	29	29

Source: Calculated from the data of Barber (1966, 1974).

ents except N, P, K, Fe, and Mn, provided sufficient nutrient concentrations are present in the soil solution.

B. Diffusion

Diffusion can be defined as the movement of molecules from a region of high concentration to a region of low concentration. When the supply of the nutrients to the root vicinity is not sufficient to satisfy the plant demand by mass flow and root interception, a concentration gradient develops and nutrients move by diffusion. Diffusion is described by Fick's first law (Barber, 1974; Mengel, 1985):

$$F = -D\left(\frac{dc}{dx}\right)$$

where F is the flux, dc/dx the concentration gradient, and D is the diffusion coefficient that generally describes the diffusivity of a homogeneous medium. Since the soil is not homogeneous, the concept of a diffusive flux in the soil medium presents difficulties. Nye (1979) proposed the following formula to calculate the diffusion coefficient for soil medium:

$$D = D_e \theta f_e \left(\frac{dC_e}{dC} \right) + D_E$$

where D is the diffusion coefficient in the whole soil medium, D_e is the diffusion coefficient in free water, θ is the fraction of the soil volume filled with solution, f_e is the impedance factor, C_e is the ion concentration in the soil solution, C is the concentration of labile forms in soil, and D_E is surface diffusion. In the process of diffusion, soil and plant factors are involved. The following equation illustrates the factors which are important in determining the rate at which a soluble nutrient diffuses to the root surface (Corey, 1973):

$$\frac{dq}{dt} = DAP \frac{C_1 - C_2}{L}$$

where dq/df represents the rate of diffusion to the root surface; D is the diffusion coefficient of the nutrient in water; A is the cross-sectional area considered, which can be assumed to represent the total absorbing surface of a plant root; P is the fraction of the soil volume occupied by water (it also includes a tortuosity factor); C_1 is the concentration of the soluble nutrient at a distance L from the root surface; C_2 is the concentration of the soluble nutrient at the root surface; and L is the distance from the root surface to where C_1 is measured.

The distance for diffusive nutrient movement through the soil to the root is usually in the range of 0.1–15 mm (Barber, 1974). Hence, only soil nutrients within this soil zone contribute to diffusive nutrient supply to roots. Diffusion coefficients for some ions in soil solution are given in Table 3.3. It can be

Table 3.3 Diffusion Coefficients for Some Ions in Soil Solution

Ion	Diffusion ($cm^2\ s^{-1}$)	Reference
NO_3^-	1×10^{-6}	Nye, 1969
NO_3^-	$10^{-6} - 10^{-7}$	Barber, 1974
NH_4^+	1.4×10^{-6}	Husted and Low, 1954
$H_2PO_4^-$	$10^{-8} - 10^{-11}$	Barber, 1974
$H_2PO_4^-$	2.4×10^{-11}	Vasey and Barber, 1963
K^+	1.4×10^{-6}	Husted and Low, 1954
K^+	$10^{-7} - 10^{-8}$	Barber, 1974
K^+	$2.1 - 9.5 \times 10^{-7}$	Baligar, 1984
Ca^{2+}	$0.9 - 4.0 \times 10^{-7}$	Baligar, 1984
Ca^{2+}	3×10^{-7}	Spiegler and Coryell, 1953
Mg^{2+}	$0.6 - 11.5 \times 10^{-7}$	Baligar, 1984
Cl^-	1.2×10^{-6}	Dutt and Low, 1962
MoO_4^{2-}	$0.5 - 8.4 \times 10^{-7}$	Lavy and Barber, 1964

Table 3.4 Soil Volume Occupied by Roots of Different Crops at 0–15 cm Soil Depth

	Soil volume occupied (%)	Reference
Soybean (*Glycine max* L. Merr.)	0.91	Dittmer, 1940
Oats (*Avena sativa* L.)	0.55	Dittmer, 1940
Rye (*Secale cereale* L.)	0.85	Dittmer, 1940
Corn (*Zea mays* L.)	0.19–1.06	Barber, 1971
Wheat (*Triticum aestivum* L.)	0.64	Barber, 1974
Alfalfa (*Medicago sativa* L.)	1.1	Barber, 1974

observed from Table 3.3 that phosphorus has the slowest diffusion rate. When an ion in the bulk soil solution is dilute and its diffusion is slow, its concentration at the root surface can be reduced very quickly (in the space of a few hours) to almost zero if plant demand is high. Table 3.2 shows that plant requirements for P and K are mostly met by the process of diffusion.

C. Root Interception

As roots grow in the soil, they push the soil aside and root surfaces come in direct contact with soil particles and plant nutrients. Nutrient interception by roots depends on the soil volume occupied by roots, root morphology, and the concentration of nutrients in the root-occupied soil volume. The amount of roots present per unit volume of soil can be measured in terms of root surface area, root length, or root volume. The root surface available for ion absorption is a function of surface area. Root density varies with soil properties, species, and management practices. Table 3.4 presents root density data obtained by different researchers for some important field crops. On the average, the soil volume occupied by roots of important food crops is around 0.7–0.9%. Table 3.2 shows the quantity of nutrients intercepted by roots of corn in a fertile silt loam soil. The only nutrient which might be supplied completely by interception is Ca, although the process may provide a significant part of the requirement for Mg, Zn and Mn.

III. ION ABSORPTION BY PLANTS

Nutrient absorption by plants is usually referred to as ion uptake or ion absorption because it is the ionic form in which nutrients are adsorbed (Hiatt and Leggett, 1974). Ion uptake by intact plants is certainly a catenary process. After reaching plant roots, ions have to enter into root cells, be transported

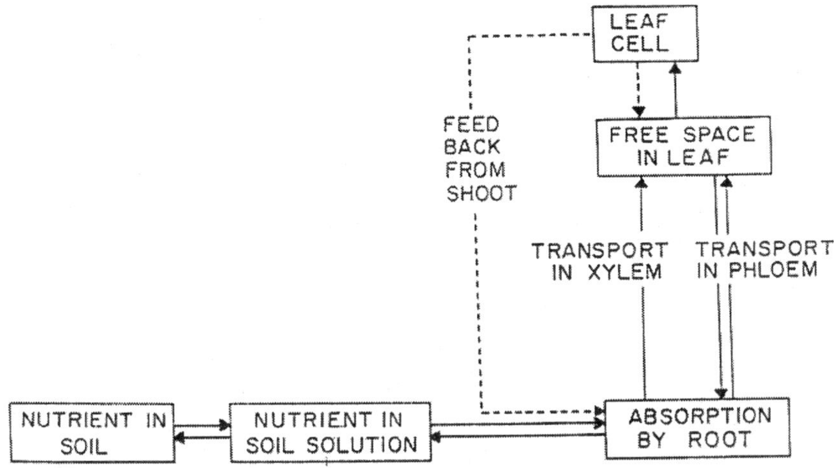

Figure 3.1 Process of nutrient uptake in soil-plant system. (Modified after Pitman, 1972.)

from cell to cell toward the xylem, be excreted into the xylem vessels, and finally be transported to the growing organs in the plant. In such a chain process, the speed of the slowest link will finally govern the reaction velocity of the whole process (Becking, 1956). A great deal of work has been done on ion uptake in plants, but the process still is not fully understood. The overall process of nutrient uptake in the soil-plant system is summarized in Figure 3.1. In ion absorption, roots constitute the dominant organ of plants; therefore, a brief discussion of root morphology is important.

A. Root Morphology

The root structure of crop plants is similar, although it may vary in size. A cross-section of a root is shown in Figure 3.2. Root structures are composed of root hairs, epidermis, cortex, endodermis, and stele. Stele is composed of xylem and phloem. Ions absorbed by root hairs move through the epidermis, cortex, endodermis, and stele, and finally to the xylem. Through xylem the ions are translocated to the shoot, and the phloem supplies photosynthate from shoot to root. There are two parallel pathways of solute movement across the cortex to reach the stele: one passing through the extracellular space, or apoplast (cell wall and intercellular spaces), and another passing from cell to cell in the symplast through the plasmodesmata, which bypasses the vacuoles (Marschner, 1995).

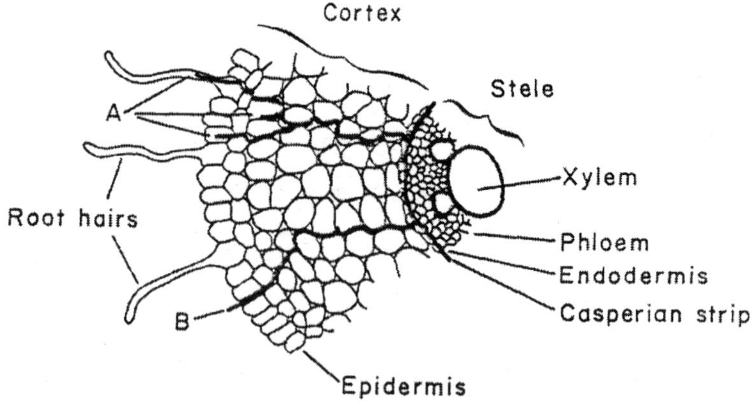

Figure 3.2 Transverse section of corn root showing the symplastic (A) and apoplastic (B) pathways of ion transport across the root. (From Marschner, 1995; copyright © Academic Press, reproduced with permission.)

B. Plant Cell and Membranes

Active nutrient uptake takes place across the plant cells, which are surrounded by membranes. Therefore, it is important to give a brief account of cell structure and membranes to facilitate understanding of nutrient uptake mechanisms. Figure 3.3 shows a simplified diagram of a mesophyll cell and its components. It has a vacuole, nucleus, chloroplasts, ribosomes, and mitochondria, all embedded in a cell wall. The plasma membrane separates the cytoplasm and cell wall, while the tonoplast membrane separates the cytoplasm and vacuole. The plasmalemma membrane forms the boundary between the cell and the outer medium, and it is this membrane—not the cell wall—which presents the effective barrier against uptake of all ions and molecules dissolved in the aqueous outer medium (Mengel and Kirkby, 1982).

Each of the plant cell organs has specific functions to facilitate plant growth and development. The vacuole plays an important role in the water economy of the cell, as well as providing a site for the segregation of water and the end products of metabolism. Chloroplasts are the organs in which light energy conversion and CO_2 assimilation take place. In the mitochondria, enzymes are present which control the various steps of the tricarboxylic acid cycle, respiration, and fatty acid metabolism. The ribosomes are supermolecular assemblies of ribosomal nuclei acid and proteins which enable the synthesis of polypeptides from free amino acids. Plasmodesmata connect the cell to other cells (Mengel and Kirkby, 1982).

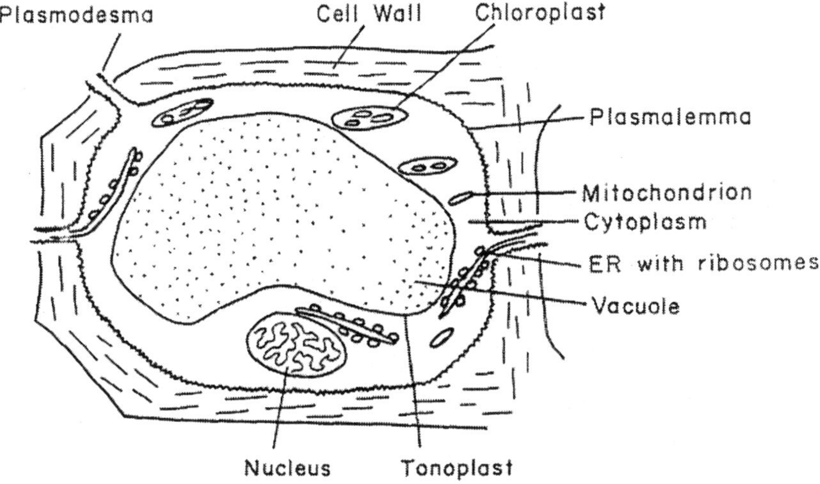

Figure 3.3 Simplified scheme of a mesophyll cell. (From Mengel and Kirkby, 1982.)

As far as membrane structure is concerned, biological membranes consist of protein and lipid molecules in approximately equal proportions and are about 7–10 nm thick (Mengel and Kirkby, 1982). Danielli and Davson (1935) proposed a unit-membrane model which presented biological membranes as a unit consisting of two lipid molecular layers in which the hydrophobic tails of the fatty acids are oriented inward (Marschner, 1995). Both outer boundaries of the lipid layer are coated with a protein layer (Figure 3.4a). This type of structure could well serve as the barrier because the protein layer would enhance rigidity and the lipidic fraction would prevent penetration of the membrane by hydrophilic particles, including hydrated inorganic ions (Mengel and Kirkby, 1982).

Singer (1972), based on electron microscopic studies, proposed a membrane model consisting mainly of a liquid amphiphilic bilayer and amphiphilic proteins (Figure 3.4b). The term amphiphilic indicates the presence of both hydrophilic (OH groups, NH_2 groups, phosphorus groups, carboxylic groups) and hydrophobic (hydrocarbon chains) regions in the membrane. Lipids and proteins may thus be bound by electrostatic bonds, hydrogen bonds, and hydrophobic bonds (Mengel and Kirkby, 1982). In the Singer model, there is no coating protein on the outer sides of the membrane such as exists in Danielli-Davson model, and globular proteins are embedded in the lipid bilayer. Some of these proteins may even extend through the membrane, forming protein channels from one side of the membrane to the other. These channels

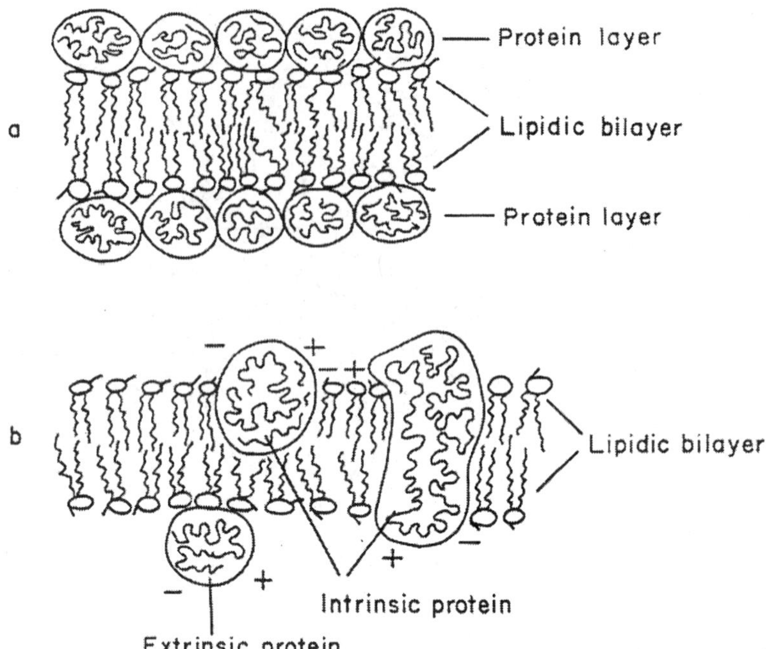

Figure 3.4 Biological membrane models: (a) Danielli-Davson and (b) Singer. (From Mengel and Kirkby, 1982.)

can be considered the hydrophilic pores through which polar solutes such as ions are transported (Walker, 1976). Biological membranes are not completely impermeable. They may allow the diffusion of hydrophilic ions and molecules, the degree of permeability depending on the components which make up the membranes. In addition, enzymes present in biological membranes may directly or indirectly be involved in the transport of ions and molecules across the membrane (Mengel and Kirkby, 1982).

C. Active and Passive Ion Transport

Ion movements in plant cells are active and passive. In the passive process, ions move from a higher to a lower concentration or down a chemical gradient of potential energy. In the case of active uptake, ions move against a concentration gradient and ion movement depends on an electrochemical potential gradient. In the active process, cations are attracted to a negative electropotential, whereas anions are attracted to a positive electropotential. Electrochemical potentials are established across membranes by unequal charge distributions. The difference between the membrane potential and the actual potential created by the nonequilibrium distribution is a measure of the quantity of energy required. A modified Nernst equation that can be used to calculate electrical charge is as follows (Ting, 1982):

$$\psi = \frac{RT}{ZF} \ln \frac{a_i}{a_o}$$

where ψ is the electrochemical potential between the root cells and the external solution in millivolts (mV), R the gas constant (8.3 J mol^{-1} K^{-1}), T the absolute temperature (K), Z the net charge on the ion (dimensionless), F the Faraday constant (96,400 J mol^{-1}), a_i the activity of the ion inside the tissue, and a_o the activity of the ion outside the tissue. For quick calculation, it is convenient to remember that RT/F = 26 mV.

Measurement of the electrochemical potential in the cell and outer medium and of the ionic concentrations in the cell and outer medium can give an indication whether the ions move actively or passively. Values from these measurements can be put into the Nernst equation to calculate electrical potential. For cations, a negative value indicates passive uptake and a positive value active uptake. For anions, negative values are indicative of active transport and positive values of passive transport (Mengel and Kirkby, 1982). These measurements are only valid when equilibrium conditions are maintained in the system, which is difficult under practical conditions. The electrical potential difference across the plasma membrane is in the range of –60 to –200 mV (cytoplasmic negative), and the electrical potential difference across the tonoplast is relatively small, only 0 to –20 mV with the cytoplasm being negative relative to the vacuole (Hodges, 1973).

D. Ion Uptake Mechanisms

The concentration of ions in the cytoplasm is often much higher than in the soil solution, in extreme cases 10,000-fold higher. Therefore, the roots must be able to take up ions against a considerable concentration gradient. Uptake against a concentration gradient or, strictly speaking, against an electrochemical gradient requires metabolic energy, and the process is commonly termed active uptake. At present, there are two principal theories of ion transport across the membrane: the carrier theory and the ion pump theory (Mengel and Kirkby, 1982).

The Carrier Theory

The term carrier is commonly used to refer to an agent responsible for transporting ions from one side of the membrane to the other. Carriers have

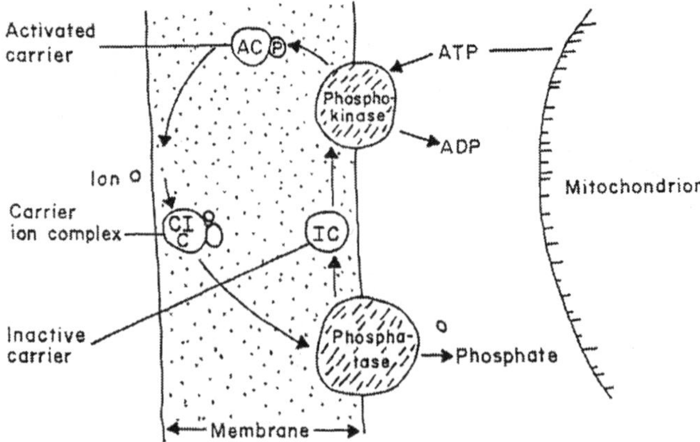

Figure 3.5 Carrier ion transport mechanism. (From Mengel and Kirkby, 1982.)

properties similar to those of enzymes, but, unlike enzymes, carriers have not been isolated and characterized. Isolation of a carrier will not necessarily entail its removal from the membrane, but there is no way of measuring its activity. Figure 3.5 shows a hypothetical scheme of the carrier ion transport across a membrane. In this transport process, a carrier meets the particular ions for which it has affinity, forms a carrier ion complex, and moves across the membrane. The enzyme phosphatase, which is located at the inner membrane boundary, splits off the phosphate group from the carrier complex, and the ion is released. In this process of transport, energy is required and involvement of ATP (adenosine triphosphate) is reported (Marschner, 1995). The ATP is generated from ADP (adenosine diphosphate) plus inorganic phosphate (P_i) via respiration (oxidative phosphorylation).

The whole uptake may be described as follows (Mengel and Kirkby, 1982):

$$\text{Carrier} + \text{ATP} \xrightarrow{\text{Kinase}} \text{carrier}Ⓟ + \text{ADP}$$

$$\text{Carrier}Ⓟ + \text{ion} \longrightarrow \text{carrier}Ⓟ - \text{ion}$$

$$\text{Carrier}Ⓟ - \text{ion} \xrightarrow{\text{Phosphatase}} \text{carrier} + P_i + \text{ion}$$

$$\text{Net: ion} + \text{ATP} \xrightarrow{\text{Transport}} \text{ion} + \text{ADP} + P_i$$

ATPase Theory of Ion Transport

Hodges (1973) proposed the ATPase theory of ion transport in plants (Figure 3.6). ATPase is a group of enzymes that have the capacity to dissociate ATP

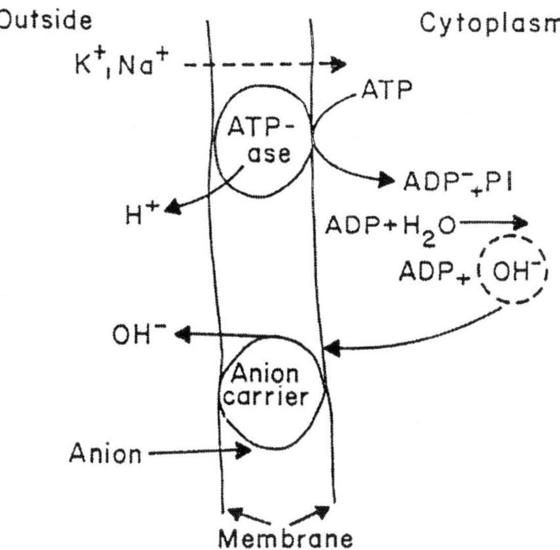

Figure 3.6 ATPase model of ion transport. (From Hodges, 1973, and Mengel and Kirkby, 1982.)

into ADP and inorganic phosphate. Energy liberated by this process can be utilized in ion transport across the membrane. In plants, the phenomenon is known as activity of ATP, which is associated with the plasmalemma and is activated by cations (Hodges et al., 1972). A detailed description of this process of ion transport is given by Hodges (1973), Mengel and Kirkby (1982), and Clarkson (1984).

E. Ion Uptake Kinetics

Epstein and Hagen (1952) formulated the enzyme kinetic hypothesis of ion transport and carrier function. A brief description of this hypothesis is given here because the enzyme kinetic hypothesis of membrane transport has been extensively reviewed in the literature (Clarkson and Hanson, 1980; Epstein, 1973; Hodges, 1973; Nissen, 1974; Nissen et al., 1980; Fageria, 1984). According to Epstein and Hagen (1952), transport of an ion into a plant cell may be analogous to the relationship between the binding of a substrate to an enzyme and the release of its products after catalysis. The overall sequence of events during an enzyme-catalyzed reaction, $S \overset{E}{\leftrightarrow} P$ may be depicted as shown below (Segal, 1968):

$$E + S \leftrightarrow ES \leftrightarrow EX \leftrightarrow EY \leftrightarrow EZ \leftrightarrow E - P \leftrightarrow P + E$$

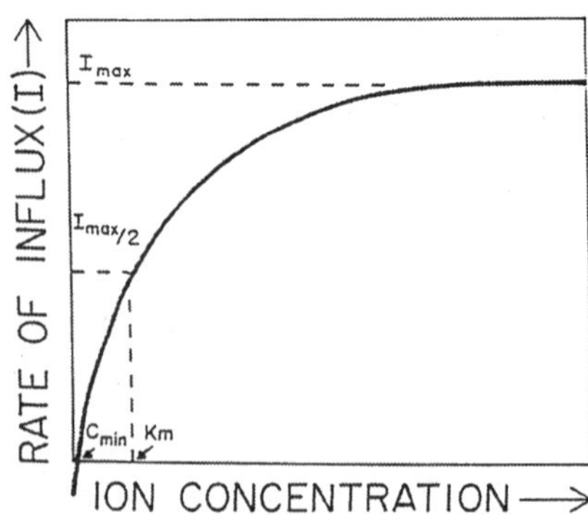

Figure 3.7 Relationship between ion concentration and influx rate.

The enzyme (E) first combines with the substrate (S) to form an enzyme-substrate complex (ES). On the surface of the enzyme, the substrate may go through one or more transitional forms (X, Y, Z) and finally be converted to the product (P). The product then dissociates, allowing the free enzyme (E) to begin again. Ion uptake by plants follows a hyperbolic relationship with increasing concentrations (up to about 200 mmol m^{-3}) in the growth medium (Ingestad, 1982) and can be explained by Michaelis-Menten kinetics. The uptake rate at a given concentration can be predicted with the help of the following equation:

$$V = \frac{V_{\max} C_i}{K_m + C_i}$$

where $V_{\max}$ is the maximum velocity, C_i the concentration of the ion in the growth medium, and K_m the Michaelis constant, equal to the substrate ion concentration giving half the maximal rate of uptake (Figure 3.7). A small value for K_m implies a high affinity between the ion and the carrier. If the uptake rate and concentration are plotted as reciprocal, a straight line is frequently obtained. Extrapolation of this line gives an intercept at $1/V_{\max}$; the concentration at half-maximal velocity corresponds to K_m. This method of calculating K_m is known as the lineweaver-Burk plot method. An alternative procedure is to plot the rate of uptake (V) against V/C and obtain $V_{\max}$ and $V_{\max}/K_m$ by extrapolation of the experimentally determined slope to the ordinate and abscissa, respectively (Clarkson, 1974). This method is known as the Hofstee plot method.

Claasen and Barber (1976) described the ion absorption isotherm in terms of $I_{\max}$, K_m, and E (Figure 3.7) and proposed the following equation:

Table 3.5 Nutrient Uptake Kinetics Values of Intact Crop Plants

Crop	Plant age (days)	Nutrient	I_{max} (nmol m^{-2} s^{-1})	K_m (μmol L^{-1})	C_{min} (μmol L^{-1})	Reference
Wheat	20–38	P	1.4	6	—	Anghinoni et al., 1981
	20–40	K	7.0	7	—	Barber, 1995
	30–40	Ca	1.6	5		Barber, 1995
	20–40	Mg	0.4	1		Barber, 1995
Corn	18–20	NO_3	10	3	4	Edwards and Barber, 1976
	14–28	P	4	3	0.2	Jungk and Barber, 1975
	18	K	40	16	1	Baligar and Barber, 1978

$$I_{max} = \frac{I_{max} C_i}{K_m + C_i} - E$$

Since the uptake rate was zero at a concentration above zero, the line was extended to the ordinate, giving negative uptake at zero ion concentration, which was termed efflux (E) from the root (Barber, 1995). Nielsen and Barber (1978) made a further modification by using the concentration in solution where the net influx reaches zero, rather than E, as the third point describing the lower end of the absorption curve. This value was termed C_{min} and the equation is written as follows:

$$I_{max} = \frac{I_{max}(C_i - C_{min})}{K_m + C_i - C_{min}}$$

Values of I_{max}, K_m, and C_{min} for some nutrients and plant species are presented in Table 3.5. Note, however, that ion uptake kinetics values vary with plant age, nutrient concentration, temperature, root morphology, plant demand for nutrients, and analytical techniques.

Hai and Laudelot (1966) and Fageria (1973, 1976) proposed a continuous-flow technique to measure nutrient uptake kinetics. The basic principle of the continuous-flow system is that the rate of nutrient uptake (U) is equal to the product of the flow rate (F) and the difference between the concentration of the solution entering the system (C_o) and of the outgoing solution (C_s). A mathematical equation can be written as follows:

$$U = C(C_o - C_s)$$

The rate of ion uptake, expressed in μg h^{-1} g^{-1} root weight (may be fresh or dry), is calculated by:

$$\text{Rate of ion uptake} = \frac{(1 - C_s/C_o) \times F \times C}{\text{root weight}}$$

where C_s is the concentration of the outgoing solution, C_o the concentration of the ingoing solution, and C the concentration of the stable ion in the nutrient solution (ppm or μmol L^{-1}).

In the continuous-flow culture technique, flow rate through the system is one of the most important parameters to be considered in ion uptake studies. Edwards and Asher (1974) discussed the significance of solution flow rate in flowing culture experiments and concluded that the actual flow rate required for a particular experiment will depend upon the nature and concentration of the ion, age of the plant, efficiency of roots in absorbing the test ion, and conditions of the experiment.

The response of the I_{max} and K_m of a transport process to physical and metabolic factors can undoubtedly provide some insight into the general nature of the processes moving ions across membranes (Clarkson, 1984). The specificity of the mechanisms involved can be inferred from the effect of competing ions on V_{max}. Studies of this kind by Epstein (1966) led to the conclusion that there were two sets of binding sites, one with affinity for K^+ and another with affinity for Na^+, as well. At low concentrations of K^+, Na^+ had no effect on V_{max} of K^+, but at higher concentrations of K^+, I_{max} was increasingly depressed by increasing Na^+.

Epstein and coworkers found the relation between ion concentration and uptake rate to be more complex when the concentration was varied over a large range. They attributed the kinetics to the simultaneous functioning of two different carriers for the same ion and suggested that both mechanisms are located in the plasmalemma (Figure 3.8). On the other hand, Laties and coworkers suggested that the mechanisms are in series, one in the plasmalemma and one in the vacuolar membrane or tonoplast. Later, Nissen (1971, 1974) suggested that uptake mechanisms remain unchanged over the entire concentration range but the characteristics of uptake change at certain discrete external concentrations. Nissen (1973) and Nissen et al. (1980) reexamined ion absorption data for many plant species and came to the conclusion that ion uptake in higher plants can be described by multiphasic uptake mechanisms and that this accounts for the apparently contradictory evidence for the parallel and the series models (Figure 3.8). Further, Nissen (1991) described kinetic models as a single diffusion model and multiphasic model. To calculate ion uptake kinetics, he gave a formula for each model (Table 3.6) and discussed

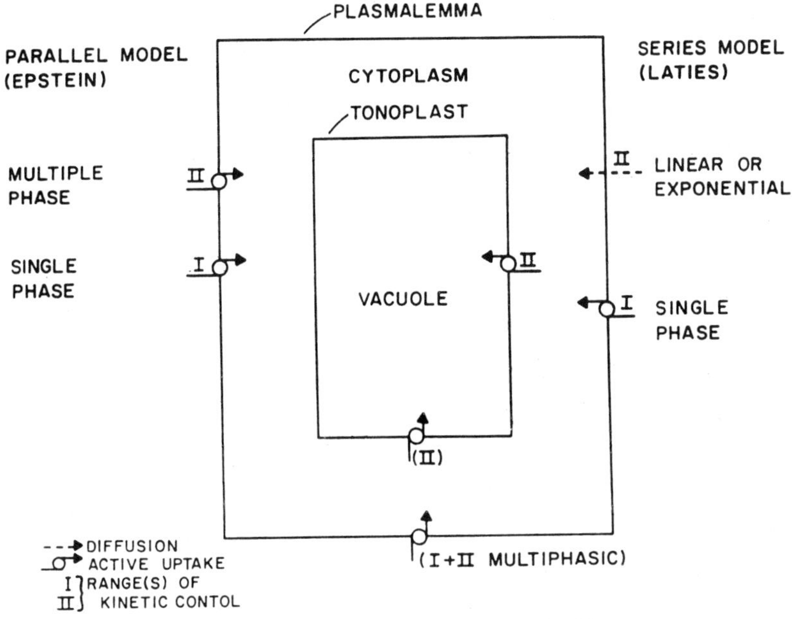

Figure 3.8 Ion transport models. (From Nissen, 1973.)

Table 3.6 Kinetic Models for Solute Uptake in Plants

Model	Formula[a]
Single + diffusion	$\underline{v} = \dfrac{\underline{V}_{\max} \cdot \underline{S}}{\underline{K}_m + \underline{S}} + \underline{k}_D \cdot \underline{S}$
Dual	$\underline{v} = \dfrac{\underline{V}_{\max_1} \cdot \underline{S}}{\underline{K}_{m_1} + \underline{S}} + \dfrac{\underline{V}_{\max_2} \cdot \underline{S}}{\underline{K}_{m_2} + \underline{S}}$
Dual + diffusion	$\underline{v} = \dfrac{\underline{V}_{\max_1} \cdot \underline{S}}{\underline{K}_{m_1} + \underline{S}} + \dfrac{\underline{V}_{\max_2} \cdot \underline{S}}{\underline{K}_{m_2} + \underline{S}} + \underline{K}_D \cdot \underline{S}$
Multiphasic	$\underline{v} = \dfrac{\underline{V}_{\max_n} \cdot \underline{S}}{\underline{K}_{m_n} + \underline{S}}$ (for phase $\underline{n}$)

[a] S, solute concentration; $\underline{v}$, uptake rate; $\underline{V}_{\max}$, maximum uptake rate; K_m, Michaelis constant; K_D, diffusion constant.
Source: Nissen (1991).

their advantages and shortcomings in ion uptake kinetic studies (Nissen, 1991). In contrast to other kinetic models, which may all be termed continuous, the multiphasic model predicts a discontinuous relationship between ion concentration and rate of uptake. Fitting of kinetic models to data for different ions and for a variety of plants and tissue shows that the fit to the multiphasic model is better than the fit to any of the continuous models which have been proposed (Nissen et al., 1980).

F. Ion Absorption Measurement

Ion uptake measurement is usually measured by tracer techniques in excised roots. Measuring rates of uptake in this way tends to ignore the large amount of ions transported across the root to the xylem and finally to the shoot. In practice, only a small fraction of the nutrients absorbed is retained in the roots and the major part is exported to the shoots (Asher and Ozanne, 1967; Loneragan and Snowball, 1969). Therefore, in ion absorption measurements, roots as well as shoots should be taken into account.

One way to measure the rates of absorption of nutrients by plant roots is to estimate the changes in nutrient content of root and shoot by chemical analysis. The values obtained by this procedure are necessarily averaged over several days but are nonetheless useful indications of the scale of the absorption process (Pitman, 1976). Equations for analyzing uptake by plants have been presented and discussed by many workers (Williams, 1948; Loneragan, 1968; Pitman, 1976). The net rate of ion uptake relative to root weight (U_R) is equal to ($1/W_R$) dm/dt, where W_R is root weight and M is the nutrient content of the plant (Pitman, 1976). An average rate of uptake is therefore

$$U_R = \frac{1}{\overline{W}_R} \frac{M_2 - M_1}{t_2 - t_1} \tag{1}$$

where $\overline{W}_R$ is average root weight for young plants growing exponentially.

$$W_R = \frac{(W_{R2} - W_{R1})}{\ln (W_{R2}/W_{R1})}$$

Hence, in this case:

$$U_R = \frac{M_2 - M_1}{(t_2 - t_1)(W_{R2} - W_{R1})} \ln \left(\frac{W_{R2}}{W_{R1}} \right) \tag{2}$$

Note that if the relative content M/W is constant (X), then:

$$U_R = X \frac{W}{W_R} \frac{\ln(W_{R2}/W_{R1})}{t_2 - t_1} = X \cdot \frac{W}{W_R} \cdot R \tag{3}$$

where R is relative growth rate, provided W/W_R is constant. Note that U_R is an average made up of uptake to each part of the plant,

$$U_R = U_{Ra} + U_{Rb} + \ldots U_{Rn} \tag{4}$$

where $a \ldots n$ refer to different parts of the plant, and if X_n is constant then

$$U_n = \frac{W_n}{W_R} X_n R_n \tag{5}$$

Alternatively, if X_n is not constant, individual U_{Rn} values can be calculated (Equation 1) and summed to give total uptake to the shoot (Pitman, 1976).

IV. SUMMARY

Nutrient absorption by plants is a dynamic and complex process and kinetics is usually more important than thermodynamics in describing this system. The rate of nutrient absorption by a root depends on the nutrient supply to the root surface, active absorption by roots, and plant demand for nutrients. Nutrients are transported to the root by mass flow, diffusion, and root interception. After reaching the root surface, nutrients move into the xylem through various root cells; from the xylem, ions are transported to the growing organs in the shoot for metabolic processes. Ion concentrations in the cell sap are often much higher than in the outside medium, and ions frequently have to move against a concentration gradient. In this process, energy is required and is supplied through root respiration. The kinetics of ion absorption in roots is similar to the kinetics of enzyme-catalyzed reactions. This kinetic information is providing insight into the nature of ion carriers. Most ion uptake studies have been done in short-duration solution culture experiments using excised roots. These studies need to be conducted for a longer duration using intact plants in soil to obtain experimental results with more practical applicability.

REFERENCES

Anghinoni, I., V. C. Baligar, and S. A. Barber. 1981. Growth and uptake rates of P, K, Ca, and Mg in wheat. J. Plant Nutr. 6: 923–933.

Asher, C. J. and P. G. Ozanne. 1967. Growth and potassium content of plants in solution cultures maintained at constant potassium concentrations. Soil Sci. 103: 155–161

Baligar, V. C. 1984. Effective diffusion coefficient of cations as influenced by physical and chemical properties of selected Indiana soils. Commun. Soil Sci. Plant Anal. 15: 1367–1376.

Baligar, V. C. and S. A. Barber. 1978. Use of K/Rb ratio to characterize potassium uptake by plant roots growing in soil. Soil Sci. Soc. Am. J. 42: 575–579.

Barber, S. A. 1966. The role of root interception, mass flow and diffusion in regulating the uptake of ions by plants from soils. Tech. Rep. Ser. Int. Atomic Energy Agency, No. 65, pp. 39–45.

Barber, S. A. 1971. The influence of the plant root system in the evaluation of soil fertility. Proc. Int. Symp. Soil Fertility Evaluation, New Delhi 1: 249–256.

Barber, S. A. 1974. Influence of the plant root on ion movement in soil, pp. 525–564. *In*: E. W. Carson (ed.). The plant root and its environment. University Press of Virginia, Charlottesville.

Barber, S. A. 1995. Soil Nutrient Bioavailability: A Mechanistic Approach. 2nd Ed. Wiley, New York.

Becking, J. H. 1956. On the mechanisms of ammonium ion uptake by maize roots. Acta Bot. Neerl. 5: 1–79.

Claasen, N. and S. A. Barber. 1976. Simulation model for nutrient uptake from soil by a growing plant root system. Agron. J. 68: 961–964.

Clarkson, D. T. 1974. Ion transport and cell structure in plants. McGraw-Hill, London.

Clarkson, D. T. 1984. Ionic relationships, pp. 319–353. *In*: M. B. Wilkins (ed.). Advanced plant physiology. Pitman, London.

Clarkson, D. T. and J. B. Hanson. 1980. The mineral nutrition of higher plants. Annu. Rev. Plant Physiol. 31: 239–298.

Corey, R. B. 1973. Factors affecting the availability of nutrients to plants. pp. 23–25. *In*: L. M. Walsh and J. D. Beaton (eds.). Soil testing and plant analysis. Soil Sci. Soc. Am., Madison, Wisconsin.

Danielli, J. F. and H. A. Davson. 1935. A contribution to the theory of the permeability of thin films. J. Cell. Comp. Physiol. 5: 495–508.

Dittmer, H. J. 1940. A quantitative study of the subterranean members of soybean. Soil Conservation 6: 33–34.

Dutt, R. and P. F. Low. 1962. Diffusion of alkali chlorides in clay-water systems. Soil Sci. 93: 233–240.

Edwards, D. G. and C. J. Asher. 1974. The significance of solution flow rate in flowing culture experiments. Plant Soil 41: 161–175.

Edwards, J. H. and S. A. Barber. 1976. Nitrogen flux into corn roots as influenced by shoot requirement. Agron. J. 68: 471–473.

Epstein, E. 1966. Dual pattern of ion absorption by plant cells and by plants. Nature (London) 212: 1324–1327.

Epstein, E. 1973. Mechanisms of ion transport through plant cell membranes. Int. Rev. Cytol. 34: 123–168.

Epstein, E. and C. E. Hagen. 1952. A kinetic study of the absorption of alkali cations by barley roots. Plant Physiol. 23: 457–474.

Fageria, N. K. 1973. Uptake of nutrient by the rice plant from dilute solutions. Doctoral Thesis, Catholic University, Louvain.

Fageria, N. K. 1976. Effect of P, Ca, and Mg concentrations in solution culture on growth and uptake of these ions by rice. Agron. J. 68: 726–732.

Fageria, N. K. 1984. Fertilization and mineral nutrition of rice. EMBRAPA-CNPAF/ Editora Campus, Rio de Janeiro.

Hai, T. V. and H. Laudelot. 1966. Phosphate uptake by intact rice plants by continuous flow method at low phosphate concentration. Soil Sci. 101: 408–417.

Hiatt, A. J. and J. E. Leggett. 1974. Ionic interactions and antagonisms in plants, pp. 101–134. *In*: E. W. Carson (ed.). The plant root and its environment. University Press of Virginia, Charlottesville.

Hodges, T. K. 1973. Ion absorption by plant roots. Adv. Agron. 25: 163–207.

Hodges, T. K., R. T. Leonard, C. E. Bracker, and T. W. Keenan. 1972. Purification of an ion-stimulated adenosine triphosphatase from plant roots association with plasma membranes. Proc. Natl. Acad. Sci. USA 69: 3307–3311.

Husted, R. F. and P. F. Low. 1954. Ion diffusion in bentonite. Soil Sci. 77: 343–353.

Ingestad, T. 1982. Relative addition rate and external concentration driving variables used in plant nutrition research. Plant Cell Environ. 5: 443–453.

Jungk, A. O. 1991. Dynamics of nutrient movement at the soil-root interface, pp. 455–481. *In*: Y. Waisel, A. Eshel, and U. Kafkafi (eds.) Plant roots: The hidden half. Marcel Dekker, New York.

Jungk, A. and N. Claassen. 1989. Availability in soil and acquisition by plants as the basis for phosphorus and potassium supply to plants. Z. Pflanzenernaehr. Bodenkd. 152: 151–157.

Jungk, A. and S. A. Barber. 1975. Plant age and the phosphorus uptake characteristics of trimmed and untrimmed corn root systems. Plant Soil 42: 227–239.

Lavy, T. L. and S. A. Barber. 1964. Movement of molybdenum in the soil and its effect on availability to the plant. Soil Sci. Soc. Am. Proc. 28: 93–97.

Loneragan, J. F. 1968. Nutrient requirements of plants. Nature (London) 188: 1307–1308.

Loneragan, J. F. and K. Snowball. 1969. Calcium requirements of plants. Aust. J. Agric. Res. 20: 465–278.

Marschner, H. 1995. Mineral nutrition of higher plants. 2nd ed. Academic Press, New York.

Mengel, K. 1985. Dynamics and availability of major nutrients in soil. Adv. Soil Sci. 2: 65–131.

Mengel, K. and E. A. Kirkby. 1982. Principles of plant nutrition. 3rd ed. International Potash Institute, Bern.

Nielsen, N. E. and S. A. Barber. 1978. Differences among genotypes of corn in the kinetics of phosphorus uptake. Agron. J. 70: 695–698.

Nissen, P. 1971. Uptake of sulfate by roots and leaf slices of barley: Mediated by single, multiphasic mechanisms. Physiol. Plant. 24: 315–324.

Nissen, P. 1973. Kinetics of ion uptake in higher plants. Physiol. Plant. 28: 113–120.

Nissen, P. 1974. Uptake mechanisms: Inorganic and organic. Annu. Rev. Plant Physiol. 25: 53–79.

Nissen, P. 1991. Uptake mechanizations, pp. 483–502. *In*: Y. Waisel, A. Eshel, and U. Kafkafi (eds.) Plant Roots: The Hidden Half. Marcel Dekker, New York.

Nissen, P., N. K. Fageria, A. J. Rayer, M. M. Hassan, and T. V. Hai. 1980. Multiphasic accumulation of nutrients by plants. Physiol. Plant. 49: 222–240.

Nye, P. H. 1969. The soil model and its application to plant nutrition, pp. 105–114. *In*: I. H. Rorison (ed.). Ecological aspects of the mineral nutrition of plants. Blackwell, London.

Nye, P. H. 1979. Diffusion of ions and unchanged solutes in soils and clays. Adv. Agron. 31: 225–272.

Pitman, M. G. 1972. Uptake and transport of ions in barley seedlings. III. Correlation between transport to the shoot and relative growth rate. Aust. J. Biol. Sci. 25: 905–919.

Pitman, M. G. 1976. Ion uptake by plant roots, pp. 95–128. *In*: U. Lüttage and M. G. Pitman (eds.). Transport in plants II. Part B. Tissues and organs. Encyclopedia of plant physiology, Vol. II, Springer-Verlag, New York.

Russell, R. S. 1977. The soil environment, pp. 143–168. *In*: Plant root systems. McGraw-Hill, London.

Segal, I. H. 1968. Biochemical calculations. Wiley, New York.

Shaviv, A. and R. L. Mikkelsen. 1993. Controlled-release fertilizers to increase efficiency of nutrient use and minimize environmental degradation—a review. Fert. Res. 35: 1–12.

Singer, S. J. 1972. A fluid lipid-globular protein mosaic model of membrane structure. Ann. N.Y. Acad. Sci. 195: 16–23.

Spiegler, K. S. and C. D. Coryell. 1953. Electromigration in cation-exchange resins. III. J. Phys. Chem. 57: 687–690.

Ting, I. P. 1982. Plant mineral nutrition and ion uptake, pp. 331–363. *In*: Plant physiology. Addison-Wesley, Reading, Massachusetts.

Vasey, E. H. and S. A. Barber. 1963. Effect of placement on the absorption of Rb^{86} and P^{32} from soil by corn roots. Soil Sci. Soc. Am. Proc. 27: 193–197.

Walker, N. A. 1976. The structure of biological membranes, pp. 3–11. *In*: U. Lüttage and M. G. Pitman (eds.). Transport in plants. Part A. Cells. Encyclopedia of plant physiology, New Ser., Vol. 2, Springer-Verlag, New York.

Williams, R. F. 1948. The effects of phosphorus supply on the rates of uptake of phosphorus and nitrogen upon certain aspects of phosphorus metabolism in gramineous plants. Aust. J. Sci. Res. Ser. B 1: 336–361.

4

Diagnostic Techniques for Nutritional Disorders

I. INTRODUCTION

A rapidly increasing world population demands ever-increasing food production. In this context modern agriculture must use every available alternative and practice to produce adequately and efficiently. Further, public concern about environmental quality and the long-term productivity of agroecysystems has emphasized the need to develop and implement management strategies that maintain soil fertility at an adequate level without degrading soil and water resources. Therefore, nutrient inputs must be supplied to replace those removed from the soil and to achieve higher yields from a limited land resource. To meet these demands, nutrient needs must be accurately identified and the nutrient applied so as to achieve maximum benefit from their use.

Diagnostic techniques for nutritional disorders are the methods for identifying nutrient deficiencies, toxicities, or imbalances in the soil-plant system. Nutrient deficiency can occur when a nutrient is insufficient in the growth medium and/or cannot be absorbed and assimilated by plants due to unfavorable environmental conditions. Table 4.1 shows soil conditions associated with

Table 4.1 Soil Conditions Inducing Nutrient Deficiencies for Crop Plants

Nutrient	Conditions inducing deficiency
N	Excess leaching with heavy rainfall, low organic matter content of soils, burning the crop residue
P	Acidic, organic, leached, and calcareous soils, high rate of liming
K	Sandy, organic, leached, and eroded soils; high liming application, intensive cropping system
Ca	Acidic, alkali, or sodic soils
Mg	Similar to calcium
S	Low organic matter content of soils; use of N and P fertilizers containing no sulfur, burning the crop residue
Fe	Calcareous soils; soils high in P, Mn, Cu, or Zn; high rate of liming
Zn	Highly leached acidic soils, calcareous soils, high levels of Ca, Mg, and P in the soils
Mn	Calcareous silt and clays, high organic matter, calcareous soils
B	Sandy soils, naturally acidic leached soils, alkaline soils with free lime
Mo	Highly podzolized soils; well drained calcareous soils

nutrient deficiencies. Similarly, nutrient or elemental toxicity may occur due to excess, imbalance, and unfavorable environmental conditions.

Nutrient or elemental disorders limit crop production in all types of soils around the world. To obtain good yields of crops, nutrient disorders must be corrected. The first step in this direction is to identify the nutritional disorder. The four most common methods of identifying nutritional deficiencies, toxicities, or imbalances are soil analysis, plant analysis, visual symptoms, and crop growth response. Each method has its advantages and disadvantages, and it is difficult to say which is best. One or a combination of all techniques can be used to identify nutritional disorders in the soil-plant system.

Soil testing probably has a greater agronomic application than plant analysis for annual row crops, but for perennial horticulture crops, plant analysis is the more significant testing procedure (Jones, 1985).

II. SOIL TESTING

Soil testing has become an essential and integrated part of soil management in present-day agricultural systems. In addition to being a useful diagnostic tool for evaluating soil fertility for fertilizer recommendations, soil testing also plays an important role in the prevention of environmental degradation by providing guidelines for minimizing loss of nutrients to surface and groundwaters.

Table 4.2 Extractable P, K, and Ca + Mg at Different Depths in an Oxisol of Central Brazil[a]

Soil depth (cm)	P ($mg\ kg^{-1}$)	K ($mg\ kg^{-1}$)	Ca + Mg ($cmol\ kg^{-1}$)
0–20	1.17	30	2.37
20–40	0.50	19	0.70
40–60	0.40	16	0.53
60–80	0.37	14	0.56

[a]Phosphorus and K were extracted by the Mehlich 1 (0.05 mol L^{-1} HCl + 0.0125 mol L^{-1} H_2SO_4) extracting solution and Ca + Mg with 1 M KCl.
Source: Fageria et al., 1991.

In a broad sense, soil testing is any chemical or physical measurement that is made on a soil (Melsted and Peck, 1973). The main objective of soil testing is to measure soil nutrient status and lime requirements in order to make fertilizer and lime recommendations for profitable farming. Soil testing includes collection of soil samples, sample preparation, laboratory analysis, calibration and interpretation of the tests, and recommendations. Since the late 1940s soil testing has been widely accepted as an essential tool in formulating a sound lime and fertilizer program. Currently, soil testing is the most widely used practice for soil fertility evaluation and management. Routine soil analysis is done for pH, P, K, Ca, and Mg, and where soil acidity is a problem, Al is also determined. Other determinations, such as organic matter and micronutrients, are performed under specific conditions but are not done routinely.

A. Soil Sampling

A reliable fertilizer or lime recommendation depends on a soil sample representative of the area from which it was taken, an accurate laboratory analysis, and correct interpretation of the laboratory results. The greatest source of error is usually the soil sample itself. Since soil is extremely heterogeneous, it is important that the sample tested be truly representative of the area sampled. A common error in soil sampling occurs when the top few centimeters of soil are dry and are not included in the normal sampling procedure. Most of the immobile nutrients remain in the top layers (Table 4.2), and if a few centimeters of the top layer are not collected, soil analysis results will show abnormally low values. During sampling, surface and subsurface soil should be kept separate for individual analysis and interpretation. Sampling

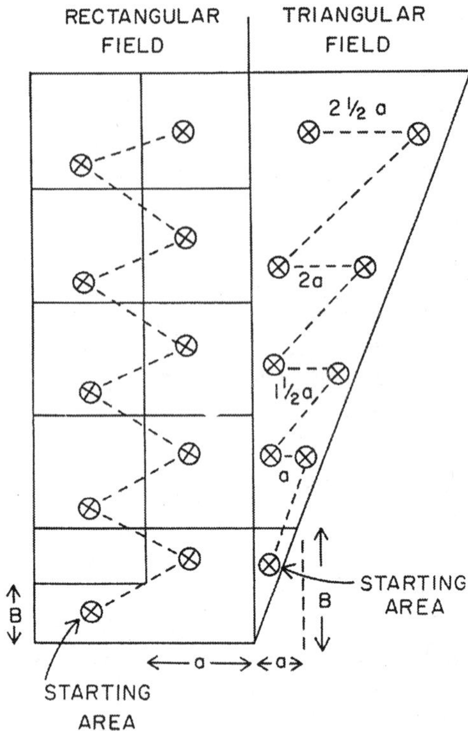

Figure 4.1 Zigzag sampling in rectangular and triangular fields. (From Sabbe and Marx, 1987.)

depth is dependent on nutrient mobility and crops to be planted. Mobile nutrients, such as NO_3^-, SO_4^{2-}, and borates should be sampled to a depth of about 60 cm and samples must be taken when biological activity is low (Sabbe and Marx, 1987). For immobile nutrients like P, K^+, Ca^{2+}, and Mg^{2+}, sampling to tillage depth can give satisfactory results. For a pasture crop a 10–cm depth is normally sufficient, whereas for field crops a depth of 15–20 cm is desired.

Sampling pattern in rectangular and triangular fields is described by Sabbe and Marx (1987) as presented in Figure 4.1. The soil samples should be taken in a zigzag pattern, and this pattern can be formed on the basis of number of samples desired in a given area. For example, in a rectangular field if 10 samples are desired, the field can be divided into 10 equal blocks. In addition, since each block has dimensions, a and $2B$, the distance and direction for the zigzag pattern are defined. The distance is $(a^2 + B^2)^{1/2}$ and the direction is given by θ = tangent (B/a). Once the sampling plan has been devised, the actual samples may be collected by the shortest route. Similarly, in a triangular field, the zigzag pattern can be evolved by defining the first direction as θ = tangent (B/a) and the distance as $(a^2 + B^2)^{1/2}$. The third sampling location is distance a from the second. The fourth sampling location is $2a$ to the east and B north from the third location. Location five is west 1.5 a. Continue the same sequence traveling north B and east the same distance as on the preceding westward move, but move an additional a east. The difference between the rectangular and the triangular fields lies in the direction and distance to the next sample location (Sabbe and Marx, 1987).

A representative soil sample is composed of 15–20 subsamples from a uniform field with no major variation in slope, drainage, or past fertilizer history. If fertilizer is applied in a band, the best way to sample such an area is, if possible, to plough the field before sampling. Alternatively, sampling should occur between the crop rows and well away from the fertilizer bands. If cultivation is intensive, sampling should be done annually to evaluate the fertility status; otherwise, sampling once in a three-year period is sufficient if the field is planted with one crop per year.

B. Sample Preparation

After collecting the soil samples, the next step is sample preparation for laboratory analysis. Samples can be dried in moving air but not in heated air. Samples can also be dried at about 50°C in a cabinet-type forced-air dryer for 24–48 h. Wet clay soils may require a longer drying period. However, at no time should samples remain in a heated drier for more than 72 h after becoming dry (less than 2% moisture). Organic soils or soil-less mixes are analyzed as received and should not be dried (Jones, 1985). After drying, the samples are ground to pass a 10–mesh (2 mm) screen.

C. Laboratory Analysis

After sample preparation, laboratory analysis is the next step in a soil testing program. It involves the extraction of nutrients and determination of their concentrations in the extract. In the laboratory analysis, nutrient values are expressed as *extractable nutrients* rather than *available nutrients* because the quantities of nutrients absorbed by plant may be close to the amount of nutrients extracted by various standard extractants. Soil testing can provide a relative index of the quantity of nutrients that plants may utilize from a soil but rarely can provide an absolute measurement (Thomas and Peaslee, 1973). That is why extractable nutrient is a better term than available nutrient. The laboratory analysis procedure should be simple, rapid, and reasonably accurate, and extracted nutrients should be correlated with growth or yield of crops under various conditions. To define a suitable extractant, soil test correlation studies must be conducted under greenhouse conditions on a large number of the more important agricultural soils. An ideal extractant would remove approximately the same amount of the element as is removed by the plant. This is never achieved in practice, but a close correlation between plant uptake and the amount of the element extracted chemically is sought.

pH and Lime Requirement

Soil pH is one of the most important chemical properties used as an index in soil management for crop production. It measures the H^+ ion activity in the soil solution. In measuring soil pH electrometrically, most laboratories use soil/water slurries of 1:1 or 1:3 (vol/vol) and allow them to stand for 1 h or less before reading (Adams, 1984). The use of 0.01 M $CaCl_2$ has also been recommended in pH determinations (Woodruff, 1967). Even though there appear to be good theoretical reasons for measuring soil pH in a 0.01 M $CaCl_2$ solution, failure to do so evidently does not generally result in serious error, as pH values of soils in 0.01 M $CaCl_2$ tend to be slightly lower than, but highly correlated with, those in water (McLean, 1982). The lower value of pH in a $CaCl_2$ solution is related to the displacement of H^+ and Al^{3+} ions from exchange sites.

According to Conyers and Davey (1988), pH values measured in water, 0.01 M $CaCl_2$, and 0.1 M KCl at dilution ratios of 5:1, 2.5:1, and 2:1 (solution: soil) are highly correlated over a wide soil pH range. The relationship between pH in water (pH_w) and pH in 0.01 M $CaCl_2$ (pH_{Ca}) at 2.5:1 solution: soil was found to be

$$pH_{Ca} = 1.05\ pH_w - 0.9$$

The most common, and in most cases the most effective, way to correct soil acidity has been by applying lime. Lime significantly increased grain yields of annual crops grown on an Oxisol (Table 4.3) and consequently improved nutrient uptake (Table 4.4).

Soil lime requirement is the amount of liming material required to adjust soil to the pH desired for crop production. The main objective of liming acid soils is to increase pH, increase exchangeable Ca and Mg, and neutralize Al (Table 4.5). Soil pH indicates whether a soil needs lime, but the quantity of lime required is not determined by soil pH. Several methods are used to determine lime requirements of soils. In the United States, for example, no single method has been officially accepted for estimating lime requirement, and the methods used vary from state to state (Table 4.6). Each method has its advantages and disadvantages, and these are described in detail by Adams (1984). In most cases, the basis of selection of the point to which the soil is to be limed is the pH giving the most favorable plant growth. However, the recommendation may be based on another criterion such as inactivation of

Table 4.3 Grain Yield of Three Annual Crops as Influenced by Liming on an Oxisol of Brazil

Lime rate ($Mg\ ha^{-1}$)	Common bean[a] ($kg\ ha^{-1}$)	Lime rate ($Mg\ ha^{-1}$)	Soybean[b] ($kg\ ha^{-1}$)	Lime rate ($Mg\ ha^{-1}$)	Maize[b] ($kg\ ha^{-1}$)
0	1566	0	470	0	2883
3	1928	1.5	1163	3	4803
6	2062	3.0	1489	6	5619
9	2264	4.5	1625	9	6219
12	2245				
Linear	**		*		*
Quadratic	NS		*		*

*, ** Significant at the 5 and 1% probability levels, respectively. NS = not significant.
Source: [a]Fageria et al., 1991; [b]Raij, 1991.

exchangeable Al. In Brazil, liming recommendations are based on soil exchangeable Al, Ca, and Mg. Lime requirement (LR) is calculated using the following formula:

$$LR\ (\text{metric tons ha}^{-1}) = \text{Al} \times 2 + [2 - (\text{Ca} + \text{Mg})]$$

The lime requirement computed from the Al, Ca, and Mg contents is based on the assumption that Al toxicity and Ca and Mg deficiencies are the most important growth-limiting factors in acid soils where legumes are not prominent in cropping systems.

To improve liming recommendations, a factor of Al × 2 should be used for species susceptible to soil acidity, Al × 1.5 for moderately tolerant species, and Al × 1 for tolerant species. Lime requirements are expressed in terms of $CaCO_3$ because, for reasons of cost and nutrient value, crushed limestone and the $CaCO_3$ therein is the most common base used for neutralizing soil acidity.

Nitrogen

So far, there is no widely accepted or reliable method that can be used to test soil for available nitrogen. This is because most of the nitrogen in soil is found in organic forms, and its mineralization from this source is dependent on soil and climatic factors that vary constantly during crop growth. Nitrogen mineralization is defined as the conversion of organic N to inorganic N as a result of microbial activity. Nitrogen immobilization is the corollary to mineralization and is defined as the conversion of inorganic N to the organic N form in microbial tissues (Soil Science Society of America, 1987). Nitrogen as NO_3^- is subject to leaching, denitrification, and immobilization by micro-

Table 4.4 Effects of Liming on Nutrient Concentration in Ear Leaves of Maize on a Acid Soil of State of São Paulo, Brazil

Lime rate ($Mg\ ha^{-1}$)	N ($g\ kg^{-1}$)	P ($g\ kg^{-1}$)	K ($g\ kg^{-1}$)	Ca ($g\ kg^{-1}$)	Mg ($g\ kg^{-1}$)	Fe ($mg\ kg^{-1}$)	Mn ($mg\ kg^{-1}$)	Cu ($mg\ kg^{-1}$)	Zn ($mg\ kg^{-1}$)
0	24.6	2.5	14.8	3.7	1.0	124	38	11	26
3	30.0	2.4	15.9	4.7	1.6	116	46	11	25
6	32.8	2.5	16.3	5.3	2.3	117	48	11	23
9	32.9	2.6	16.0	5.3	3.0	126	43	10	20
12	30.5	2.7	15.0	5.0	3.6	144	31	10	15
Sig.	Q**	L*	Q**	Q*	L**	NS	Q**	NS	L**

L = linear and Q = quadratic effects (*P 0.05 and **P 0.01).
NS = not significant.
Source: Quaggio et al., 1991.

Table 4.5 Influence of Dolomitic Lime on pH, Ca + Mg, and Al During Cultivation of Upland Rice in an Oxisol of Central Brazil[a]

Lime level ($t\ ha^{-1}$)	Before lime application	18 days after sowing	67 days after sowing	After first crop harvest	After second crop harvest
pH					
0	5	5.0	5.0	5.0	5.0
3	5	5.3	5.4	5.6	5.3
6	5	5.6	5.6	5.8	5.4
9	5	5.9	5.8	6.0	5.7
12	5	6.0	6.0	6.1	5.9
Ca + Mg ($cmol\ Kg^{-1}$)					
0	0.82	1.44	1.49	1.28	1.29
3	0.82	2.17	2.69	2.01	2.10
6	0.82	2.89	3.62	3.31	2.76
9	0.82	3.49	4.29	3.83	3.42
12	0.82	4.00	5.26	3.99	3.92
Al ($cmol\ kg^{-1}$)					
0	0.63	0.62	0.52	0.47	0.49
3	0.63	0.28	0.26	0.15	0.22
6	0.63	0.13	0.11	0.09	0.13
9	0.63	0.10	0.07	0.05	0.06
12	0.63	0.08	0.04	0.04	0.05

[a] Lime was applied about 160 days before sowing the rice crop. First harvest was about 280 days after lime application and second harvest about 1 year after first harvest. Soil samples were taken from 0 to 20 cm soil depth.

Source: Fageria et al., 1991.

Table 4.6 Lime Requirement Methods Used by State Soil Testing Laboratories of the United States

State	Method
Alabama	Adams-Evans buffer
Arkansas	Soil pH, exchangeable Ca^{2+}, soil texture, crop
Florida	Adams-Evans buffer
Georgia	Adams-Evans buffer
Kentucky	SMP buffer
Louisiana	$Ca(OH)_2$ titration
Mississippi	Modified Woodruff buffer
North Carolina	Mehlich buffer
Oklahoma	Modified SMP buffer
South Carolina	Adams-Evans buffer
Tennessee	Adams-Evans buffer
Texas	Soil pH, soil texture
Virginia	Soil pH, soil texture

Source: Adams, 1984.

Table 4.7 Average Recovery Efficiency (%) of Nitrogen for Five Major Cereals in Various Continents. Number of Observations Between Brackets

Continent	Millet	Sorghum	Maize	Rice	Wheat
Europe	–	–	40(3)	–	48(34)
Africa	40(25)	45(6)	51(11)	28(30)	39(4)
Asia	40(4)	38(9)	41(12)	39(66)	45(22)
North America	–	18(4)	29(22)	53(1)	51(4)
South America	–	32(9)	32(42)	55(17)	34(32)
Australia	–	35(9)	45(3)	50(9)	43(12)

Source: Duivenbooden et al., 1996.

organisms, and this further complicates the development of a soil test for available nitrogen (Dahnke and Vasey, 1973). Due to nitrogen losses by various processes in the soil-plant system, recovery of this nutrient by annual crops is quite low (Table 4.7). This unrecovered N constitutes a potential contributor to groundwater contamination, eutrophication, acid rain, global warming, and farm insolvency. A primary goal of predicting N mineralization from soils and other organic N sources is to increase the overall efficiency of N-use in crop production (Honeycutt, 1994). Various biological and chemical methods have been proposed for estimating the organic N-supplying capacities of soils, but none of these methods is sufficiently reliable to warrant its routine use in soil testing laboratories (Stanford, 1982). Nitrogen recommendations are therefore made on the basis of long-term crop response data (Cope and Evans, 1985). Accurate recommendations for N fertilizer require knowledge of plant N requirements, external N sources, and N losses from the soil system. These components are represented by the equation (Rice et al., 1995):

$$N_f = N_c - (N_{\text{sources}} - N_{\text{losses}})$$

where:

N_f = N fertilizer
N_c = N needed by the crop
N_{sources} = N sources
N_{losses} = N losses

Phosphorus

In selecting an extractant for a particular soil and nutrient, the important consideration is the degree of correlation between the extracted nutrient and plant growth. Thomas and Peaslee (1973) discussed the characteristics of suitable extractors of P. These are:

1. Rapidly dissolve and/or desorb soil P and be time-independent after 30 minutes or less.
2. Maintain organic matter and soil clay in a flocculated state.
3. Avoid reprecipitation of dissolved P and/or hydrolysis of organic P.
4. Contain no excess salts, buffers, or ions that interfere with analytical determinations.
5. Extract meaningful quantities of other nutrient ions as well as P.
6. Easy to prepare, store, and dispose.

There are several extracting reagents used for P extraction from the soils (Table 4.8). However, the most common extractants used in routine analysis of P in various parts of the world are Mehlich 1 (0.05 N HCl + 0.025 N H_2SO_4), Bray 1 (0.03 N NH_4F + 0.025 N HCl), and Olsen (0.5 N $NaHCO_3$,

Table 4.8 Reagents Used for Extraction of Available P

Extracting reagents	Soil/reagent ratio	Name of procedure
0.025*N* HCl + 0.03*N* NH_4F	1:10	Bray 1
0.1*N* HCl + 0.03*N* NH_4F	1:17	Bray 2
0.5*M* $NaHCO_3$, pH 8.5	1:20	Olsen
0.05*N* HCl + 0.025*N* H_2SO_4	1:4	Mehlich 1
0.2*N* CH_3COOH + 0.2*N* NH_4Cl + 0.015*N* NH_4F + 0.012*N* HCl	1:10	Mehlich 2
0.2*N* CH_3COOH + 0.25*N* NH_4NO_3 + 0.015*N* NH_4F + 0.013*N* HNO_3 + 0.001 MEDTA	1:10	Mehlich 3
0.002*N* H_2SO_4 buffered at pH 3 with $(NH_4)_2SO_4$	1:100	Truog
0.54*N* HOAc + 0.7 NaOAc, pH 4.8	1:10	Morgan
0.02*N* Ca-lactate + 0.02*N* HCl	1:20	Egner
1% Citric acid	1:10	Citric acid

Source: Tan, 1996.

pH 8.5). These extractants cover a broad range of soil conditions ranging from acid to alkaline, low to high cation exchange capacity (CEC), and arid to humid.

The Mehlich 1 extractant is suitable for low-CEC, highly weathered, acidic soils. It is a good extractant of calcium phosphates, and extracts some Al-P but not as effectively as Ca-P. In soils where rock phosphate has been applied, care should be taken in using this extractor because it will dissolve some unreacted rock phosphate, resulting in overestimation of the P supply and underestimation of the fertilizer P requirements (Cope and Evans, 1985).

Table 4.9 compares Mehlich 1 and Bray 1 extractants when used to extract P in an Oxisol of central Brazil treated with different P fertilizers. It is clear from this table that, compared to Bray 1, the Mehlich 1 extractor gave higher P values for rock phosphate-treated soil but a lower P value when no fertilizer was applied or the soil was treated with a soluble source of P such as triple superphosphate and partially acidulated phosphate rock. These results suggest that Mehlich extractant was dissolving unreacted phosphate rock and overestimating P availability. Therefore, separate calibration curves are needed for soluble and PR-based fertilizers.

There are two soil tests that show promise as suitable tests in soils fertilized with soluble as well as PR-based fertilizers. These are the iron oxide impregnated paper (P_i) test and the ion exchange resin paper test. In both cases, the strips act as a sink for P mobilized in a soil solution, and P measured

Table 4.9 The Effect of P sources and Rates on P Level (mg kg^{-1}) in the soil (0–20 cm)

P added	TSP		APPA		PPPA		PAC		PC		PJ		PPM		PA	
(kg ha^{-1})	Bray 1	Meh. 1	Bray 1	Meh. 1	Bray 1	Meh. 1	Bray 1	Meh. 1	Bray 1	Meh. 1	Bray 1	Meh. 1	Bray 1	Meh. 1	Bray 1	Meh. 1
First crop (rice)																
Control	2.7	1.3	—	—	—	—	—	—	—	—	—	—	—	—	—	—
87	4.1	1.6	5.1	4.9	4.5	5.8	3.5	5.3	3.2	4.5	3.1	3.4	3.1	8.1	3.4	1.6
174	7.9	4.0	7.1	7.3	9.1	21.2	3.8	15.2	3.1	7.5	3.4	7.8	3.6	16.0	3.3	1.8
262	14.1	5.3	11.1	16.3	10.6	25.0	3.9	26.1	3.7	9.9	3.4	9.0	4.0	33.5	4.9	2.4
Linear	**	NS	**	**	**	**	NS	**	NS	NS	NS	*	NS	**	NS	NS
Quadratic	*	NS	NS	NS	*	NS	NS	NS	NS	NS	NS	NS	NS	*	NS	NS
Statistical significance																
Sources	**	**														
P rates	**	**														
S x P	**	**														

TSP = Triple superphosphate; APPA = Arafertil phosphate partially acidulated; PPPA = Phosphate of Patos partially acidulated; PAC = Phosphate of Araxa concentrated; PC = Phosphate of Catatão; RJ = Phosphate of Jacupiranga; PPM = Phosphate of Patos de Minas; and PA = Phosphate of Abaete.

*, ** Significant at the 5% and 1% probability levels, respectively; NS = nonsignificant.

Source: Fageria et al., 1991.

depends only on the concentration of P mobilized in the solution and not on the source of P or properties of the soil. Both tests somewhat simulate the sorption of P by plant roots without disturbing the chemical equilibrium, unlike other tests that extract P by the destructive dissolution of specific soil P compounds. In both cases, P measured from soils fertilized with PR-based fertilizers has shown very good correlation with plant response. Field calibration with crops under different pedological and agroecological regimes is needed for using these soil tests in developing fertilizer recommendations (Menon and Chien, 1995).

Bray 1 extractant is good for medium- to high-CEC soils, is a strong extractant for Al-P, and extracts some Ca-P due to acid decomposition. This extractant may also dissolve some rock-P, and therefore its use on soils where rock phosphate is applied is not recommended. According to Cope and Evans (1985), one possible solution to this problem is to increase the HCl to 0.1 M, as in the Bray 2 extractant, and to do two extractions on the same sample. The Bray 2 extractant has the same concentration of NH_4F (0.03 M) as the Bray 1, but the HCl has been increased to 0.1 M to give it increased capacity to extract less soluble Ca-P. The Olsen extracting solution (0.5 M $NaHCO_3$, pH 8.5) is normally used for alkaline soils of semiarid regions which have high CEC and high base saturation and, very often, free $CaCO_3$. Nowadays some state laboratories in the United States are also using the Mehlich 3 extractant (0.2 N CH_3COOH + 0.25 N NH_4NO_3 + 0.015 N NH_4F + 0.013 N HNO_3+ 0.001 M EDTA) for P extraction (Sims, 1993).

Potassium, Calcium, and Magnesium

The commonly used extractants for K, Ca, and Mg are double acid (0.05 M HCl in 0.0125 M H_2SO_4), 1 M NH_4OAc at pH 7, and 1 M NaOAc at pH 4.8. The double-acid extractant is suitable for sandy soils and acid soils with low CEC, whereas ammonium extractants are suitable for a wide range of soils and are most commonly used in routine soil testing laboratories. Detailed descriptions of sample size, volume of extractant, shaking time, and sensitivity are given by Jones (1979). Where the Bray 1 extractant works well for P, it is commonly used to extract K at the same time. In Brazil, 1 M KCl is used to extract Ca and Mg from acid soils in routine soil testing programs.

Micronutrients

Use of micronutrients for agronomic and horticultural crops has increased markedly in recent years. Increased use is related to higher nutrient demands from more intensive cropping practices and also from farming marginal lands. Most of the fertilizers used to correct micronutrient deficiencies are water-soluble inorganic sources or soluble organic products such as synthetic chelates or natural organic complexes. These fertilizers may react with soil to decrease their availability to plants. The rates of such chemical reactions may differ considerably with micronutrient fertilizer and soil environment (Mortvedt, 1994). Crop recovery values for micronutrients generally range from only 5–10%. Some of the reasons for the low eficiency of micronutrient fertilizers are poor distribution of the low rates applied to soils, fertilizer reactions with soil to form unavailable reaction products, and low mobility to soil, especially of the cationic micronutrients (Cu, Fe, Mn, and Zn).

Various methods have been developed for determining the availability of micronutrients in soils (Cox and Kamprath, 1972). Lindsay and Norvell (1978) developed a DTPA soil test to identify near-neutral and calcareous soils with insufficient available Zn, Fe, Mn, or Cu for maximum yields of crops. The extractant consists of 0.005 M diethylenetriaminepentaacetic acid (DTPA), 0.1 M triethanolamine (TEA), and 0.01 M $CaCl_2$ with a pH of 7.3. The soil test consists of shaking 10 g of air-dried soil with 20 ml of extractant for 2 h. The leachate is filtered, and Zn, Fe, Mn, and Cu are measured in the filtrate by atomic absorption spectrophotometry.

The pH of 7.3 buffered with TEA is used to prevent excess dissolution of the trace metals, a process which is highly pH dependent. The presence of 0.01 M $CaCl_2$ enables the extractant to obtain equilibrium with CO_2, which minimizes the dissolution of $CaCO_3$ from calcareous soils (Baker and Amacher, 1982). In a preliminary soil test calibration study, Edlin et al. (1983) evaluated four Zn extraction methods (DTPA, EDTA, 1 M NH_4OAc, and 0.1 M HCl) for their ability to separate responding from nonresponding soils in a growth chamber experiment using alfalfa as an indicator plant. The DTPA test (Lindsay and Norvell, 1978) was the most suitable index of Zn-deficient soils, with a critical value of 0.55 mg DTPA-Zn kg^{-1} soil.

Cox and Wear (1977) reported critical Zn soil test levels for corn and rice for three methods (DTPA, 0.1 M HCl, and 0.05 M HCl + 0.0125 M H_2SO_4) based on experiments conducted as a part of a joint regional project that included the southern states in the United States, Pennsylvania, and Puerto Rico. For corn the critical soil test levels were 0.5 mg Zn kg^{-1} for DTPA, 1.4 mg Zn kg^{-1} for 0.1 M HCl, and 0.8 mg Zn kg^{-1} for the double acid. For rice the values were 0.7, 1.8, 1.4 mg Zn kg^{-1}, respectively. Similarly, Singh et al. (1987) successfully used DTPA-extractable Zn to predict Zn deficiency in a large number of soil samples taken across Saskatchewan, Canada. In acid soils, the Mehlich 1 (0.05 mol L^{-1} HCl + 0.0125 mol L^{-1} H_2SO_4) has been satisfactory for Zn and Mn determinations (Weaver and Evans, 1968; Cox, 1968). The DTPA method was developed for use on calcareous soils, but inclusion of soil pH in the interpretation of results makes the method useful for estimating trace metal availability in acid soils well below the pH range for which the test was originally intended (Haq and Miller, 1972). The acid extractants are not recommended for calcareous soils.

For boron, the most satisfactory extractor is hot water. As far as Mo is concerned, water, normal NH_4OAc, strong acid, strong base, and ammonium oxalate (pH 3.3) have been investigated to relate extractable soil Mo to plant response. Many soil factors other than extractable Mo levels affect plant uptake of Mo, so the general usefulness of extractable soil Mo measurements remains quite limited (Cox and Kamprath, 1972; Kubota and Cary, 1982).

Recommended micronutrient rates have been based on results of numerous experiments, and these rates vary with crop, soil, and other factors. The usual application rates (on an elemental basis) range from 1 to 10 kg ha^{-1} for Cu, Fe, Mn, and Zn; < 1 kg ha^{-1} for B; and < 100 g ha^{-1} for Mo (Mortvedt, 1994).

Aluminum

Aluminum toxicity is one of the most important growth-limiting factors in acid soils in many regions of the world (Fageria et al., 1988). Aluminum ions are most commonly displaced with unbuffered salt solution, such as 1 M KCl, $CaCl_2$, and $BaCl_2$ (Barnhisel and Bertsch, 1982). Extraction techniques employing unbuffered salt solutions are probably best suited for estimating truly exchangeable Al, at least as a first approximation. Other complexing agents and acid salt solutions may extract Al from both exchangeable and nonexchangeable sources, including structural oxyhydroxy polymeric interlayers, organically bound species, and other noncrystalline measurable forms (Barnhisel and Bertsch, 1982).

D. Calibration and Interpretation

Soil test values have no meaning if they are not calibrated against crop response for appropriate interpretation and fertilizer recommendations. After a soil test procedure has been selected on the basis of thorough greenhouse and laboratory experimentation, it is necessary to calibrate the test on a large number of sites under field conditions. Calibration should be done for each crop and each agroclimatic region. The objective of soil calibration is to determine the amount of nutrient that must be added to a specific soil at different soil test levels of that nutrient to obtain maximum yield. In soil test calibration studies, the following are important considerations:

1. Soils should be deficient in the nutrient for which the calibration study is conducted.
2. All other nutrients except the one under study should be applied in adequate amounts.
3. Other factors that affect growth, such as water deficiency, diseases, insects, and weeds, should not be limiting or their presence should be properly documented.

4. It is more desirable to have a large number of experiments at different locations than to have more replications at a few sites.
5. Under field conditions, soil and climatic variables cannot be controlled effectively. It is necessary to repeat field calibration studies for 3–5 years before definitive conclusions can be made.
6. Nutrient levels selected for calibration studies should cover a wide range from deficiency to sufficiency.

For interpretation of soil test results, crop yields are plotted against soil test values. Absolute yield values should be transformed into relative yield values by:

$$\text{Relative yield (\%)} = \frac{\text{yield of control or treated plot}}{\text{maximum yeild of treated plot}} \times 100$$

According to Evans (1987), a wide scattering of absolute yields may occur as a result of factors other than soil fertility. This scattering of absolute yields does not necessarily mean that there is poor correlation, but a better relationship may be obtained by using a relative yield to eliminate some of the climate and site influences.

The use of soil analysis as a fertilizer recommendation method is based on the existence of a functional relationship between the amount of nutrient extracted from the soil by chemical methods and the crop yield. When a soil analysis test shows a low level of a particular nutrient in a given soil, application of that nutrient is expected to increase crop yield. Figure 4.2 shows a

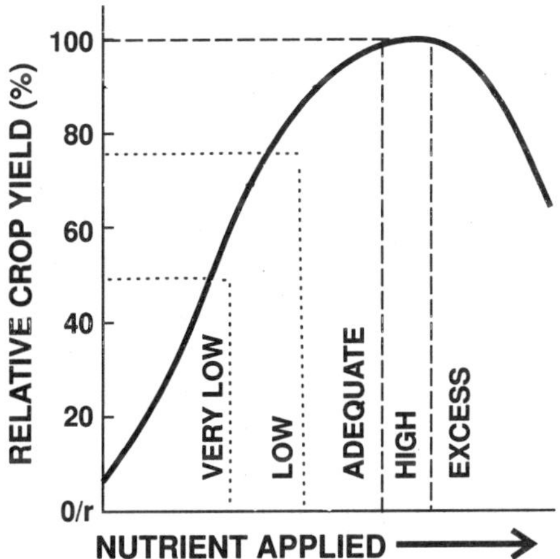

Figure 4.2 Theoretical relationship between nutrient applied and relative yield.

theoretical relationship between nutrient applied and relative crop yield. Nutrient analysis has been arbitrarily classified as very low, low, adequate, high, and excess. Under the very low nutrient level, relative crop yield is expected to be less than 50%, and a larger application of fertilizer for soil-building purposes is required. After the application of the nutrient, growth response is expected to be dramatic and profitable. Under the low fertility level, relative yield is expected to be 50–75%. Under this situation, annual application of fertilizer is necessary to produce maximum response and increase soil fertility. The increased yield justifies the cost of fertilization. When a soil analysis test shows an adequate level, relative crop yield is expected to be 75–100%. Normal annual applications to produce maximum yields are recommended. In this case, more fertilizer may increase yields slightly, but the added yield would not pay back the expense of the additional fertilizer. Under the high nutrient level, there is no increase in yield. Under this situation, a small application is necessary in order to maintain the soil level. The amount suggested may be doubled and applied in alternate years. When the soil test shows very high or excess levels of a nutrient, the yield may be reduced due to toxicity or imbalances of nutrients. Under this situation, there is no need to apply a nutrient until the level drops back into the low range. To get such nutrient level and yield relationship, it is necessary to conduct fertilizer yield trials in several locations in a given agro-ecological region for different crops.

Figure 4.3 shows the relationship between soil test P and relative yield of common bean (*Phaseolus vulgaris* L.) in an Oxisol of central Brazil. Based on relative yield, soil test P values are classified as very low (< 50% relative yield), low (50–75% relative yield), medium (75–100% relative yield), and high (> 100% relative yield). When soil test values are very low for a particular nutrient, heavy fertilizer applications are needed for soil-building purposes. When yield is 50–75% of its potential, annual applications are preferred to produce maximum response and slowly increase soil fertility. With medium soil fertility ratings, normal annual fertilizer applications are used to produce maximum yield. When soil fertility is classified as high, small applications to maintain the soil test level are essential.

In calibration studies, the cation exchange capacity of soil plays an important role. Soils with high CEC release less P to chemical extractants at comparable levels of adequacy than do the lower-CEC soils, and these should therefore be classified differently for making P fertilizer recommendations (Cope, 1981). The reverse is true for K, Ca, and Mg, as demonstrated for K in Figure 4.4 (Cope and Evans, 1985).

Tables 4.10 and 4.11 show typical interpretations of soil tests for P, K, Mg, and micronutrients.

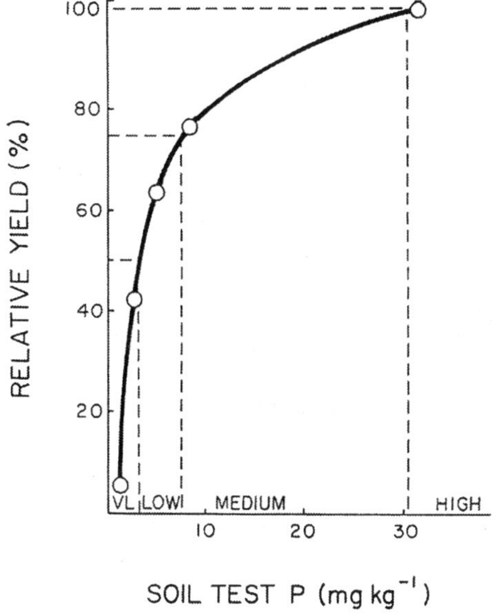

Figure 4.3 Relationship between soil test P and relative yield of common bean.

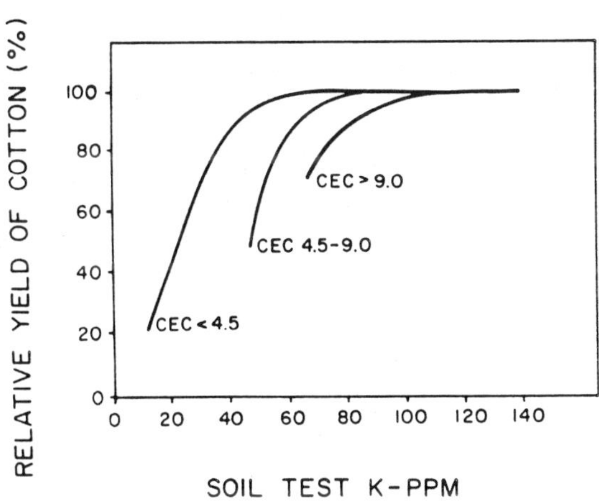

Figure 4.4 Relationship between soil test K and relative yield of cotton. (From Cope and Evans, 1985.)

Table 4.10 Interpretation of Soil Test Values for P, K, and Mg

Relative rating	Extractable element (mg kg^{-1})				
	Mehlich 1			Bray 1	
	P	K	Mg	P	K
Very low	0–6	0–14	0–17	0–12	0–60
Low	7–13	15–35	18–25	13–25	61–90
Medium	14–23	36–37	26–59	26–37	91–150
High	23–44	68–133	60–126	38–75	151–250

Source: Lorenz and Maynard, 1980.

Table 4.11 Interpretation of Soil Test Values for Micronutrients

Micronutrient	Critical level (mg kg^{-1})		
	Range	Mean	Extractant
Iron	2.5–5	4.0	DTPA
Manganese	1–2	1.4	DTPA
	4–8	7.0	Mehlich 1
Zinc	0.25–2	0.8	DTPA
	0.5–3	1.1	Mehlich 1
Boron	0.1–2	0.7	Hot H_2O
Copper	0.12–2.5	0.8	DTPA
	0.1–10	3.0	Mehlich 1
Molybdenum	—	0.2	$(NH_4)_2C_2O_4$

Source: Compiled from Cox, 1987, and Reisenauer et al., 1973.

E. Fertilizer Recommendations

The last phase in soil testing is making fertilizer recommendations. After calibration and interpretation, soil test data need to be associated with appropriate fertilizer rates. Each of the four common interpretative categories (very low, low, medium, and high) signifies different fertilizer rates. The very low level requires large amounts of fertilizer, the low level needs annual application, the medium level requires normal application, and the high level needs little or no fertilizer. The economic loss to farmers is not as great with excess fertilization as with underfertilization by the same proportion. Thus, it would seem more profitable to be sure that the optimum amount is recommended even if there is a chance that the rate would be more than optimum, as in unfavorable growth years (Barber, 1973). This holds especially true for nutrients not easily leached from the soil.

In making fertilizer recommendations, not only the soil test but also climate, disease, insects, previous crops, previous fertilizer application, and soil yield potential should be considered. Maximum crop yield is obtained by fertilizer application only if all growth factors are at optimum levels. Besides optimum fertilizer application, other cultural practices such as timely planting, adequate plant density, effective fertilizer application, and efficient harvesting also contribute to increased yield. If a nutrient is mobile, as nitrogen is, large amounts will be needed as potential yield increases.

Cope (1973) described the index system now in use at the Auburn University Soil Testing Laboratory for making fertilizer recommendations. It utilizes the concept of percentage of nutrients in the soil as introduced by Mitscherlich and further discussed by Bray (1944). The index expresses the percent sufficiency of a nutrient based on the amount necessary to produce 100% relative yield. Rouse (1968) has proposed use of such a system in combination with ratings defined in terms of relative yield. This rating system is presented in Table 4.12.

The Auburn University Laboratory makes fertilizer recommendations for 53 different crops or crop rotation (Cope, 1972). Alabama crops are divided into three categories based on P and K requirements. The P_1K_1 group with the lowest requirement includes perennial grasses, corn and small grains, temporary summer grasses, and peanuts. Crops in the intermediate P_1K_2 category include cotton, annual and perennial legumes, and tree crops such as peaches, apples, and pecans. The P_2K_2 requirement level includes gardens, lawns, shrubs, golf courses, athletic fields, and truck crops (Cope, 1973). The relation between fertility index, soil test ratings, and soil test values for P and K are shown in detail in Table 4.13 and for Mg and Ca in Table 4.14. Figure 4.5 shows a sequence for soil testing through fertilizer recommendations.

Table 4.12 Relationships Among Fertility Index, Soil Test Ratings, Relative Yield, and Recommendations Based on Soil Tests

Fertility index[a]	Soil test rating	Relative yield of cotton or legumes (%)	Recommendations
0–50	Very low	< 50	Large applications for soil-building purposes.
60–70	Low	50–75	Annual application to produce maximum response and increase soil fertility.
80–100	Medium	75–100	Normal annual applications to produce maximum yields.
110–200	High	100	Small applications to maintain soil level. Amount suggested may be doubled and applied in alternate years.
210–400	Very high	100	None until level drops back into high range. This rating permits growers, without risk of loss in yields, to benefit economically from high levels added in previous years. Where no P or K is applied, soils should be resampled in 2 years.
410–999	Extremely high	100 ?	Used for P on samples from lawns, gardens, and shrubs. A warning is added that P is excessive and further additions may cause Fe or Zn deficiency.

[a] Fertility index is percent sufficiency expressed to the nearest 10%.

Source: Cope, 1973.

Table 4.13 Fertility Index, Soil Test Rating, and Soil Test P and K for All Soils and Crops of Alabama

Fertility index	Phosphorus				Potassium				
	Rating		Soil test P(kg ha^{-1})		Rating		Soil test K (kg ha^{-1})		
(% sufficiency)	P_1K_1, P_1K_2	P_2K_2	CEC 0–10	CEC > 10	P_1K_2, P_2K_2	P_1K_1	CEC 0–5	CEC 5–10	CEC > 10
0	Very low	Very low	0	0	Very low	Very low	0–22	0–34	0–45
10	Very low	Very low	1–2	1	Very low	Low	23–25	35–37	46–49
20	Very low	Very low	3–5	2	Very low	Low	26–27	38–40	50–54
30	Very low	Very low	6–8	3	Very low	Low	28–29	41–44	55–58
40	Very low	Very low	9–11	4–6	Very low	Low	30–32	45–47	59–64
50	Very low	Very low	12–14	7–8	Very low	Low	33–34	48–51	65–67
60	Low	Very low	15–21	9–12	Low	Low	35–45	52–67	68–90
70	Low	Very low	22–28	13–17	Low	Medium	46–67	68–101	91–135
80	Medium	Low	29–38	18–24	Medium	Medium	68–90	102–135	136–179
90	Medium	Low	39–48	25–29	Medium	High	91–112	136–168	180–224
100	Medium	Low	49–56	30–34	Medium	High	113–135	169–202	225–269
110–130	High	Medium	57–73	35–45	High	High	136–179	203–269	270–359
140–200	High	Medium	74–112	46–67	High	Very high	180–269	270–403	360–538
210–400	Very high	High	113–224	68–135	Very high	Very high	270–538	404–807	539–1075
410–800	Extremely high	Very high	225–448	136–269	Extremely high	Extremely high	539+	808+	1076+
810+	—	Extremely high	449+	270+	Extremely high	Extremely high	—	—	—

Source: Compiled from Cope, 1973.

Table 4.14 Fertility Index, Soil Test Ratings, and Soil Test Mg and Ca for All Soils and Crops of Alabama

Fertility index	Magnesium (all crops) soil test Mg (kg ha^{-1})			Calcium (all soils)		
(% sufficiency)	Rating	CEC 0–5	CEC > 5	Rating	Peanuts (kg ha^{-1})	Tomatoes (kg ha^{-1})
0	Low	0	0	Low	0	0
10	Low	1–2	1–6	Low	1–28	1–45
20	Low	3–6	7–11	Low	29–56	46–90
30	Low	7–8	12–17	Low	57–84	91–135
40	Low	9–11	18–23	Low	85–112	136–179
50	Low	12–14	24–28	Low	113–140	180–224
60	Low	15–17	29–34	Low	141–168	225–280
70	Low	18–19	35–39	Low	169–196	281–336
80	Low	20–23	40–45	Medium	197–241	337–409
90	Low	24–25	46–51	Medium	242–286	410–482
100	Low	26–28	52–56	Medium	287–336	483–560
110–130	High	29–37	57–73	High	337–437	561–728
140–200	High	38–56	74–112	High	438–672	729–1120
210–400	High	57–112	113–224	High	673–1344	1121–2240
410–800	High	113–224	225–448	High	1345–2688	2241–4480
810+	High	225+	449+	High	2689+	4481+

Source: Compiled from Cope, 1973.

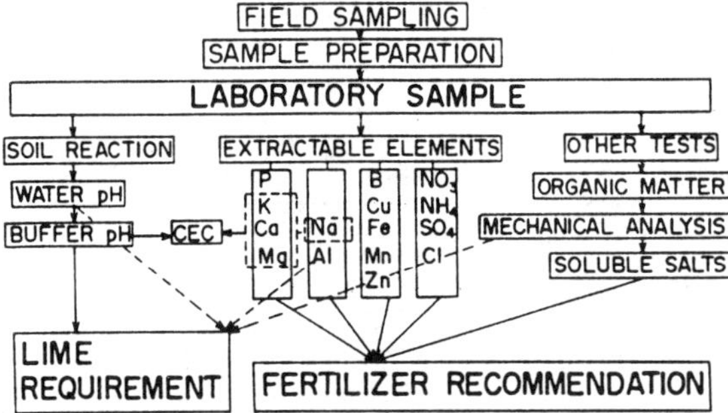

Figure 4.5 Sequence for conducting a soil test, from field sampling through laboratory analysis to lime and fertilizer recommendations. (From Jones, 1988.)

III. PLANT ANALYSIS

Plant analysis in a narrow sense is the determination of the concentration of an element or extractable fraction of an element in a sample taken from a particular part or portion of a crop at a certain time or stage of morphological development (Munson and Nelson, 1973). The concentration is generally expressed on a dry weight basis. Plant analysis considerations include collection of the plant samples, preparation of the samples for analysis, interpretation of analytical results, and recommendations. Plant analysis has many applications, such as: 1) diagnosis of nutrient deficiencies, toxicities, or imbalances; 2) measurement of the quantity of nutrients removed by a crop to replace them in order to maintain soil fertility; 3) estimating overall nutritional status of the region or soil types; 4) monitoring of the effectiveness of the fertilizer practices adopted; 5) prediction of crop yields; and 6) estimation of nutrient levels in diets available to livestock (Smith, 1986). With the development of instruments such as atomic absorption spectrophotometers, spark emission spectrometers, and inductively coupled plasma (ICP) emission spectrometers, plant analysis has become more sensitive and simplified.

After identifying a nutrient deficiency in a growing crop through plant analysis, its correction depends on the nutrient and growth stage of the crop. If the deficiency is identified in an early growth stage and the nutrient is a micronutrient, it is easy to correct, and the farmer may get a favorable return on his investment. The most common method of correcting nutrient deficiencies in a growing crop is foliar application of the deficient nutrient. For field crops, major nutrients should not be applied by foliar application. To satisfy plant requirements for major nutrients, several foliar applications are necessary for beneficial results, and this is rarely economically feasible.

Table 4.15 Plant Parts and Number of Plants to be Sampled for Field Crops at Different Growth Stages

Crop	Growth stage	Plant part to sample	No. of plants to sample
Wheat	Seedling stage	All the aboveground portion	50–100
Rice, barley	Prior to heading	The four uppermost leaves	50–100
Corn	Seedling stage	All the aboveground portion	20–30
	Prior to tasseling	The entire leaf fully developed below the whorl	15–25
	From tasseling and shooting to silking	The entire leaf at the ear node or immediately below or above it	15–25
Soybean or other beans	Seedling stage	All the aboveground portion	20–30
	Prior to or during initial flowering	Two or three fully developed leaves at the top of the plant	20–30
Sugarbeets	Midseason	Fully extended and mature leaves midway between the younger center leaves and the oldest leaf whorl on the outside	30–40
Peanuts	Prior to or at bloom stage	Mature leaves from the main stems and either cotyledon lateral branch	40–50
Cotton	Prior to or at bloom or when first squares appear	Youngest fully mature leaves on main stem	30–40
Alfalfa	Prior to or one-tenth bloom stage	Mature leaf blades taken about one-third of the way down the plant	40–50
Clover	Prior to bloom	Mature leaf blades taken about one-third of the way down from the top of the plant	40–50
Sugarcane	Up to 4 months	Third or fourth fully developed leaf from top	15–25
Sorghum	Prior to or at heading	Second leaf from top of plant	15–25

Source: Compiled from Jones and Steyn, 1973.

A. Collecting the Plant Sample

Probably no other single aspect of the plant analysis technique can have as much effect on the final result as the procedure used to collect the sample for analysis (Jones, 1985). The primary consideration in deciding what part of the plant to sample is the degree to which nutrient concentrations in a given part reflect the nutrient status of the plant. In particular, it is important to use a plant part that gives a sharp transition from a concentration that reflects a deficiency of the nutrient to one that indicates an adequate supply of the nutrient (Ulrich et al., 1959). Jones et al. (1971) and Jones and Steyn (1973) proposed a scheme indicating which plant parts should be sampled at different growth stages. This scheme for important field crops is presented in Table 4.15. When sampling, precautions should be taken to avoid soiled, diseased, and insect-damaged or mechanically damaged plants and to exclude dying or dead tissue. Plant samples should be collected in paper bags rather than plastic bags.

B. Sample Preparation

After collecting plant tissues, the next step is preparation of the sample, which may include washing, drying, grinding, and storage. Plant material should be transported from the field to the analytical laboratory as quickly as possible to avoid respiratory and evaporation losses in weight and enzymatic activity, both of which produce corresponding errors in nutrient determination (Reuter et al., 1986). The plant material destined for micronutrient analysis then should be washed in deionized water to remove dust deposits accumulated from pesticide or nutrient foliar sprays. If dirt cannot be removed by deionized water, a 0.1–0.3% detergent solution can be used. If samples cannot be washed immediately, material should be stored in a refrigerator at 5°C to minimize respiratory losses and plant spoilage. After washing, the samples can be dried to a constant weight in a forced-draft oven at 70–75°C. Prolonged drying at temperatures in excess of 80°C can promote thermal decomposition and appreciable loss of volatile constituents (Grundon and Asher, 1981). Dried material can be ground to manageable sizes and mixed to provide homogeneous samples for chemical analysis. However, grinding is not essential for small-sized samples because the whole sample can be crushed by hand and weighed for analysis (Reuter et al., 1986). Grinding should be performed in mills with stainless steel grinding surfaces in order to minimize contamination. Ground material can be stored in glass or polycarbonate containers under cool, dry, or refrigerated conditions for future analysis. Stored samples should be redried for about 8–12 h just prior to weighing for analysis if the relative humidity in the laboratory is high.

Table 4.16 Comparison of Dry and Wet Oxidation Methods Used for Plant Digestion

Dry ashing	Wet oxidation
Time consuming	Comparatively rapid
Requires higher temperature, and chances of nutrient loss by volatilization	Low temperature and volatilization losses are less
Generally more sensitive to nature of sample	Generally less sensitive to nature of sample
Requires less supervision	Requires more supervision
Reagent blank smaller	Reagent blank larger
Can handle large samples	Difficult to handle large samples

Source: Compiled from Gorsuch, 1976.

C. Chemical Analysis

Detailed analytical procedures for tissue analysis are available in several publications (e.g., Ramirez-Munoz, 1968; Norrish and Hutton, 1977; Liegel et al., 1980; Jones, 1981, 1985; and Munter et al., 1984. However, a brief discussion of digestion methods is appropriate. There are two common digestion methods—dry ashing and wet oxidation—and both have advantages and disadvantages. A comparison of these two methods is given in Table 4.16. The basic digestion reagents used in wet oxidation are: 1) nitric and sulfuric acids, 2) sulfuric acid and hydrogen peroxide, and 3) mixtures containing perchloric acid (Jones, 1985). Among these, nitric and sulfuric acid digestion is suitable for a wide range of sample types. Problems may occur with samples high in Ca when calcium sulfate precipitation causes losses due to coprecipitation. Sulfuric acid and hydrogen peroxide digestion is a vigorous procedure with potential for losses of some elements in the presence of chlorides. A mixture of nitric (sometimes sulfuric) and perchloric acids is a widely used digestion reagent, although its use requires extreme care and a specially designed hood (Jones, 1985). Tolg (1974) gave characteristics of common wet oxidation procedures (Table 4.17). It is now a common practice to use block digestors in wet oxidation processes so that temperature can be regulated and thereby avoid losses due to rapid boiling and excessive temperatures.

In the dry ashing oxidation procedure, care should be taken regarding the vessels used and temperature. Silica (quartz) is probably one of the best materials for the ashing vessel, although well-glazed, acid-washed porcelain high-form crucibles are equally suitable vessels for most uses (Jones, 1985).

Table 4.17 Wet Oxidation Digestion Reagents and Their Applicability

Reagents	Applicability to organic matrix	Remarks
H_2SO_4/HNO_3	Vegetable origin	Most commonly used
H_2SO_4/H_2O_2	Vegetable origin	Not very common
HNO_3	Biological origin	Easily purified reagent, short digestion time, temperature 350°C
$H_2SO_4/HClO_4$	Biological origin	Suitable only for small samples; danger of explosion
$HNO_3/HClO_4$	Protein, carbohydrate (no fat)	Less explosive
$H_2SO_4/HCO_3/HClO_4$	Universal (also fat and carbon black)	No danger with exact temperature control

Source: Tolg, 1974.

Ashing temperature should not exceed 500°C to avoid volatilization losses for most elements.

Besides appropriate oxidation procedures, some other precautions are necessary to achieve accuracy and precision in plant chemical analysis. These precautions include checking the purity of the reagents by including blanks in routine testing, use of appropriate methods to account for known interferences, inclusion of reference plant material with each batch of samples to monitor analytical performance, regularly checking the authenticity of standard solutions used for establishing calibration curves and replacing them at defined intervals, and exchanging samples with other laboratories on a regular basis as a means of verification of analytical procedures (Reuter et al., 1986).

Since 1956 the Department of Soil Science, University of Wageningen, Netherlands, has been comparing chemical analytical results from laboratories all over the world (Houba et al., 1986). Houba and coworkers compared the analytical results of 23 elements, using data from the collaborative interlaboratory study (Table 4.18). In particular, the relationship between content level and coefficient of variation (*CV*) was examined. Usually, a constant *CV* was found at high content levels, with a sharp increase in *CV* at low levels. The precision found for N, P, K, Ca, Mg, Zn, and nitrate was high enough ($CV < 20\%$) to yield reasonably comparable content values. Comparison of analytical results for B, Cu, Fe, Cd, Mn, and Na may be difficult, since a *CV* of about 20% was reached at the levels usually present in plant material. The analytical results for Al, Co, Cr, Mo, Ni, Pb, S, Se, and sulfate varied con-

Table 4.18 Lowest Measurable Level of Some Nutrients for a Chosen Interlaboratory Variability Level of 20% C.V. Compared to Normal Nutrient Values in the Literature[a]

Nutrient	Lowest measurable concentration[b]	Normal concentration[c]
NO_3	15	—
Ca	30	188–338
Mg	15	80–180
Na	30	—
Cl	40	60–600
B	10	9–75
Cu	7.5	4–17
Fe	200	50–200
Mn	25	30–200
Zn	15	22–100

[a] Values of NO_3, Ca, Mg, Na, and Cl are in mmol kg^{-1} and values of B, Cu, Fe, Mn, and Zn are in mg kg^{-1}.
Sources: [b]Houba et al, 1986; [c]Average values obtained from Fink, 1968; Jones, 1972; and Mengel and Kirby, 1982

siderably, irrespective of the content level, which means that results are very difficult to reproduce with these elements.

The data presented by Houba et al. (1986) can be used to assess the possibilities and limitations of plant analysis for the evaluation of the nutrient status of plants (Table 4.18). No analytical problems are to be expected with the determination of N, P, and K, since the normal levels are much higher than the values that correspond to 20% *CV*, a maximum acceptable value chosen by these authors. From Table 4.18 it is apparent that the situation for Ca, Cl, Mg, and Zn is also favorable. Comparison of results is difficult for B, Cu, Fe, and Mn, since values that correspond with 20% *CV* fall in the range of normal values. Elements like Al, Co, Cr, Mo, Ni, Pb, S, Se, and sulfate pose problems with respect to the comparability of analytical results from different laboratories. These results suggest that for many elements plant analysis techniques still need to be refined.

D. Interpretation of Results

The basis for plant analysis as a diagnostic technique is the relationship between nutrient concentration in the plant and growth and production response. This relation should be significant to have complete interpretation in terms of deficient, adequate, and excess nutrient concentrations in the plant.

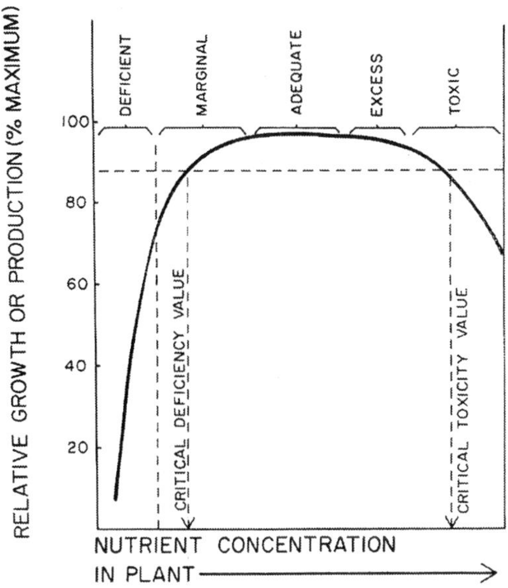

Figure 4.6 Relationship between yield and nutrient concentration in plant.

Curves representing the relationship between nutrient concentration and growth response vary in shape and character depending on both the nutrient concentration in the growth medium and the plant species. A hypothetical curve showing this relationship is shown in Figure 4.6. In the literature, several different terminologies have been used in classifying nutrient concentrations in plant tissue.

In one scheme (based on Figure 4.6), nutrient concentrations can be classified as deficient, marginal, excess, and toxic. When nutrients are in the deficiency range, plant growth and yield are significantly reduced and foliar deficiency symptoms appear. In this range, application of the nutrient results in a sharp increase in growth with very little change in nutrient concentration in the plant. In the marginal range, growth or yield is reduced, but plants do not show deficiency symptoms, and both nutrient concentrations and growth increase as more nutrient is absorbed. Sometimes the marginal range is also called the transition zone (Ulrich and Hills, 1973). Within the marginal or transition zone lies the critical level or concentration. The critical level can be defined as that concentration at which the growth or yield begins to decline significantly. This, of course, is a matter of judgment and how one interprets the term (Munson and Nelson, 1973). However, under most practical farming conditions, a decrease of 20% from optimum yield can be considered as a

Table 4.19 Average Concentration of Essential Nutrients in Dry Matter Sufficient for Adequate Growth of Field Crops[a]

Nutrient	$g\ kg^{-1}$ or $mg\ kg^{-1}$	$\mu mol\ g^{-1}$
Carbon	450	37,500
Oxygen	450	38,125
Hydrogen	60	60,000
Nitrogen	14	1,000
Phosphorus	1.9	60
Potassium	9.8	250
Calcium	5.0	125
Magnesium	1.9	80
Sulfur	1.0	30
Iron	112	2
Manganese	55	1
Zinc	20	0.3
Copper	6	0.1
Boron	22	2
Molybdenum	0.1	0.001
Chlorine	106	3

[a] Values of macronutrients are in $g\ kg^{-1}$ and values of micronutrients in $mg\ kg^{-1}$

Sources: Bergmann and Neubert, 1976; Salisbury and Ross, 1985.

significant decrease in crop yield. This is an arbitrary value based on practical experience of the authors. The critical value is normally estimated on the basis of a 5, 10, or 20% reduction in maximum yield (Ulrich and Hills, 1973; Reuter and Robinson, 1986; Fageria, 1987). In determining critical nutrient concentrations experimentally, it is important that plant growth not be limited by factors other than supply of the nutrient being studied. The third range or zone is the adequate zone, in which there is no increase in growth but nutrient concentration increases. This classification is also known as satisfactory, normal, or sufficient concentration. The high classification range represents the range of concentrations between the adequate and toxic ranges. Fertilizer use on crops with values in this range should be reduced until the nutritional status of plants lies in the adequate range (Reuter and Robinson, 1986). The fourth (toxic) range is the range of nutrient in which there is reduction in growth and yield but concentration of nutrients continues to increase. In this range plants start showing toxicity symptoms. The critical toxicity value lies in this range (Figure 4.6). Table 4.19 shows average adequate concentrations

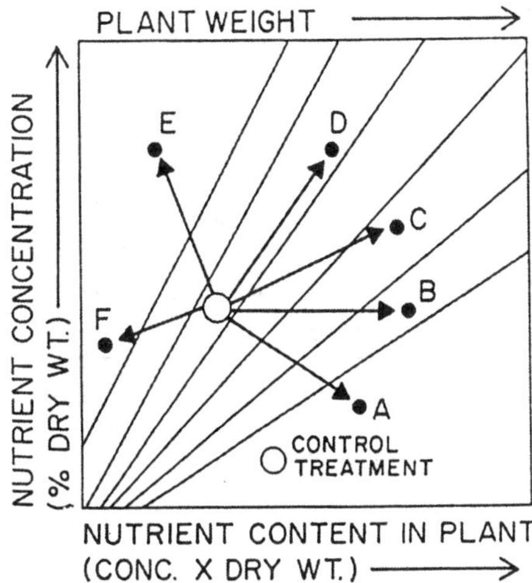

Figure 4.7 Interpretation of directional relationship between foliar concentration and absolute content of an element following treatment. (From Timmer, 1985.)

for essential nutrients in crop plants. No doubt these values vary with soil, climate, crop, and management practices, and it is very difficult and not logical to make generalizations. Generalized values (Table 4.19) give the reader some idea about what adequate levels of nutrients are in crop plants. Specific values for the adequacy range are presented in chapters devoted to individual crops.

Timmer and Stone (1978) suggested a different approach to the interpretation of plant analysis results (Figure 4.7). In their interpretation method, plant nutrient content (concentration × dry matter) is plotted on the x axis and nutrient concentration (present dry weight) on the y axis, and a series of diagonal lines from the origin represent unit increases in yield (Figure 4.7). Comparisons involve an untreated or control plot (open symbol) and fertilized plots (shaded symbol) in the same stand. Based on data from controls and treated plots, points above or below the line (D) would indicate a gain or loss, respectively, in plant weight. The direction and magnitude of the resulting changes in all three parameters can be designated by a single arrow. Diagnostic interpretations depend on the change in nutrient concentrations, content, and plant dry weight or yield following fertilization or any other treatment. Shifts along the diagonal depict no real change in weight; those to the right or left signify gains or losses, respectively (Timmer and Morrow, 1984). The arrows in Figure 4.7 indicate dilution (A), sufficiency (B), deficiency (C), luxury (D), toxicity (E), and possible antagonisms (F). Table 4.20

Table 4.20 Interpretation of Plant Analysis Results[a]

Direction of shift	Plant weight	Nutrient concentration	Nutrient content	Interpretation	Nutrient diagnosis
A	+	–	+	Dilution	Nonlimiting
B	+	0	+	Sufficiency	Nonlimiting
C	+	+	+	Deficiency	Limiting
D	0	+	+	Luxury consumption	Nontoxic
E	–	+	±	Excess	Toxic
F	–	–	–	Excess	Antagonistic

[a] +, –, and 0 mean response positive, response negative, and no response, respectively.
Source: Timmer, 1985.

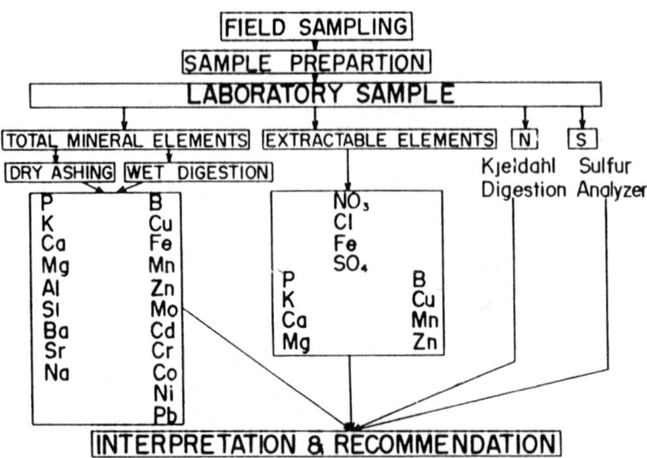

Figure 4.8 Sequence for conducting a plant analysis, from field sampling through laboratory analysis to interpretation and recommendations. (From Jones, 1988.)

summarizes the interpretation and diagnosis of plant mineral nutrition analysis results by this technique. Figure 4.8 shows the various steps from plant sampling through interpretation of analysis results and recommendations.

E. Factors Affecting Nutrient Concentrations

Nutrient concentrations in plants are affected by several factors including species and cultivar within species, plant age, interaction with other nutrients, and environmental factors such as moisture supply, humidity, and light.

Genotypic Differences

Differences in nutrient absorption and utilization among field crops and cultivars within a crop are well established (Lafever, 1981; Vose, 1984; Gerloff, 1987; Siddiqui et al., 1987; Baligar et al., 1987; Saric, 1987; Itamar et al., 1987). Nutrient concentration and uptake by different plant genotypes are the most important criteria used in recent years for identifying the existing genetic specificity of plant mineral nutrition (Saric, 1987). Nutrient uptake values are transformed into a nutrient use efficiency ratio, and this ratio is one of the best parameters for comparing the different plant species or cultivars in terms of nutrient utilization (Gerloff and Gabelman, 1983). The nutrient efficiency ratio is the milligrams of dry matter produced per milligram of element absorbed by a plant or present in a plant part. The ratio should be compared under stress and adequate nutrient supply to verify plant species or cultivar differences in nutrient utilization under suboptimum and optimum conditions.

Table 4.21 summarizes various soil and plant mechanisms and processes and other factors that influence genotypic differences in plant processes and other factors that influence genotypic differences in plant nutrient efficiency. Similarly, a summary of morphological and physiological traits in plants associated with major nutrient use efficiency is presented in Table 4.22. Figure 4.9 shows the mechanisms of plant adaptation to high and low nutrient levels. A detailed description of this subject can be found in Fageria and Baligar (1993) and Baligar and Fageria (1996).

Plant Age

Growth and development of a plant make a difference in nutrient concentration in plant organs. Normally, with advancement of plant age, nutrient concentration expressed per unit dry weight decreases. This is viewed as a dilution effect (Jarrell and Beverly, 1981). Phosphorus decreases in bean plant tops with advancement of age under different levels of P supplied in oxisols of central Brazil in a greenhouse experiment (Table 4.23). The decrease in P concentration occurred throughout the growing season where P was applied.

Table 4.21 Soil and Plant Mechanisms and Processes and Other Factors that Influence the Genotypical Differences in Nutrient Efficiency in Plant at Nutrient Stress Conditions

A. Nutrient acquisition
   1. Diffusion and mass flow (buffer capacity, ionic concentration, ionic properties, tortuosity, soil moisture, bulk density, temperature)
   2. Root morphological factors (number, length, root hair density, root extension, root density)
   3. Physiological (root: shoot, root microorganisms such as VAM fungi, nutrient status, water uptake, nutrient influx and efflux, rate of nutrient transport in root and shoot, affinity to uptake Km, threshold concentration C_{min}.
   4. Biochemical (enzyme secretion as phosphatase, chelating compounds, siderophore), proton exudate, organic acid such as citric, trans-aconite, melic acid exudate

B. Nutrient movement in root
   1. Transfer across endodermis and transport within root
   2. Compartmentalization/binding within roots
   3. Rate of nutrient release to xylem

C. Nutrient accumulation and remobilization in shoot
   1. Demand at cellular level and storage in vacuoles
   2. Retransport from older to younger leaves and from vegetative to reproductive parts
   3. Rate of chelates in xylem transport

D. Nutrient utilization and growth
   1. Metabolism at reduced tissue concentration of nutrient
   2. Lower element concentration in supporting structure, particularly the stem
   3. Elemental substitution (e.g., Na for K function)
   4. Biochemical (nitrate reductase for N-use efficiency, glutamate dehydrogenase for N metabolism, peroxidase for Fe efficiency, pyruvate kinase for K deficiency, metallothionein for metal toxicities

E. Other factors
   1. Soil factors
      a. Soil solution (ionic equilibria, solubility precipitation, competing ions, organic ions, pH, phytotoxic ions)
      b. Physicochemical properties of soil (organic matter, pH, aeration, structure, soil moisture)
   2. Environmental effects
      a. Intensity and quality of light (solar radiation)
      b. Temperature
      c. Moisture (rainfall, humidity)
   3. Plant diseases, insects, and allelopathy.

Source: Compiled from various sources by Baligar and Fageria, 1996.

Table 4.22 Efficiency Traits and External Factors that Determine Nutrient Efficiency in Plants

Morphological and physiological traits	External factors that improve efficiency
Nitrogen	
Root length density, ratio of lateral/primary roots	Use high-yielding, N-efficient cultivars
Higher efficiency for uptake, incorporation, utilization	Fertilizers—right level, time, and depth of placement
Higher dry matter production/unit N; harvest index (H), nutrient harvest index (NHI)	Suitable N source use of ammonification/ nitrification inhibitors
Higher physiological efficiency index for N (PEN = Kg grain produced / Kg N absorbed)	Incorporate crop residue
Nitrate reductase levels (uptake)	Reduce leaching, denitrification, volatilization losses
Glutamate dehydrogenase (metabolism)	Maintain adequate moisture and other essential nutrients
Arginine residue (uptake/transport)	Control weeds, insects, and diseases
NO_3/NH_4 nutrition	
Phosphorus	
Root-Fibrous roots, high root density	Use of P-efficient cultivars
High density and length of root hair	Use of phosphate rock and inorganic forms
Exudates, piscidic acid	Fertilizer band/strip placement
High acquisition capacity	Incorporation of organic matter
More dry matter/unit P absorbed	Lime addition
Partitioning organic/inorganic P	Adequate supply of moisture and other nutrients
High levels of sucrose to glucose, fructose, and phosphatase enzyme	VA-mycorrhiza-increased root surface
High phytic phosphate in grain	Control of weeds, diseases, and insects
Potassium	
High efficiency (uptake, incorporation, utilization	Use of K-efficient cultivars
High dry matter/unit K absorbed	Topdress—light-textured soil
K uptake and transport	Incorporate crop residue
Pyruvate kinase (low K status)	Control of weeds, diseases, insects
Membrane mechanisms	Addition of other nutrients (NP)
	Reduce leaching and run-off losses
Calcium/Magnesium	
High dry matter/unit of nutrient	High-yielding crops
High efficiency (uptake, incorporation, utilization)	Adequate levels of moisture and other nutrients
	Type and quality of liming materials and incorporation
	Control of weeds, diseases, and insects

Source: Compiled from various sources by Baligar and Fageria, 1996.

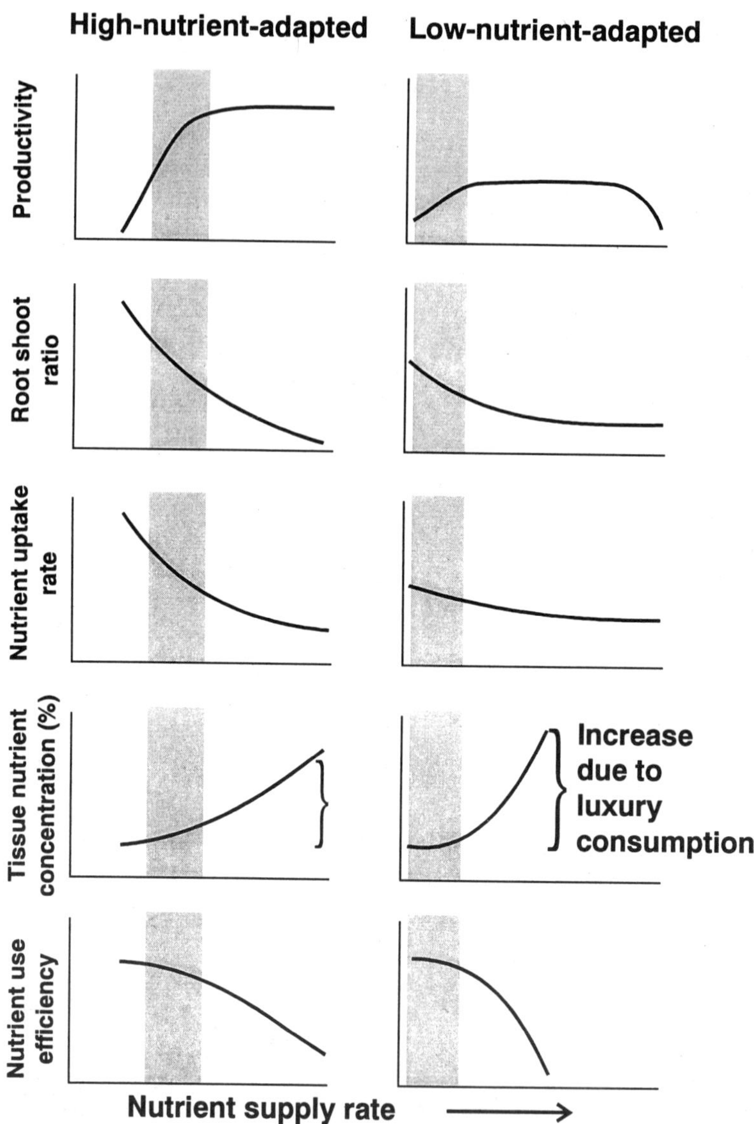

Figure 4.9 Mechanisms of high and low nutrient adaptation by crop plants. (From Chapin, 1987.)

Table 4.23 Concentration of P in Bean Plant Tops at Different Growth Stages

P level	Concentration ($g\ kg^{-1}$)				
($mg\ kg^{-1}$)	17 days	31 days	45 days	60 days	80 days
0	3.5	2.0	1.1	1.2	1.3
25	4.5	3.0	2.5	2.3	1.4
50	4.0	3.3	3.1	2.8	1.9
75	4.3	3.3	3.2	2.6	2.0
100	4.3	3.5	3.3	2.7	2.1
125	4.8	3.5	3.3	2.4	2.3
150	4.3	3.5	3.4	2.4	2.2
175	4.0	3.5	3.4	2.6	2.5
200	4.8	3.0	3.2	2.5	2.0

Source: Fageria, 1989.

Nutrient Interaction

Interaction between nutrients in crop plants occurs when the supply of one nutrient affects the absorption and utilization of other nutrients. This type of interaction is most common when one nutrient is in excess concentration in the growth medium. Nutrient interactions can occur at the root surface or within the plant. According to Robson and Pitman (1983), nutrient interactions can be classified into two major categories. In the first category are interactions which occur between ions because the ions are able to form a chemical bond. Interactions in this case are due to formation of precipitates or complexes. For example, this type of interaction occurs where the liming of acid soils decreases the concentration of almost all micronutrients except molybdenum. But this decrease varies from nutrient to nutrient. For example, Cu is more strongly complexed by soluble organic matter than Zn (Hodgson et al., 1966), and effects of increasing soil pH are more marked on Zn uptake than on Cu uptake by plants (Robson and Pitman, 1983). The second form of interaction is between ions whose chemical properties are sufficiently similar that they compete for site of adsorption, absorption, transport, and function on plant root surfaces or within plant tissues. Such interactions are more common between nutrients of similar size, charge, geometry of coordination, and electronic configuration (Robson and Pitman, 1983). This type of interaction is common among Ca^{2+}, Mg^{2+}, K^+, and Na^+. Figure 4.10 shows the processes in soil and plants where interactions between nutrients may occur.

In crop plants, the nutrient interactions are generally measured in terms of growth response. When plant growth is taken into consideration, interactions

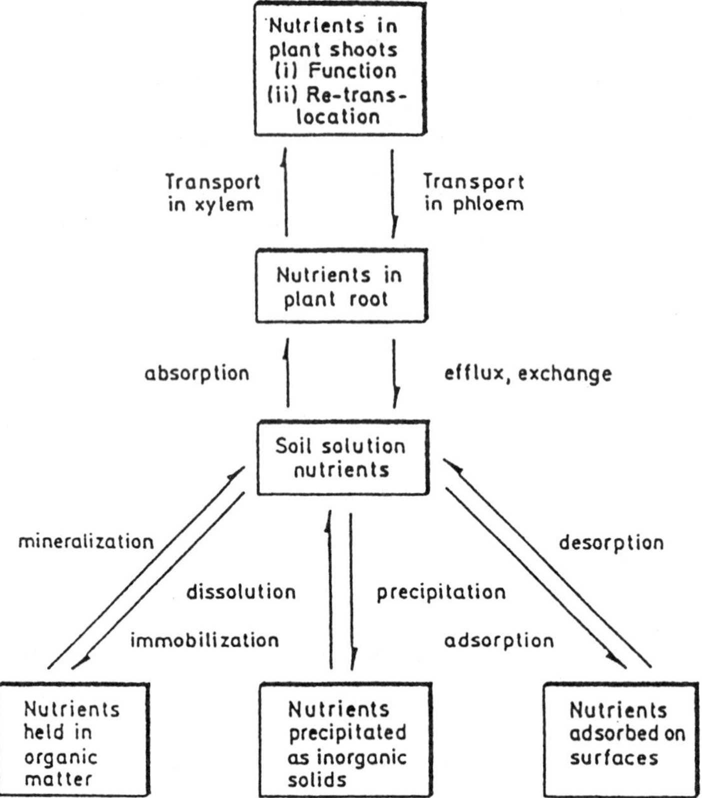

Figure 4.10 Processes in soil and plants where interactions between nutrients may occur. (From Robson and Pitman, 1983.)

may be positive or negative. When nutrients in combination result in a growth response that is greater than the sum of their individual effects, the interaction is positive; when the combined effect is less, the interaction is negative. In the former case the nutrients are synergistic, whereas in the latter they are antagonistic. Additivity indicates the absence of interaction (Sumner and Farina, 1986).

Environment

Environmental factors such as moisture supply, temperature, and light affect nutrient concentrations in plants. Since weather varies from year to year in a given agroclimatic region, nutrient concentrations vary in crop plants from year to year. This variation is more significant in annual shallow-rooted crops due to greater variability in moisture supply (Bates, 1971). Soil moisture plays an important role in movement of nutrients to roots and hence absorption and concentration in the plants. Fisher (1980), working with *Stylosanthes humilis* in Australia, found that P concentrations in shoots were reduced from 0.20% to as low as 0.08% by water stress during early vegetative growth and from 0.22% to 0.15% during late vegetative growth.

When temperature is below the optimum, plant growth is reduced. As a result the nutrient concentration in the dry matter tends to increase (Bouma, 1983). In subterranean clover grown at three temperatures (15°C day/10°C night, 21°C/16°C, 27°C/22°C), and at two P levels, the response in growth to increasing P level was least at the lowest temperature. The P concentration in all plant parts was highest at the lowest temperature and decreased with rising temperature (Bouma and Dowling, 1969). There is a potential limit at which further decreases in temperature will not cause a further rise in nutrient concentration or may even cause a decrease (Bouma, 1983). Light influences nutrient concentrations in a similar manner.

IV. VISUAL SYMPTOMS

Identification of nutritional disorders in crop plants through visual symptoms is the cheapest diagnostic technique. However, considerable experience is needed on the part of the observer to identify nutritional disorders through visual symptoms. Visual symptoms are sometimes confused with disease, insect, and drought stress. The first step in identifying nutrient disorder by visual symptoms is observation of the growth and development of the plant. Growth can be stunted by deficiency or toxicities of all elements, but nitrogen and phosphorus are two important nutrients whose deficiency brings about growth reduction. The second step is to note what plant part is affected. Whether the foliar symptoms appear on the lower leaves or older leaves or on the stem, flowers, and growing points of the plant is nutrient dependent. After identifying the part of the plant, the third step is recognition of the nature of the symptoms—whether chlorotic, necrotic, or deformed.

If symptoms appear on the lower leaves, the possibility exists that there is a deficiency of mobile nutrients such as N, P, K, Mg, and Zn. Mobile nutrients are those which can be retranslocated within plants. They move from the original site of deposition (older leaves) to the organ where deficiency occurs. As a result, deficiency will first occur on older leaves on the lower portion of the plant. When a shortage of the immobile element occurs, the element is not translocated to the growing region of the plant but remains in the older leaves where it was originally deposited. Deficiency symptoms, therefore, first appear on the upper young leaves of the plant. The immobile elements include

Table 4.24 General Description of Nutrient Deficiency Symptoms in Field Crops

Nutrient	Symptoms
N	Chlorosis starts in old leaves; in cereals tillering is reduced; under field conditions, if deficiency is severe, whole crop appears yellowish and growth is stunted
P	Growth is stunted; purple orange color of older leaves; new leaves dark green; in cereals tillering is drastically reduced
K	Older leaves may show spots or marginal burn starting from tips; increased susceptibility to diseases, drought, and cold injury
Ca	New leaves become white; growing points die and curl
Mg	Marginal or interveinal chlorosis with pinkish color of older leaves; sometimes leaf-rolling like drought effect; plants susceptible to winter injury.
S	Chlorosis of younger leaves; under severe deficiency whole plant becomes chlorotic and similar to appearance in N deficiency
Zn	Rusting in strip of older leaves with chlorosis in fully matured leaves; leaf size is reduced
Fe	Interveinal chlorosis of younger leaves; under severe deficiency whole leaf becomes first yellow and finally white
Mn	Similar to iron deficiency; at advanced stage necrosis develops instead of white color
Cu	Chlorosis of young leaves, rolling, and dieback
Mo	Mottled pale appearance in young leaves; bleaching and withering of leaves
B	Pale green tips of blades, bronze tint; death of growing points

Ca, Fe, S, B, Mn, and Mo. Tables 4.24 and 4.25 describe some important key points in the identification of symptoms of nutrient deficiency or toxicity symptoms in crop plants. For further information, there are several publications which describe the foliar symptoms of nutritional disorders in crop plants and illustrate them with colored photography (Wallace, 1961; Wilcox and Fageria, 1976; Fageria, 1984; Bould et al., 1983; Chapman, 1966; Bergmann and Neubert, 1976).

V. CROP GROWTH RESPONSE

Soil and plant analyses are the common practices for identifying nutritional deficiencies in crop production. The best criterion, however, for diagnosing nutritional deficiencies in annual crops is through evaluating crop responses to applied nutrients. If a given crop responds to an applied nutrient in a given soil, this means that the nutrient is deficient for that crop. Relative decrease

Table 4.25 General Descriptions of Toxicity Symptoms in Field Crops

Nutrient/element	Symptoms
Nitrogen	Plants usually dark green in color with abundant foliage, but usually with a restricted root system; NO_3 toxicity shows marginal burn of older leaves followed by interveinal collapse; NH_4^+ toxicity produces blackening around tips of older leaves and necrosis.
Phosphorus	Necrosis and tip dieback; interveinal chlorosis in younger leaves; marginal scorch of older leaves.
Potassium	Excess K may lead to Mg and possibly Mn, Zn, and Fe deficiency.
Sulfur	Reduction in growth and leaf size; sometimes interveinal yellowing or leaf burning.
Magnesium	Excess Mg can induce K deficiency.
Iron	Common in flooded rice plant on acid soils; bronzing of older leaves; induced P, K, and Zn deficiency.
Boron	Interveinal necrosis.
Manganese	Yellowing beginning at the leaf edge of older leaves; uneven chlorophyll distribution; interveinal bronze-yellow chlorosis in beans.
Chlorine	Burning of leaf tips or margins; reduced leaf size, sometimes chlorosis.
Zinc	Excess zinc induces iron chlorosis in plants.
Copper	Stunting, reduced branching, induced iron chlorosis.
Molybdenum	Rarely observed.
Aluminum	Yellowing with white interveinal stripe on older leaves.

in yield in the absence of a nutrient, as compared to an adequate soil fertility level, can give an idea of the magnitude of nutrient deficiency. This can be done by conducting experiments under greenhouse and/or field conditions.

Fageria (1994) conducted a greenhouse experiment that provided evidence of which major nutrient is most limiting for five important annual crops in an Oxisol of central Brazil (Figure 4.11). Phosphorus was the most yield-limiting nutrient among the three nutrients evaluated for all the crops tested. This means that (in an Oxisol) phosphorus deficiency is the primary yield-limiting nutrient for annual crop production. Among the crops tested, the upland rice growth was the lowest without P treatment, as compared to the treatment which received N, P, and K. These results also suggest that there is a difference among crop species in relation to P deficiency and tops growth susceptibility. Crops can be classified for P deficiency susceptibility in the order of rice > wheat > common bean > corn > soybean.

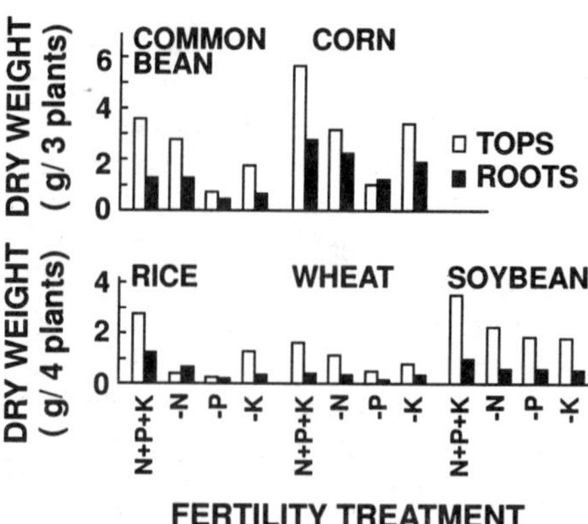

Figure 4.11 Comparison among five crops in dry matter production of tops and roots under different fertility treatments. (From Fageria, 1994.)

VI. CORRECTION OF NUTRIENT DISORDER

After having diagnosed a nutrient disorder, the next step is to correct it in order to improve crop production. Methods of correcting nutrient deficiencies or toxicities vary according to agroclimatic regions, the socioeconomic situation of the region, the magnitude of the disorder, and the nutrient or element involved. A generalized description of these methods is presented in Tables 4.26 and 4.27. Use of efficient or tolerant cultivars in combination with fertilizers or amendments may be the best solution for correcting nutrient disorders in field crops, but this will vary according to the situation. Fertilizer recommendations are usually based on results of field trials in which crop response to various rates of fertilizer application is determined. Such response curves provide relationships between yield and the amount of fertilizer required for a particular crop grown in a particular agroclimatic region. From such curves, economic rates of fertilizer use can be derived. In such calibration studies, it is important that all other controllable crop production factors are at optimum levels.

VII. SUMMARY

Soil testing, plant analysis, visual foliar symptoms, and crop growth response are the most common guides to the fertilization of field crops. Among these diagnostic techniques, visual symptoms are the least expensive, but soil analysis is widely used for soil fertility evaluation. Quantities of fertilizer and lime are determined on the basis of soil test calibration studies for each crop

Table 4.26 Methods of Correcting Nutrient Deficiency and Al and Mn Toxicity

Nutrient/element	Corrective measures
N	Addition of organic matter to the soil; application of N fertilizers, including legumes in the crop rotation; use of foliar spray of 0.25–0.5% solution of urea
P	Adjustment of extreme pH; application of phosphorus fertilizers
K	Application of potassium fertilizers, incorporation of crop residues
Ca	Liming of acid soils; addition of gypsum or other soluble calcium source where lime is not needed; foliar spray in acute cases with 0.75–1% calcium nitrate solution
Mg	Application of dolomitic limestone; foliar application of 2% magnesium sulfate solution
S	Use of fertilizer salt containing sulfur such as ammonium sulfate and single superphosphate; application of gypsum or elemental sulfur
Zn	Addition of zinc sulfate to soil; foliar spray of 0.1–0.5% solution of zinc sulfate
Fe	Foliar spray of 2% iron sulfate or 0.02–0.05% solution of iron chelate; use of efficient cultivars
Cu	Soil application of copper source of fertilizer or foliar spray of 0.1–0.2% solution of copper sulfate
B	Soil application of boron source or foliar spray of 0.1–0.25% solution of borox
Mo	Liming of acid soils, soil application of sodium or ammonium molybdate; foliar spray of 0.07–0.1% solution of ammonium molybdate
Mn	Foliar application of 0.1% solution of manganese sulfate
Al/Mn	Application of lime; use of tolerant species or cultivars

Table 4.27 Tolerance of Plant Foliage to Mineral Nutrient Sprays

Nutrient	Formulation or salt	kg per 400 L[a] of water
Nitrogen	Urea	3–5
	NH_4NO_3, $(NH_4)_2HPO_4$, $(NH_4)_2SO_4$	2–3
	NH_4Cl, $NH_4H_2PO_4$	2–3
Phosphorus	H_3PO_4, others see N above	1.5–2.5
Potassium	KNO_3, K_2SO_4, KCl	3–5
Calcium	$CaCl_2$, $Ca(NO_3)_2$	3–6
Magnesium	$MgSO_4$, $Mg(NO_3)_2$	3–12
Iron	$FeSO_4$	2–12
Manganese	$MnSO_4$	2–3
Zinc	$ZnSO_4$	1.5–2.5
Boron	Sodium borate	0.25–1
Molybdenum	Sodium molybdate	0.1–0.15

[a] 400 L of solution is sufficient to spray on 1 ha of field crop.
Source: Compiled from Wittwer, 1967.

in a given agroclimatic region. One of the greatest values of tissue analysis is in the prevention of deficiencies rather than their correction after they appear. Thus, trends in tissue analysis over a period of years may be studied in relation to the fertilizer programs to determine whether the supply of one or more elements is deficient, adequate, or excessive in a particular soil for a particular crop. Soil analysis, plant analysis, and visual symptoms are all useful and complementary in nutritional diagnosis of crop plants. The three techniques provide information for evaluating the nutrient status of the soil-plant environment and for establishing the basis for fertilizer and lime applications.

REFERENCES

Adams, F. 1984. Crop response to lime in the southern United States, pp. 211–265. *In*: F. Adams (ed.) Soil acidity and liming, 2nd ed. Monogr. No. 12, Am. Soc. Agron., Madison, Wisconsin.

Baker, D. E. and M. C. Amacher. 1982. Nickel, copper, zinc, and cadmium, pp. 323–336. *In*: A. L. Page (ed.). Methods of soil analysis, Part 2, 2nd Ed. Monogr. No. 9, Am. Soc. Agron., Madison, Wisconsin.

Baligar, V. C. and N. K. Fageria. 1996. Nutrient use efficiency in acid soils: Nutrient management and plant use efficiency. Paper presented at 4th Int. Symp. on Plant-Soil Interactions at Low pH, March 17–24, 1996, Belo Horizonte, Brazil.

Baligar, V. C., R. J. Wright, T. B. Kinraide, C. D. Foy, and J. H. Elgin, Jr. 1987. Aluminum effects on growth, mineral uptake, and efficiency ratios in red clover cultivars. Agron. J. 79: 1038–1044.

Barber, S. A. 1973. The changing philosophy of soil test interpretations, pp. 201–211. *In*: L. M. Walsh and J. D. Beaton (eds.). Soil testing and plant analysis. Soil Sci. Soc. Am., Madison, Wisconsin.

Barnhisel, R. and P. M. Bertsch. 1982. Aluminum, pp. 275–300. *In*: A. L. Page (ed.). Methods of soil analysis, Part 2, 2nd ed., Am. Soc. Agron., Madison, Wisconsin.

Bates, T. E. 1971. Factors affecting critical nutrient concentrations in plants and their evaluation. A review. Soil Sci. 112: 116–130.

Bergmann, W. and P. Neubert. 1976. Plant diagnosis and plant analysis. Vebgustav Fischer Verlag, Jena.

Bould, C., E. J. Hewitt, and P. Needham. 1983. Diagnosis of mineral disorder in plants. Vol. 1: Principles. HMSQ, London.

Bouma, D. 1983. Diagnosis of mineral deficiencies using plant tests, pp. 120–146. *In*: A. Lauchli and R. L. Bieleski (eds.). Inorganic mineral nutrition. Encyclopedia of plant physiology, Vol. 15A. Springer-Verlag, New York.

Bouma, D. and E. J. Dowling. 1969. Effects of temperature on growth and mineral uptake in subterranean clover during recovery from phosphorus stress. II. Phosphorus uptake and distribution. Aust. J. Biol. Sci. 22: 515–522.

Bray, R. H. 1944. Soil plant relations. I. The quantitative relation of exchangeable potassium to crop yields and to crop response to potash additions. Soil Sci. 58: 305–324.

Chapin, F. S. 1987. Adaptation and physiological responses of wild plants to nutrient stress, pp. 15–25. *In*: W. H. Gabelman and B. C. Loughman (eds.). Genetic aspects of plant nutrition. Kluwer Academic Publishers, Dordsecht, The Netherlands.

Chapman, H. D. 1966. Diagnostic criteria for plants and soils. University of California, Division of Agricultural Science, Riverside.

Conyers, M. K. and B. G. Davey. 1988. Observations on some routine methods for soil pH determination. Soil Sci. 145: 29–36.

Cope, J. T. 1972. Fertilizer recommendations and computer program key. Auburn Univ. Agric. Exp. Stn. Circ. 176 (revised).

Cope, J. T. 1973. Use of a fertility index in soil test interpretation. Comm. Soil Sci. Plant Anal. 4: 137–146.

Cope, J. T. 1981. Effects of 50 years of fertilization with phosphorus and potassium on soil test levels and yield at six locations. Soil Sci. Soc. Am. J. 45: 342–347.

Cope, J. T. and C. E. Evans. 1985. Soil testing. Adv. Soil Sci. 1: 201–228.

Cox, F. R. 1968. Development of a yield response prediction and manganese soil test interpretation for soybeans. Agron. J. 60: 521–524.

Cox, F. R. 1987. Micronutrient soil test: Correlation and calibration, pp. 97–117. *In*: J. R. Brown (ed.). Soil testing: Sampling, correlation, calibration, and interpretation. Spec. Publ. No. 21, Soil Sci. Soc. Am., Madison, Wisconsin.

Cox, F. R. and E. J. Kamprath. 1972. Micronutrients soil tests, pp. 289–318. *In*: J. J. Mortvedt, P. M. Giordano, and W. L. Lindsay (eds.). Micronutrients in agriculture. Soil Sci. Soc. Am., Madison, Wisconsin.

Cox, F. R. and J. I. Wear. 1977. Diagnosis and correction of zinc problems in corn and rice production. South. Coop. Ser. Bull. 222 for cooperative regional research project S-80.

Dahnke, W. C. and E. H. Vasey. 1973. Testing soils for nitrogen, pp. 97–114. *In*: L. M. Walsh and J. D. Beaton (eds.). Soil testing and plant analysis. Soil Sci. Soc. Am., Madison, Wisconsin.

Duivenbooden, N. Van, C. T. de Wit, and H. Van Keulen. 1996. Nitrogen, phosphorus and potassium relations in five major cereals reviewed in respect to fertilizer recommendations using simulation modelling. Fert. Res. 44: 37–49.

Edlin, W. M., R. E. Karamanos, and E. H. Halstead. 1983. Evaluation of soil extractants for determining Zn and Cu deficiencies in Saskatchewan soil. Commun. Soil Sci. Plant Anal. 14: 1167–1179.

Evans, C. E. 1987. Soil test calibration, pp. 23–29. *In*: J. R. Brown (ed.). Soil testing: sampling, correlation, calibration, and interpretation. Spec. Publ. No. 21, Soil Sci. Soc. Am., Madison, WI.

Fageria, N. K. 1984. Fertilization and mineral nutrition of rice. EMBRAPA-CNPAF/ Editora Campus, Rio de Janeiro.

Fageria, N. K. 1987. Variation in the critical level of phosphorus in rice plant at different growth stages. R. Bras. Ci. Solo. 11: 77–80.

Fageria, N. K. 1989. Effects of phosphorus on growth, yield and nutrient accumulation in common bean. Trop. Agric. 66: 249–255.

Fageria, N. K. 1994. Soil acidity affects availability of nitrogen, phosphorus and potassium. Better Crops International 10: 8–9.

Fageria, N. K. and V. C. Baligar. 1993. Screening crop genotypes for mineral stresses, pp. 142–159. Proc. Workshop on Adaptation of Plant to Soil Stresses, August 1–4, 1993. University Nebraska, Lincoln. INTSORMIL Publ. No. 94–2.

Fageria, N. K., V. C. Baligar, and R. J. Wright. 1988. Aluminum toxicity in crop plants. J. Plant Nutr. 11: 303–319.

Fageria, N. K., V. C. Baligar, and R. J. Wright. 1991. Influence of phosphate rock sources and rates on rice and common bean production in an Oxisol, pp. 539–546. *In*: R. J. Wright, V. C. Baligar, and R. P. Murrmann (eds.) Plant-soil interactions at low pH. Kluwer Academic Publisher, Dordrecht, The Netherlands.

Fageria, N. K., R. J. Wright, V. C. Baligar, and J. R. P. Carvalho. 1991. Response of upland rice and common bean to liming on an Oxisol, pp. 519–528. *In*: R. J. Wright, V. C. Baligar, and R. P. Murrmann (eds.) Plant-soil interactions at low pH. Kluwer Academic Publisher, Dordrecht, The Netherlands.

Finck, A. 1968. Grenzwere der Nahrelementgehalte in Pflanzen und ihre Auswertung zur ermittlung de dungerbedarfs. Z. Pflanzenernaehr. Dung. Bodenk. 119: 197–208.

Fisher, M. J. 1980. The influence of water stress on nitrogen and phosphorus uptake and concentration in Townsville Stylo (*Stylosanthes humilis*). Aust. J. Exp. Agric. Anim. Husb. 20: 175–180.

Gerloff, G. C. 1987. Intact-plant screening for tolerance of nutrient deficiency stress. Plant and Soil. 99: 3–16.

Gerloff, G. C. and W. H. Gabelman. 1983. Genetic basis of inorganic plant nutrition, pp. 453–480. *In*: A. Lauchli and R. L. Bieleski (eds.). Inorganic plant nutrition. Encyclopedia of plant physiology, Vol. 15A. Springer-Verlag, New York.

Gorsuch, T. T. 1976. Dissolution of organic matter, pp. 491–508. *In*: P. D. Lafleur (ed.). Accuracy in trace analysis: Sampling, sample handling, analysis, Vol. 1. Spec. Publ. 422, National Bureau of Standards, Washington, DC.

Grundon, N. J. and C. J. Asher. 1981. Volatile losses of sulphur from oven drying plant material. Commun. Soil Sci. Plant Anal. 12: 1181–1194.

Haq, A. U. and M. H. Miller. 1972. Prediction of available Zn, Cu, and Mn using chemical extractants. Agron. J. 64: 779–782.

Hodgson, J. F., W. L. Lindsay, and J. F. Trierweiler. 1966. Micronutrient cation complexing in soil solution. II. Complexing of zinc and copper in displaced solution from calcareous soils. Soil Sci. Soc. Am. Proc. 30: 723–726.

Honeycutt, C. W. 1994. Linking nitrogen mineralization and plant nitrogen demand with thermal units, pp. 49–79. In: L. P. Wilding (ed.) Soil testing: Prospect for improving nutrient recommendations. SSSA Special Publication No. 40, SSSA, ASA, Madison, WI.

Houba, V. J., I. Novozamsky, and J. J. Lee. 1986. Inorganic chemical analysis of plant tissue: Possibilities and limitations. Neth. J. Agric. Sci. 34: 449–456.

Itamar, I. P., M. Thung, J. Kluthcouski, H. Aidar, and J. R. P. Carvalho. 1987. Screening bean cultivars in relation to higher efficiency of phosphorus use. Pesq. Agropec. Bras., Brasilia. 22: 39–45.

Jarrell, W. M. and R. B. Beverly. 1981. The dilution effect in plant nutrition studies. Adv. Agron. 34: 197–224.

Jones, J. B., Jr. 1972. Plant tissue analysis for micronutrients, pp. 319–347. *In*: J. J. Mortvedt, P. M. Giordano, and W. L. Lindsay (eds.). Micronutrients in agriculture. Soil Sci. Soc. Am., Madison, Wisconsin.

Jones, J. B., Jr. 1979. Soil tests, pp. 514–521. *In*: R. W. Fairbridge and C. W. Finkl, Jr. (eds.). The encyclopedia of soil science, Part 1. Dowden, Hutchinson and Ross, Stroudsburg, Pennsylvania.

Jones, J. B., Jr. 1981. Analytical techniques for trace element determination in plant tissue. J. Plant Nut. 3: 77–92.

Jones, J. B., Jr. 1985. Soil testing and plant analysis: Guides to the fertilization of horticulture crops. Hortic. Rev. 7: 1–67.

Jones, J. B., Jr. 1988. Soil testing and plant analysis: Procedures and use. Tech. Bull. 109, Food and Fertilizer Technological Center, Taipei City, Taiwan. 14 p.

Jones, J. B., Jr., and W. J. A. Steyn. 1973. Sampling, handling and analyzing plant tissue samples, pp. 249–270. *In*: L. M. Walsh and J. D. Beaton (eds.). Soil testing and plant analysis. Soil Sci. Soc. Am., Madison, Wisconsin.

Jones, J. B., Jr., R. L. Large, D. B. Pfleiderer, and H. S. Klosky. 1971. How to properly sample for a plant analysis. Crops Soils. 23: 15–18.

Kubota, J. and E. E. Cary. 1982. Cobalt, molybdenum, and selenium, pp. 485–500. *In*: A. L. Page (ed.). Methods of soil analysis, Part 2, 2nd Ed. Monogr. No. 9, Am. Soc. Agron., Madison, Wisconsin.

Lafever, H. N. 1981. Genetic differences in plant response to soil nutrient stress. J. Plant Nutr. 4: 89–109.

Liegel, E. A., C. R. Simson and E. E. Schulte. 1980. Wisconsin procedure for soil testing, plant analysis and feed forage analysis. Department of Soil Science, University of Wisconsin, Madison.

Lindsay, W. L. and W. A. Norvell. 1978. Development of a DTPA soil test for zinc, iron, manganese, and copper. Soil Sci. Soc. Am. J. 42: 421–428.

Lorenz, O. A. and D. N. Maynard. 1980. Knott's handbook for vegetable growers, 2nd Ed. Wiley, New York.

McLean, E. O. 1982. Soil pH and lime requirement, pp. 199–224. *In*: A. L. Page (ed.). Methods of soil analysis, Part 2. Monogr. No. 9, Am. Soc. Agron., Madison, Wisconsin.

Melsted, S. W. and T. R. Peck. 1973. The principles of soil testing, pp. 13–21. *In*: L. M. Walsh and J. D. Beaton (eds.). Soil testing and plant analysis. Soil Sci. Soc. Am., Madison, Wisconsin.

Mengel, K. and E. A. Kirkby. 1982. Principles of plant nutrition, 3rd ed. International Potash Institute, Bern.

Menon, R. A. and S. H. Chien. 1995. Soil testing for available phosphorus in soils where phosphate rock-based fertilizers are used. Fert. Res. 41: 179–187.

Mortvedt, J. J. 1994. Need for controlled-availability micronutrient fertilizers. Fert. Res. 38: 213–221.

Munson, R. D. and W. L. Nelson. 1973. Principles and practices in plant analysis, pp. 223–248. *In*: L. M. Walsh and J. D. Beaton (eds.). Soil testing and plant analysis. Soil Sci. Soc. Am., Madison, Wisconsin.

Munter, R. C., T. C. Halverson, and R. D. Anderson. 1984. Quality assurance for plant tissue analysis by ICP-AES. Commun. Soil Sci. Plant Anal. 15: 1285–1322.

Norrish, K. and J. T. Hutton. 1977. Plant analysis by x-ray spectrometry. 1. Low atomic number elements Na-Ca. X-ray Spectrom. 6: 6–11.

Quaggio, J. A., V. J. Ramos, P. R. Furlani, and M. L. C. Carelli. 1991. Liming and molybdenum effects on nitrogen uptake and grain yield of corn, pp. 327–332. *In*: R. J. Wright, V. C. Baligar, and R. P. Murrmann (eds.) Plant-soil interactions at low pH. Kluwer Academic Publ., Dordrecht, The Netherlands.

Raij, B. Van. 1991. Fertility of acid soils, pp. 159–167. *In*: Plant-soil interactions at low pH. Kluwer Academic Publ., Dordrecht, The Netherlands.

Ramirez-Munoz, J. 1968. Atomic absorption spectroscopy and analysis by atomic absorption flame photometry. Elsevier, Amsterdam.

Reisenauer, H. M., L. M. Walsh, and R. G. Hoeft. 1973. Testing soils for sulphur, boron, molybdenum and chlorine, pp. 173–200. *In*: L. M. Walsh and J. B. Beaton (eds.). Soil testing and plant analysis. Soil Sci. Soc. Am., Madison, Wisconsin.

Reuter, D. J. and J. B. Robinson. 1986. Plant analysis: An interpretation manual. Inkata Press, Melbourne.

Reuter, D. J., J. B. Robinson, K. I. Peverill, and G. H. Price. 1986. Guidelines for collecting, handling and analyzing plant materials, pp. 20–23. *In*: D. J. Reuter and J. B. Robinson (eds.). Plant analysis: An interpretation manual. Inkata Press, Melbourne.

Rice, C. W., J. L. Havlin, and J. S. Schepers. 1995. Rational nitrogen fertilization in intensive cropping systems. Fert. Res. 42: 89–97.

Robson, A. D. and M G. Pitman. 1983. Interactions between nutrients in higher plants, pp. 147–180. *In*: A. Lauchli and R. L. Bieleski (eds.). Inorganic plant nutrition. Encyclopedia of plant physiology, Vol. 15A. Springer-Verlag, New York.

Rouse, R. D. 1968. Soil test theory and calibration for cotton, corn, soybean and coastal bermudagrass. Auburn Univ. Agric. Exp. Stn. Bull. No. 375.

Sabbe, W. E. and D. B. Marx. 1987. Soil sampling: spatial and temporal variability, pp. 1–14. *In*: J. R. Brown (ed.). Soil testing: Sampling, correlation, calibration, and interpretation. Spec. Publ. No. 21, Soil Sci. Soc. Am., Madison, Wisconsin.

Salisbury, F. B. and C. W. Ross. 1985. Mineral nutrition, pp. 96–113. *In*: Plant physiology, 3rd ed. Wadsworth, Belmont, California.

Saric, M. R. 1987. Progress since the first international symposium: "Genetic Aspects of Plant Mineral Nutrition," Beograd, 1982, and perspectives of future research. Plant Soil. 99: 197–209.

Siddiqui, M. Y., A. D. M. Glass, A. I. Hsiao, and A. N. Minjas. 1987. Genetic differences among wild oat lines in potassium uptake and growth in relation to potassium supply. Plant Soil 99: 93–105.

Sims, J. T. 1993. Environmental soil testing for phosphorus. J. Prod. Agric. 6: 501–507.

Singh, J. P., R. E. Karamanos and J. W. B. Stewart. 1987. The zinc fertility of Saskatchewan soils. Can. J. Soil Sci. 67: 103–116.

Smith, F. W. 1986. Interpretation of plant analysis: Concepts and principles, pp. 1–12. *In*: D. J. Reuter and J. B. Robinson (eds.). Plant analysis: An interpretation manual. Inkata Press, Melbourne.

Soil Science Society of America. 1987. Glossary of Soil Science Terms. SSSA, Madison, Wisconsin.

Stanford, G. 1982. Assessment of soil nitrogen availability, pp. 651–688. *In*: F. J. Stevenson (ed.). Nitrogen in agricultural soils. Monogr. 22, Am. Soc. Agron., Madison, Wisconsin.

Sumner, M. E. and M. P. W. Farina. 1986. Phosphorus interactions with other nutrients and lime in field cropping systems. Adv. Soil Sci. 5: 201–236.

Tan, K. H. 1996. Soil sampling, preparation, and analysis. Marcel Dekker, New York.

Thomas, G. W. and D. E. Peaslee. 1973. Testing soils for phosphorus, pp. 115–132. *In*: L. M. Walsh and J. D. Beaton (eds.). Soil testing and plant analysis. Soil Sci. Soc. Am., Madison, Wisconsin.

Timmer, V. R. 1985. Response of a hybrid popular clone to soil acidification and liming. Can. J. Soil Sci. 65: 727–735.

Timmer, V. R. and L. D. Morrow. 1984. Predicting fertilizer growth response and nutrient status of jack pine by foliar diagnosis, pp. 335–351. *In*: E. L. Stone (ed.). Forest soils and treatment impacts. Proc. 6th North American Forage Conference, Knoxville.

Timmer, V. R. and E. L. Stone. 1978. Comparative foliar analysis of young balsam fir fertilized with nitrogen, phosphorus, potassium, and lime. Soil Sci. Soc. Am. J. 42: 125–130.

Tolg, G. 1974. The basis of trace analysis, pp. 698–710. *In*: F. Korte (ed.). Methodicium chimicum, Vol. 1. Analytical method. Part B. Micromethods, biological methods, quality control, automation. Academic Press, New York.

Ulrich, A. and F. J. Hills. 1973. Plant analysis as an aid in fertilizing sugar crops. Part 1. Sugarbeet, pp. 271–288. *In*: L. M. Walsh and J. D. Beaton (eds.). Soil testing and plant analysis. Soil Sci. Soc. Am., Madison, Wisconsin.

Ulrich, A., D. Ririe, F. J. Hills, A. G. George, and M. D. Morse. 1959. Plant analysis, a guide for sugarbeet fertilization. Calif. Agric. Exp. Stn. Bull. No. 766.

Vose, P. B. 1984. Effects of genetic factors on nutritional requirements of plants, pp. 67–114. *In*: P. B. Vose and S. G. Blixt (eds.). Crop breeding: A contemporary basis. Pergamon, London.
Wallace, T. 1961. The diagnosis of mineral deficiencies in plants by visual symptoms. A color atlas and guide. HMSQ, London.
Weaver, J. I. and C. E. Evans. 1968. Relationship of zinc uptake by corn and sorghum to soil zinc measure by three extractants. Soil Sci. Soc. Am. Proc. 32: 543–546.
Wilcox, G. E. and N. K. Fageria. 1976. Identification and correction nutrients deficiency in bean. EMBRAPA/CNPAF, Tech. Bull. 5.
Wittwer, S. H. 1967. Foliar application of nutrients—part of the chemical revolution in agriculture. Plant Food Rev. No. 2, National Plant Food Inst., Washington, DC.
Woodruff, C. M. 1967. Crop response to lime in the midwestern United States. *In*: R. W. Pearson and F. Adams (eds.). Soil acidity and liming. Monogr. 12, Am. Soc. Agron., Madison, Wisconsin.

5

Nutrient Management of Degraded Soils

I. INTRODUCTION

The world's arable land resources are finite and nonrenewable on a human time scale. Arable land is the primary medium for food and fiber production for mankind. The world total land area is about 13.4 billion ha (Table 5.1). Crops are planted on an area of about 1.36 billion ha, and another 0.11 billion ha are under perennial crops around the world. In addition to this, about 3.3 billion ha are under pastures for animal production. These areas together represent about 36% of the world's total land area. Further, about 31% of the total land area is under forests, with the remainder devoted to populated areas, roads, and recreational areas.

Soil degradation affects about 35% of the earth's land surface (Mabbutt, 1984). Larson (1986) suggested that historically more land has been forced out of crop production because of soil degradation than the amount of land in crop production at the present time. It has been estimated that 0.3–0.5% (4–7 million ha) of the world's crop land is taken out of production each year and that the rate of degradation is accelerating. It is projected that, by the end of the century, 10 million ha (0.7%) will be lost each year (FAO, 1983).

Table 5.1 Land Use in Different Regions of the World in the Year 1989 (billion ha)

Land use	Africa	North and Central America	South America	Asia	Europe	Oceania	USSR	World
T. area[a]	3.03	2.24	1.78	2.76	0.49	0.851	2.24	13.391
L. area[b]	2.96	2.14	1.75	2.68	0.47	0.842	2.23	13.072
Ara. land and P. crops	0.19	0.27	0.14	0.45	0.14	0.051	0.23	1.470
Ara. land[c]	0.17	0.26	0.11	0.42	0.13	0.049	0.22	1.359
Perm. crops[d]	0.02	0.01	0.03	0.03	0.01	0.001	0.01	0.111
Perm. past.[e]	0.89	0.37	0.48	0.68	0.08	0.432	0.37	3.302
F. W. land[f]	0.68	0.72	0.89	0.54	0.16	0.157	0.95	4.097
Other land[g]	1.20	0.78	0.24	1.01	0.09	0.202	0.68	4.202

[a] T. area refers to the total area of the continent, including area under inland water bodies.
[b] L. area refers to total area excluding area under inland water bodies. The definition of inland water bodies generally includes major rivers and lakes.
[c] Ara. land refers to land under temporary crops (double-cropped areas are counted only once), temporary meadows for mowing or pasture, land under market and kitchen gardens (including cultivation under glass, and land temporarily fallow (less than five years). The abandoned land resulting from shifting cultivation is not included in this category.
[d] Perm. crops refers to land cultivated with crops that occupy the land for long periods and need not be replanted after each harvest, such as cocoa, coffee, and rubber; it includes land under shrubs, fruit trees, nut trees, and vines but excludes land under trees grown for wood or timber.
[e] Perm. past. refers to land used for meadows and pasture and permanently (five years or more) for herbaceous forage crops either cultivated or growing wild (wild prairie or grazing land).
[f] F. W. land refers to land under natural or planted stands of trees, whether productive or not, and includes land from which forests have been cleared but that will be reforested in the foreseeable future.
[g] Other land refers to any other land not specifically listed under items 3 through 6. It includes built-on areas, roads, barren land, etc.
Source: FAO (1991).

Soil productivity in developing countries may be reduced by one-fifth (Dudal, 1982). If these projections are approximately correct, the amount of land lost to crop production may approach the amount of new lands that can be brought into productivity. Archaeological evidence suggests that many ancient civilizations vanished because of a decline in soil productivity due to degradation (Lal, 1989). The decrease in productivity depends on the stage of soil degradation (Figure 5.1). There is always a threshold of soil degradation that causes a reduction in crop or animal productivity.

Mengel (1993) stated that global resources are declining and environmental pollution is increasing in an exponential manner. According to Buringh (1982), within 100 years no potential cropland will be left unused. In some countries all potential cropland has already been used. In the United States all reserves

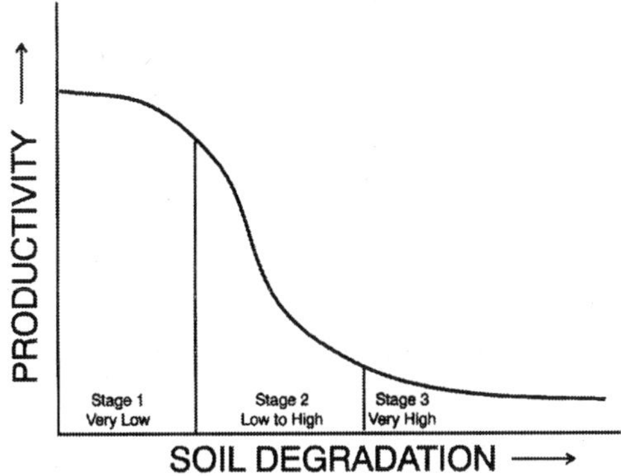

Figure 5.1 Relationship between soil degradation and productivity (Modified from Lal et al., 1989).

will be used within a period of 25 years (USDA, 1981). Schnepf (1979) reported that the production potential of American land is limited and that there is a real need to protect prime land (i.e., good agricultural land) because it is one of the most important resources of the country. Continuing deterioration of water quality is now seen as resulting mainly from pollution, such as water interaction with soil (Warkentin, 1994). Boyle et al. (1989) and Smith and Elliott (1990) described cropping practices in the last several decades that degraded the soil resource and caused ever-diminishing crop yields.

Progress of the present as well as future generations depends on conservation of our valuable soil resources. Due to continuously increasing world demand for more food and fiber, farmers have to use the best existing technologies, and agricultural scientists have to develop new technologies to arrest or minimize future degradation of arable lands and restore already degraded lands for crop production.

II. DEFINITIONS OF SOIL DEGRADATION AND SUSTAINABLE SOIL MANAGEMENT

Before discussing soil degradation processes or causes, it is important to define soil degradation and sustainable soil management. The United Nations Environment Program (UNEP; 1982) defines soil degradation as the decline in soil quality caused by its use by humans. According to FAO (1978), soil degradation is the diminution of the current and/or potential capability of soil to produce (quantitative or qualitative) goods or services as a result of one or more degradation processes. Lal et al. (1989) defined soil degradation as diminution of soil quality and/or reduction in its ability to be a multipurpose resource due both to natural and man-induced causes. In the agricultural sense, soil degradation leads to loss of sustainable production (Lal et al., 1989). Recently Doran and Parkin (1994) defined soil quality as the capability of a soil to function within ecosystem boundaries to sustain biological productivity, maintain environmental quality, and promote plant and animal health. Thus there are several definitions of soil degradation in the literature. Based on all these definitions, we propose a simple definition of soil degradation. For our purposes, soil degradation is the deterioration of soil physical, chemical, and biological properties due to disturbances in its original environment which limit its ability to sustain efficient farming systems. According to Smith et al. (1993), soil quality may be defined in several different ways including productivity, sustainability, environmental quality, and effects on human nutrition. To quantify soil quality, specific soil indicators need to be measured spatially. These indicators are mainly soil properties whose values relate directly to soil quality but may also include policy, economic, or environmental considerations (Smith et al., 1993).

Similarly, many definitions for sustainable land management have been proposed (Bouma, 1994). However, we feel that at present the most ideal definition is that proposed by an International Working Group for the development of an international framework for evaluating sustainable land management.

> Sustainable land management combines technologies, policies and activities aimed at integrating socioeconomic principles with environmental concerns so as to simultaneously maintain or enhance production and services; reduce the level of production risks; achieve environmental stability by preserving soil and water quality; and be economically viable and socially acceptable (Dumanski et al., 1991).

III. PROCESSES AND/OR FACTORS OF SOIL DEGRADATION

A. Physical Degradation

Physical soil degradation is related to changes in soil physical, mechanical, hydrological, and rheological properties which have a negative effect on crop and animal production, farm income, and environmental quality (Lal et al., 1989). These important physical processes are deterioration in soil structure leading to soil compaction, crusting, accelerated erosion, and hardsetting. In addition, physical degradation processes may lead to an imbalance in the water/air ratio, which may cause wetness or drought. Deforestation, burning of vegetation, monoculture, and overgrazing of pasture lands may lead to soil degradation.

Deterioration of Soil Structure

Soil structure refers to the arrangement of primary soil particles into secondary particles or aggregates. Soil structure, per se, is a qualitative concept. Soil structure can be defined operationally as the interaction between soil solids and soil pores (Gupta et al., 1989). Decline in soil structure eventually leads to surface sealing, crusting, compaction, hardsetting, reduction in permeability, poor aeration, and waterlogging.

Surface Sealing or Crusting. Dispersion and subsequent illuviation of fine particles into pores has often been suggested as a major process causing formation of soil crusts (Bresson and Cadot, 1992). The concept of washing-in (i.e., plugging of large pores by washed-in fine material) was introduced by McIntyre (1958a,b) to describe crust formation under rainfall. Surface sealing increases the probability of runoff and soil erosion and occurs on a variety of soils worldwide (Ewing and Gupta, 1994). Rainfall and sprinkler irrigation often form a surface seal or crust, especially when the soil surface is bare. Surface seals generally range from 2 to 3 mm thick and are sometimes overlain by a skin seal (0.1 mm in thickness) composed almost entirely of fine particles (Tarchitzky et al., 1984). This seal reduces infiltration of water into the soil and thus increases the probability of runoff, erosion, and surface water pollution. Surface sealing from rains that occur after planting and before seedling emergence also hamper stand development. The formation of a thin, dense surface layer is a product of the combined effects of aggregate breakdown and soil particle rearrangement due to raindrop impact. The mechanisms of seal formation have not been fully delineated, but some contributing processes have been identified. The underlying cause appears to be dispersion of clay (Shainberg et al., 1989), which weakens the soil. Southard et al. (1988) reported that clay disperses because of a decrease in electrolyte concentration of the soil solution during rainfall. LeBissonats (1990) stated that in dry soils the initial aggregate breakdown is caused by slaking, while in wet soils raindrop impact is the cause.

The flocculation of suspended soil colloids plays an important role in the processes of surface crust formation (Southard et al., 1988). Dispersed soil particles have a negative impact on soil structure and contribute to soil erosion and contaminant movement. Flocculation at a given soluble bivalent cation charge fraction increased as the organic C content of the soil colloids decreased (Goldberg et al., 1990). Lack of organic matter content and a high proportion of silt are responsible for crust formation. The FAO (1978) developed this index to characterize soils with these properties:

$$\text{Crusting index} = \frac{\%\ \text{fine silt} + \%\ \text{coarse silt}}{\%\ \text{clay}}$$

This index will exceed 2.5 for soils prone to intense crusting. An index based on soil organic matter content also used by FAO (1978) is

$$\text{Crusting index} = \frac{1.5(\%\ \text{fine silt}) + 0.75(\%\ \text{coarse silt})}{\%\ \text{clay} + 10\ (\text{organic matter})}$$

This index will exceed 2 for soils prone to intense crusting.

Management Strategies for Surface Sealing. Many factors have been related to seal formation including soil texture, aggregate stability, organic matter content, surface coverage by residue, cropping and tillage systems, and rainfall percolation (Chiang et al., 1993). Organic matter appears to be a dominant factor controlling soil particle flocculation, and consequently crust formation. Available experimental evidence suggests that organic matter stabilizes mineral particles in suspension (Goldberg et al., 1990; Miller et al., 1990; Ryan and Gschwend, 1990). The effect of organic matter on the flocculation of soil particles is often explained in terms of particle charge. Dixit (1982) concluded that adsorption of organic matter onto clay particles increases their negative charge and, therefore, suspension stability. Goldberg et al. (1990) concluded that organic matter decreased the flocculation of soil particles through an effect on particle charge. Thus improving or maintaining adequate levels of organic matter in the soil is an important strategy to reduce soil sealing or crusting. The important management practice which can reduce this problem is maintaining a vegetative cover on the soil surface.

Soil Compaction. Soil compaction refers to the compression of unsaturated soils, during which an increase in the density of the soil body and a simultaneous reduction in fractional air volume occurs. Soil compaction is often characterized in terms of bulk density, void ratio, or total porosity (Gupta et al., 1989). Compaction effects on soil degradation are due to changes in physical, chemical, and biological processes. These processes, in turn, are dependent on the soil structure. In addition to chemical toxicity, physical barriers in the soil profile can also limit root penetration and proliferation (Alcordo and Recheigl, 1993). These include natural hardpans, dense B horizons, and tillage pans formed by heavy machinery (Bowen, 1981).

Correcting Strategies for Soil Compaction. Soil compaction can be reduced by improving soil organic matter content, keeping the soil surface covered with vegetation, use of conservation or minimum tillage, and preparing soil at an appropriate moisture level. Sumner et al. (1990) presented mechanical impedance and aggregate stability data that demonstrated that application of both mined gypsum and phosphogypsum to highly weathered soils improved penetration of subsoil hardpans by roots. Similarly, Radcliffe et al. (1986) showed that mechanical impedance was lower on gypsum-treated soils that have been cropped for years as opposed to fallowed plots. They concluded that gypsum increased subsoil root activity, which in turn reduced subsoil mechanical impedance.

Soil Erosion. Loss of topsoil by wind and water erosion caused by poor soil management is by far the largest single factor contributing to deterioration of soil physical, chemical, and biological properties and to the decline in productivity of most cropland soils. The magnitude of the effect of erosion on yields also varies among soils, crops, and management practices (Lal, 1987). Soil erosion by water depends primarily on soil detachment by raindrop impact (splash) and the transport capacity of the sheet flow. Erosion models have separated water erosion into two components: rill and interrill erosion. Runoff from the soil surface may concentrate in small erodible channels known as rills. In rill erosion, soil loss is due mainly to detachment of soil particles by flowing water (Ben-Hur et al., 1992). In interrill erosion, soil detachment is caused by raindrop impact, and soil transport is due to raindrop splash and runoff flow (Watson and Laflen, 1986). The detachment capacity of interrill flow is small because of its low velocity (Young and Wiersma, 1973). Raindrop detachment capacity is high because the kinetic energy of raindrops has been estimated to be 260 times that of surface flow (Hudson, 1971). However, most of the sediment removed from the interrill area is transported by runoff flow (Young and Wiersma, 1973). In addition to soil detachment, the beating action of raindrops causes the development of a seal at the soil surface (Levy et al., 1994). Seal formation in soils exposed to raindrop impact is due to two mechanisms (Agassi et al., 1981): 1) physical disintegration of soil aggregates and their compaction, and 2) physicochemical dispersion and movement of clay particles into a region of 0.1–0.5 mm depth, where they lodge and clog the conducting pores. The two mechanisms act simultaneously as the first enhances the latter. The seals formed are layers less than 2–3 mm thick that have a greater density, higher shear strength, and lower saturated conductivity than the underlying soil (Levy et al., 1994). Seal strength, as inferred from surface pitting by impacting raindrops, decreases with an increase in clay content and is inversely related to soil erosion (Levy et al., 1994).

Soil loss depends on the inherent susceptibility of the soil to erosion and is called soil erodibility (Wischmeier and Smith, 1978). Slope steepness is an important factor governing water erosion. For nonerodible soils, soil loss doubles as slope steepness increases from 5 to 30% (Watson and Laflen, 1986). For erodible soils, increasing slope from 5 to 30% increases erosion by several fold (Warrington et al., 1989). Nowak et al. (1985) estimated the mean annual soil loss in the United States (average of sheet, rill, and wind erosion) to be 15.3 t ha^{-1} yr^{-1} for cropland, 5.9 t ha^{-1} yr^{-1} for pasture, and 2.7 t ha^{-1} yr^{-1} for forest land. Such quantitative information is not available

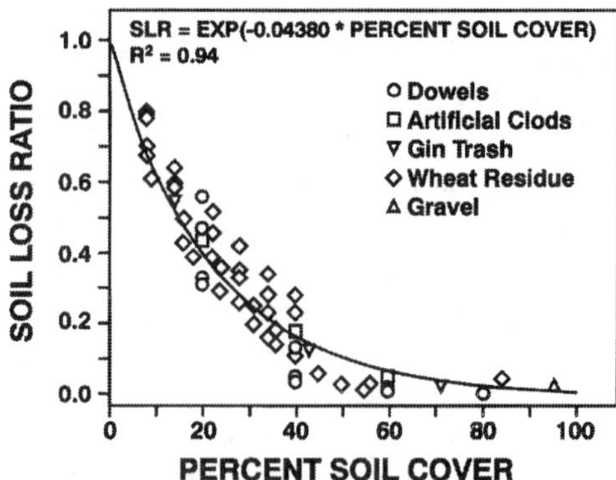

Figure 5.2 Soil loss ratio (SLR) as a function of percent of soil covered by non-erodible material (From Bilbro and Fryrear, 1994).

for other countries, especially those in the tropical and subtropical region where demographic pressures are high and where soils are highly susceptible to erosion and are of low inherent fertility (Lal, 1987).

The U.S. Department of Agriculture (USDA) has assigned a soil loss tolerance value (T-value) for most cultivated soils. This value defines the maximum rate of soil loss that still permits sustained crop production. These T-values never exceed 11.2 t ha^{-1} yr^{-1}, and some are less, depending on the soil depth and other factors (Larson, 1981). Nationally, soil erosion by water alone exceeded the T-value on more than 45.3 million hectares (27.1%) of cropland (Larson, 1981). Water caused annual sediment discharge of 15.9 t ha^{-1} in the cultivation of winter wheat in the Southern Great Plains (Smith et al., 1991). Water erosion and runoff are serious factors in soil degradation of the loamy soils of northern and western Europe. The main damage is related to the accumulation of excess surface water and the concentration of runoff in rills and gullies during rain storms (Courault et al., 1993).

Control Measures for Soil Erosion. Some management practices, such as maintaining or increasing soil organic matter, in turn will reduce compaction and improve other soil properties such as infiltration, water retention, and aeration. Conservation tillage systems offer tremendous potential for erosion control as well as for conserving water and increasing organic matter content of some soils. Dickey et al. (1984) reported that no-tillage systems reduced water erosion by 95% during the fallow period of a wheat-fallow rotation. Conservation tillage systems generally range from practices that retain a minimum of 30% surface cover after planting to a complete lack of mechanical

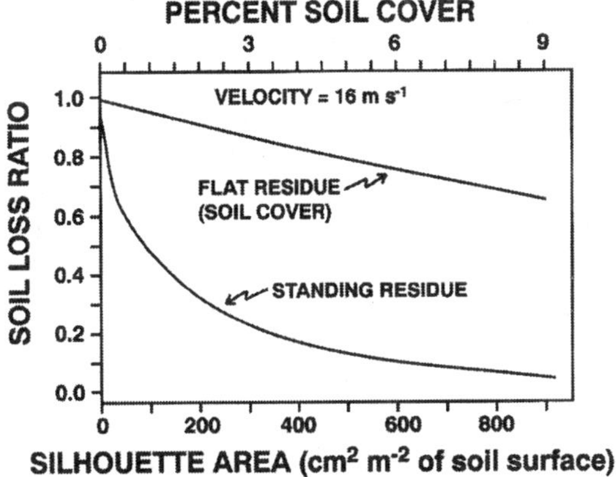

Figure 5.3 Soil loss ratios for flat and standing residue (From Bilbro and Fryrear, 1994).

tillage that preserves most plant residues on the soil surface after harvest of plant parts of economic value (Conservation Technology Information Center, 1988). Conservation tillage systems are extensively used in major U.S. agroecosystems for soil erosion control (Dao, 1993). During 1991, 7% of U.S. acreage was planted no-till while another 20% was planted with full-width tillage that left 30% surface cover after planting (Conservation Technology Information Center, 1991). By 1995 no-till and mulch-till acreage was expected to double (Stephens and Johnson, 1993).

Various types of soil cover have been shown to be effective in reducing water as well as wind erosion potential (Figure 5.2). When 60% of the soil is covered by nonerodible material, SLR (SLR = soil loss from protected soil/soil loss from flat, bare soil) is nearly zero. Even a small amount of soil cover is very important. For example, if a bare field had a potential soil erosion loss of 20 t ha^{-1} yr^{-1}, 10% soil cover on that field would reduce potential soil losses to 12.9 t ha^{-1} yr^{-1}. A 20% cover would reduce potential loss to 8.3 t ha^{-1} yr^{-1}, and a 30% cover would reduce losses to 5.4 t ha^{-1} yr^{-1} (Bilbro and Fryrear, 1994). After harvesting, much of the residue of many crops such as sorghum, corn, millet, and rice is left standing. Standing residue is superior to flat residue in decreasing potential wind erosion (Figure 5.3). Figure 5.4 shows the effect of vegetative cover in intercepting rainfall energy and soil erosion control. When 20% of the soil surface was covered with vegetative cover, the water erosion rate was only 30%; and when 60% of the surface was covered with vegetative cover, water erosion was practically zero.

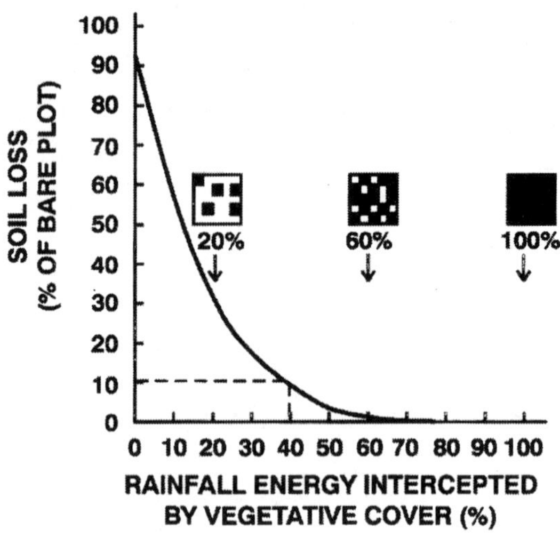

Figure 5.4 Relationship between rainfall energy interception by vegetative cover and soil erosion by water (From Shaxson, 1993).

Furrow diking or basin tillage is the practice of constructing small earthen dams within furrows to increase surface water retention, thus preventing runoff and increasing infiltration (Jones and Stewart, 1990). Deep tillage is another water conservation method which increases soil permeability and reduces erosion. Detailed descriptions of soil erosion control measures are given by Lal (1984, 1986).

Drought

Water deficiency is one of the most important soil degradation factors in arid and semi-arid regions. Arid climate is defined as receiving 25 cm or less average annual precipitation, while semi-arid regions receive between 25 and 50 cm of precipitation annually (Stephens, 1994). Together these two regions comprise about 35% of the earth's surface, excluding the polar deserts (Potter, 1992). Occurrence of drought is the common phenomenon in arid and semi-arid regions. Drought can also occur in humid regions. For example, in the central part of Brazil, the average annual rainfall is about 150 cm, mainly concentrated during October to March. This amount is sufficient to produce two annual crops. However, a two- to three-week drought during the rainy season is very common in this region (Fageria, 1980). Sometimes drought occurs at a sensitive crop growth stage (e.g., flowering) and crop yields are significantly reduced. This means that rainfall distribution during the crop growth cycle is just as important as total rainfall.

Water controls vegetation growth on the land surface. In low rainfall regions, only drought-tolerant plants can grow and survive. Under these situations, if population and animal pressure is high, existing vegetation will disappear, which leads to desertification.

Management Strategies for Drought

Supplementary irrigation during the rainy season and permanent irrigation during the dry period provide the best solution for coping with drought stress, thereby reducing soil degradation. According to 1979 statistics (FAO, 1991), irrigated land only accounts for 16% of the cultivated land under permanent crops worldwide but produces 33% of the world's food. China and India have approximately 45 and 43 million hectares of irrigated land, respectively, representing about 40% of the world's irrigated area. Irrigation is one of the most important technological factors that made these countries selfsufficient in food supply, allowing them to support about 37% of the world population. Irrigation in some areas makes farming possible; in others it supplements rainfall.

However, it is not possible in all drought regions to provide irrigation facilities to growers due to water scarcity or economic reasons. Some other management practices, such as improving the infiltration rate of soil through deep plowing, use of mulching to reduce evaporation, planting drought-resistant plant species or cultivars, and other complementary solutions, may reduce the impact of drought in crop production. Use of appropriate crop rotations can increase the efficiency of water use. When grown in rotation, one wheat and one sorghum crop are produced in a three-year period, and a fallow period of about 330 days follows each crop in parts of Southern and Central Great Plains of the United States. These fallow periods afford an opportunity for increasing storage of precipitation as soil water, thus reducing dependence on irrigation and providing some additional water for achieving more favorable yields under limited irrigation or dryland conditions (Unger, 1992). The percentage of precipitation stored as soil water increases with increasing amounts of crop residue retained on the soil surface during fallow (Unger, 1984; Wilhelm et al., 1986). Similarly, Groenevelt et al. (1989) pointed out that relatively thin surface layers of gravel and coarse sands can reduce evaporation to 10–20% of that occurring from recently wetted, unmulched soil surfaces.

Lower ridges (shallow furrows) should increase the wetted areas, thus increasing the potential for water conservation through increased infiltration and more effective wetting of the ridge (Unger, 1992). Additional water conservation is possible by capturing potential runoff water from precipitation in blocked furrows (Jones and Stewart, 1990).

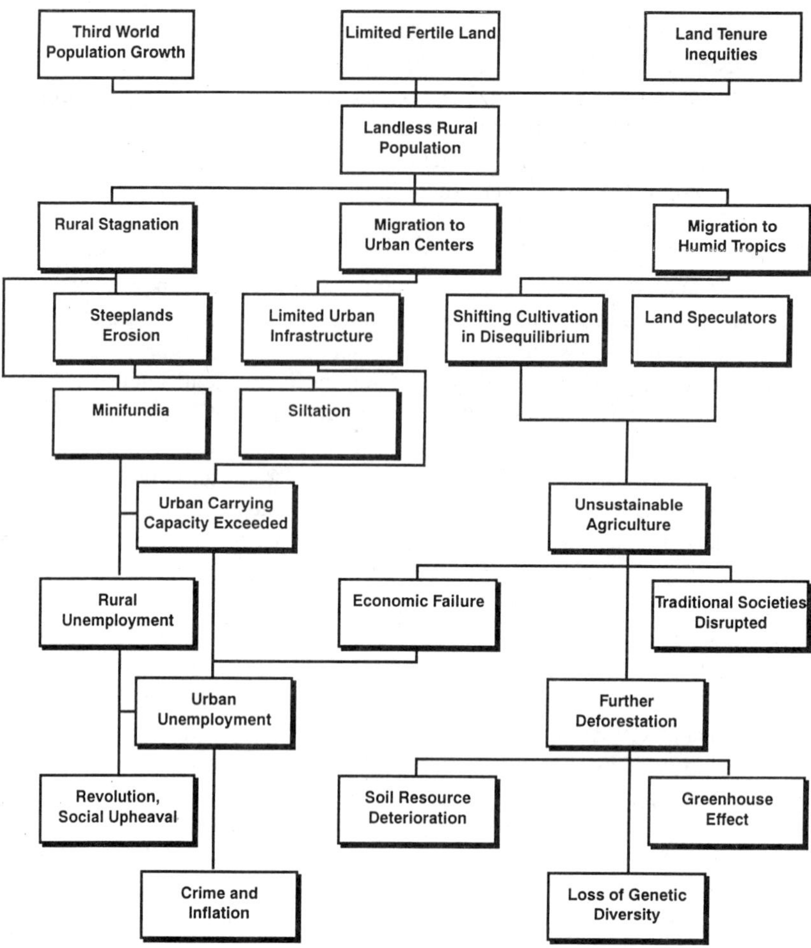

Figure 5.5 Cause-effect relationship related to tropical deforestation in developing countries (From Sanchez et al., 1991).

Deforestation

Expansion of cropland areas to meet the needs of rapidly rising human and livestock populations has resulted in increased deforestation (Kang et al., 1990). Deforestation and forest conversion are major factors of soil degradation (Harrison, 1984). Each year, 14 million hectares of primary forest disappear (Sanchez et al., 1991). Shifting cultivation is responsible for almost 70% of deforestation in tropical Africa (Kang et al., 1990). In tropical South America, after clearing the land, upland rice is generally planted followed by pasture. The incidence of forest decline in Europe and North America has rapidly increased during past decades (Pitelka and Raynal, 1989). The unfavorable effects of deforestation and cropping on soil microclimate, soil physical and chemical properties, and biotic components have been widely reviewed and investigated (Lal, 1986; Kang et al., 1990). Recent estimates indicate that about 18% of all global warming is caused by clearing of tropical rain forests (Sanchez et al., 1991). Deforestation is also decimating the world's largest depository of plant and animal diversity. Tropical deforestation is driven by a complex set of demographic, biological, social, and economic forces (Figure 5.5).

Deforestation is one of the main reasons for the global net release of CO_2 from soil to the atmosphere (Veldkamp, 1994). Soil-vegetation systems can act as a CO_2 sink or a CO_2 source, depending on decomposition rate and rate of soil organic carbon (SOC) formation (Van Breemen and Feijtel, 1990). When forest is cleared, the soil turns into a CO_2 source. The annual relative increase in global atmospheric CO_2 concentration is 0.5%, which corresponds with 3.6 Pg C yr^{-1} (Bouman, 1990). Estimates of global CO_2 release caused by deforestation are between 1.0 and 3.2 Pg C yr^{-1}. Recent concern for global climate change has created a heightened interest in the role of forested ecosystems in the global C cycle. Deforestation followed by 25 years of pasture caused a net loss of 21.8 Mg ha^{-1} in soil organic C for an Eutric Hapludand and 1.5 Mg ha^{-1} for an Oxic Humitropep (Veldkamp, 1994). Detwiler (1986) estimated that cropping of tropical forest soils reduced their C content by 40%; the use of these soils for pastures reduced their C content by about 20%.

Globally, 1576 Pg of C is stored in soils, with 506 Pg (32%) of this in soils of the tropics. It is also estimated that 40% of the C in soils of the tropics is in forest soils. Other studies have shown that deforestation can result in 20–50% loss of this stored C, largely through erosion (Eswaran et al., 1993a).

Management Strategies for Deforestation

Land management options that improve the economic status of subsistence farmers, maintain agricultural productivity on deforested lands, and recuperate productivity of degraded lands are urgently needed. These options must be compatible with the various socioeconomic needs in the region so that they are readily and widely adopted. Sanchez et al. (1991) describe some of the main management options.

1. Land clearing that does not damage the soil.
2. Using low-input cropping systems on infertile arid soils such as Oxisols and Ultisols.

3. Using agroforestry for much of the humid tropics because it can be adapted to a wide range of socioeconomic and soil-landscape conditions.
4. Developing grass-legume pastures which can provide sustainability for cattle production.
5. Using effective crop rotations and the judicious application of lime and fertilizers.
6. Maximizing nutrient cycling in order to minimize the need for external nutrient inputs. The management of crop and root residues is essential in this regard.

Wetlands

Types of soil wetness include short-term wetness caused by excessive rainfall or flooding, groundwater table rise caused by irrigation and canal seepage, perched shallow water tables caused by soil compaction, groundwater table rise due to land surface management, and impeded surface drainage due to construction of highways (Fausey and Lal, 1990). Measurements of hydrology, vegetation, and soils constitute the three-parameter approach that is currently favored by the Federal government for delineating wetlands in the United States (Federal Interagency Committee for Wetland Delineation, 1989). A wetland protected by law must meet the criteria listed for hydric soils by the Soil Conservation Service (1991) and Megonigal et al. (1993). A hydric soil is one that is saturated, flooded, or ponded long enough during the growing season to develop anaerobic conditions in the upper part (Soil Conservation Service, 1991). The most accurate methods for demonstrating hydric conditions involve monitoring soil moisture, water table fluctuations, soil O_2 content, reduction-oxidation potential, or Fe^{2+} activity (Faulkner and Patrick, 1992; Megonigal et al., 1993).

In soil taxonomy, wet soils are identified by an aquic moisture regime at the suborder level or by properties used to define an aquic soil moisture regime. A soil that is saturated with water and essentially depleted of oxygen is defined as having an aquic moisture regime. A reducing environment, a result of stagnant water, persists for a sufficient time for aerobic microorganisms to deplete soil oxygen. As the reducing environment is extended, organisms extract chemically bound oxygen.

The annual flooding of the Paraguay River and its meandering tributaries such as the Cuiba, Itiquira, Taquari, and Miranda is responsible for the name "Pantanal," a large swampland in Portuguese. During the November through March rainy season, the flooding plains begin filling and ephemeral lakes take shape. Permanent lakes are connected one to another and to tributaries by temporary streams. The Pantanal's total area is about 15 million ha, covering much of the states of Mato Grosso and Mato Grosso de Sul of Brazil and some adjoining territory of Paraguay and Bolivia.

The Everglades represent a unique and complex composite of ecosystems forming a vast wetland covering a large portion of Southeastern Florida in the United States. The original Everglades encompassed an area of about 1 million ha (Debusk et al., 1994).

In Southeast Asia, millions of hectares of land in coastal swamp areas have been and are being reclaimed for agriculture (Konsten et al., 1994). A large part of the 25 million ha of coastal swamp soils in Indonesia has only limited potential for agriculture (Nedeco/Euroconsult, 1984). In their original waterlogged state, many of the soils contained pyrite. If the soils are reclaimed and drained, pyrite oxidation results in acidification of soil and water; acid sulfate soils then develop (Dent, 1986). Low nutrient content, low pH, and relatively high contents of dissolved Al and Fe, organic acids, and H_2S are responsible for poor crop performance on many young sulfate soils (Konsten et al., 1994).

There are many diverse effects of soil wetness. One of these is anaerobiosis, a very significant effect from an agronomic or biological perspective. The major physical change that can be defined as soil degradation associated with soil wetness is loss of soil strength. The chemical effects of soil wetness that can be associated with soil degradation are accumulation of salts at or near the surface in semiarid or arid regions under high water table conditions and changes in solubility and chemical form of nutrients under anaerobic conditions (Fausey and Lal, 1990).

Reclamation of Wetlands

Among reclamation measures, adequate drainage is one of the most effective, but expensive, control measures. The drainage system may be surface or subsurface. The use of drainage to ameliorate excessive soil wetness has long been a subject of study (USDA, 1987). Drainage is not a guarantee against soil degradation by wetness, but drainage can minimize periods of anaerobiosis, improve trafficability, aid in flushing salts, and reduce soil erosion (Fausey and Lal, 1990). The use of soil ridges or ridge tillage has been reported as an alternative tillage method to alleviate high soil water content and low soil temperatures on poorly drained soils (Radke et al., 1993).

Crop Monoculture Versus Crop Rotation

Monoculture is growing a single crop at one time. Monoculture usually requires large amounts of inputs such as fertilizers, pesticides, and machinery. Repeatedly growing the same crop on the same land can develop "soil sickness," which is thought to be caused by a combination of soil pathogens, mineral depletion, change in soil structure, and accumulation of toxic substances. Johnson et al. (1992) reported that spore populations of mycorrhizal fungi, which proliferated in corn monoculture, generally correlated negatively

Table 5.2 Upland Rice Grain Yield (kg/ha) During Three Consecutive Years in Monoculture on an Oxisol of Central Brazil (kg ha^{-1})

P added	First year	Second year	Third year
Control	529	674	402
87	1958	1509	488
174	2739	1897	672
262	2217	1588	555
F test	**	NS	NS
Linear	NS	NS	NS
Quadratic	**	NS	NS

Source: Fageria et al. (1991a).

with yield and tissue mineral concentrations of corn, but were positively correlated with the yield and tissue mineral concentrations of soybean. Continuous monoculture of both corn and soybean generally had lower yields and tissue concentrations of P, Cu, and Zn than first-year crops. Table 5.2 shows results of a field experiment with upland rice on an Oxisol of Central Brazil (but the decreased productivity could have been caused by drought, insects, etc.) There is no comparison with rotations.

Detailed crop rotation experiments in The Netherlands demonstrated that yields of wheat, and especially of potato, decreased over the years and stabilized at different levels depending on the frequency of crop rotation. Potatoes grown in the same plots every fourth (1: 4) or every third (1: 3) year yielded 10–15 lbs less, in general, than potatoes grown every sixth year (a 1: 6 potato-cropping frequency). Yields were even lower (30% less) in fields where potatoes had been cropped every second year or every year (Schippers et al., 1987).

Legume and meadow-based rotations and conservation tillage systems often maintain more favorable soil properties compared to monocultures and plow-based methods (Lal et al., 1994). In addition to decreasing the incidence of disease, crop rotations may also improve soil physical and nutritional properties. Meadow and leguminous cover crops are believed to improve soil structure and increase soil fertility (Power, 1990; Lal et al., 1990). Dick et al. (1986) observed that rotation significantly improved corn grain yield, with corn yields in a corn-oat-meadow rotation averaging 1.22 Mg ha^{-1} more than in continuous corn and 0.96 Mg ha^{-1} more than in a corn-soybean rotation.

Havlin et al. (1990) found that increasing the frequency of corn and sorghum with soybean increased surface organic C and N, especially under no-tillage. After only 3.5 yr of no-tillage, wheat-corn-millet-fallow rotations had greater surface organic C and N concentrations and potential mineralization than wheat-fallow rotations on soils that had previously been managed under conventional tillage for 50 or more years (Wood et al., 1990). Crop rotation increased yields, increased profitability via diversification, and decreased environmental risks due to reduced chemical inputs (Pierce and Rice, 1988).

Although crop rotation may change soil mineral status, particularly N, there may also be a rotation effect beyond that which can be explained by soil mineral status alone (Copeland and Crookston, 1992). At Urbana, Illinois, high rates of limestone and N, P, and K fertilizers did not substitute for rotation, which increased yield of corn rotated with soybean by 16% over corn grown in monoculture (Welch, 1976). In rotation studies at Lancaster, Wisconsin, in which N was not limiting, the eight-year average yields for continuous corn were less than the average yield for first-year corn (Higgs et al., 1976).

Overgrazing

Grazing is an important element in most pasture ecosystems since it interacts with and determines the structure and composition of vegetation (Anderson, 1990; Hobbs et al., 1991). Grazing tends to reduce root growth and rhizome carbohydrate reserves (Turner et al., 1993). The decrease of below-ground C inputs could result in a reduced C/N ratio of below-ground plant biomass, reduced microbial growth, and reduced potential for N immobilization (Holland and Detling, 1990).

Overgrazing led to denudation of vegetation and caused severe erosion by wind and water in both semiarid and arid regions. Livestock numbers in Africa increased from 295 million animals compared with 219 million people in 1950, to 520 million animals compared with 515 million people in 1983 (Lal, 1988). Although a high proportion of these livestock animals are of poor quality and low productivity, they have caused widespread degradation of grasslands. Existing herds often exceed the carrying capacity of the land, thereby accelerating degradation (Lal, 1988). Similarly, in the Thar desert of Rajasthan, India, a large number of animals (especially sheep, goats, and camels) denuded the scanty vegetation, causing soil degradation by wind erosion in the extremely dry periods of May and June.

Management Strategies for Overgrazing

The best management strategy is rational and controlled grazing of pastures. Research can determine how many animals should be grazed per unit area and time. During the rainy season adequate fertilization should be applied to allow pasture species to recover, and grazing animals should be prohibited. If a pasture is adequately managed, it not only supports good animal production but also reduces environmental pollution. For example, the internal N cycle of grassland and prairie ecosystems has been shown to accumulate little NO_3 in the soil profile because of high plant and microbial activities (Jackson et al., 1989; Schimel et al., 1989). Jackson et al. (1989) studied short-term N turnover in a grassland ecosystem to demonstrate that microbes actually out-compete plants for available NO_3 and NH_4. The ability of microorganisms to rapidly immobilize available NO_3 and NH_4 is dependent on the presence of a readily available C source. Schimel (1986) described a higher level of N immobilization in prairie soils than in cultivated soils and attributed this difference to the quality and quantity of C substrate available for microbial use in the prairie soil.

Surface Mining

Exploration and surface mining of precious metals and minerals degrade sizable land areas in various parts of the world. It has been estimated that 4 million ha of land in the United States will be disturbed as a result of surface mining for coal (Doolittle and Hossner, 1988). Texas, the nation's sixth largest coal-producing state, will have more than 400,000 ha disturbed by surface mining for lignite (Clarke and Baen, 1980). The extensive volume of overburden material generated from mining is generally disposed of on the land surface near the site. This spoil material, composed of regolith and bedrock, degrades the soils and presents a challenge for reclamation.

Management Strategies for Surface Mining

Successful reclamation of surface-mined land can be accomplished by using a mixed overburden topsoil substitute. Mixed overburden can form mine-soils that have better chemical and physical properties than pre-mine-soils (Bearden, 1984). When properly managed, mine-soils from mixed overburden have an excellent yield potential (Doolittle et al., 1994).

Schuman and Sedbrook (1984) demonstrated that abandoned bentonite mine spoils could be successfully reclaimed using sawmill wastes (wood chips, bark, and sawdust) as a spoil amendment that enabled immediately improved water infiltration, which allowed successful reestablishment of perennial vegetation. Schuman and Meining (1993) reported that surface-applied gypsum at the rate of 56 Mg ha^{-1} effectively ameliorated bentonite mine spoil under natural rainfall conditions in a semiarid environment.

Colonization of vegetation by vesicular arbuscular mycorrhizal fungi may enhance plant growth and P cycling during reclamation of mined lands (White et al., 1992). Sutton and Dick (1987) gave a detailed description of management processes adopted for the reclamation of acid mined lands in humid areas.

Vegetation Burning

Wild fires, prescribed burning, and slash burning are common in various parts of the world. For example, burning pasture and forest land for crop planting is a very common practice in South America in the dry period of July, August, and September. In tall grass prairies, a major ecosystem in the United States, fire is a common disturbance (Garcia and Rice, 1994). Vegetation burning changes physical, chemical, and biological properties of a soil and enhances the risk of degradation. However, deterioration depends on intensity and duration of fire. Subjective methods for classifying burns based on litter and on soil appearance after fire have been described by Chandler et al. (1983). Low-intensity or lightly burned areas are characterized by black ash, scorched litter, low plant mortality, and maximum surface temperature during burning of 100–250°C. Moderate burning produces surface temperatures of 300–400°C and consumes most of the plant material, thus exposing the underlying soil which otherwise is not altered. High-intensity or severe burning produces surface soil temperatures in excess of 500°C and is recognized by white ash remaining after the complete combustion of heavy fuel and by reddening of the soil.

Prescribed burning is a vegetation management tool that removes plant cover which has variable effects on infiltration, surface runoff, and erosion (Knight et al., 1983; Lloyd-Reilley et al., 1984). Vegetation cover is an important factor controlling surface runoff and erosion (Emmerich and Cox, 1994). Prescribed burning of vegetation may increase the potential for surface runoff and erosion. Immediately after a rangeland burn, runoff, and sediment production may be unchanged, but within one year, significant increases can occur, probably due to soil surface morphological changes during the one-year time period.

Soil texture changes have been observed in response to both fires and laboratory heating. Ulery and Graham (1993) found a significant decrease in the clay content of severely burned soils and a corresponding increase in sand, suggesting the aggregation of clay-sized particles into stable sand-sized secondary particles. Aggregation and decomposition of clay minerals have been reported as the mechanism for shifts in particle-size distribution on heating, but no evidence other than mechanical analysis has been shown. Ulery and Graham (1993) also reported that in the reddened layers, organic C was reduced by 90–100%, while in the blackened layers, it was reduced by 15–68%, compared with the unburned surface soil. Sertsu and Sanchez (1978) noted almost total elimination of organic C when soils were heated in the laboratory to 400°C.

Schwertmann and Fetcher (1984) suggested that goethite, a yellowish-brown Fe oxide that is nearly ubiquitous in soils, can be transformed to maghemite and hematite during a fire, especially if organic matter is present, which contributes to an O_2–deficient environment.

Litter and aboveground standing biomass serve to maintain high infiltration rates by protecting the soil surface aggregates and structure from destruction by raindrop impact, thus reducing crust formation (Thurow et al., 1986; Smith et al., 1990). Reduction in soil aggregate sizes after a burn has been shown to occur and persist for more than five years (Ueckert et al., 1978).

Fire induces numerous chemical changes in the soil (Raison, 1979). The intensive heat produced by fires acts on soil organic matter and plant materials and influences soil elemental pools. Water-soluble substances, including Ca^{2+}, Mg^{2+}, K^+, NH_4^+, Cl^-, and SO_4^{2-}, increase after fires (Khanna and Raison, 1986; Grove et al., 1986). Intense soil heating can also lead to the formation of hydrophobic compounds, influencing water infiltration (DeBano et al., 1976), formation of organometallic cement on soil mineral particles (Giovannini and Lucchesi, 1983), and structural alteration of the soil humic fraction (Almendros et al., 1988).

Blank et al. (1994) reported that compared with unburned soils, significant decreases in NO_3^- and orthophosphate and significant increases in SO_4^{2-}, acetate, formate, oxalate, and glycolate occurred immediately after fire in the surface 5 cm of soil under shrub growth. Concentrations of organic acids in burned soils increased significantly in the weeks following a wildfire. Elevated concentrations of organic acids may influence seed germination, plant establishment, and mineral nutrition. The nature and magnitude of soil chemical changes are a function of a number of factors, including the nature and density of the vegetation community, fire temperature, length of burn, soil water content, and soil texture (Wright and Bailey, 1982).

Some short-term beneficial effects have also been reported by burning vegetation. For example, burning increases the photosynthetic capacity of postburn plant growth and results in changes in soil temperature, water, and nutrient status (Knapp and Seastedt, 1986). Ojima (1987) reported that annual burning resulted in lower soil organic matter but higher plant productivity compared with no burning. This apparent contradiction may be explained by: 1) synchronization of nutrient release with plant uptake and microbial activity, 2) extension of the growing season because of earlier soil warming, 3) changes in the rate of ecosystem processes that allow for recovery of volatilized nutrients (N_2 fixation), 4) changes in the utilization of available nutrients, or 5) a combination of all these reasons (Knapp and Seastedt, 1986; Ojima, 1987).

These short-term effects should not cause us to overlook important long-term effects of vegetation burning. Lal (1986) concluded that repeated fires resulted in a change in climax vegetation from forest to savanna. Large trees of the forest give way to fire-resistant shrubs and grasses, resulting in the widespread occurrence of derived savanna in Africa, Asia, and Latin America where rainfall and climatic factors would normally support forest vegetation.

B. Chemical Degradation

Chemical degradation is the change in soil chemical properties from a favorable to an unfavorable state which results in decreasing soil productivity. The process of degradation may occur due to soil acidity, inadequate fertilization, flooding, practicing monoculture for long periods on the same field, soil erosion, accumulation of salts in harmful concentrations, release of allelochemicals, and indiscriminate use of pesticides. Some of these degradation processes are also related to deterioration of soil physical properties through soil erosion, monoculture, and flooding of wetlands.

Nutrient Stress

Nutrient stresses refer to deficiency of essential plant nutrients as well as toxicities (Fageria and Baligar, 1993). Nutrient deficiencies are more common than toxicity in many arable lands around the world. If this stress is not alleviated, crop yields are decreased and soils are not in a position to support adequate plant growth. Under these situations soil degradation starts. If land continues to be used for crop production, crop yield becomes so low that farmers have to abandon the degraded areas. Approximately one-fourth of the earth's soils are considered to produce some kind of mineral stress (Dudal, 1976). Figure 5.6 shows the importance of N, P, and K nutrient supply for

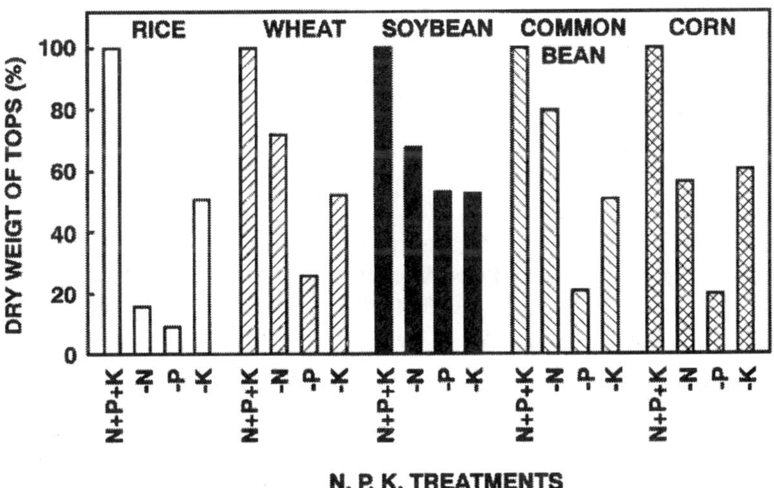

Figure 5.6 Comparison among five crops in dry matter production of tops under different fertility treatments (From Fageria, 1994).

the growth of rice, wheat, corn, soybean, and common bean in an Oxisol of Central Brazil. P is the most important yield-limiting nutrient in this, like most other Oxisols.

Management Strategies for Nutrient Stress. Nutrient stresses can be alleviated with adequate fertilizer application. Periodic soil testing helps determine if current nutrient management practices are sustainable. Soil testing is a vital component of sustainable farming programs that are profitable, efficient, and environmentally responsible. A good procedure for making fertilizer recommendations for a given crop on any one soil is to carry out a series of fertilizer trials, fit a response function, and substitute that function into a profit equation to calculate optimal fertilizer levels according to economic variables. Adequate soil fertility reduces soil degradation for the following reasons:

1. Adequate fertility protects the soil from erosion by giving the crop early vigor for quick canopy cover and helps build strong root systems that hold the soil in place.
2. Adequate fertility results in more residues remaining after harvest to protect the soil against wind and water erosion while building organic matter levels and increasing long-term production potential.
3. Adequate soil fertility improves nutrient use efficiency, which is good both economically and environmentally. Balancing nitrogen with phosphate and potassium improves nitrogen use efficiency and leaves less nitrate in the soil for potential leaching losses and effects on groundwater.
4. Adequate soil fertility conserves water by reducing the amount required per unit of dry matter production.
5. Adequate soil fertility interacts positively with other production inputs such as tillage practices, variety selection, pest control, and plant population to get the most out of the crop being grown.

In addition to adequate rate, placement of fertilizers is an integral part of efficient crop management. Placement can affect both crop yield and nutrient use efficiency (Mahler et al., 1994). Most nutrients such as P and K are banded, and even band application of N is generally preferred over broadcast (Mahler et al., 1994) because: 1) fertilizer is placed where small seedling root systems can more readily utilize the nutrients, 2) the amount of fertilizer needed per unit area is lower than with broadcasting, 3) fertilizer is positionally more available to the crop than to germinating weeds, 4) only one operation is needed with planting, and 5) there is less loss of N due to erosion and immobilization. An integrated nutrient management system is presented in Figure 5.7. If this system is implemented in any cropping system, nutrient management can be optimized and soil degradation can be reduced to a minimum value.

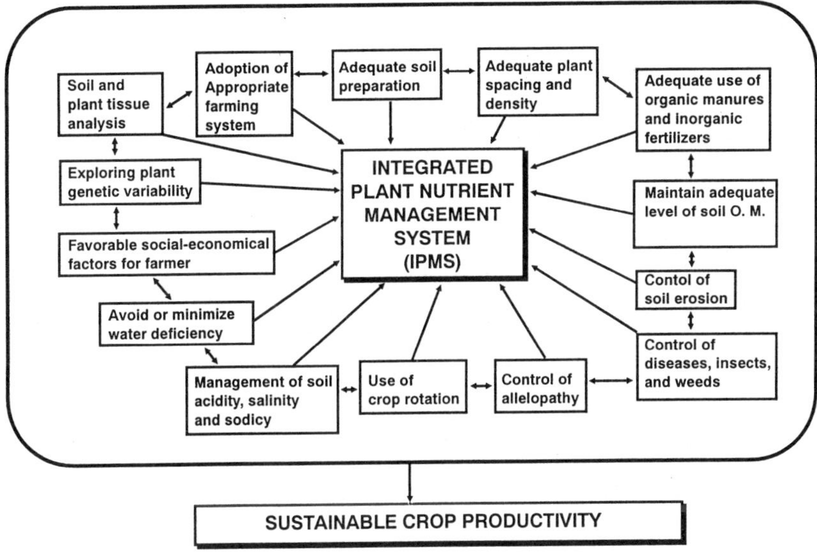

Figure 5.7 Integrated plant nutrient management system for sustainable environment.

The transformation of extensive rangelands to productive farmland and improved pastureland in the central Brazilian Cerrado has been a quiet social revolution since the early 1970s. Management strategies to overcome chemical constraints (nutrient deficiencies and/or elemental toxicities) through appropriate liming, fertilizer rates, and cultivating adapted plant species have developed out of two decades of research on Savannah Oxisols in South America and are coincident with the opening of these lands for agricultural production on a large scale.

In addition to these measures, exploitation of plant genetic variability in absorption and utilization of nutrients is another important strategy to improve nutrient deficiencies and reduce soil degradation. Utilization of plant genetic resources has been the foundation for improvement of agronomic crops. Agricultural scientists around the world use plant genetic resources to develop new crop cultivars that are more productive under environmental stresses. One example of using plant genetic resources is the U.S. National Plant Germplasm System (NPGS). The NPGS maintains more than 415,000 accessions of various crop species (Table 5.3). The ten International Agricultural Research Centers involved with crops are key institutions for the collection, preservation, and distribution of many agronomically important crops. These centers are CIAT (Centro Internacional de Agricultur Tropical; Colombia), CIMMYT

Table 5.3 Germplasm Collection at the National Plant Germplasm System Genetic Resources, USA

Genus	Species	Crop	Number of accessions
Arachis	*hypogaea*	Peanut	7,943
Avena	*sativa*	Oat	6,580
Avena	*sterilis*, etc.	Oat relatives	13,419
Cajanus	*cajan*	Pigeonpea	4,156
Capsicum	*annuum*	Pepper	2,313
Carthamus	*tinctorius*	Safflower	2,218
Cicer	*arietinum*	Chickpea	3,962
Cucumis	*melo*	Melon	3,374
Glycine	*max*	Soybean	14,316
Gossypium	*hirsutum*	Cotton	4,746
Helianthus	*annuus*	Sunflower	2,607
Hordeum	*vulgare*	Barley	28,612
Lens	*culinaris*	Lentil	2,618
Linum	*usitatissimum*	Flax	2,722
Lycopersicon	*esculentum*	Tomato	8,601
Malus	*domestica*	Apple buds	163
Medicago	*sativa*	Alfalfa	3,454
Oryza	*sativa*	Rice	18,213
Phaseolus	*vulgaris*	Bean	10,448
Pisum	*sativum*	Peas	3,590
Secale	*cereale*	Rye	2,618
Solanum	*tuberosum*	Potato	5,486
Sorghum	*bicolor*	Sorghum	34,480
Triticum	*aestivum*	Wheat	34,391
Triticum	*durum*	Durum wheat	6,831
Vigna	*unguiculata*	Cowpea	3,958
Zea	*mays*	Corn	28,376
Other			155,710
		TOTAL	415,905

Source: Eberhart (1993).

(Centro Internacional de Mejoramento de Maiz Y Trigo; Mexico), CIP (Ce. Internacional de la Papa, Peru), ICARDA (International Center for Agricultural Research in the Dry Area, Syria), ICRISAT (International Crops Research Institute for the Semi-Arid Tropics; India), IITA (International Institute of Tropical Agriculture; Nigeria), ILCA (International Livestock Center for Africa, Ethiopia), IRRI (International Rice Research Institute; Philippines), WARDA (West Africa Rice Development Association, Ivory Coast), and AVRDC (Asian Vegetable Research Development Center, Taiwan). These centers together have collections of 498,615 crop germplasm accessions (Eberhart, 1993).

Soil Acidity

Soil acidification refers to a complex set of processes that results in the formation of an acid soil (pH less than 7.0) (Robarge and Johnson, 1992). According to Krug and Frink (1983), however, soil acidification in the broadest sense can be considered as the summation of natural and anthropogenic processes that lower measured soil pH. Soil pH is a measure of the activity of H^+ ions in the soil solution, which measures the degree of acidity or alkalinity of a soil. A soil with a pH of 5.0 is 10 times more acid than one with a pH of 6.0 and 100 times more acid than one with pH 7.0. Soil acidification cannot be quantitatively described by a single index or parameter, even though it is often assumed that soil pH is such a parameter (Matzner, 1989). Other changes in soils that may occur during soil acidification include loss of nutrients due to leaching, loss or reduction in the availability of certain plant nutrients (such as P, Ca, Mg, and Mo), an increase in the solubility of toxic metals (such as Al and Mn) which may influence root growth and nutrient and water uptake, and a change in microbial populations and activities (Binkley et al., 1989; Myrold, 1990). Such changes will often be accompanied by changes in overall soil pH, but the degree of change will be dependent on a combination of properties within a given soil (Robarge and Johnson, 1992).

Acidity is a major degrading factor of soils and affects extensive areas both in the tropics and in temperate zones. Acid soils are reported to occupy about 3.0 billion hectares, of which over 89% are Oxisols and Ultisols situated in the tropics (Eswaran et al., 1993b). The global distribution of acid soils is 1.17 billion ha Oxisols, 1.13 billion ha Ultisols, 0.48 billion ha Spodosols, and 0.25 billion ha Aridisols (Eswaran et al., 1993b).

In tropical South America, 85% of the region as a whole has acid soils, and there are approximately 850 million hectares of under-utilized acid soils, undoubtedly the most extensive area of acid tropical soils in the world (Cochrane, 1991). Soil acidity may result from parent materials that were acid and naturally low in basic cations (Ca^{2+}, Mg^{2+}, K^+, and Na^+) or because these elements have been leached from the soil profile by heavy rains (Fageria et al., 1990a). Soil acidity may also develop from exposure to the air of mine spoils containing iron pyrite (FeS_2) or other sulfides. Crop fertilization with ammonia or ammonium fertilizers can result in soil acidification. In addition, acidity may also be produced by the decomposition of plant residues or organic wastes into organic acids. This process is of particular importance in many forest soils. The important yield-limiting factors in acid soils are: toxicities of H, Al, and Mn; deficiencies of N, P, K, Ca, Mg, Mo, and Zn; and reduced activity of beneficial microorganisms (Fageria et al., 1989; Fageria et al., 1990a). Soil acidity constraints in crop production are very complex, and sometimes it is difficult to separate one constraint from another. Therefore we assume that three important factors limit plant growth in acid soil: deficiency of nutrients, toxicity of aluminum, and low activities of beneficial microorganisms.

Phosphorus Deficiency. Phosphorus deficiency is a major limitation to crop production in acid, infertile soils. Sanchez and Salinas (1981) estimated that P deficiency was a constraint to plant growth over 96% of the total area of acid soils in tropical America (23°N to 23°S). Sanchez (1987) estimated that P deficiency was a constraint over 90% of the total land area of the Amazon Basin. Fageria (1994) studied responses of five crop species to N, P, and K in an Oxisol of Central Brazil (Figure 5.6). Results of this study suggested that P was the most yield-limiting nutrient in this acid soil. The two major reasons for the occurrence of P deficiency in acid soils are low native soil P content and high P-fixation capacity. The amount of P fixed (not recovered by Mehlich 1 extractant) in an Oxisol from central Brazil increased from 45 to 268 kg P ha^{-1} when the P application rate was increased from 50 to 400 kg P ha^{-1} (Fageria and Barbosa Filho, 1987). High phosphorus fixation capacity is related to clay and Fe and Al contents of these soils (Van Riemsdijk et al., 1984). To obtain good yields, sufficient P fertilization is a prerequisite in these soils (Fageria et al., 1990a).

Aluminum Toxicity. Aluminum is not an essential element for plant growth. However, soluble Al is deleterious to plant growth in acid soils (Chaudhary et al., 1987). The interaction between aluminum and various inorganic components in the soil is complex. The soil pH at which aluminum toxicity is expected is therefore ill-defined. The initial soil pH at which Al becomes soluble or exchangeable in toxic concentrations depends on many soil factors, including the predominant clay minerals, organic matter levels, concentrations of other cations, anions and total salts, and particularly the plant species or cultivar (Foy, 1984; Fageria et al., 1988a). Generally, aluminum toxicity may be observed at any pH below 5.5, but more commonly it is observed at soil pH below 5.0 (Foy, 1984; Foy, 1992). Much of the poor root development and drought stress on soils with strongly acidic (pH below 5.0) subsoil layers is probably due primarily to Al toxicity which limits both rooting depth and degree of root branching. Aluminum toxicity is a complex disorder which may be manifested as a deficiency of P, Ca, or Mg, or as drought stress (Fageria and Carvalho, 1982; Baligar et al., 1987, 1989; Foy, 1984, 1988;

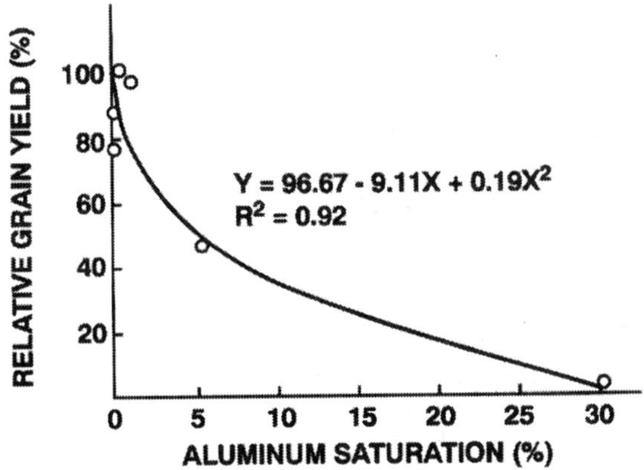

Figure 5.8 Relationship between aluminum saturation and common bean grain yield in a "Varzea" soil of Goias State of Brazil.

Kamprath and Foy, 1985; Hai et al., 1989). Excess Al has even induced Fe deficiency symptoms in rice, sorghum, and wheat (IRRI, 1974; Clark et al., 1981; Foy and Fleming, 1982). Excess Al in the growth medium influences several physiological and biochemical processes in plants which in turn affect their growth and development.

In general, a more useful predictor of Al toxicity is the percentage of the cation exchange capacity (CEC) occupied by Al (Cregan, 1980; Farina and Channon, 1990; Kamprath and Foy, 1985). To be most effective, the percentage Al saturation must be applied within a rather narrowly defined set of conditions because the critical Al saturation associated with toxicity varies with soil type and with plant species and genotypes (Foy, 1987; Foy, 1992).

In Brazil there are about 30 million hectares of lowland areas known locally as Varzea. These soils have high potential for producing two to three crops per year due to availability of water throughout the year. These lowland areas are located near rivers or small natural streams. In the rainy season these soils are good for flooded rice cultivation, and in the dry season other crops can be planted provided there is adequate drainage. Aluminum toxicity is one of the important chemical constraints to crop production in Varzea soils (Fageria et al., 1994). Figure 5.8 shows the relationship between Al saturation in Varzea soil of Brazil and relative grain yield of common bean.

Manganese Toxicity. Manganese toxicity is another important plant growth-limiting factor in acid soils. Unlike Al, Mn seems to affect plant tops more directly than roots, but root damage follows when the toxicity is severe

(Fageria et al., 1988a). The solubility and potential toxicity of Mn to a given crop depend upon many soil properties, including total Mn content, pH, organic matter content, aeration, and microbial activity (Foy, 1992). There is a strong antagonism between Fe and Mn. Iron-deficient snapbean grown in nutrient solution absorbed excess Mn and developed toxicity symptoms (Fleming, 1989). In Brazilian Oxisols excess Mn creates Fe deficiency in upland rice (Fageria et al., 1990b). However, Wright et al. (1987) found no evidence of Mn toxicity in snapbeans grown in 55 horizons from 14 acidic Appalachian soils. Growth was most closely related to Al/Ca ratios in soil solutions, and Wright et al. (1988) concluded that Mn toxicity was not the major limiting factor in the growth of subclover and switchgrass on acid soils in the Appalachian region.

Adverse Effects on Beneficial Microorganisms. The growth of beneficial microorganisms is affected by the pH of the environment. Soil pH is an important factor in determining the amounts and activities of microorganisms involved in organic matter transformations (Alexander, 1980). By regulating microbial activity, pH affects the mineralization of organic matter and the subsequent availabilities of N, P, S, and micronutrients to higher plants (Kamprath and Foy, 1985). In general, organic matter, whether natural or added, decomposes more rapidly in neutral soils than in acid soils. In some strongly acid soils, Al toxicity as well as H^+ ion toxicity may limit microbial breakdown of organic matter.

Inhibited growth of legumes depending on rhizobial N_2 fixation can be attributed to a direct effect of soil acidity on the growth of the host plant itself (Munns et al., 1981; Jarvis and Robson, 1983), to effects on establishment of the legume-*Rhizobium* symbiosis, or to inhibition of nodule development and/or function (Franco and Munns, 1982). Schubert et al. (1990) reported that at low pH (pH 4.7 and 5.4), dry matter production, seed yield, and N_2 fixation were significantly lower than at the higher pH levels (pH 6.2 and 7.0) in broadbean. Buerkert et al. (1990) also reported that in common bean, liming resulted in 40% greater shoot and 18% greater root dry weight, and also improved nodule weight per plant by 110% at early flowering. According to Glenn and Dilworth (1991), *Rhizobium* and *Bradyrhizobium* grow best at around pH 6.5–7.

Management Strategies for Soil Acidity. Soil acidity is a combination of soil conditions that limit plant growth, and its management requires manipulation of various soil and plant factors in favor of better plant growth or crop production. These management practices may vary with severity of acidity, type of soil, type of farming practices, and socioeconomic conditions of the farmer. However, the important soil acidity constraints discussed earlier can be improved in favor of better plant growth by adopting appropriate management strategies. An example of successful soil acidity management can be found in the "Cerrado," an acid Savanna ecoregion of Brazil considered unsuitable for agricultural crop production as recently as the 1970s. The Cerrado covers 205 million ha, of which 175 million ha are in central Brazil. Approximately 112 million ha of the Cerrado are considered adequate for developing sustainable agricultural production in central Brazil (Schaffert, 1993). The soils of the Cerrado are commonly characterized by low pH, low phosphorus availability, low fertility, and toxicity of Al and Mn (Goedert, 1983). Today 12 million hectares of the Brazilian Cerrado are in crop production, producing 25% of Brazil's rice, maize, and soybean, 20% of its coffee, and 15% of its common beans. Another 35 million hectares of improved pastures have been developed in the Cerrado, carrying 53 million head of cattle and producing 40% of Brazil's meat and 12% of its milk (Schaffert, 1993). All of this progress has been achieved by the adoption of acid soil management strategies.

Improving P-Use Efficiency. Various management practices can be used to improve P-use efficiency in P-deficient soils. An important practice is band application of P based on soil conditions and the cropping system. Field trials in combination with soil tests for available phosphorus are required to recommend fertilizer P requirements for representative soil types, crops, and climatic conditions (Mengel, 1993). Phosphorus use efficiency depends on soil structure, since a good structure favors root growth and thus the capacity of plant roots to exploit soil phosphate. In this situation adequate phosphorus fertilization is an important cultural practice to improve soil structure. Moreover, phosphorus deficiency in high P-sorbing soils can be corrected by an initial application of a large quantity of P, or a combination of an initial broadcast application and repeated band applications (Yost et al., 1979). Liming is another practice to improve P uptake by plants in acid soils through precipitation of Fe and Al hydroxides up to a pH of about 6.5 (Fageria, 1984, 1992). Beyond that pH, fixation of P occurs through formation of calcium phosphate. Addition of organic manures and inoculation with mycorrhizal fungi are additional management strategies to improve P-use efficiency in P-deficient acid soils. All these practices are related to modification of the soil environment to improve P-utilization efficiency by crops. However, farmers are facing difficulties with the high costs of phosphorus fertilizers, especially in developing countries where a large percentage of acid soils are located.

An integrated fertilization-plant breeding approach seems likely to give more economically viable and practical results in the future. The possibility of exploiting genotypic differences in absorption and utilization of P to improve efficiency of P fertilizer use or to obtain higher productivity on P-deficient soils has received considerable attention in recent years (Fageria et al., 1988b; Baligar et al., 1990; Clark, 1990). Fageria (1994) studied differences among crop species in relation to decrease in dry weights of tops under P stress and adequate supply of P in an acid soil of Brazil. The P susceptibility order among crop species was rice > corn > common bean > wheat > soybean. This means among the five crop species, rice was most sensitive to P deficiency and soybean was least sensitive. Plant species differ in their ability to acquire soil P (Hanway and Olson, 1981) due to differences in root morphology (Barley, 1970), uptake characteristics (Fohse et al., 1991), mycorrhizal associations (Bolan, 1991), and the effect of the plant root on soil chemistry and P solubility (Ibrikci et al., 1994).

Increased fertilizer P uptake efficiency has been noted when P has been applied with N fertilizers. The effect of added N may be physiological enhancement of P uptake with added NH_4, enhanced P uptake when applied in a band with N, or NH_4^+-induced acidification near the root and an increase in the concentration of $H_2PO_4^-$ compared with HPO_4^{2-} (Miller et al., 1990; Riley and Barber, 1971). A decrease in precipitation of fertilizer P when banded with $(NH_4)_2SO_4$ was attributed to NH_4^+-induced acidification (Miller et al., 1970). Fan and Mackenzie (1994) reported that total N and P uptake by corn was increased by banding urea with triple superphosphate or monoammonium phosphate, and fertilizer P-use efficiency increased from 40 to 80%.

Addition of organic matter can also improve P uptake. In addition to providing a reservoir of organic matter, dissolved organic matter could increase the availability of P (Chien et al., 1987) and reduce P fixation (Mikeni and Mackenzie, 1987).

Liming. The most common and, in most cases, the most effective way to correct soil acidity is by applying lime. Liming is the practice of adding liming materials to acid soils for the purpose of increasing soil pH and maintaining a favorable soil environment for plant growth. A more favorable root environment may be a consequence of:

1. desirable soil pH
2. decreasing the toxicity of Al and Mn
3. increasing Ca and Mg supplies
4. enhancing the availability of P and Mo
5. improving mineralization of organic compounds, thereby improving N, S, and P uptake
6. improving soil biological activity, such as nitrogen fixation

The quantity of lime added depends on type of soil, liming material, crop species, cultivar (within species), and economic considerations. The formula to compute lime requirement by this method is given in Chapter 4. In some parts of Brazil, base saturation is being used as a parameter to calculate lime requirement using the following formula (Raij, 1991):

$$\text{Lime (T ha}^{-1}) = \frac{\text{EC(B}_2 - \text{B}_1)}{\text{TRNP}} \times \text{df}$$

where *EC* is the total exchangeable cations (Ca^{2+} + Mg^{2+} + K^+ + H^+ + A^{3+}) in cmol kg^{-1}, B_2 is the optimum base saturation, B_1 is the existing soil base saturation, *TRNP* is the total relative neutralizing power of liming material, and *df* is the soil depth factor (1.0 for 20 cm, 1.5 for 30 cm depth). For Brazilian Oxisols the optimum base saturation for most cereals is considered to be in the range of 50–60%; for most legumes, it is in the range of 60–70% (Raij, 1991).

Cochrane et al. (1980) developed the following equation for liming acid mineral soils to compensate crop aluminum tolerance while accounting for the levels of exchangeable Ca and Mg in the soil:

$$\text{Lime required (CaCO}_3\text{, equiv. t ha}^{-1}) = 1.8\left[\frac{\text{Al} - \text{CAS (ECEC)}}{100}\right]$$

where:

CAS = critical aluminum saturation or required aluminum saturation of the effective cation exchange capacity

Al Sat = Al/ECEC

ECEC = effective cation exchange capacity, which is the sum of exchangeable Al, Ca, Mg, and K in cmol kg^{-1} of soil in 1M KCl extractant at original soil pH

For most agronomic crops, Al^{3+} concentration or activity in the soil solution or Al saturation in the exchange complex appears to be the best single measure to assess potential Al toxicity for a given soil (Kamprath, 1971). When Al saturation of exchange capacity exceeds 60%, appreciable amounts of Al^{3+} start to get into the soil solution (Nye et al., 1961). At this point Al toxicity, caused by soil acidity, can occur. Table 5.4 shows critical Al saturation for important field crops. Heavy fertilization could induce Al toxicity even in soils with a relatively low Al saturation (Kamprath, 1970, 1971). Intensive cropping and use of acid-forming fertilizers without proper liming will aggravate the situation (Beverly and Anderson, 1987).

Use of gypsum ($CaSO_4$ $2H_2O$) or phosphogypsum is another management strategy to reduce soil acidity. Phosphogypsum is a by-product of wet acid production of phosphoric acid from rock phosphate. It is essentially hydrated $CaSO_4$ with small proportions of P, F, Si, Fe, Al, several minor elements, heavy metals, and radionucleides as impurities (Alcordo and Recheigl, 1993). Worldwide production of phosphoric acid, estimated at 11 million Mg of P annually (Lin et al., 1990), also results in the production of approximately 125 million Mg of phosphogypsum. With only about 4% of the world's phosphogypsum production being used in agriculture and in gypsumboard and cement

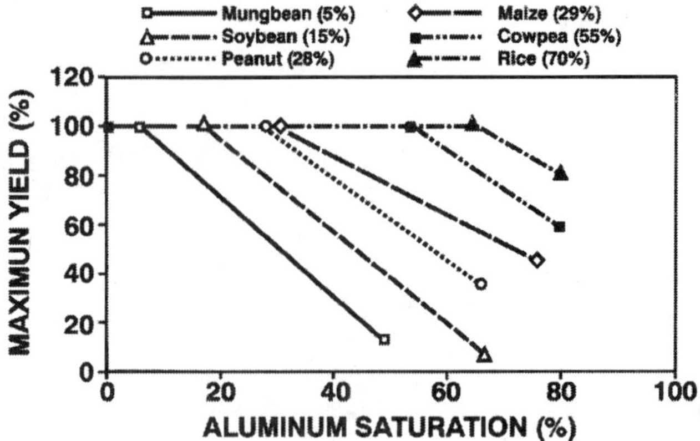

Figure 5.9 Aluminum tolerance of crop species (From Wade et al., 1989).

industries, about 120 million Mg of phosphogypsum accumulates annually. Most of this excess is stockpiled, and some is stored in abandoned quarries or, in certain countries, dumped into waterways (Alcordo and Recheigl, 1993).

Field calibration data would provide the best approach for recommending gypsum for field crops. However, Malavolta (1991) proposed the following equations to apply gypsum for field crops in Cerrado soils of Brazil (mostly Oxisols and Ultisols):

$$\text{Rate of gypsum (t ha}^{-1}) = (0.4 \times \text{ECEC} - \text{Ca}) \times 2.5$$

or

$$\text{Rate of gypsum (t ha}^{-1}) = (\text{Al} - 0.2 \times \text{ECEC}) \times 2.5$$

where values of effective cation exchange capacity (ECEC) and Al should be expressed in cmol kg^{-1} of soil. In a much simpler approach, Souza (1988) suggested that for highly weathered Oxisols, such as the soils of the Cerrado, gypsum rates should be based on clay content. He proposed 0.5, 1.0, 1.5, and 2.0 t ha^{-1} gypsum for soils of sandy texture, medium texture, clayey texture, and very clayey texture, respectively.

For the management of Al toxicity with the use of lime or gypsum materials, the Ca-Al balance index of Noble et al. (1988) should be helpful. Most of the studies showed that surface-applied gypsum or phosphogypsum ameliorated subsoil Al toxicity, acidity, and infertility in shorter time periods than surface-applied liming materials (Alcordo and Recheigl, 1993). In addition, phosphogypsum is a good source of S and Ca for crops.

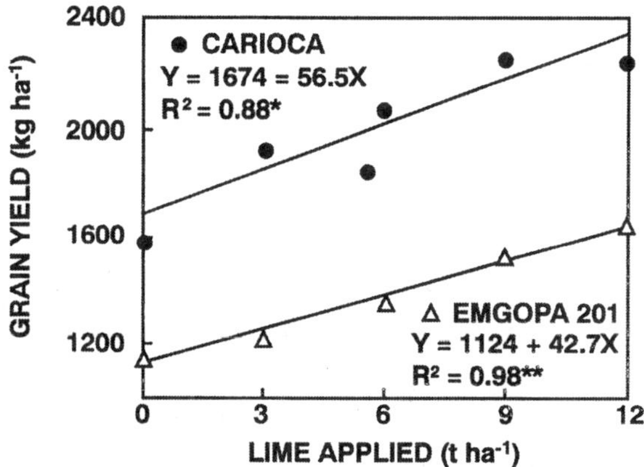

Figure 5.10 Relationship between lime applied and grain yield of two common bean cultivars in an Oxisol of central Brazil (From Fageria et al., 1991b).

Use of Tolerant Species and/or Cultivars. In addition to liming and gypsum application, use of Al- and Mn-tolerant crop species and cultivars can be a complementary solution for crop production on acid soils. Aluminum tolerance of some important crop species is presented in Figure 5.9. Among six crop species, rice is most tolerant to Al toxicity and mungbean is most sensitive. Other acid-tolerant crop species are cassava, potato, millet, and pigeonpea. Differences among cultivars of important food crops have been widely reported for Ca and Mg deficiencies and Mn toxicities (Fageria et al., 1989; Foy, 1984; Kamprath and Foy, 1985). Figure 5.10 shows similar responses of two common bean cultivars to liming in an Oxisol of central Brazil. The cultivar Carioca produced much higher grain yield at low as well as higher lime levels as compared to the cultivar EMGOPA 201. Table 5.4 shows critical Al saturation of various crop species.

In addition to the above management practices, acid soils amended with large quantities of organic residues give low Al^{3+} concentrations in soil solution and permit good growth of crops under conditions where toxicities would otherwise occur. Deep tillage to lower the concentration of toxic elements in the surface soil and selection of plants with deep rooting characteristics are additional methods to enhance crop production on acid soils.

Saline-Sodic Soils

Many soils have been adversely modified for growth of most crop plants by the presence of high soluble salts or exchangeable sodium, or both. A saline sodic soil has saturation extract conductivity more than 2 ds m^{-1} at 25°C and

Table 5.4 Critical Al Saturation for Important Field Crops at 90–95% of Maximum Yield

Crop	Type of soil	Critical Al saturation (%)	Reference
Cassava	Oxisol/Ultisol	80	Howeler, 1991
Upland rice	Oxisol/Ultisol	70	Wade et al., 1989
Cowpea	Oxisol/Ultisol	55	Wade et al., 1989
Cowpea	Oxisol	42	Smyth and Cravo, 1991
Peanut	Oxisol/Ultisol	65	Foster et al., 1980
Peanut	Xanthic Hapludox	54	Smyth and Cravo, 1992
Soybean	Oxisol	19	Smyth and Cravo, 1991
Soybean	Xanthic Hapludox	27	Smyth and Cravo, 1992
Soybean	Oxisol/Ultisol	15	Wade et al., 1989
Soybean	Not given	< 20	Kamprath, 1970
Soybean	Ultisol	20–25	Sartain and Kamprath, 1975
Soybean	Histosol	10	Mengel and Kamprath, 1978
Soybean	Ultisol	20	Pearson et al, 1977
Corn	Oxisol	19	Smyth and Cravo, 1991
Corn	Xanthic	27	Smyth and Cravo, 1992
Corn	Oxisol/Ultisol	29	Wade et al., 1989
Corn	Oxisol/Ultisol	25	Foster et al., 1980
Corn	Oxisol	28	Sanchez, 1976
Mungbean	Oxisol/Ultisol	15	Foster et al., 1980
Mungbean	Oxisol	5	Wade et al., 1989
Coffee	Oxisol	60	Sanchez, 1976
Sorghum	Oxisol	20	Sanchez, 1976
Common bean	Oxisol	10	Howeler, 1991
Common bean	Oxisol	8–10	Abruna et al., 1975
Common bean	Oxisol	23	Salinas and Sanchez, 1977
Cotton	Not given	< 10	Kamprath, 1970

exchangeable sodium percentages of more than 15. The pH of such soil is less than 8.5. Salt-affected soils are distributed worldwide. The total area of these degraded soils is about 0.9 billion hectares (Lal et al., 1989). The global distribution of saline soils is about 0.34 billion hectares and of sodic soils is about 0.56 billion hectares. Salt-affected soils are common in arid and semi-arid regions where evaporation is higher than precipitation. As a result, salts are not leached from the soil and accumulate in amounts and types detrimental to plant growth. Soils are also salinized in coastal areas due to tides. Salts generally originate from native soil and irrigation water. Use of inappropriate levels of fertilizers with inadequate management practices can create saline conditions even in humid regions.

Sodic soils (which have a Na absorption ratio greater than 15 in saturation extracts) form when annual fluctuations in soil water conditions cause two opposing processes: Na concentration and dissipation (Seelig and Richardson, 1994). For example, lateral subsurface flow and annual fluctuation of a shallow water table created an area of sodic soils in Alberta, Canada (Fullerton and Pawluk, 1987). Sodic soils typically occur on landscape positions with high evaporative discharge that concentrates Na salts near the soil surface. Evaporative discharge is defined as a loss of water from groundwater to atmosphere by evapotranspiration (Seelig and Richardson, 1994). Reduced productivity occurs as a result of decreased yields on land that is presently cultivated. About one-third of all irrigated land is considered to be affected by salt (Epstein et al., 1980). Salt problems also restrict significant agricultural expansion into areas that presently are not cultivated. In the United States, salinity is a major limiting factor to agricultural productivity, and as the quality of irrigation water continues to decline, this problem will become more acute.

Management Strategies for Saline-Sodic Soils

Successful crop production on salt-affected soils depends on soil, water, and plant management. Technological approaches to cope with salinity and sodicity include water and soil management, irrigation methodology, and perhaps desalinization. Saline-sodic soils can be reclaimed by providing a source of Ca^{2+}, such as gypsum, to replace Na^{+} from cation exchange sites, a process that requires the flow of water through the soil (Ilyas et al., 1993). The cost of soil reclamation is frequently so high that it is not possible to reclaim such soils for crop production. Under these circumstances, biological approaches include the identification of halophytes that are potential crop plants and, if necessary, the introduction of more desirable horticultural or agronomic traits into them (i.e., the introduction of salt-tolerance characteristics into crop plants, the majority of which are glycophytic, or the manipulation of glycophytic crop plants to adjust and produce under conditions of moderate or low levels of salinity) (Hasegawa et al., 1986). Table 5.5 shows differences in crop species in relation to salinity tolerance. With recent developments in biotechnology, there is also potential for obtaining salt-tolerant crop genotypes by the use of somatic cell selection or protoplast fusion methodologies or by gene transformation using recombinant DNA methodologies (Hasegawa et al., 1986).

Allelopathy

Allelopathy, the direct or indirect effect of one plant on another through the production of chemical compounds that escape into the environment (Rice,

Table 5.5 Relative Tolerance of Important Field Crops to Salinity

Tolerant	Moderately tolerant	Moderately sensitive	Sensitive
Barley	Sorghum	Corn	Rice
Cotton	Wheat	Peanut	Sesame
Oats	Pearl millet	Chickpea	Blackgram
Rye	Soybean	Sugarcane	Pigeonpea
Triticale	Sunflower	Alfalfa	Common bean
Sugar beet	Cowpea	Broadbean	Mungbean
Guar	Winged bean	Ladino clover	Carrot
Canola or rapeseed		Common vetch	Onion
		Cassava	
		Eggplant	
		Garlic	
		Pea	
		Pepper	
		Potato	
		Radish	
		Sweet potato	
		Tepary bean	
		Tomato	

Source: Compiled from Maas (1993).

1974, 1979; Smith and Martin, 1994), occurs widely in natural plant communities and is postulated to be one mechanism by which soil degradation occurs. Actual and potential roles of allelopathy in agriculture have been extensively reviewed (Putnam and Duke, 1978; Rice, 1974, 1979; Smith, 1991). In general, allelopathy has been related to problems with crop production on certain types of soil, with stubble mulch farming, with certain types of crop rotations, with crop monoculture, and with forest site replanting (Putnam and Duke, 1978). Some chemical compounds implicated as effective allelopathic agents include simple phenolic acids, coumarins, terpenoids, flavonoids, alkaloids, cyanogenic glycosides, and glucosinolates. Secondary compounds implicated in biochemical interactions among plants are also involved in several protective or defensive functions for the plant (Putnam and Duke, 1978). Release of these chemical compounds into the environment occurs: 1) by oxidation of volatile chemicals from living plant parts, 2) by leaching of water-soluble toxins from aboveground parts in response to the action of rain, fog, or dew, 3) by exudation of water-soluble toxins from below ground parts, 4) by release of toxins from nonliving plant parts through leaching of toxins from litter, sloughed root cells, or 5) as microbial by-products resulting from litter decomposition. Once these chemicals are released into the immediate environment, they must accumulate in sufficient quantity to affect other plants, persist for some period of time, or be constantly released in order to have lasting effects (Putnam and Duke, 1978).

Minimizing Allelochemical Effects

Allelochemical effects can be reduced by adopting certain management practices. Among these practices are use of appropriate crop rotations, improving organic matter content of the soil, leaving cropped areas fallow for certain periods of time to allow decomposition of allelochemicals, and planting resistant cultivars or plant species. The use of companion plants that contribute organic matter or inoculation with microorganisms which can readily metabolize the toxins might prove useful in perennial crop ecosystems (Putnam and Duke, 1978).

Indiscriminate Use of Pesticides

Indiscriminate use of pesticides such as insecticides, fungicides, and herbicides has adverse effects on soil biology. Changes in the soil microflora may have significant implications regarding soil productivity and sustainability. Excessive use of pesticides may also pollute groundwater. Although the application of agricultural chemicals has greatly increased the productivity of modern agriculture, deleterious effects of these chemicals on soil biology and water quality have been reported (Li and Ghodrati, 1994). Many agricultural chemicals have been detected in groundwater in many areas of the United States (Holden, 1986). Extensive groundwater monitoring and experimental data, however, show that preferential flow is likely to be one of the principal mechanisms responsible for accelerated movement of these chemicals in many field soils (Ellsworth et al., 1991). Preferential flow refers to any transport process through which water and solutes gain enhanced movement (Luxmoore, 1991). The rise of soil degradation due to excessive use of pesticides can be avoided by advising farmers to use these materials rationally, as shown by experimental evidence for each agro-ecological region.

C. Biological Degradation

Biological degradation refers to the loss of organic matter, reduction in biomass carbon, and decline in the biotic activity of soil fauna (Lal, 1989). Biological degradation factors also include infestation of crop plants with diseases, insects, and weeds which reduce crop growth and yield.

Decrease in Soil Organic Matter Content

Soil organic matter is composed of a series of fractions from very active to passive (Schimel et al., 1989). These fractions have been conceptualized in mathematical models as kinetically defined pools with different turnover rates (Van Veen et al., 1984; Parton et al., 1987). Stable organic matter usually has a C/N ratio between 10 and 17 (Stevenson, 1986; Vaughan and Ord, 1991). The C stored in soils is nearly three times that in the aboveground biomass and approximately double that in the atmosphere (Eswaran et al., 1993a).

The literature on soil organic matter is replete with references to the positive effects of organic matter on soil chemical, physical, and biological properties that, in turn, contribute to improved crop yields (Stevenson, 1982a, 1982b; Bauer and Black, 1994). As a chemical reservoir, there is universal acknowledgment that soil organic matter is the main indigenous source of soil available N, that it contains as much as 65% of the total soil P, and that it provides significant amounts of sulfur and other nutrients essential for plant growth. It is also universally accepted that the C fraction is used by soil microorganisms as a major energy source for metabolic activity, in the process altering nutrient availability and soil structure (Paul, 1991). Because organic matter affects soil structure, soil physical properties can also be altered (Bauer and Black, 1994). The contribution of 1 Mg organic matter ha^{-1} to soil productivity (across the range of 64–142 Mg organic matter ha^{-1}) was calculated as equivalent to 35.2 kg ha^{-1} for spring wheat total dry matter and 15.6 kg ha^{-1} for grain yield. Loss of productivity associated with depletion of soil organic matter in the Northern Great Plains of the United States is primarily a consequence of a concomitant loss of fertility (Bauer and Black, 1994). Soil erosion in the Northern Great Plains is deemed to diminish soil productivity through loss of soil organic matter. The contribution to productivity of a unit quantity of soil organic matter is the quotient of the difference in the crop yield grown on sites differing in soil organic matter content and the difference in the soil organic matter content of these sites in the plow layer (Bauer and Black, 1994). Kaolinitic soils of temperate as well as tropical regions are characterized by high dispersibility, a factor that may increase the susceptibility of cultivated soils to aggregate disruption, surface crusting, reduced infiltration, and erosion (Sumner, 1992). In warm, humid climate regions, these factors can contribute to a rapid loss of soil organic matter and a decline in the productivity of agricultural soils (Bruce et al., 1990).

Long-term cultivation alters soil structure and increases losses of soil organic matter (Dalal and Mayer, 1986). Apart from soil type and climate variables, the magnitude of these effects depends on the intensity of cultivation, in particular the type and frequency of tillage and the quantity and quality of fertilizers and organic residues returned to the soil (Rasmussen and Collins,

1991). Elliott (1986) and Gupta and Germida (1988) attributed much of the soil organic matter lost during cultivation of grassland soils to mineralization of soil organic matter binding microaggregates into macroaggregates. Macroaggregates are much less stable than microaggregates (Oades, 1984; Beare et al., 1994a), probably because of the nature of the binding agents involved. Macroaggregates are also more susceptible to the disruptive forces of cultivation and to the dispersion that results from rapid wetting or raindrop impact (Tisdall and Oades, 1982).

The decomposition ratio of soil organic matter is doubled for every 10°C increase in mean temperature. The climate index used by FAO to assess loss of organic matter is (Lal, 1989):

$$K = \frac{1}{12}\sum_{1}^{12} e^{0.1065t} \times \frac{P}{PET} \qquad (\text{with } P < PET)$$

where t is mean air temperature during the month, P is precipitation, PET is potential evapotranspiration, and K is rate of humus decay in percent by year. If P is greater than PET, P divided by $PET = 1$, and if t is less than 0, $t = 0$.

Management Strategies for Soil Organic Matter Content

Agricultural management practices influence the amount of organic matter present in soils and cause changes in the rate of soil organic matter turnover (Cambardella and Elliott, 1994). Because soil organic matter is composed of a series of fractions, management practices also influence the distribution of organic C and N among soil organic matter pools. Cultivation of soils results in the disruption of soil aggregates and the loss of soil organic matter compared with native sod and pasture soils (Elliott, 1986; Kay, 1990). Therefore, one strategy for improving soil organic matter content, either fallowing or pasture, should be included in the farming system. Little inorganic N is accumulated in a prairie or in a grassland ecosystem, and thus little is lost to either leaching or denitrification (Goodroad and Keeney, 1984). Although inorganic N is continuously mineralized from soil organic matter, it rarely accumulates in a grassland soil due to rapid immobilization by microbial activity (Woodmansee et al., 1981; Jackson et al., 1989). Conservation tillage is another practice to improve soil organic matter content of cultivated land. With no-tillage management, the soil is not plowed, and crop residues accumulate on the soil surface as a mulch. Several studies have shown that if residues are not removed, no-tillage or minimum tillage can improve soil aggregation and reduce losses of soil organic matter that result from cultivation (Havlin et al., 1990; Carter, 1992; Beare et al., 1994b). Conventional tillage practices disrupt soil aggregates, exposing more organic matter to microbial attack. Organic matter may be protected from microbial attack by adsorption to clay minerals (Oades, 1984; Ladd et al., 1985) and the formation of microaggregates (Gregorich et al. 1989) by isolation in micropores (Foster, 1981), and by physical protection within stable macroaggregates (Elliott, 1986; Gupta and Germida, 1988).

According to Beare et al. (1994b), macroaggregates in no-tillage soils provide an important mechanism for the protection of soil organic matter that may otherwise be mineralized under conventional tillage. Soil organic matter is enriched by return of crop residues, growing green manures, and application of farmyard manures and other materials containing organic substances (e.g., peat and humic materials). Simultaneous additions of organic materials and N fertilizers can increase total soil organic matter content (Stevenson, 1986). The result of long-term additions of organic materials to soil is increased soil organic matter, crop productivity, and soil biological activity (Collins et al., 1992). High rates of animal manures can sustain crop yields (Bouldin et al., 1984). However, studies utilizing typical farm scale management practices have shown that replacement of inorganic N with organic (animal or green manure) N sources resulted in unacceptable yield reduction related to N availability during the first few years of transition (Doran et al., 1987). A successful transition from inorganic to organic N sources, therefore, can be achieved by adding a small amount of inorganic fertilizer to supplement the manures. Grassland soils tend to lose 30–50% of their original soil organic matter in the first 40–50 years of cultivation (Tiessen et al., 1982; Mann, 1985). Rapid depletion of the easily mineralizable organic matter is the primary reason for the initial C loss (Bowman et al., 1990). Organic matter changes thereafter largely become a function of soil management and erosion (Rasmussen and Parton, 1994). Use of green manuring, crop residues, and livestock manures are some specific practices to add or improve soil organic matter content.

Green Manuring. Green manure crops historically were used to supply plant nutrients and organic matter for increasing soil fertility. China has a 3,000–year history of using green manure to increase yields of cereal crops and to maintain and increase soil fertility (Lizhi, 1988). As agriculture became more specialized, especially in developed countries, use of green manure in the cropping systems was practically eliminated. Research at many experiment stations in the United States and abroad showed that concentrated fertilizer nitrogen could replace legumes and manures as a readily available source of nitrogen for nonlegume crops. The result was a rapid intensification of cropping based on fertilizer nitrogen. In recent years, the high cost of nitrogen fertilization and water quality problems associated with inappropriate use of nitrogen fertilizers have generated a renewed interest in legume green manures as an alternative source of nitrogen. In addition to fixing nitrogen, legume green manures maintain ground cover, reduce erosion, and provide weed control. They improve soil physical conditions and promote mycorrhizae on the roots of succeeding crops, increasing soil phosphorus availability. They may also suppress plant pests such as nematodes. Studies in Brazil have shown that some legume green manures are very effective in suppressing soybean nematodes, resulting in increased yields (Sharma et al., 1982). Green manures, especially those which have tap roots or root deeply into the soil, may also prevent or help alleviate compaction in intensively cultivated soils (Taylor, 1974). Partial fallow replacement with legumes reduces the risk of erosion and nutrient leaching and minimize the hazard of salinization and eutrophication of downstream ecosystems (Biederbeck and Bouman, 1994).

A vast array of legume species have potential as green manures. In the temperate region many legumes are used mainly as forage for livestock. These include alfalfa (*Medicago sativa* L.), several clovers (*Trifolium* spp.), vetch (*Vicia* spp.), and other less familiar species (Lathwell, 1990). Annual dry matter accumulation by these legumes varies from 1 Mg/ha to over 10 Mg/ha under ideal growing conditions. Quantities of N accumulated in the aboveground dry matter range from 20 kgN/ha to as much as 300 kgN/ha annually. There are several hundred species of tropical legumes, but only a fraction of these have been studied for their potential as green manures. Some important tropical species used as green manures are mucuna or velvet bean (*Mucuna aterrima*), crotalaria (*Crotalaria striata*), zornia (*Zornia latifolia*), jackbean (*Canavalia ensiformes*), pigeonpea (*Cajanus cajan*), tropical kudzu (*Pueraria phaseoloides*), guar (*Cymopsis tetragonoloba*), mung bean (*Vigna radiata*), and cowpea (*Vigna unguiculata*). Effects on corn production by some tropical green manure crops in Oxisols of the Cerrado region of Brazil are presented in Table 5.6. Dry matter produced by the legume green manures ranged from 6.5 Mg/ha with *Pueraria* to over 13 Mg/ha with *Crotalaria*. The total N accumulated in the aboveground dry matter varied from 112 to 180 kgN/ha. The grain yield of corn after incorporation of green manure crops in some cases was comparable to that obtained with the application of 200 kg N/ha as urea (Table 5.6).

Mughogho et al. (1982) reported that in Trinidad, cowpeas produced about 3.5 Mg/ha of dry matter, with about 1.8 Mg/ha as grain. Grain contained from 45 to 50 kg N/ha, whereas the residue contained from 15 to 20 kg N/ha. Results from the International Institute of Tropical Agriculture (IITA) in Nigeria were similar to those found in Trinidad (Eaglesham, 1982). In an experiment at the International Crop Research Institute for the Semi-Arid Tropics in India, pigeonpea produced 6 Mg/ha of dry matter, of which 1.6 Mg was grain (Kumar et al., 1982). Estimated N content was about 40 kgN/ha in the grain and 50 kgN/ha in the residue. George et al. (1994) reported that mungbean

Table 5.6 Dry Matter and N Added by Several Green Manure Legumes and Subsequent Grain Yield and N Uptake by Maize.

Treatment	Dry matter added (mg/ha)	Total N added (kg/ha)	Grain yield (mg/ha)	Total N aboveground (kg/ha)
None	—	—	3.3	69
Canavalia brasiliensis	10.0	228	6.6	159
Cajanus	10.4	229	7.0	149
Canavalia ensiformes	8.9	231	6.4	177
Calopogonium	7.3	142	6.1	137
Crotalaria	13.7	306	6.6	180
Mucuna	6.8	152	6.3	130
Pueraria	6.5	116	5.7	112
Urea (fertilizer N)	—	200	7.2	198

Source: Carsky (1989).

Table 5.7 Effect of Maize Straw Management on K Balance in a Long-Term Trial in Togo

	Straw removed		Straw returned	
Parameter	K_0	K_2	K_0	K_2
Mean yield (t ha^{-1} $year^{-1}$)	2.9	5.5	2.1	4.6
Input (kg ha^{-1} yr^{-1} K) fertilizer	0	57	0	86
Output (kg ha^{-1} yr^{-1} K)				
Grain	13	22	9	20
Straw	51	85	23	38
Leaching	2.5	4	2.5	4
Balance (kg ha^{-1} yr^{-1} K)	–66	–72	–35	+23

Source: Fardeau et al. (1992).

fixed from 37 to 63 kg N/ha and *Sesbania rostrata* fixed 68–154 kg N/ha as green manures in lowland rice cropping systems on Alfisols in the Philippines. There are several desirable characteristics that can contribute to the effective use of green manure crops. Some of these characteristics are: 1) short duration, fast growing, and high nutrient accumulation ability; 2) wide ecological adaptability; 3) efficient water use; 4) pest and disease resistance; 5) ease in incorporation; and 6) high N sink in underground plant parts. Certainly, green manure crops have some distinct advantages in some climates, on some soils, and in some socioeconomic situations.

In certain climates, green manure could have definite physicochemical advantages; in other climates they face major constraints. For example, in temperate climates, low temperatures can hinder organic decomposition which could allow build-up of toxicity. In addition, fertilizer nitrogen is relatively easy to transport and apply, and farmers can readily adjust the timing and rate of application to meet crop requirements. Legume green manures, on the other hand, require careful management. This means that the use of green manures in crop production should be carefully evaluated according to each situation.

Use of Crop Residue. Crop residues can play an important role in controlling the rate of change in soil C and N content (Uhlen, 1991; Paustian et al., 1992). In general, as much crop residue as possible should be returned to the soil. Compared with clean tillage, minimum tillage cropping systems leave more residue on the soil surface, resulting in reduced soil erosion, less farm energy use, and greater water conservation (Unger and McCalla, 1980). Up to 70% of the total potassium accumulated in crops is found in crop residues (Fageria et al., 1990a). Complete removal of crop residues (e.g., for use as fuel, roofing, manufacturing) from the land causes considerable loss of nutrients and presents a great challenge to low-input farmers to develop suitable residue management techniques.

The impact of maize stover treatment on K balance was studied in a long-term (11 years) field experiment on a Ferrasol in Togo. The results (Table 5.7) clearly indicated that K balance was strongly negative if straw was removed during the study. Mineral fertilizer application could not prevent depletion of soil K on this soil. Growing continuous maize, K balance is maintained only if K fertilizer is applied and all is stover retained. This is convincingly reflected by the measured contents of available soil K at the end of the experiment: 36 and 199 mg K/kg soil, without and with stover.

In recent years, crop residue management has been widely promoted for soil and water conservation purposes (Unger, 1994). When adequate residues are retained on the soil surface, soil water storage increased compared with residue incorporation with soil (Unger, 1984).

Optimum residue management strategies must consider the role of residues on the soil water balance, as well as their role in nutrient management. Greater water availability to buried residues enhances their decomposition and nutrient release compared with surface residue placement (Schomberg et al., 1994). However, maximum soil protection requires that substantial residue amounts remain on the soil surface, so practices that promote rapid residue drying, such as leaving residues standing, should be encouraged on highly erodible lands. Nutrient management under these conditions may become more critical. The influence of residue quality should also be considered (i.e., residues that decompose rapidly protect soils for shorter periods but accelerate the return of nutrients to the soil) (Schomberg et al., 1994).

Use of Livestock Manures. Manure from livestock is an important source of N and P for crop production in many areas, but efficient management of manure is critical to improve the economics of manure use and to minimize the impact on water quality (Jokela, 1992). Animal manures are generally required in large quantity due to their low nutrient content. Therefore, the best strategies should be to apply livestock manures in combination with inorganic fertilizers.

Soil microbial biomass is a source and sink for plant nutrients and an active participant in nutrient cycling (McGill et al., 1986). Soils managed with organic amendments generally have larger and more active microbial populations than those managed with mineral fertilizers (Bolton et al., 1985; Dick et al., 1988; Alef et al., 1988; Fauci and Dick, 1994).

Diseases, Insects, and Weeds

Diseases, insects, and weeds continue to cause major problems in agriculture throughout the world, reducing yield and quality of crops by competing for light, nutrients, and water. These biotic factors are also responsible for substantial reduction in crop yields and soil degradation. Walker (1975) suggested that overall average crop loss due to insects is 14%, to diseases 12%, and to weeds 9%. These figures are considerably less than Russell's (1978) estimate that more than half of the world's potential crop production is lost by the action of pests. The resultant losses in economic terms are impossible to estimate accurately because the severity of pests varies greatly from place to place and from year to year due to changes in environmental factors. Nematodes, weeds, and defoliating insects account for more than 85% of economic losses from biotic factors in soybean in the United States (Hammond et al., 1991).

The importance of diseases, insects, and weeds is reflected by the rapid growth of the use of pesticides in agriculture throughout the world. Weeds are the most economically important of all pests with respect to use of pesticides and limitation to crop yield (Levesque and Rahe, 1992). An indirect measure of the impact of weeds compared with other pests in cropping systems may be the quantity of herbicide applied. The projected pesticide use in 1992 on major field crops in the United States was 219 million kg active ingredient. Herbicides accounted for 84% of total pesticide use while insecticides comprise 14%, and fungicides 2% (USDA, 1992). Herbicide sales represent almost one half of the \$21 billion worldwide pesticide market (Belcher, 1989). Herbicides comprise 84% of the total amount of pesticides applied to wheat in the United States (Young et al., 1994).

Control Measures for Diseases, Insects, and Weeds

The best management strategies for control of diseases, insects, and weeds are by employing all the technologies available rather than by relying entirely on one or two measures. This strategy has been encompassed by the concept of integrated pest management. This means all measures such as cultural, chemical, and genetic should be applied in order to achieve effective pest control. An integrated disease, insect, and weed management system must take all aspects of a cropping system into account, from sowing to harvest. In addition, climate and edaphic factors, such as soil moisture and soil fertility, must be taken into account. The impact of pests (diseases, insects, and weeds) on crop productivity may vary during different phases of crop development, and an effective management system should take into account the dynamic nature of crop-pest interactions throughout the growing season (Tollenaar et al., 1994a). The exploitation of beneficial effects of natural enemies by altering the soil environment is an important strategy in pest control. According to Sayre and Walter (1991), there are three types of biological control agents: 1) natural, where an agent has increased to concentrations that suppress the pest population without having been specifically introduced; 2) the augmentation of a crop or soil with an agent; and 3) the introduction of an exotic enemy with the hope that it may become established and increase to densities that have an economic impact on a pest.

Management decisions require a great deal of information regarding pest population levels and expected losses in crop value. Integrated pest management strategies are designed to employ pesticides only when pest populations reach economically damaging numbers. Furthermore, selective pesticides applied at proper times can control insects, but have minimal impact on beneficial insects (Funderburk et al., 1994).

The relative competitive ability of maize can be enhanced by increasing plant density. Tollenaar et al. (1994a) reported that increasing maize plant density from 4 to 10 plants m^{-2} reduced weed biomass up to 50%. Ghafar and Watson (1983) reported that biomass of yellow nutsedge (*Cyperus es-*

culentus L.) was significantly reduced when maize density was increased from 3 to 13 plants m^{-2}, and Weil (1982) reported a negative correlation between maize plant density and weed dry matter. The leaf area index (LAI) of maize usually increases when plant density is increased (Tollenaar, 1992), which reduces the transmission of irradiance by the maize canopy. The LAI also may influence the transmitted irradiance qualitatively (i.e., the red to infrared ratio; Ballare et al., 1990), which may affect weed growth and development. Plant-suppressive rhizobacteria have the potential to be used as biological control agents to combat weeds (Johnson et al., 1993). One bacterium, *Pseudomonas fluorescens* strain D_7, is a promising biological control agent, having been screened in laboratory bioassays for benign interaction with wheat roots and an inhibitory effect on downy brome roots (Kennedy et al., 1991). Application of the bacterium to winter wheat fields resulted in reduced downy brome competition and enhanced wheat yield (Kennedy et al., 1991).

Use of genetic-resistant crop species and cultivars within species is an important pest control strategy. Genetic yield improvement in maize has been associated with increased stress tolerance (Tollenaar, 1994b), which may suggest that modern maize hybrids are more tolerant of diseases, insects, and weed interference than earlier hybrids.

D. Socioeconomic Factors

Among socioeconomic factors which are responsible for soil degradation, high population pressure and land tenure systems are the most important. The expanding human population, in its search for food, fiber, and fuel for today, puts tomorrow's sustainable agriculture production and natural resources preservation in jeopardy in many areas of the world (Lal, 1991). The world population today is about 5.7 billion people, and it is expected to be about 8.5 billion in the year 2025 and more than 10 billion in 2050. Most of the population increase will be in developing countries. African countries have to import food even to maintain minimum nutritional standards, and by the turn of the century the food deficit in sub-Saharan countries will total some 50 million tons if current farming practices are not improved. The main problems in many of these areas are high population growth and low and fluctuating crop yields (International Potash Institute, 1993).

E. Management Strategies for Socioeconomic Factors

High population growth can be controlled by an effective family planning system. An effective family planning system involves improving the education standard of the people, improving the economic situation, and providing efficient health service at the local level. In developing countries, lack of trained medical personnel, faulty transport and communication systems, and lack of well-equipped health facilities still complicate the situation. However, with the help of some international agencies and developed countries, the situation could be improved.

IV. SUMMARY

Soil degradation, the loss or decline in those properties of soil that affect its life-supporting processes, is influenced by physical, chemical, and biological processes. All these processes are accelerated in favor of soil degradation by man's activities such as deforestation, expansion of agricultural planting on steep and marginal lands, monoculture, use of poor quality irrigation water, use of excess agrochemicals, overgrazing of pasture lands, and indiscriminate use of mechanization. Soil degradation is responsible for decreasing soil fertility or nutrient-supplying capacity of soils. By adopting appropriate soil management practices, it is possible to restore the productivity of degraded soils and, consequently, soil fertility.

REFERENCES

Abruna, F., R. W. Pearson, and R. Perez-Escolar. 1975. Lime response of corn and beans grown on typical Oxisols and Ultisols of Puerto Rico, pp. 261–286. *In*: Soil management in tropical America. North Carolina State University, Raleigh.

Agassi, M., I. Shainberg and J. Morin. 1981. Effect of electrolyte concentration and soil sodicity on infiltration rate and crust formation. Soil Sci. Soc. Am. J. 45: 848–851.

Alcordo, I. S. and J. E. Recheigl. 1993. Phosphogypsum in agriculture: A review. Adv. Agron. 49: 55–118.

Alef, K. T., Beck, L. Zelles, and D. Kleiner. 1988. A comparison of methods to estimate microbial biomass and N-mineralization in agriculture and grassland soils. Soil. Biol. Biochem. 20: 561–565.

Alexander, M. 1980. Introduction to soil microbiology. Wiley, New York.

Almendros, G., F. Martin and F. J. Gonzalez-Villa. 1988. Effects of fire on humic and lipid fractions in a Dystric Xerochrept in Spain. Geoderma 42: 115–127.

Anderson, R. C. 1990. The historic role of fire in the North American grassland, pp. 8–18. *In*: S. L. Collins and L. L. Wallace (eds.) Fire in North American tallgrass prairies. Univ. Oklahoma Press, Norman.

Baligar, V. C., R. R. Duncan, and N. K. Fageria. 1990. Soil-plant interaction on nutrient use efficiency in plants: An overview, pp. 351–373. *In*: V. C. Baligar and R. R. Duncan (eds.) Crops as enhancers of nutrient use. Academic Press, San Diego, California.

Baligar, V. C., J. H. Elgin, Jr., and C. D. Foy. 1989. Variability in alfalfa for growth and mineral uptake and efficiency ratios under aluminum stress. Agron. J. 81: 223–229.

Baligar, V. C., R. J. Wright, T. B. Kinraide, C. D. Foy, and J. H. Elgin, Jr. 1987. Aluminum effects on growth, mineral uptake, and efficiency ratios in red clover cultivars. Agron. J. 79: 1038–1044.

Ballare, C. L., A. L. Scopel, and R. A. Sanchez. 1990. Far infrared radiation reflected from adjacent leaves: An early signal of competition in plant canopies. Science 248: 329–331.

Barley, K. P. 1970. The configuration of the root system in relation to nutrient uptake. Adv. Agron. 22: 159–201.

Bauer, A. and A. L. Black. 1994. Quantification of the effect of soil organic matter content on soil productivity. Soil Sci. Soc. Am. J. 58: 185–193.

Bearden, E. D. 1984. A comparison of variability of undisturbed and surface mined soils in Freestone county, Texas. M.S. thesis, Texas A & M Univ., College Station.

Beare, M. H., M. L. Cabrera, P. F. Hendrix, and D. C. Coleman. 1994a. Aggregate-protected and unprotected organic matter pools in conventional and no-tillage soils. Soil Sci. Soc. Am. J. 59: 787–795.

Beare, M. H., P. F. Hendrix, and D. C. Coleman. 1994b. Water stable aggregates and organic matter fractions in conventional and no-tillage soils. Soil Sci. Soc. Am. J. 58: 777–786.

Belcher, J. E. 1989. Monsanto Company. Duff and Phelp, Inc., Chicago. 31 pp.

Ben-Hur, M., R. Stern, A. J. Vander Merwe, and I. Shainberg. 1992. Slope and gypsum effects on infiltration and erodibility of dispersive and nondispersive soils. Soil Sci. Soc. Am. J. 56: 1571–1576.

Beverly, R. B. and D. L. Anderson. 1987. Effects of acid source on soil pH. Soil Sci. 143: 301–303.

Biederbeck, V. O. and O. T. Bouman. 1994. Water use by annual green manure legumes in dryland cropping systems. Agron. J. 86: 543–549.

Bilbro, J. D. and D. W. Fryrear. 1994. Wind erosion losses as related to plant silhouette and soil cover. Agron. J. 86: 550–553.

Binkley, D., C. T. Driscoll, H. L. Allen, P. Schoeneberger, and D. Mcavoy. 1989. Acidic deposition and forest soils: Context and case studies of the Southeastern United States. Ecol. Study. Vol 72. Springer-Verlag, Berlin.

Blank, R. R., F. Allan, and J. A. Young. 1994. Extractable anions in soils following wildfire in a sagebrush-grass community. Soil Sci. Soc. Am. J. 58: 564–570.

Bolan, N. S. 1991. A critical review on the role of mycorrhizal fungi in the uptake of phosphorus by plants. Plant Soil 134: 189–207.

Bolton, H., L. F. Elliott, R. I. Papendick, and D. F. Bezdicek. 1985. Soil microbial biomass and selected soil enzyme activities: Effect of fertilization and cropping practices. Soil Biol. Biochem. 17: 297–302.

Bouldin, D. R., S. W. Klausner, and W. S. Reid. 1984. Use of nitrogen from manure, pp. 221–245. *In*: R. D. Hauck (ed.) Nitrogen in crop production. ASA, CSSA, and SSSA. Madison, Wisconsin.

Bouma, J. 1994. Sustainable land use as a future focus for pedology. Soil Sci. Soc. Am. J. 58: 645–646.

Bouman, A. F. 1990. Soils and the greenhouse effect. Wiley, Chichister, England.

Bowen, H. D. 1981. Alleviating mechanical impedance, pp. 21–57. *In*: G. F. Arkin and H. M. Taylor (eds.) Modifying the root environment to reduce crop stress. Monogr. 4, Am. Soc. Agric. Eng., St. Joseph, Michigan.

Bowman, R. W., J. D. Reeder, and R. W. Lober. 1990. Changes in soil properties in a Central Plains rangeland soil after 3, 20, and 60 years of cultivation. Soil Sci. 150: 851–857.

Boyle, M., W. T. Frankenburger, Jr., and L. H. Stolzy. 1989. The influence of organic matter on soil aggregation and water infiltration. J. Prod. Agric. 2: 290–299.

Bresson, L. M. and L. Cadot. 1992. Illuviation and structural crust formation on loamy temperate soils. Soil Sci. Soc. Am. J. 56: 1565–1570.

Bruce, R. R., G. W. Langdale, and L. T. West. 1990. Modification of soil characteristics of degraded soil surface by biomass input and tillage affecting soil water regime, pp. 4–9. *In*: 14th Trans. Int. Congr. Soil Sci., Kyota, Japan, August 12–18, 1990. Wageningen, The Netherlands.

Buerkert, A., K. G. Cassman, R. Piedra, and D. N. Munns. 1990. Soil acidity and liming effects on stand nodulation and yield of common bean. Agron. J. 82: 749–754.

Buringh, P. 1982. Potentials of world soils for agricultural production, pp. 33–41. *In*: Managing soil resources. 12th Int. Soil Sci. Congr., New Delhi.

Cambardella, C. A. and E. T. Elliott. 1994. Carbon and nitrogen dynamics of soil organic matter fractions from cultivated grassland soils. Soil Sci. Soc. Am. J. 58: 123–130.

Carsky, R. J. 1989. Estimating availability of nitrogen from green manure to subsequent maize crops using a buried bag technique. Ph.D. thesis, Correll Univ., Ithaca, New York.

Carter, M. R. 1992. Influence of reduced tillage systems on organic matter, microbial biomass, macroaggregate distribution and structural stability of the surface soil in a humic climate. Soil Tillage Res. 23: 361–372.

Chandler, C. P., P. Cheney, P. Thomas, L. Trabaud, and D. Williams. 1983. Fire in forestry. Vol. 1. Forest fire behavior and effects. Wiley, New York.

Chaudhary, M. A., S. Yoshida, and B. S. Vergara. 1987. Induced mutations for aluminum tolerance after N-methyl-N-nitrosourea treatment of fertilized egg cells in rice. Envir. and Exp. Botany 27: 37–43.

Chiang, S. C., D. E. Radcliffe, and W. P. Miller. 1993. Hydraulic properties of surface seals in Georgia soils. Soil Sci. Soc. Am. J. 57: 1418–1426.

Chien, S. H., D. Somlongse, J. Hensao, and D. R. Hellums. 1987. Greenhouse evaluation of P availability from compacted phosphate rocks with urea or urea and triple superphosphate. Fert. Res. 14: 245–256.

Clark, R. B. 1990. Physiology of cereals for mineral nutrient uptake, use, and efficiency, pp. 131–209. *In*: V. C. Baligar and R. R. Duncan (eds.) Crops as Enhancers of Nutrient Use. Academic Press, San Diego.

Clark, R. B., H. A. Pier, D. Knudsen, and J. W. Maranville. 1981. Effect of trace element deficiencies and excesses on mineral nutrients in sorghum. J. Plant Nutr. 3: 357–374.

Clarke, N. P. and S. R. Baen. 1980. Forward, p. 3. *In*: L. R. Hossner (ed.) Reclamation of surface mined lignite spoil in Texas. Texas Agric. Exp. Stn. Bull. RM-10.

Cochrane, T. T. 1991. Understanding and managing acid soils of tropical South America, pp. 113–122. *In*: P. Deturck and F. N. Ponnamperuma (eds.) Rice production on acid soils of the tropics. Institute of Fundamental Studies, Kandy, Sri Lanka.

Cochrane, T. T., J. G. Salinas, and P. A. Sanchez. 1980. An equation for liming acid mineral soils to compensate crop aluminum tolerance. Trop. Agric. Trinidad 57: 133–140.

Collins, H. P., P. E. Rasmussen, and C. L. Douglas. 1992. Crop rotation and residue management effects on soil carbon and microbial biomass dynamics. Soil Sci. Soc. Am. J. 56: 783–788.

Conservation Technology Information Center. 1988. Conservation tillage adoption: A survey of research and education needs. Conservation Technology Information Center, West Lafayette, Indiana.

Conservation Technology Information Center. 1991. National Survey of Conservation Tillage Practices. Conservation Technology Information Center, West Lafayette, Indiana.

Copeland, P. J. and R. K. Crookston. 1992. Crop sequence affects nutrient composition of corn and soybean grown under high fertility. Agron. J. 84: 503–509.

Courault, D., P. Bertuzzi, and M. C. Girard. 1993. Monitoring surface changes of bare soils due to slaking using spectral measurements. Soil Sci. Soc. Am. J. 57: 1595–1601.

Cregan, P. D. 1980. Soil acidity and associated problems. Guideline for farmer recommendations. Agric. Bull. Wagga Wagga Dept. of Agric., New South Wales, Australia, October 7, 1980.

Dalal, R. C. and R. J. Mayer. 1986. Long-term trends in fertility of soils under continuous cultivation and cereal cropping in Southern Queensland. II. Total organic carbon and its rate of loss from the soil profile. Aust. J. Soil Res. 24: 281–292.

Dao, T. H. 1993. Tillage and winter wheat residue management effects on water infiltration and storage. Soil Sci. Soc. Am. J. 57: 1586–1595.

DeBano, L. F., S. M. Savage, and D. A. Hamilton. 1976. The transfer of heat and hydrophobic substances during burning. Soil Sci. Soc. Am. J. 40: 779–782.

Debusk, W. F., K. R. Reddy, M. S. Koch, and Y. Wang. 1994. Spatial distribution of soil nutrients in a Northern Everglades marsh: Water Conservation Area 2A. Soil Sci. Soc. Am. J. 58: 543–552.

Dent, D. 1986. Acid sulfate soils: A baseline for research and development. IILRI Publ. 39. Int. Inst. Land Reclamation and Improvement, Wageningen, The Netherlands.

Detwiler, R. P. 1986. Land use change and the global carbon cycle: The role of tropical soils. Biogeochemistry 2: 67–93.

Dick, R. P., P. E. Rasmussen, and E. A. Kerle. 1988. Influence of long-term residue management on soil enzyme activities in relation to soil chemical properties of a wheat-fallow system. Biol. Fertil. Soils 6: 159–164.

Dick, W. A., D. M. VanDoren, Jr., G. B. Triplett, Jr., and J. E. Henry. 1986. Influence of long-term tillage and rotation combinations on crop yields and selected soil parameters. *In*: Results obtained for a Typic Fragiudalt soil. Res. Bull. 1181, Ohio Agric. Res. Dev. Center, Wooster, Ohio.

Dickey, E. C., C. R. Fenster, R. H. Mickelson, and J. M. Laflen. 1984. Tillage and erosion in a wheat fallow rotation, pp. 183–195. *In*: Proc. of the Great Plains Conservation Tillage Symposium. North Platte, Nebraska, Aug. 21–23, 1984. GPE-2 Conservation Tillage Task Force, North Platte.

Dixit, S. P. 1982. Influence of pH on electrophoretic mobility of some soil colloids. Soil Sci. 133: 144–149.

Doolittle, J. J. and L. R. Hossner. 1988. Resources recoverable by surface mining, pp. 1–17. *In*: L. R. Hossner (ed.) Reclamation of surface-mined lands. Vol. 1. CRC Press, Boca Raton, Florida.

Doolittle, J. J., L. R. Hossner, and L. P. Wilding. 1994. Simulated aerobic pedogenesis in pyritic overburden with a positive acid-base account. Soil Sci. Soc. Am. J. 57: 1330–1336.

Doran J. W., D. G. Fraser, M. N. Culik, and W. C. Liebhardt. 1987. Influence of alternative and conventional agricultural management on soil microbial processes and nitrogen availability. Am. J. Altern. Agric. 2: 99–106.

Doran J. W. and T. B. Parkin. 1994. Defining and assessing soil quality, pp. 3–21. *In*: J. W. Doran, D. C. Coleman, D. F. Bezdicek, and B. A. Stewart (eds.). Defining soil quality for a sustainable environment. SSSA Spec. Publ. 35, Madison, Wisconsin.

Dudal, R. 1976. Inventory of major soils of the world with special reference to mineral stress hazards, pp. 3–13. *In*: M. J. Wright (ed.) Plant adaptation to mineral stress in problem soils. Cornell Univ. Press, Ithaca, New York.

Dudal, R. 1982. Land degradation in a world perspective. J. Soil Water Conserv. 37: 245–250.

Dumanski, J., H. Eswaran, and M. Latham. 1991. A proposal for an international framework for evaluating sustainable land management, pp. 25–49. *In*: J. Dumanski et al. (eds.) Evaluation for sustainable land management in the developing world. Vol. 2. Technical papers. Proc. Conf. 12, Bangkok, Thailand, Sept. 15–21, 1991. Int. Board Soil Res. Manag. (IB SRAM), Bangkok.

Eaglesham, A. R. J. 1982. Assessing the nitrogen contribution of cowpea in monoculture and intercropped, pp. 641–656. *In*: P. H. Graham and S. C. Harris (ed.) Biological nitrogen fixation technology for tropical agriculture. Paper presented at a workshop held at CIAT, March 9–13, 1981, CIAT, Cali, Colombia.

Eberhart, S. A. 1993. Plant genetic resources, pp. 51–61. *In*: INTSORMIL (ed.) Proceedings of a workshop on adaptation of plants to soil stresses. August 1–4, 1993. University of Nebraska, Lincoln.

Elliott, E. T. 1986. Aggregate structure and carbon, nitrogen, and phosphorus in native and cultivated soils. Soil Sci. Soc. Am. J. 50: 627–633.

Ellsworth, T. R., W. A. Jury, F. R. Ernstard, and P. J. Shouse. 1991. A three-dimensional field study of solute transport through unsaturated, layered porous media. I. Methodology, mass recovery and mean transport. Water Resour. Res. 27: 951–965.

Emmerich, W. E. and J. R. Cox. 1994. Changes in surface runoff and sediment production after repeated rangeland burns. Soil Sci. Soc. Am. J. 58: 199–203.

Epstein, E., J. D. Norlyn, D. W. Rush, R. W. Kingsbury, D. B. Kelley, G. A. Cunningham, and A. F. Wrona. 1980. Saline culture of crops: A genetic approach. Science 210: 399–404.

Eswaran, H., E. V. D. Berg, and P. Reich. 1993a. Organic carbon in soils of the world. Soil Sci. Soc. Am. J. 57: 192–194.

Eswaran, H., N. Bliss, D. Lytle, and D. Lammers. 1993b. Major soil regions of the world. USDA-SCS. U.S. Govt. Printing Office, Washington, DC.

Ewing, R. P. and S. C. Gupta. 1994. Pore-scale network modeling of compaction and filtration during surface sealing. Soil Sci. Soc. Am. J. 58: 712–720.

Fageria, N. K. 1980. Upland rice response to phosphate fertilization as affected by water deficiency in Cerrado soils. Pesq. Agropec. Bras. 15: 259–265.

Fageria, N. K. 1992. Maximizing Crop Yields. Marcel Dekker, New York.

Fageria, N. K. 1984. Response of rice cultivars to liming in Cerrado soil. Pesq. Agrop. Bras. 19: 883–889.

Fageria, N. K. 1994. Soil acidity affects availability of nitrogen, phosphorus and potassium. Better Crops International 10: 8–9. Atlanta, Georgia.

Fageria, N. K. and V. C. Baligar. 1993. Screening crop genotypes for mineral stresses, pp. 142–159. INTSORMIL (ed.) Proceedings of a workshop on adaptation of plants to soil stresses. August 1–4, 1993. University Nebraska, Lincoln.

Fageria, N. K., V. C. Baligar, and D. G. Edwards. 1990a. Soil-plant nutrient relationships at low pH stress, pp. 475–507. *In*: V. C. Baligar and R. R. Duncan (eds.) Crops as enhancers of nutrient use. Academic Press, San Diego.

Fageria, N. K., V. C. Baligar, and R. J. Wright. 1990b. Iron nutrition of plants: An overview on the chemistry and physiology of its deficiency and toxicity. Pesq. Agropec. Bras. 25: 553–570.

Fageria, N. K., V. C. Baligar, and R. J. Wright. 1989. Growth and nutrient concentration of alfalfa and common bean as influenced by soil acidity. Plant Soil 119: 331–333.

Fageria, N. K., V. C. Baligar, and R. J. Wright. 1988a. Aluminum toxicity in crop plants. J. Plant Nutr. 11: 303–319.

Fageria, N. K., V. C. Baligar, and R. J. Wright. 1991a. Influencing phosphate rock sources and rates on rice and common bean production in an Oxisol. Plant Soil 134: 137–144.

Fageria, N. K. and M. P. Barbosa Filho. 1987. Phosphorus fixation in Oxisol of Central Brazil. Fert. Agric. 94: 33–37.

Fageria, N. K., M. P. Barbosa Filho, and F. J. P. Zimmermann. 1994. Characterization of physical and chemical properties of lowland soils of some states of Brazil. Pesq. Agropec. Bras. 29: 267–274.

Fageria, N. K. and J. R. P. Carvalho. 1982. Influence of aluminum in nutrient solutions on chemical composition in upland rice cultivars. Plant Soil 69: 31–44.

Fageria, N. K., R. J. Wright, and V. C. Baligar. 1988b. Rice cultivar evaluation for phosphorus use efficiency. Plant Soil 111: 105–109.

Fageria, N. K., R. J. Wright, V. C. Baligar, and J. R. P. Carvalho. 1991b. Response of upland rice and common bean to liming on an Oxisol, pp. 519–525. *In*: R. J. Wright, V. C. Baligar, and R. P. Murrmann (eds.) Plant–soil interaction at low pH. Kluwer Academic Publishers, Dordrecht, The Netherlands.

Fan, M. X. and A. F. Mackenzie. 1994. Corn yield and phosphorus uptake with banded urea and phosphate mixtures. Soil Sci. Soc. Am. J. 58: 249–255.

FAO. 1978. A provisional methodology for soil degradation assessment. Rome. 84 p.

FAO. 1983. Guidelines for the control of soil degradation. Rome.

FAO. 1991. The production year book for 1990. Vol. 44. Rome.

Fardeau, J. C., P. Poss, and H. Saragoni. 1992. Effect of potassium fertilization on K-cycling in different agroecosystems. Paper presented at 23rd Colloquium of the Int. Potash Institute in Prague, Czech. Republic, October 11–16, 1992.

Farina, M. P. W. and P. Channon. 1990. Acid subsoil amelioration. I. A comparison of several mechanical procedures. Soil Sci. Soc. Am. J. 52: 169–175.

Fauci, M. F. and R. P. Dick. 1994. Soil microbial dynamics: Short and longterm effects of inorganic and organic nitrogen. Soil Sci. Soc. Am. J. 58: 801–806.

Faulkner, S. P. and W. H. Patrick, Jr. 1992. Redox processes and diagnostic wetland soil indicators in bottomland hardwood forests. Soil Sci. Soc. Am. J. 56: 856–865.

Fausey, N. R. and R. Lal. 1990. Soil wetness and anaerobiosis. Adv. Soil Sci. 11: 173–186.

Federal Interagency Committee for Wetland Delineation. 1989. Federal manual for identifying all jurisdictional wetlands. USACE, USEPA, USDA-FWS, and USDA-SCS. Cooperative Tech. Publ., U.S. Govt. Print. Office, Washington, DC.

Fleming, A. L. 1989. Enhanced Mn accumulation by snapbean cultivars under Fe stress. J. Plant Nutr. 12: 715–731.

Fohse, D., N. Classen, and A. Junk. 1991. Phosphorus efficiency of plants: II. Significance of root radius, root hairs and cation-anion balance for phosphorus influx in seven plant species. Plant Soil 132: 261–272.

Foster, R. C. 1981. Polysaccharides in soil fabrics. Science 214: 655–667.

Foster, H. L., A. R. Ahmad, and A. A. Ghani. 1980. The correction of soil acidity in Malaysian soils, pp. 157–171. *In*: Proceedings of the Conference on Soil Science and Agricultural Development in Malaysia. E. Pushparajah and C. S. Lock (eds.) Malaysian Society of Soil Science, Kula Lumpur, Malaysia.

Foy, C. D. 1984. Physiological effects of hydrogen, aluminum, and manganese toxicities in acid soils, pp. 57–97. *In*: F. Adams (ed.) Soil acidity and liming. 2nd ed. ASA, Madison, Wisconsin.

Foy, C. D. 1987. Acid soil tolerances of two wheat cultivars related to soil pH, KCl extractable aluminum, and aluminum saturation. J. Plant Nutr. 10: 609–623.

Foy, C. D. 1988. Plant adaptation to acid, aluminum toxic soils. Commun. Soil Sci. Plant Anal. 19: 959–987.

Foy, C. D. 1992. Soil chemical factors limiting plant root growth. Adv. Soil Sci. 19: 97–149.

Foy, C. D. and A. L. Fleming. 1982. Aluminum tolerance of two wheat cultivars related to nitrate reductase activities. J. Plant Nutr. 5: 1313–1333.

Franco, A. A. and D. N. Munns. 1982. Acidity and aluminum restraints on nodulation, nitrogen fixation, and growth of *Phaselous vulgaris* in solution culture. Soil Sci. Soc. Am. J. 46: 296–301.

Fullerton, S. and S. Pawluk. 1987. The role of seasonal salt and water fluxes in the genesis of solonetzic B horizons. Can J. Soil Sci. 67: 719–730.

Funderburk, J. E., F. M. Rhoads, and I. D. Feare. 1994. Modifying soil nutrient level affects soybean insect predators. Agron. J. 86: 581–585.

Garcia, F. O. and C. W. Rice. 1994. Microbial biomass dynamics in tallgrass prairie. Soil Sci. Soc. Am. J. 58: 816–823.

George, T., J. K. Ladha, D. P. Garrity, and R. J. Buresh. 1994. Legumes as nitrate catch crops during the dry-to-wet season transition in lowland rice based cropping systems. Agron. J. 86: 267–272.

Ghafar, Z. and A. K. Watson. 1983. Effect of corn seeding rate on growth of yellow nutsedge. Weed Sci. 31: 572–575.

Giovannini, G. and S. Lucchesi. 1983. Effects of fire on hydrophobic and cementing substances of soil aggregates. Soil Sci. 136: 231–235.

Glenn, A. R. and M. J. Dilworth. 1991. Soil acidity and growth of bacteria in low pH, pp. 567–579. *In*: R. J. Wright, V. C. Baligar, and R. D. Murrmann (eds.) Plant-soil interactions at low pH. Kluwer Academic Publishers, Dordrecht, The Netherlands.

Goedert, W. J. 1983. Management of Cerrado soils of Brazil: A review. J. Soil Sci. 34: 405–428.

Goldberg, S., B. S. Kapoor, and J. D. Rhoades. 1990. Effect of aluminum and iron oxides and organic matter on flocculation and dispersion of arid zone soils. Soil Sci. 150: 588–593.

Goodroad, L. L. and D. R. Keeney. 1984. Nitrous oxide emissions from soils during thawing. Can. J. Soil Sci. 64: 187–194.

Gregorich, E. G., R. G. Kachanoski, and R. P. Voroney. 1989. Carbon mineralization in soil size fractions after various amounts of aggregate disruption. J. Soil Sci. 40: 649–659.

Groenevelt, P. H., P. Vanstraaten, V. Rasiah, and J. Simpson. 1989. Modification in evaporation parameters by rock mulches. Soil Technol. 2: 279–285.

Grove, T. S., A. M. O'Connell, and G. M. Dimmock. 1986. Nutrient changes in surface soils after an intense fire in jarrah (*Eucalyptus marginata*) forest. Aust. J. Ecol. 11: 303–317.

Gupta, V. S. R. and J. J. Germida. 1988. Distribution of microbial biomass and its activity in different soil aggregate size classes as affected by cultivation. Soil Biol. Biochem. 20: 777–786.

Gupta, S. C., P. P. Sharma, and S. A. Defranchi. 1989. Compaction effects on soil structure. Adv. Agron. 42: 311–338.

Hai, T. V., T. T. Nga, and H. Laudelout. 1989. Effect of aluminum on the mineral nutrition of rice. Plant Soil 114: 173–185.

Hammond, R. R., R. A. Higgins, T. P. Mack, L. P. Pedigo, and E. J. Bechinski. 1991. Soybean pest management, pp. 341–472. *In*: D. Pimentel (ed.) Handbook of pest management in agriculture. 2nd ed., CRC Press, Boca Raton, Florida.

Hanway, J. J. and R. A. Olson. 1981. Phosphate nutrition of corn, sorghum, soybeans and small grains, pp. 671–692. *In*: F. E. Khasawneh (ed.) The role of phosphorus in agriculture. ASA, GSSA, and SSA. Madison, Wisconsin.

Harrison, P. H. 1984. Population, climate and future food supply. Ambio. 13: 161–167.

Hasegawa, P. M., R. A. Bressan, and A. K. Handa. 1986. Cellular mechanisms of salinity tolerance. Hort. Sci. 21: 1317–1324.

Havlin, J. L., D. E. Kissel, L. D. Maddux, M. M. Claassen, and J. H. Long. 1990. Crop rotation and tillage effects on soil carbon and nitrogen. Soil Sci. Soc. Am. J. 54: 448–452.

Higgs, R. L., W. H. Paulsen, J. W. Pendleton, A. F. Peterson, J. A. Jacobs, and W. D. Shrader. 1976. Crop rotations and nitrogen: Crop sequence comparisons on soils of the driftless area of Southwestern, Wisconsin. 1967–1974. Univ. Wisconsin College of Agric. and Life Sci. Res. Bull. R. 2461.

Hobbs, N. T., D. S. Schimel, C. E. Owensby, and D. S. Ojima. 1991. Fire and grazing in the tallgrass prairie: Contingent effects on nitrogen budgets. Ecology 72: 1374–1382.

Holden, P. W. 1986. Pesticides and groundwater quality: Issues and problems in four states. National Academic Press, Washington, DC.

Holland, E. A. and J. K. Detling. 1990. Plant response to herbivory and below ground nitrogen cycling. Ecology 71: 1040–1049.

Howler, R. H. 1991. Identifying plants adaptable to low pH conditions, pp. 885–904. *In*: R. J. Wright, V.C. Baligar, and R. P. Murrmann (eds.) Plant-soil interactions at low pH. Kluwer Academic Publishers, Dordrecht, The Netherlands.

Hudson, N. 1971. Soil Conservation. Cornell Univ. Press, Ithaca, New York.

Ibrikci, H., N. B. Comerford, E. A. Hanlon, and J. E. Rechcigl. 1994. Phosphorus uptake by Bahiagrass from Spodosols: Modeling of uptake from different horizons. Soil Sci. Soc. Am. J. 58: 139–143.

Ilyas, M., R. W. Miller, and R. H. Qureshi. 1993. Hydraulic conductivity of saline-sodic soil after gypsum application and cropping. Soil Sci. Soc. Am. J. 57: 1580–1585.

International Potash Institute. 1993. Food security and the role of fertilizers in sub-Saharan Africa 34: 1–5. Basel, Switzerland.

IRRI. 1974. International Rice Research Institute Report for 1973. IRRI, Los Banos, Philippines.

Jackson, L. E., J. P. Schimel, and M. K. Firestone. 1989. Short-term partitioning of ammonium and nitrate between plants and microbes in an annual grassland. Soil Biol. Biochem. 21: 409–415.

Jarvis, S. C. and A. D. Robson. 1983. A comparison of the cation/anion balance of ten cultivars of *Trifolium subterranean* L. and their effects on soil acidity. Plant Soil 75: 235–243.

Johnson, B. N., A. C. Kennedy, and A. G. Ogg, Jr. 1993. Suppression of downy brome growth by a Rhizobacterium in controlled environments. Soil Sci. Soc. Am. J. 57: 73–77.

Johnson, N. C., P. J. Copeland, R. K. Crookston, and F. L. Pfleger. 1992. Mycorrhizae: Possible explanation for yield decline with continuous corn and soybean. Agron. J. 84: 387–390.

Jokela, W. G. 1992. Nitrogen fertilizer and dairy manure effects on corn yield and soil nitrate. Soil Sci. Soc. Am. J. 56: 148–154.

Jones, O. R. and B. A. Stewart. 1990. Basin tillage. Soil Tillage Res. 18: 249–265.

Kamprath, E. J. 1970. Exchangeable aluminum as a criterion for liming leached mineral soils. Soil Sci. Soc. Am. Proc. 34: 252–284.

Kamprath, E. J. 1971. Potential detrimental effects from liming highly weathered soils to neutrality. Proc. Soil Crop Sci. Fla. 31: 200–203.

Kamprath, E. J. and C. D. Foy. 1985. Lime fertilizer plant interactions in acid soils, pp. 91–151. *In*: O. P. Englesford (ed.) Fertilizer technology and use. Third Ed. Soil Sci. Soc. Am., Madison, Wisconsin.

Kang, B. T., L. Reynolds, and A. M. Atta-krah. 1990. Alley farming. Adv. Agron. 43: 315–359.

Kay, B. D. 1990. Rates of change of soil structure under different cropping systems. Adv. Soil Sci. 12: 1–52.

Kennedy, A. C., L. F. Elliott, F. L. Young, and C. L. Douglas. 1991. Rhizobacteria suppressive to the weed downy brome. Soil Sci. Soc. Am. J. 55: 721–727.

Khanna, P. K. and R. J. Raison. 1986. Effect of fire intensity on solution chemistry of surface soil under a Eucalyptus Pauciflora forest. Aust. J. Soil Res. 24: 423–434.

Knapp, A. K. and R. R. Seastedt. 1986. Detritus accumulation limits productivity of tall grass prairie. Bioscience 36: 662–668.

Knight, R. W., W. H. Blackburn, and C. J. Scifus. 1983. Infiltration rates and sediment production following herbicide/fire brush treatments. J. Range Manag. 36: 154–157.

Konsten, C. J., N. V. Breemen, S. Suping, I. B. Aribawa, and J. E. Groenenberg. 1994. Effects of flooding on pH of rice producing, acid sulfate soils in Indonesia. Soil Sci. Soc. Am. J. 58: 871–883.

Krug, E. C. and C. R. Frink. 1983. Acid rain on acid soil: A new perspective. Science 221: 520–525.

Kumar, J. V. D. K., P.-J. Dart, and P. V. S. Subrahmanya. 1982. Residual effects of pigeonpea, pp. 659–663. *In*: P. H. Graham and S. C. Harris (eds.) Biological nitrogen fixation technology for tropical agriculture. Paper presented at a workshop held at CIAT, March 9–13, 1981, CIAT, Cali, Colombia.

Ladd, J. N., M. Amato, and J. M. Oades. 1985. Decomposition of plant material in Australian soils. III. Residual organic and microbial biomass. C and N from isotopic labeled plant material and organic matter decomposing under field conditions. Aust. J. Soil Res. 23: 603–611.

Lal, R. 1984. Soil erosion from tropical arable lands and its control. Adv. Agron. 37: 183–248.

Lal, R. 1986. Soil surface management in the tropics for intensive land use and high and sustained production. Adv. Soil Sci. 5: 2–109.

Lal, R. 1987. Effects of soil erosion on crop productivity. CRC Critical Rev. in Plant Sci. 5: 303–367.

Lal, R. 1988. Soil degradation and the future of agriculture in sub-Saharan Africa. J. Soil Water Cons. 43: 444–451.

Lal, R. 1989. Soil degradation in relation to climate, pp. 257–268. *In*: IRRI (ed.) Climate and food security. IRRI, Los Banos, Philippines.

Lal, R. 1991. Strategies for sustainable management of soil and water resources to enhance per capita productivity in the tropics. Scientific Liaison Officers to Int. Agric. Res. Centers, A.I.D., Washington, DC.

Lal, R., G. F. Hall, and F. P. Miller. 1989. Soil degradation. I. Basic Processes. Land Degradation and Rehabilitation 1: 51–69.

Lal, R., T. J. Logan, and N. R. Fausey. 1990. Long-term tillage effects on a Mollic Orchraqualf in northwest Ohio. III. Soil nutrient profile. Soil Tillage Res. 15: 371–382.

Lal, R., A. A. Mahboubi, and N. R. Fausey. 1994. Long-term tillage and rotation effects on properties of a central Ohio soil. Soil Sci. Soc. Am. J. 58: 517–522.

Larson, W. E. 1981. Protecting the soil resource base. J. Soil Water Cons. 36: 13–16.

Larson, W. E. 1986. The adequacy of world soil resources. Agron. J. 78: 221–225.

Lathwell, D. J. 1990. Legume green manures. Tropsoils Bulletin 90–01. North Carolina State University, Raleigh.

LeBissonais, Y. 1990. Experimental study and modeling of soil surface crusting processes, pp. 13–28. *In*: R. B. Bryant (ed.) Soil erosion experiments and models. Catena Suppl. 17. Catena Verlag, Cremlingen-Destedt, Germany.

Levesque, C. A. and J. E. Rahe. 1992. Herbicide interactions with fungal root pathogens with special reference to glyphosate. Annu. Rev. Phytopathol. 30: 579–602.

Levy, G. J., J. Levin, and I. Shainberg. 1994. Seal formation and interrill soil erosion. Soil Sci. Soc. Am. J. 58: 203–209.

Li, Y. and M. Ghodrati. 1994. Preferential transport of nitrate through soil columns containing root channels. Soil Sci. Soc. Am. J. 58: 653–659.

Lin, K. T., C. I. Lai, N. Ghafoori, and W. F. Chang. 1990. High-strength concrete utilizing industrial byproduct, pp. 456–472. *In*: W. F. Chang (ed.) Proceedings, 3rd Int. Symposium on Phosphogypsum. Publ. 01–060–083. Florida.

Lizhi, C. 1988. Green manure cultivation and use for rice in China, pp. 63–70. *In*: IRRI (ed.) Green manure in rice farming. Los Banos, Philippines.

Lloyd-Reilley, J., C. J. Scifres, and W. H. Blackburn. 1984. Hydrologic impacts of brush management with tibuthiuron and prescribed burning on post oak Savannah watersheds in Texas. Agric. Ecosys. Environ. 11: 213–224.

Luxmoore, R. J. 1991. Preferential flow and its measurement, pp. 113–121. *In*: T. J. Gish and A. Shirmohammadi (eds.) Preferential Flow. Proc. Symp. Chicago. Dec. 16–17, 1991. ASAE, St. Joseph, Michigan.

Maas, E. V. 1993. Testing crops for salinity tolerance, pp. 234–247. *In*: INTSORMIL (ed.) Proceedings of a Workshop on Adaptation of Plants to Soil Stresses. August 1–4, 1993. University Nebraska, Lincoln.

Mabbutt, J. A. 1984. The impact of desertification as revealed by mapping. Environ. Conserv. 11: 103–113.

Mahler, R. L., F. E. Koehler, and L. K. Lutcher. 1994. Nitrogen source, timing of application and placement: Effects on winter wheat production. Agron. J. 86: 637–642.

Malavolta, E. 1991. Agricultural gypsum for plant nutrition and environments. Questions and answers. Mimeograph. Piracicaba, Brazil.

Mann, L. K. 1985. A regional comparison of carbon in cultivated and uncultivated Alfisols and Mollisols in the central United States. Geoderma 36: 241–253.

Matzner, E. 1989. Acidic precipitation: Case study Solling, pp. 39–83. *In*: D. C. Adriano and M. Havas (eds.) Acidic precipitation. Vol. 1. Springer-Verlag, Berlin.

McGill, W. B., K. R. Cannon, J. A. Robertson, and F. D. Cook. 1986. Dynamics of soil microbial biomass and water soluble organic C in Bretonl after 50 years of cropping to two rotations. Can. J. Soil Sci. 66: 1–19.

McIntyre, D. S. 1958a. Permeability measurement of soil crusts formed by raindrop impact. Soil Sci. 85: 185–189.

McIntyre, D. S. 1958b. Soil splash and the formation of surface crusts by raindrop. Soil Sci. 85: 261–266.

Megonigal, J. P., W. H. Patrick, Jr., and S. P. Faulkner. 1993. Wetland identification in seasonally flooded forest soils: Soil morphology and redox dynamics. Soil Sci. Soc. Am. J. 57: 140–149.

Mengel, D. B. and E. J. Kamprath. 1978. Effect of soil pH and liming on growth and nodulation of soybeans in Histosols. Agron. J. 70: 959–963.

Mengel, K. 1993. Impact of intensive agriculture on resources and environment, pp. 613–617. *In*: M. A. C. Fragosa and M. L. Van Beusichem (eds.) Optimizing of plant nutrition. Kluwer Academic Publishers, Dordrecht, The Netherlands.

Mikeni, P. N. S. and A. F. Mackenzie. 1987. Effect of added organic residues in four Quebec soils. Plant Soil 104: 163–167.

Miller, M. H., C. P. Mamaril, and G. J. Blair. 1970. Ammonium effects on phosphorus absorption through pH changes and phosphorus precipitated at the soil-root interface. Agron. J. 62: 524–527.

Miller, W. P., H. Frenkel, and K. D. Newman. 1990. Flocculation concentrations and sodium/calcium exchange of kaolinitic soil clays. Soil Sci. Soc. Am. J. 54: 346–351.

Mughogho, S. K. J. Awai, H. S. Lowendorf, and D. J. Lathwell. 1982. The effects of fertilizer nitrogen and *Rhizobium* inoculation on yield of cowpea and subsequent crops of maize, pp. 297–301. *In*: P. H. Graham and S. C. Harris (eds.) Biological nitrogen fixation technology for tropical agriculture. Paper presented at a workshop held at CIAT, March 9–13, 1981, Cali, Columbia.

Munns, D. N., J. S. Hohenberg, T. L. Righetti, and D. J. Lauter. 1981. Soil acidity tolerance of symbiotic and nitrogen-fertilized soybeans. Agron. J. 73: 407–410.

Myrold, D. D. 1990. Effects of acidic deposition on soil organisms, pp. 163–187. *In*: A. A. Lucier and S. G. Haines (eds.) Mechanisms of forest response to acidic deposition. Springer-Verlag, Berlin.

Nedeco/Euroconsult. 1984. Nationwide study of coastal and near coastal swamp land in Sumatra, Kalimanton, and Irianjaya. Final report, Vol. 1; Main report. Nedeco/Euroconsult, Arnheim, The Netherlands.

Nobel, A. D., M. V. Fey, and M. E. Sumner. 1988. Calcium aluminum balance and the growth of soybean roots in nutrient solutions. Soil Sci. Soc. Am. J. 52: 1651–1658.

Nowak, P. J., J. Timmons, J. Carlson, and R. Miles. 1985. Economic and social perspectives on T-values related to soil erosion and crop productivity. *In*: R. F. Follett and B. A. Stewart (eds.) Soil erosion and crop productivity. ASA, Madison, Wisconsin.

Nye, P. D., D. Craig, N. T. Coleman, and J. L. Ragland. 1961. Ion exchange equilibrium involving aluminum. Soil Sci Soc. Am. Proc. 25: 14–17.

Oades, J. M. 1984. Soil organic matter and structure stability: Mechanisms of implications for management. Plant Soil 76: 319–337.

Ojima, D. S. 1987. The short-term and long-term effects of burning on tallgrass ecosystem properties and dynamics. Ph.D. Diss., Colorado State Univ., Ft. Collins (Diss. Abstr. 87–25646).

Parton, W. J., D. S. Schimel, C. V. Cole, and D. S. Ojima. 1987. Analysis of factors controlling soil organic matter levels in Great Plains grassland. Soil Sci. Soc. Am. J. 51: 1173–1179.

Paul, E. A. 1991. Decomposition of organic matter. *In*: J. Lederburg (ed.) Encyclopedia of microbiology. Academic Press, San Diego, California.

Paustian, K., W. J. Parton, and J. Persson. 1992. Modeling soil organic matter in organic-amended and nitrogen-fertilized long-term plots. Soil Sci. Soc. Am. J. 56: 476–488.

Pearson, R. W., R. P. Escolar, F. Abruna, Z. F. Lund, and E. J. Brenes. 1977. Comparative responses of three crop species to liming several soils of the Southeastern United States and of Puerto Rico. J. Agric. Univ. Puerto Rico 61: 361–382.

Pierce, F. J. and C. W. Rice. 1988. Crop rotation and its impact on efficiency of water and nitrogen use, pp. 21–42. *In*: W. L. Hargrove (ed.) Cropping strategies for efficient use of water and nitrogen. ASA Special Publ. No. 51. ASA, CSSA, and SSSA, Madison, Wisconsin.

Pitelka, L. F. and D. J. Raynal. 1989. Forest decline and acidic deposition. Ecology 70: 2–10.

Potter, L. D. 1992. Desert characteristics as related to waste disposal, pp. 21–56. *In*: C. C. Reith and B. M. Thomsom (eds.) Deserts and dumps. Univ. New Mexico, Press.

Power, J. F. 1990. Legumes and crop rotations, pp. 178–204. *In*: C. A. Francis et al. (eds.) Sustainable agriculture in temperate zone. Wiley, New York.

Putnam, A. R. and W. B. Duke. 1978. Allelopathy in agroecosystems. Ann. Rev. Phytopathol. 16: 431–451.

Radcliffe, D. E., J. L. Clark, and M. E. Sumner. 1986. Effect of gypsum and deep-rooting perennials on subsoil mechanical impedance. Soil Sci. Soc. Am. J. 50: 1566–1570.

Radke, J. K., D. C. Reicosky, and W. B. Voorhees. 1993. Laboratory simulation of temperature and hydraulic head variations under a soil ridge. Soil Sci. Soc. Am. J. 57: 652–660.

Raij, B. V. 1991. Soil fertility and fertilization. Editoria Agronomica Ceres, São Paulo.

Raison, R. J. 1979. Modification of the soil environment by vegetation fires, with particular reference to nitrogen transformation. A review. Plant Soil 51: 73–108.

Rasmussen, P. E. and H. P. Collins. 1991. Long-term impacts of tillage, fertilizer and crop residue on soil organic matter in temperate semiarid regions. Adv. Agron. 45: 93–134.

Rasmussen, P. E. and W. J. Parton. 1994. Long-term effects of residue management in wheat-fallow. I. Inputs, yield and organic matter. Soil Sci. Soc. Am. J. 58: 523–530.

Rice, E. L. 1974. Allelopathy. Academic Press. New York.

Rice, E. L. 1979. Allelopathy: An update. Bot. Rev. 45: 15–109.

Riley, D. and S. A. Barber. 1971. Effect of ammonium and nitrate fertilization on phosphorus uptake as related to root-induced pH changes at the root-soil interface. Soil Sci. Soc. Am. Proc. 35: 301–306.

Robarge, W. P. and D. W. Johnson. 1992. The effects of acidic deposition on forested soils. Adv. Agron. 47: 1–83.

Russell, G. E. 1978. Plant breeding for pest and disease resistance. Butterworths, London.

Ryan, J. N. and P. M. Gschwend. 1990. Colloid mobilizing in two Atlantic Coastal Plain aquifers: Field studies. Water Resour. Res. 26: 307–322.

Salinas, J. G. and P. A. Sanchez. 1977. Tolerance to Al-toxicity and low available P, pp. 115–137. *In*: Agronomic-economic research in the tropics. Annual report for 1976–1977. North Carolina State Univ., Raleigh.

Sanchez, P. A. 1976. Properties and management of soils in the tropics. Wiley, New York.

Sanchez, P. A. 1987. Management of acid soils in the humid tropics of Latin America, pp. 63–107. *In*: Management of acid tropical soils for sustain-able agriculture. IBSRAM Proc. No. 2, Bangkok.

Sanchez, P. A., C. A. Palm, and T. J. Smyth. 1991. Alternative to tropical deforestation, pp. 242–246. *In*: Tropsoils technical report 1988–1989. North Carolina State Univ., Raleigh.

Sanchez, P. A. and J. G. Salinas. 1981. Low-input technology for managing Oxisols and Ultisols in tropical America. Adv. Agron. 34: 279–406.

Sartain, J. B. and E. J. Kamprath. 1975. Effects of liming a highly Al-saturated soil on the top and root growth and soybean nodulation. Agron. J. 67: 507–510.

Sayre, R. M. and D. E. Walter. 1991. Factors affecting the efficacy of natural enemies of nematodes. Ann. Rev. Phytopathol. 29: 149–166.

Schaffert, R. E. 1993. Discipline interaction in the quest to adapt plants to soil stresses through genetic improvement, pp. 1–13. *In*: Proceedings of the workshop on adaptation of plants to soil stresses. August 1–4, 1993, Univ. Nebraska, Lincoln.

Schimel, D. S. 1986. Carbon and nitrogen turnover in adjacent grassland and cropland ecosystem. Biogeochemistry 2: 345–357.

Schimel, D. S., D. C. Coleman, and K. A. Horton. 1989. Soil organic matter dynamics in prairie rangeland and cropland toposequences in North Dakota. Geoderma 36: 201–214.

Schippers, B., A. W. Bakker, and M. M. Bakker. 1987. Interactions of deleterious and beneficial rhizosphere microorganisms and the effect of cropping practices. Ann. Rev. Phytopathol. 25: 339–359.

Schnepf, M. 1979. Farmland, food and the future. Soil Cons. Soc. Am., Ankeny, Iowa.

Schomberg, H. H., J. L. Steiner, and P. W. Unger. 1994. Decomposition and nitrogen dynamics of crop residues. Residue quantity and water effects. Soil Sci. Soc. Am. J. 58: 372–381.

Schubert, E., K. Mengel, and S. Schubert. 1990. Soil pH and calcium effect on nitrogen fixation and growth of broadbean. Agron. J. 92: 969–972.

Schuman, G. E. and J. L. Meining. 1993. Short-term effects of surface-applied gypsum on revegetated sodic bentonite spoils. Soil Sci. Soc. Am. J. 57: 1083–1088.

Schuman, G. E. and T. A. Sedbrook. 1984. Sawmill wood residue for reclaiming bentonite spoils. For. Prod. J. 34: 65–68.

Schwertmann, U. and H. Fetcher. 1984. The influence of aluminum on iron oxides: XI. Aluminum substituted maghemite in soils and its formation. Soil Sci. Soc. Am. J. 48: 1462–1463.

Seelig, B. D. and J. L. Richardson. 1994. Sodic soil toposequence related to focused water flow. Soil Sci. Soc. Am. J. 58: 156–163.

Sertsu, S. M. and P. A. Sanchez. 1978. Effects of heating on some changes in soil properties in relation to an Ethiopian land management practice. Soil Sci. Soc. Am. J. 42: 940–944.

Shainberg, I., M. E. Sumner, W. P. Miller, M. P. W. Farina, M. A. Pavan, and M. V. Fey. 1989. Use of gypsum on soils: A review. Adv. Soil Sci. 9: 1–111.

Sharma, R. D., J. Pereira, and D. V. S. Resck. 1982. Green manure efficiency in nematodes control associated with soybean in Cerrado. EMBRAPA-CPAC, Res. Bull. 13, Planaltina, Brazil.

Shaxson, T. F. 1993. Improving productive potential of tropical soils. Paper presented at 24th Brazilian Soil Sci. Congress, July 25–31, 1993, Golante-GO, Brazil.

Smith, A. E. 1991. The potential importance of allelopathy in the pasture ecosystem: A review. Adv. Agron. 1: 27–37.

Smith, A. E. and L. D. Martin. 1994. Allelopathic characteristics of three cool-season grass species in the forage ecosystem. Agron. J. 86: 243–246.

Smith, H. J. C., G. J. Levy, and I. Shainberg. 1990. Water droplet energy and soil amendments: Effect on infiltration and erosion. Soil Sci. Soc. Am. J. 54: 1084–1087.

Smith J. L., J. J. Halvorson, and R. I. Papendick. 1993. Using multiple-variable indicator kriging for evaluating soil quality. Soil Sci. Soc. Am. J. 57: 743–749.

Smith, L. L. and L. F. Elliott. 1990. Tillage and residue management effects on soil organic matter dynamics in semiarid region. Adv. Soil Sci. 13: 69–80.

Smith, S. J., A. N. Sharpley, J. W. Naney, W. A. Berg, and O. R. Jones. 1991. Water quality impacts associated with wheat culture in the Southern Plains. J. Environ. Qual. 20: 244–249.

Smyth, T. J. and M. S. Cravo. 1991. Continuous cropping experiment in Manaus, pp. 252–254. *In*: Tropsoils Technical Report, 1988–1989. North Carolina State University, Raleigh.

Smyth, T. J. and M. S. Cravo. 1992. Aluminum and calcium constraints to continuous crop production in a Brazilian Amazon Oxisol. Agron. J. 84: 843–850.

Soil Conservation Service. 1991. Hydric-soils of the United States. U.S. Govt. Printing Office, Washington, DC.

Southard, R. J., I. Shainberg, and M. J. Singer. 1988. Influence of electrolyte concentration on the micromorphology of artificial depositional crust. Soil Sci. 145: 278–286.

Souza, D. M. G. 1988. Use of gypsum as fertilizer. Paper presented at the seminar on sugarcane agricultural industry of Pernambuco, Paraiba, and Rio Grande do Norte, Natal, Brazil.

Stephens, D. B. 1994. A perspective on diffuse natural recharge mechanisms in areas of low precipitation. Soil Sci. Soc. Am. J. 58: 40–48.

Stephens, L. E. and R. R. Johnson. 1993. Soil strength in the seed zone of several planting systems. Soil Sci. Soc. Am. J. 57: 481–489.

Stevenson, F. J. 1982a. Humus chemistry: Genesis, Composition, Reactions. Wiley, New York.

Stevenson, F. J. 1982b. Origin and distribution of nitrogen in soil. *In*: F. J. Stevenson (ed.) Nitrogen in agricultural soils. Agron. Monogr. 22. ASA, CSSA, and SSSA, Madison, Wisconsin.

Stevenson, F. J. 1986. Cycles of Soil Carbon, Nitrogen, Phosphorus, Sulfur, Micronutrients. Wiley, New York.

Sumner, M. E. 1992. The electric double layer and clay dispersion. *In*: M. E. Sumner and B. A. Stewart (eds.) Soil crusting: Chemical and physical processes. Lewis Publ., Chelsea, Michigan.

Sumner, M. E., D. F. Radcliffe, M. McCray, and R. L. Clark. 1990. Gypsum as an ameliorant for subsoil hardpans. Soil Technol. 3: 253–258.

Sutton, P. and W. A. Dick. 1987. Reclamation of acid mind lands in humid areas. Adv. Agron. 41: 377–405.

Tarchitzky, J., A. Banin, J. Morin, and Y. Chen. 1984. Nature, formation, and effects of soil crusts formed by water drop impact. Geoderma 33: 135–155.

Taylor, H. M. 1974. Root behavior as affected by soil structure and strength, pp. 271–291. *In*: E. W. Carson (ed.) The plant root and its environment. Univ. Press of Virginia, Charlottesville.

Thurow, T. L., W. H. Blackburn, and C. A. Taylor, Jr. 1986. Hydrologic characteristics of vegetation types as affected by livestock grazing systems. Edwards Plateau, Texas. J. Range Manag. 39: 505–509.

Tiessen, H., J. W. B. Stewart, and J. R. Betany. 1982. Cultivation effects on the amounts and concentration of carbon, nitrogen, and phosphorus in grassland soils. Agron. J. 74: 831–835.

Tisdall, J. M. and J. M. Oades. 1982. Organic matter and water stable aggregations in soils. J. Soil Sci. 33: 141–161.

Tollenaar, M. 1992. Is low plant density a stress in maize. Maydica. 37: 305–311.

Tollenaar, M., A. A. Dibo, A. Aguilera, S. F. Weise, and C. J. Swanton. 1994a. Effect of crop density on weed interference in maize. Agron. J. 86: 591–595.

Tollenaar, M., S. P. Nissanka, A. Aguilera, S. F. Weise, and C. J. Swanton. 1994b. Effect of weed interference and soil nitrogen on four maize hybrids. Agron. J. 86: 596–601.

Turner, C. L., T. R. Seastedt, and M. I. Dyer. 1993. Maximizing of above-ground grassland production: The role of defoliation frequency, intensity, and history. Ecol. Appl. 3: 175–186.

Ueckert, D. N., T. L. Whigham, and B. M. Spears. 1978. Effects of burning on infiltration, sediment, and other soil properties in a mesquite-tobosa-grass community. J. Range Manag. 31: 420–425.

Uhlen, G. 1991. Long-term effects of fertilizers, manure, straw and crop rotation on total-N and total-C in soil. Acta. Agric. Scand. 41: 119–127.

Ulery, A. L. and R. C. Graham. 1993. Forest fire effects on soil color and texture. Soil Sci. Am. J. 57: 135–140.

Unger, P. W. 1984. Tillage and residue effects on wheat, sorghum, and sunflower grown in rotation. Soil Sci. Soc. Am. J. 48: 885–891.

Unger, P. W. 1992. Ridge height and furrow blocking effects on water use and grain yield. Soil Sci. Soc. Am. J. 56: 1609–1614.

Unger, P. W. 1994. Residue management for winter wheat and grain sorghum production with limited irrigation. Soil Sci. Soc. Am. J. 58: 537–542.

Unger, P. W. and T. M. McCalla. 1980. Conservation tillage systems. Adv. Agron. 33: 1–58.

United Nations Environment Program (UNEP). 1982. Worlds Soil Policy. Nairobi, Kenya.

USDA. 1981. National agricultural land study. Final Report of Council on Environmental Quality. Washington, DC.

USDA. 1987. Farm drainage in the United States: History and prospects. G. A. Pavelis (ed.) Misc. Publ. No. 1455.

USDA. 1992. Agricultural resources situation and outlook. AR-25. Resources and Technol. Div. Economic Res. Serv., Washington, DC.

Van Breemen, N. and T. C. J. Feijtel. 1990. Soil processes and properties involved in the production of greenhouse gases with special relevance to soil taxonomic systems, pp. 195–220. *In*: A. F. Bouman (ed.) Soils and the greenhouse effect. Wiley, Chichester, England.

Van Riemsdijk, W. H., L. J. M. Bauman, and F. A. M. Haan. 1984. Phosphorus sorption by soils. I. A model for phosphate reaction with metal-oxide in soil. Soil Sci. Soc. Am. J. 48: 537–541.

Van Veen, J. A. I., J. H. Ladd, and M. J. Frissel. 1984. Modelling of C and N turnover through the microbial biomass in soil. Plant Soil 76: 257–274.

Vaughan, D. and B. G. Ord. 1991. Influence of natural and synthetic humic substances on the activity of urease. J. Soil Sci. 42: 17–23.

Veldkamp, E. 1994. Organic carbon turnover in three tropical soils under pasture after deforestation. Soil Sci. Soc. Am. J. 58: 175–180.

Wade, M. K., D. W. Gill, M. Subagjo, M. Sudjadi, and P. A. Sanchez. 1989. Overcoming soil fertility constraints in a transmigration area of Indonesia. Tropsoils Bull. 88–01, North Carolina State University, Raleigh.

Walker, P. T. 1975. Pest control problems causing major losses in world food supply. FAO Plant Protection Bull. 23, pp. 70–77.

Warkentin, B. P. 1994. The discipline of soil science: How should it be organized. Soil Sci. Soc. Am. J. 58: 267–268.

Warrington, D., I. Shainberg, M. Agassi, and J. Morin. 1989. Slope and phosphogypsum effects on runoff and erosion. Soil Sci. Soc. Am. J. 53: 1201–1205.

Watson, D. A. and J. M. Laflen. 1986. Soil strength, slope and rainfall intensity effects on interrill erosion. Trans. ASAE 29: 98–102.

Weil, R. R. 1982. Maize-weed competition and soil erosion in unweeded maize. Trop. Agric. (Trinidad) 59: 207–213.

Welch, L. F. 1976. The Morrow plots: Hundred years of research. Ann. Agron. 27: 881–890.

White, J. A., G. J. Depuit, J. L. Smith, and S. E. Williams. 1992. Vesicular arbuscular mycorrhizal fungi and irrigated mined land reclamation in Southwestern Wyoming. Soil Sci. Soc. Am. J. 56: 1464–1469.

Wilhelm, W. W., J. W. Doran, and J. F. Power. 1986. Corn and soybean yield response to crop residue management under no-tillage production systems. Agron. J. 78: 184–189.

Wischmeier, W. H. and D. D. Smith. 1978. Predicting rainfall erosion losses. A guide to conservation planning. USDA Agric. Handbook No. 537, U.S. Govt. Printing Office, Washington, DC.

Wood, C. W., D. G. Westfall, G. A. Peterson, and I. C. Burke. 1990. Impacts of cropping intensity on potential C and N mineralization in no-till dryland agroecosystems. Agron. J. 82: 1115–1120.

Woodmansee, R. A., I. Vallis, and J. J. Mott. 1981. Grassland nitrogen. Ecol. Bull. (Stockholm) 33: 443–462.

Wright, H. A. and A. W. Bailey. 1982. Fire Ecology, United States and Southern Canada. Wiley, New York.

Wright, R. J., V. C. Baligar, , and S. F. Wright. 1988. Estimation of plant available manganese in acidic subsoil horizons. Comm. Soil Sci. Plant Anal. 19: 643–662.

Wright, R. J., V. C. Baligar, and S. F. Wright. 1987. The influence of acid soil factors on the growth of snapbeans in major Appalachian soils. Commun. Soil Sci. Plant Anal. 18: 1235–1252.

Yost, R. S., E. J. Kamprath, E. Lobato, and G. C. Naderman. 1979. Phosphorus response of corn on an Oxisol as influenced by rates and placement. Soil Sci. Soc. Am. J. 43: 338–343.

Young, F. L., A. G. Ogg, Jr., R. I. Papendick, D.C. Thill, and J.R. Alldredge. 1994. Tillage and weed management affects winter wheat yield in an integrated pest management system. Agron. J. 86: 147–154.

Young, R. A. and J. L. Wiersma. 1973. The role of rainfall impact in soil detachment and transport. Water Resour. Res. 9: 1629–1636.

6

The Role of Essential Nutrients on Plant Diseases

I. INTRODUCTION

Changing tillage practices, certain climatic conditions, monoculture cropping systems, and year-round cultivation provide for an environment conducive to pathogenic organism build-up, leading to significant disease influence upon crop yield and/or crop quality. Protection measures vary widely in availability, effectiveness, and economic feasibility. In some cases pesticides are available and can be economically utilized. Crop rotations are possible in some cases. Genetic resistance to specific disease organisms exists for some crops. Natural plant resistance due to balanced nutrition is another alternative (Usherwood, 1980).

Plant diseases are greatly influenced by environmental factors, including deficiencies and/or toxicities of essential nutrients. The effects of nutrients on disease may be attributed to: 1) effects on plant growth that can influence the microclimate in a crop and thereby affect infection and sporulation of the pathogen, 2) effects on cell walls and tissues, as well as on the biochemical composition of the host, 3) the rate of growth of the host, which may enable

Table 6.1 Classification of Crop Losses in the Economic and Social Spheres as Affected by Plant Disease

Direct loss		
Primary	Secondary	Indirect loss
Yield	Contamination of sowing and plant material	Farm
Quality	Soilborne diseases	Rural municipality
Costs of control	Weakening of trees by premature defoliation	Exporters
Extra cost of harvesting	Costs of control	Traders
Extra cost of grading		Consumers
Costs of replanting		Government
Less profitable replacement crops		Environment

Source: Zadoks and Schein, 1979.

seedlings to escape infection in their most susceptible stage, and 4) effects on the pathogen through alterations in the soil environment (Colhoun, 1973).

Balanced nutrition has an important role in determining plant resistance or susceptibility to diseases. The capacity of plants to be protected from diseases is influenced by the health of the plant and its stage of phenological development. A severely nutrient-stressed plant is often more susceptible to disease than one at a nutritional optimum, yet plants receiving a large excess of a required mineral may become predisposed to disease (Piening, 1989). Mineral elements are directly involved in all mechanisms of plant defense as integral components of cells, substrates, enzymes, and electron carriers, or as activators, inhibitors, and regulators of metabolism (Huber, 1980).

Crop loss is the reduction of quantity and/or quality of yield. Crop losses resulting from diseases, insects, and weeds probably constitute the most significant constraint worldwide to increasing food production. Losses are the result of changes in structure or condition of a crop to an extent that restoration of yield and/or quality is irreversible. Zadoks and Schein (1979) suggested a classification that describes the complexity and interdependence of losses at all strata of society (Table 6.1). Figure 6.1 shows the potential yield level in an optimum environment and a lower yield level under environmental stress. Diseases and essential nutrients are the two important factors for this difference in yield. Preharvest losses caused by diseases, insects, and weeds have been estimated at 35%, whereas postharvest losses and wastage are

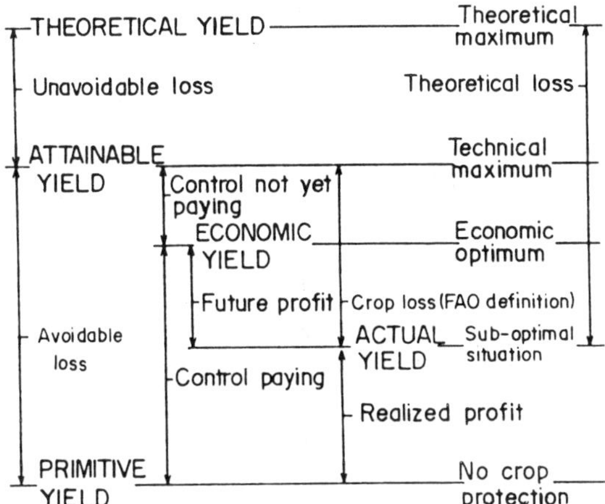

Figure 6.1 Yield levels used to explain the conceptual basis for crop losses. (From Zadoks and Schein, 1979.)

reported at 30% (James, 1980). This means that, potentially, most world food problems could be eliminated just by reducing the amount of loss caused by diseases, insects, and weeds.

By effecting changes in growth pattern, plant morphology and anatomy, and particularly chemical composition, mineral nutrients may either increase or decrease the resistance of plants to pathogens (Marschner, 1995; see also Huber and Watson, 1974; Perrenoud, 1977; Huber, 1980; Graham, 1983; Huber and Arny, 1985; and Huber and Wilhelm, 1988). Information provided herein will help in planning better strategies of disease control and, consequently, will improve crop production.

II. NITROGEN

The evidence that excessive nitrogen fertilization increases susceptibility of host plants to obligate pathogens (e.g., rust, powdery mildew, club root) seems to be conclusive, although the form of nitrogen available to the plant may also be significant (Huber and Watson, 1974; Kiraly, 1976). However, high levels of nitrogen usually increase resistance to facultative pathogens in fresh, green, young plant tissues (Kiraly, 1976). Table 6.2 shows the effects of low and high nitrogen on diseases caused by obligate and facultative pathogens. These differences in response result from differences in the nutritional requirements of the two types of pathogens. Obligate pathogens rely on assimilates supplied by living cells. Facultative pathogens, on the other hand, are

Table 6.2 Effects of Low and High N Concentrations on Disease Severity[a]

Pathogen or disease		Low N	High N
Obligate parasites	*Puccinia graminis*	*	***
	Erysiphe graminis	*	***
	Plasmodiophora brassicae	*	*
	Tobacco mosaic virus	*	***
Facultative parasites	*Xanthomonas vesicatoria*	***	*
	Alternaria solani	***	*
	Fusarium oxyporum	***	*

[a] Disease severity: ***, high; *, low.
Source: Kiraly, 1976.

semisaprophytes which prefer senescing tissue or which release toxins in order to damage or kill the host plant cells. As a rule, all factors which support the metabolic and synthetic activities of host cells and delay senescence of the host plant also increase resistance to facultative pathogens (Marschner, 1995).

The form of nitrogen (N) has an important role in plant disease. A critical NH_4^+-N: NO_3^--N ratio for take-all suppression of 3:1 was estimated from data in the literature (Christensen and Brett, 1985). Take-all severity was negatively correlated (r^2 = 0.84) with the length of time the NH_4^+-N: NO_3^--N ratio remained above the estimated critical ratio.

Urea applied to the foliage of wheat reduced levels of *Septoria tritici*, *S. nodorum*, powdery mildew (*Erysiphe graminis*), and brown rust (*Puccinia recondita*) in some experiments (Gooding et al., 1988; Peltonen et al., 1991). Increased leaf nitrogen content can improve resistance to *Septoria* spp. infection (Zadoks and Schein, 1979), and this is consistent with the finding that increasing soil applications of granular ammonium nitrate in the spring can also sometimes lead to lower levels of late-season *S. tritici* (Gooding and Davies, 1992). Additionally, spore germination and colony growth of *S. nodorum* on agar is inhibited by urea addition (6%), and scanning electron micrographs also suggest inhibition of spore germination on the surface of wheat leaves (Peltonen et al., 1991). Furthermore, foliar urea applications influence microflora populations on the surface of cereal leaves, and these can interact with cereal pathogens.

Sometimes foliar urea applications have increased the severity of *Botrytis cinerea* (Gooding et al., 1988) and *S. nodorum* infection. It is suggested that damage by urea sprays to leaf tissue is likely to encourage secondary invasion of the scorched areas by certain pathogens, where the subsequent microclimate encourages these infections. Increased severity of *S. nodorum* appears to be more likely when applications of urea are made subsequent to infection (Gooding and Davies, 1992). Table 6.3 shows the influence of soil pH and nitrogen form on plant diseases.

Various factors influence the effect that a specific form of N will have on disease. Because no one form of N controls all diseases or favors disease control on any group of plants, each disease must be considered individually. Disease control achieved with a specific form of N may depend on several factors, including host response, previous crop, nitrogen rate, residual N, time of application, soil microflora, ratio of NH_4^+–N to NO_3^-–N, and the disease complex present (Huber and Watson, 1974). This subject is very complex, and more research is needed to arrive at definite conclusions.

III. PHOSPHORUS

Although phosphorus (P) is involved in organic compounds and metabolic processes vitally important to the plant, its role in disease resistance is variable and seemingly inconsistent (Kiraly, 1976). Application of P is most beneficial in reducing seedlings and other fungal diseases where vigorous root development permits plants to escape disease. Phosphate fertilization of wheat has almost eliminated economic losses from *Pythium* root rot in the central wheat growing area in the United States (Huber, 1980). In a survey of Illinois field research, fertilizer P reduced cob rot of corn grown on soils deficient in P. This influence was noted where the casual organism was *Fusarium*. Other studies reveal that P can diminish the incidence of boil smut of corn (Potash and Phosphate Institute, 1988). Application of P reduced bacterial leaf blight on rice; downy mildew, blue mold, and leaf curl virus diseases in tobacco; pod and stem blight in soybean; yellow dwarf virus disease of barley; brown stripe disease in sugarcane; and blast disease in rice (Potash and Phosphate Institute, 1988). In contrast, applications of P may increase the severity of diseases caused by *Sclerotinia* on many garden plants, *Bremia* on lettuce, and flag smut on wheat (Huber, 1980).

IV. POTASSIUM

Although there is a large amount of literature on the relationship between potassium (K) and plant diseases, there is little quantitative information available on the concentration of K in soil or plant tissues that result in the observed effects on disease expression (Huber and Arny, 1985). Generally, K fertilization reduces the intensity of several infectious diseases, and this occurs with diseases caused by obligate as well as by facultative parasites (Table 6.4). Potassium frequently reduces the incidence of, or damage from, various diseases such as bacterial leaf blight; sheath blight, stem rot, and sesamum leaf spot in rice;

Table 6.3 Influence of Soil pH and Nitrogen Form on Plant Diseases

		pH		Nitrogen form	
Plant	Disease	Low	High	Ammonia	Nitrate
Cereals	Take-all	Decrease	Increase	Decrease	Increase
Cotton	*Phymatotrichum* root rot	Decrease	Increase	Decrease	Increase
Cotton	*Verticillium* wilt	Decrease	Increase	Decrease	Increase
Eggplant	*Verticillium* wilt	Decrease	Increase	Decrease	Increase
Potato	*Verticillium* wilt	Decrease	Increase	Decrease	Increase
Potato	*Streptomyces* scab	Decrease	Increase	Decrease	Increase
Tobacco	*Thielaviopsis basicola*	Decrease	Increase	Decrease	Increase
Tomato	*Verticillium* wilt	Decrease	Increase	Decrease	Increase
Turf	Take-all	Decrease	Increase	Decrease	Increase
Wheat	*Fusarium* root rot	Decrease	Increase	—	Increase
Avocado	Black rot	Increase	Decrease	—	Increase
Bean	*Fusarium* root rot	—	—	Increase	Increase
Brassicae	Club-root	Increase	Decrease	—	—
Cotton	*Fusarium* wilt	Increase	Decrease	Increase	Decrease
Many	*Sclerotium* collar rot	Increase	Decrease	Increase	Decrease
Pea	*Aphanomyces euteiches*	Increase	Decrease	—	—
Peach	Bacterial canker	Increase	Decrease	—	—

Source: Compiled from various sources by Huber and Wilhelm, 1988.

Table 6.4 Effects of K on Plant Diseases[a]

Pathogen or disease	Low K	High K
Puccinia graminae	***	*
Alternaria solani	***	*
Fusarium oxysporum	***	*
Xanthomonas oryzae	***	*
Tobacco mosaic virus	***	*

[a] Disease severity: ***, high; *, low.
Source: Kiraly, 1976.

black rust in wheat; sugary disease in sorghum; bacterial leaf blight in cotton; *Cercospora* leaf spot in cassava; tikka leaf spot in peanut, red rust in tea, *Cercospora* leaf spot in mungbean (*Vigna radiata* L.); and seedling rot caused by *Rhizoctonia solani* in mungbean and cowpea (Tandon and Sekhon, 1989). Perrenoud (1977) reviewed the relation between K fertilization and plant diseases in 534 references, and Usherwood (1980) drew the following conclusions from Perrenoud's review:

1. Potassium improved plant health in 65% of the studies and was deleterious 23% of the time.
2. Potassium reduced bacterial and fungal diseases 70% of the time, insects and mites 60% of the time, and the effects of nematodes and viruses in a majority of the cases.
3. Fungal disease infestation was reduced by K an average of 48% in soils tested low in K and 14% where soil test levels were unknown.
4. The influence of K on crop yield varied according to the parasite group. The average increase in yield or growth was 48% for fungal diseases, 99% for viruses, 14% for insects and mites, and 70% for bacteria.
5. The mode of action is primarily through plant metabolism and morphology. The K-deficient plants have impaired protein synthesis and accumulate simple N compounds (e.g., amides) which are good nutrient sources for invading pathogens. Tissue hardening and stomatal opening patterns are closely related to infestation intensity.
6. Crop response is not consistently different for different sources of K.
7. The balance between N and K affects disease susceptibility of plants.
8. The benefits of additional K were more frequent in the field than in laboratory and greenhouse experiments.

The nature of the action of K in controlling the severity of plant diseases is still not understood. It may relate, in part, to the effect of K in promoting the development of thicker outer walls in epidermal cells, thus preventing disease attack. In addition, plant metabolism is much influenced by K, and some plant diseases may be favored by changes in metabolism associated with low K contents in the plant (Mengel and Kirkby, 1982).

V. CALCIUM, MAGNESIUM, AND SULFUR

Calcium (Ca) sometimes increases host resistance; however, in other cases, it renders the pathogen more virulent, thereby increasing the severity of disease symptoms. Calcium has important roles in the integrity of all membranes and cell walls. These have been invoked as mechanisms for the resistance that Ca often confers against *Pythium*, *Sclerotium*, *Botrytis*, and *Fusarium* (Graham, 1983). Table 6.5 shows the severity of some diseases at high and low levels of Ca. Hancock and Miller (1965) showed that the Ca mobilized in lesions caused by *Colletotrichum trifolii* in alfalfa supports fungus growth by stimulating the macerating action of pectolytic enzyme, polygalacturonic acid transeliminase (Kiraly, 1976). Many physiological disorders of storage organs, fruits, certain vegetables, roots, and young enclosed leafy structures are related to the Ca content of the respective tissues. For example, adequate soil Ca is needed to protect peanut pods from diseases caused by *Rhizoctonia* and *Pythium*. Increasing the tissue content of Ca normally diminishes the occurrence of such diseases (Bangerth, 1979).

There is little information on the effects of sulfur (S) and magnesium (Mg) nutrition on plant diseases. Sulfur is commonly applied to reduce the severity

Table 6.5 Effects of Ca on Plant Diseases[a]

Pathogen or disease	Low Ca	High Ca
Erwinia phytophthora	***	*
Rhizoctonia solani	***	*
Sclerotium rolfsii	***	*
Botrytis cinerea	***	**
Fusarium oxsporum	***	*
Jonathan spot (nonparasitic)	***	**
Bitter pit (nonparasitic)	***	**

[a] Disease severity: ***, high; *, low.
Source: Kiraly, 1976.

of potato scab (Huber, 1980). Magnesium decreases the Ca content of peanut pods and may predispose them to pod breakdown by *Rhizoctonia* and *Pythium*.

VI. MICRONUTRIENTS

A. Zinc

Since there are few research data, the role of zinc (Zn) in the defense mechanisms of higher plants is far from clear. There are reports of beneficial effects, no effects, and negative effects of Zn application on plant diseases. Zinc application often increases host resistance to mildew and leaf spot and has suppressive effects on soil-borne diseases and bacterial and viral diseases (Graham, 1983). With Zn deficiency, a leakage of sugars onto the leaf surface of *Hevea brazilensis* increased the severity of infection with *Oidium* (Bolle-Jones and Hilton, 1956). Zinc deficiency inhibits protein synthesis and produces high concentrations of nonprotein N, including amino acids (Graham, 1983). Such accumulations may favor invading heterotrophs.

Zinc consistently stimulates germination of fungal spores (Graham, 1983), but whether this effect is significant in pathogenesis is not well known. Stimulation of plant disease infestations by Zn application may be related to its antagonism of Mn and Cu uptake. These two elements generally increase plant resistance to diseases, as discussed later in this section.

B. Boron

Boron has been used as a fertilizer for more than 400 years, but it was not shown to be an essential element until this century (Mengel and Kirkby, 1978). It seems possible that its earlier use was associated with disease control. Dicotyledonous plants require higher tissue boron concentrations than monocots, a difference apparently related to the greater elaboration of secondary metabolites in the former (Graham, 1983). However, greater susceptibility to disease in boron-deficient plants occurs at least as frequently in monocots as in dicots. In boron-deficient wheat plants, the rate of powdery mildew infection was sevenfold higher than in boron-sufficient plants, and the fungus also spread more rapidly over the plant (Schutte, 1967).

C. Manganese

The availability of manganese (Mn) in soil is extremely complex and depends on factors such as pH, moisture, nutrients, chloride, nitrification inhibitors, organic matter, and microbial activity (Graham, 1983). Differences in the concentration of Mn in plant tissue have been correlated with susceptibility to various pathogens. However, Mn may not have a casual role in disease

Table 6.6 Reported Effects of Mn on Plant Diseases

Host plant	Disease	Pathogen	Effect of Mn
Barley	Aphid	*Rhopalosiphum maidis*	Decrease
Barley	Leaf spot	*Helminthosporium*	Increase
Barley	Mildew	*Erysiphe graminis*	Decrease
Barley	Mildew	*Erysiphe graminis*	Increase
Bean	Virus	Tobacco mosaic virus	Decrease
Cotton	Damping-off	*Rhizoctonia solani*	Decrease
Cotton	Wilt	*Fusarium oxysporium*	Decrease
Cotton	Wilt	*Verticillium alboatrum*	Decrease
Cowpea	Mildew	*Erysiphe polygone*	Decrease
Cowpea	Virus	Chlorotic mottle virus	Increase
Lentil	Wilt	*Fusarium oysporium*	Decrease
Oats	Bacterial blight	*Pseudomonas* spp.	Decrease
Onion	Rot	Storage fungi	Decrease
Pigeonpea	Wilt	*Fusarium udum*	Decrease
Potato	Late blight	*Phytophthora infestans*	Decrease
Potato	Stem canker	*Rhizoctonia solani*	Decrease
Potato	Scab	*Streptomyces scabies*	Decrease
Potato	Wilt	*Verticillium dahliae*	Decrease
Rice	Bacterial blight	*Xanthomonas oryzae*	Decrease
Rice	Bacterial blight	*Xanthomonas oryzae*	Increase
Rice	Blast	*Pyricularia oryzae*	Decrease
Rice	Brown spot	*Helminthsporium oryzae*	Decrease
Rice	Leaf spot	*Helminthsporium sigmoidum*	Decrease
Sorghum	Downy mildew	*Peronosclerospora sorghi*	Decrease
Soybean	Blight	*Pseudomonas glycinea*	Decrease
Sugarbeet	Leaf spot	*Cercospora* spp.	Decrease
Sugarbeet	Insect	Root borer	Decrease
Sugarcane	Whip smut	*Ustilago scitaminca*	Decrease
Sweet potato	Root rot	*Streptomyces ipomoea*	Decrease
Wheat	Mildew	*Erysiphe graminis*	Decrease
Wheat	Rust	*Puccinia* spp.	Decrease

Source: Compiled by Huber and Wilhelm from various sources, 1988.

infection and development. Differences in susceptibility may reflect a host response to infection or an inherent difference in uptake efficiency or metabolic utilization of Mn, and they may indicate involvement of Mn through a mechanism that conditions resistance, susceptibility, or predisposition (Huber and Wilhelm, 1988).

Manganese concentration is usually lower in tissues susceptible to fungal, viral, and bacterial pathogens than in resistant tissues (Kudelova et al., 1978; Cordrey and Bergman, 1979; Anderson and Dean, 1986). Huber and Wilhelm (1988) reviewed the effects of Mn on plant diseases and concluded that out of 62 references, 85% reported decreases in fungal, bacterial, and viral diseases with Mn application. The greater resistance of paddy-grown rice to blast (*Pyricularia oryzae*) compared to upland rice has been attributed to the increased uptake of Mn by flooded rice under reduced soil where Mn^{2+} concentrations may even increase to toxic levels (Chou and Chiou, 1979).

Similarly, brown spot (*Helminthosporium oryzae*) on rice is severe on Mn-deficient soils irrespective of pH. Kaur and Padmanabhan (1974) reported that application of Mn to soil or water decreased the severity of brown spot in highly susceptible rice varieties, and even resistant varieties became susceptible when the Mn concentration in host tissue dropped below 2.5 mg kg^{-1}. Several mechanisms may be involved in the effects of Mn on disease, including a direct effect of Mn on pathogen toxicity or virulence, modification of host plant resistance, or a combination of these effects (Huber, 1980, 1981; Graham, 1983). Table 6.6 shows the effects of Mn on plant diseases.

D. Copper

Use of copper (Cu) as a fungicide has been documented over more than a century. Copper oxychloride and various Bordeaux mixtures have long been used for the control of fungal diseases. It has now become clear that Cu-deficient plants are frequently more susceptible to airborne fungal diseases than plants with adequate Cu. A number of other diseases have been reported to be more severe on Cu-deficient plants, such as *Alternaria* on sunflower, *Gaeumannomyces graminis* on wheat, *Claviceps pururea* on rye and barley, *Heterodera* on sugar beet, *Puccinia tritica* on wheat, *Pyricularia grisea* on rice, and *Selerotinia* on peanuts (Graham, 1983). Poor lignification, impaired phenol metabolism, accumulation of soluble carbohydrates, and delay in leaf senescence are probably the main reasons for the higher susceptibility of Cu-deficient plants (Marschner, 1986).

E. Iron

In iron (Fe)-deficient soils, foliar application of Fe increases the resistance of apples and pears to *Spaeropsis inalorum*, wheat to smut, and turf grass to

Fusarium patch diseases. Wallace and North (1962) reported that Fe amendments can either correct or mask some virus symptoms in camellia. Increased Fe supply results in tolerance of cabbage to *Olpidium brassicae.*

Recent papers on siderophore production by bacteria in the soil reflect a new level of sophistication about the involvement of Fe in resistance to disease. Competition between host and pathogen for Fe is considered to be an important defense mechanism in animals, and plant pathologists are looking for similar systems in the plant kingdom (Graham, 1983). The fluorescent pseudomonads are a group of agriculturally important gram-negative, rod-shaped bacteria which are present in greater numbers in soils suppressive of "take all" disease of wheat (Graham, 1983). According to Graham (1983) the fluorescent pseudomonads suppress disease by producing siderophores in the soil to compete with rhizosphere pathogens for Fe.

F. Chlorine

The application of fertilizer containing chloride (Cl^-) in Cl^--deficient soils reportedly suppressed plant diseases (Christensen and Brett, 1985). Suppression of take-all root rot of wheat caused by the fungus *Gaeumannomyces 3graminis* (Sacc.) Arx and Oliv. var. *tritici* Walker is maximized when Cl^- is applied in conjunction with other management practices to reduce disease severity (e.g., the use of ammonium N (NH_4^+-N), band-applied P, moderate soil acidity, and delayed planting; Taylor et al., 1983). Christensen et al. (1981) reported that the susceptibility of winter wheat plants to *Gaeumannomyces graminis* var. *tritici* colonization may be reduced by lowering the chemical potential of water in the plant. Because the osmotic potential in wheat plants changes readily with Cl^- application, fertilization with Cl^- salts provides an opportunity to actively manage plant water potential components for the express purpose of take-all root rot suppression.

Fixen (1987) reviewed the Cl^- work and reported that one of the most frequently reported effects of chloride fertilization is disease suppression (Table 6.7). To date, at least 15 different foliar and root diseases on 10 different crops have been significantly reduced in severity with the addition of chloride. Yield increases in Oregon and North Dakota are completely due to suppression of take-all common root rot, respectively.

It is theorized that chloride acts as a nitrification inhibitor, forcing wheat plants to take up more nitrogen as ammonium than as nitrate when ammoniacal fertilizers are applied. This, in turn, causes plant roots to give off hydrogen ions, which increases the acidity at the root surface. The take-all fungus is inhibited by microorganisms which thrive in the more acidic root zone, and disease severity decreases. These studies also have shown that the effect is negligible if soil pH is above approximately 6.1 (Fixen, 1987).

Table 6.7 Diseases Suppressed by Chloride Fertilizers

Location	Crop	Suppressed disease
Oregon	Winter wheat	Take-all
Germany	Winter wheat	Take-all
North Dakota	Winter wheat	Tanspot
Oregon	Winter wheat	Stripe rust
Great Britain	Winter wheat	Stripe rust
South Dakota	Spring wheat	Leaf rust
South Dakota	Spring wheat	Tanspot
South Dakota	Spring wheat	Septoria
North Dakota	Barley	Common root rot
North Dakota	Barley	Spot blotch
Montana	Barley	Fusarium root rot
North Dakota	Durum	Common root rot
New York	Corn	Stalk rot
India	Pearl millet	Downy mildew
Philippines	Coconut palm	Gray leaf spot
Oregon	Potatoes	Hollow heart
Oregon	Potatoes	Brown center
California	Celery	Fusarium yellows

Source: Fixen, 1987.

Common root rot effects on barley in North Dakota have a slightly different mechanism. In this theory, chloride fertilization decreases nitrate concentrations in the plant, due either to nitrification inhibition as discussed with take-all, or to general competition between chloride and nitrate for uptake. It is believed that a plant having low nitrate concentrations is less likely to develop a severe case of common root rot (Fixen, 1987).

G. Silicon

Silicon (Si) is the most abundant element in the lithosphere after oxygen, and soil contains approximately 32% silicon by weight (Lindsay, 1979). In the pH range 2–9, Si is available to plants as the monosilicic acid $Si(OH)_4$. Above pH 9 it occurs as the silicate ion. The accessibility of Si to plants, however, depends largely on how rapidly weathering takes place in bringing Si into soil solution.

Although Si is not an essential nutrient, its role in plant disease control is well known. For rice, a typical silica (SiO_2)-demanding plant, this element is

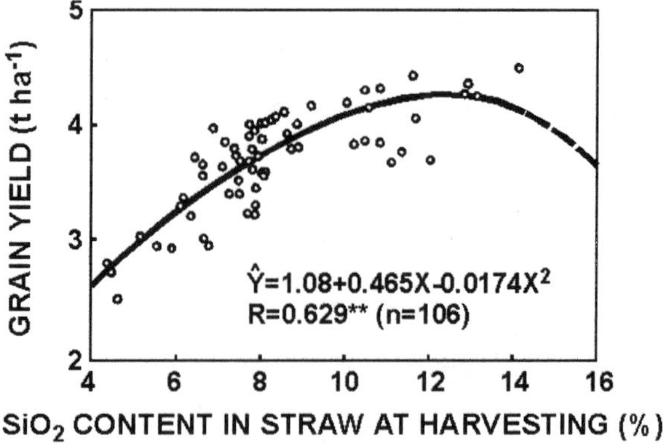

Figure 6.2 Relationship between SiO_2 content in the rice straw and grain yield. (From Park, 1970.)

essential for improving yield and structural strength. There was a significant yield increase of rice with increasing Si content in the straw (Figure 6.2).

In certain locations where organic Histosols of the Florida Everglades are low in plant-available Si, silicon fertilization has been shown to be beneficial to rice crops (Snyder et al., 1986; Deren et al., 1992). When these soils are amended with Si as calcium silicate slag, rice yields increase significantly, due in part to a reduction in disease severity (Datnoff et al., 1991). On Si-deficient mineral soils, diseases such as brown spot [*Bipolaris oryzae* (Breda de Haan) Shoemaker] and blast [*Pyricularia grisea* (Cooke) Sacc.] are greatly reduced with the addition of plant-available Si (Yamauchi and Winslow, 1987).

Another possible effect of Si on rice grown on Everglades Histosols may be that it counteracts some of the detrimental effects of N, which is nonlimiting for rice, sugarcane, and other crops grown on these soils. High rates of N fertilization are associated with droopy leaves, lodging, generally poor plant architecture, and increased disease (Tisdale et al., 1985). Silicon applications have improved rice plant architecture in the Everglades and reduce disease. Si-fertilized plants may gain the maximum benefit of the ample available N, but the Si reduces the detrimental effects, promoting increased photosynthesis (Ishizuka, 1971). Although the mechanisms by which Si increases yield and disease tolerance are not well understood, clearly there is benefit in increasing plant silicon concentration on low-Si organic soils.

In West and Central Africa, rice is often grown on freely drained upland soils in high-rainfall forest areas. These are predominantly highly weathered Ultisols and Oxisols that are largely desilicated and probably have insufficient Si to satisfy the requirements of the crop (Winslow, 1992). Application of Si to an upland Ultisol soil at Onne, Nigeria, increased rice yield and reduced damage from diseases (Yamauchi and Winslow, 1989).

The association between Si deficiency and diseases raises the possibility of increasing disease resistance by breeding cultivars with higher Si content. Some authors report higher blast resistance in high-Si genotypes (Deren et al., 1992; Winslow, 1992).

Deren et al. (1994) also reported that rice genotypes differed for Si concentration and disease severity at each location and for each Si treatment. Among genotypes, disease severity was negatively correlated with Si concentration in plant tissue. Increases in yield with added Si were attributable to a greater number of grains per panicle, whereas weight per 100 seed and panicles per square meter exhibited less change.

With poor silicification of epidermal cells, the rice plant becomes susceptible to such fungal diseases as rice blast and *Helminthosporium* leaf spot. Many workers feel that silicates in the epidermal layer prevent physical penetration of fungi (Ou, 1972). For healthy growth, the SiO_2/N ratio should be wide (Matsubayashi et al., 1963). Plants given large amounts of N are found to have less silicated epidermal cells and, thus, lower blast resistance (Ou, 1972). Silicification of cell walls seems to be linked with K nutrition. According to Nogushi and Sugawara (1966), K deficiency reduces the accumulation of SiO_2 in the epidermal cells of the leaf blades, thus increasing the susceptibility to rice blast.

H. Organic Manures

The value of plant nutrition in reducing the incidence and severity of plant pathogens has been recognized for many years. Although most metabolic or physiological mechanisms involved in host–pathogen interactions are not clearly understood, specific nutrients are known to reduce disease severity by affecting virulence of the pathogen, enhancing resistance of the plant, compensating for pathogenic damage, or activating indigenous biological control mechanisms. The source of nutrients may be either inorganic or organic, and nutrients from either source generally have comparable effects. However, organic sources, as complex mixtures of nutrients, may have a much more complex relationship to disease incidence and severity than many inorganic sources (Huber, 1981).

Huber and Watson (1970) concluded that organic amendments and crop rotation probably influence the severity of soilborne diseases by: 1) increasing the biological buffering capacity of the soil; 2) reducing pathogen numbers during anaerobic decomposition of organic matter; 3) affecting nitrification which influences the form of nitrogen predominating in the soil, and 4) denying the pathogen a host during the interim of unsuitable species. The specific form of nitrogen available to the plant and soil microflora, in turn, influences specific microbial associations and host physiology. These factors should be considered in opening new lands to cultivation to avoid costly results of land pollution from establishment of soilborne plant pathogens.

Crop residues and animal manures can provide an environment where pathogens can flourish. For example, farmyard manure can increase *Rhizobium solani* disease incidence in cotton and cowpea seedlings (Bandyopadhyay et al., 1982). Kataria and Grover (1987) reported that both farmyard manure and green manure aggravated mungbean seedling rot by *R. solani*. Crop residues in conservation tillage can support higher pathogen and insect populations and thus prompt increased pesticide use (Sojka et al., 1991). Weeks (1993) applied crop residues (peanut hay and peanut hulls) to heavy metal-contaminated land to increase adsorption of the metals and reduce their uptake by peanut, but spring hay application increased late leaf-spot disease scores. Organic amendments to soil have many positive effects; however, they can also increase disease incidence, thus increasing the need for insecticide application (Davis, 1994).

I. Plant Diseases and Salt-Affected Soils

Little information is available on the role of soil salinity and sodicity in plant diseases. The effects of soil salts on plant growth are physiological, whether nutritional, osmotic, or directly toxic (Bernstein, 1975). No direct effects of soil salinity or sodicity on pathogens are known, but some indirect effects have been observed. Generally, saline soils have higher moisture contents than nonsaline soils under given meteorological conditions (Bernstein, 1975). Wet soil conditions may favor *Phytophthora* root rot and other fungal infections (Bernstein and Francois, 1973). None of these diseases is specifically salinity induced. Improving the water regime by better irrigation or drainage may prevent the fungus diseases associated with wetter soil conditions, even without a decrease in soil salinity (Bernstein, 1975).

VII. SUMMARY

Production losses that result from infestation of crop plants by diseases are considerable, and balanced nutrition plays an important role in determining plant resistance or susceptibility to diseases.

1. A well nourished crop plant is generally more tolerant of disease than one with suboptimum nutrition.
2. Nitrogen differently affects diseases caused by obligate and facultative pathogens. High N fertilization increases susceptibility of host plants to obligate pathogens, but it decreases their susceptibility to facultative pathogens. Application of K, Ca, Mn, Fe, B, Cu, and Si to deficient

soils usually increases resistance, but the effects of P and Zn are variable. There is not sufficient information on Mg and S to reach definite conclusions.

3. Macronutrients increase resistance to disease, if at all, only in the deficiency range. Very high tissue concentrations of nutrients do not provide further protection and, in some cases, may be detrimental.
4. Correction of a micronutrient deficiency generally increases the tolerance and/or resistance of plants to diseases.
5. Copper, B, and Mn influence the synthesis of lignin and simple phenols. Silicon appears to affect physical barriers to invasion.
6. The greatest benefits from nutrients are found with moderately susceptible or partially resistant cultivars. In highly resistant and highly susceptible cultivars, the nutritional status of the plant has little influence on the severity of disease.
7. Nutritional balance is as important as the level of a single nutrient in disease control.
8. No nutrient controls all diseases or favors disease control on any one group of plants. Therefore, all control practices should be integrated for optimum plant growth and production.
9. Some nutrients appear to affect disease severity simply through greater tolerance, for example P and S. Others, including N and K, alter specific host-plant resistance mechanisms.

REFERENCES

Anderson, D. L. and J. L. Dean. 1986. Relationship of rust severity and plant nutrients in sugarcane *Saccharum* spp. Phytopathology 76: 581–585.

Bandyopadhyay, R., J. P. S. Yadav, H. R. Kataria, and R. K. Grover. 1982. Fungicidal control of *Rhizoctonia solani* in soil amended with organic manures. Annu. Appl. Bot. 101: 251–259.

Bangerth, F. 1979. Calcium-related physiological disorders of plants. Annu. Rev. Phytopathol. 17: 97–122.

Bernstein, L. 1975. Effects of salinity and sodicity on plant growth. Annu. Rev. Phytopathol. 13: 295–312.

Bernstein, L. and L. E. Francois. 1973. Comparisons of drip, furrow, and sprinkler irrigation. Soil Sci. 115: 73–86.

Bolle-Jones, E. W. and R. N. Hilton. 1956. Zinc-deficiency of *Hevea braziliensis* as a predisposing factor to *Oidium* infection. Nature (London) 177: 619–620.

Chou, C. H. and S. J. Chiou. 1979. Auto intoxication mechanisms of *Oryza sativa*. 2. Effects of culture treatments on the chemical nature of paddy soil and on rice productivity. J. Chem. Ecol. 5: 839–860.

Christensen, N. W. and M. Brett. 1985. Chloride and liming effects on soil nitrogen form and Take-all of wheat. Agron. J. 77: 157–163.

Christensen, N. W., R. G. Taylor, T. L. Jackson, and B. L. Mitchell. 1981. Chloride effects on water potentials and yield of winter wheat infected with Take-all root rot. Agron. J. 73: 1053–1058.

Colhoun, J. 1973. Effects of environmental factors on plant disease. Annu. Rev. Phytopathol. 11: 343–364.

Cordrey, T. D. and E. L. Bergman. 1979. Influence of cucumber mosaic virus on growth and elemental composition of susceptible (*Capsicum annuum* L.) and resistant (*Capsicum frutescens* L.) peppers. J. Am. Soc. Hortic. Sci. 104: 505–510.

Datnoff, L. F., R. N. Raid, G. H. Snyder, and D. B. Jones. 1991. Effect of calcium silicate on blast and brown spot intensities of rice. Plant Dis. 75: 729–732.

Davis, J. G. 1994. Managing plant nutrients for optimum water use efficiency and water conservation. Adv. Agron. 53: 85–120.

Deren, C. W., L. E. Datnoff, and G. H. Snyder. 1992. Variable silicon content of rice cultivars grown on Everglades Histosols. J. Plant Nutr. 15: 2363–2368.

Deren, C. W., L. F. Datnoff, G. H. Snyder, and F. G. Martin. 1994. Silicon concentration, disease response, and yield components of rice genotypes grown on flooded organic Histosols. Crop Sci. 34: 733–737.

Fixen, P. E. 1987. Recent research gives new answers. Crops and Soils Magazine 39: 14–16.

Gooding, M. J. and W. P. Davies. 1992. Foliar urea fertilization of cereals: A review. Fert. Res. 32: 209–222.

Gooding, M. J., P. S. Kettlewell, and W. P. Davies. 1988. Disease suppression by late season urea sprays on winter wheat and interaction with fungicide. J. Fert. Issues 5: 19–23.

Graham, R. D. 1983. Effects of nutrient stress on susceptibility of plants to disease with particular reference to the trace elements. Adv. Bot. Res. 10: 221–276.

Hancock, J. G. and R. L. Miller. 1965. Relative importance of polygalacturonate transeliminase and other pectolytic enzymes in southern anthrachose, spring black stem, and *Stemphylium* leaf spot of alfalfa. Phytopathology 55: 346–355.

Huber, D. M. 1980. The role of mineral nutrition in defense, pp. 381–406. *In*: J. G. Horsfall and E. B. Cowling (eds.). Plant pathology. An advanced treatise. Academic Press, New York.

Huber, D. M. 1981. The use of fertilizers and organic amendments in the control of plant disease, pp. 357–394. *In*: D. Pimental (ed.). Handbook of pest management in agriculture. CRC Press, Boca Raton, Florida.

Huber, D. M. and D. C. Arny. 1985. Interactions of potassium with plant disease, pp. 467–488. *In*: R. D. Munson (ed.). Potassium in agriculture. Am. Soc. Agron., Madison, Wisconsin.

Huber, D. M. and R. D. Watson. 1970. Effect of organic amendment on soil-borne plant pathogens. Phytopathology 60: 22–26.

Huber, D. M. and R. D. Watson. 1974. Nitrogen form and plant disease. Annu. Rev. Phytopathol. 12: 139–165.

Huber, D. M. and N. S. Wilhelm. 1988. The role of manganese in resistance to plant diseases, pp. 154–173. *In*: R. D. Graham, R. J. Hannam, and N. C. Wren (eds.). Manganese in soils and plants. Kluwer Academic Publishers, London.

Ishizuka, Y. 1971. Physiology of the rice plant. Adv. Agron. 23: 241–315.

James, W. C. 1980. Economic, social and political implications of crop losses: A holistic framework for loss assessment in agricultural systems, pp. 10–16. *In*: Crop loss assessment. E. C. Stakman Commemorative Symposium. Misc. Publ. F, Agric. Exp. Stn., University of Minnesota, St. Paul.

Kataria, H. R. and R. K. Grover. 1987. Influence of soil factors, fertilizers and manures on pathogenicity of *Rhizoctonia solani* on *Vigna* species. Plant Soil 103: 57–66.

Kaur, P. and S. Y. Padmanabhan. 1974. Control of *Helminthesporium* disease of rice with soil amendments. Current Sci. 43: 78–79.

Kiraly, Z. 1976. Plant disease resistance as influenced by biochemical effects on nutrients in fertilizers, pp. 33–46. *In*: International Potash Institute (ed.). Fertilizer use and plant health. Bern, Switzerland.

Kudelova, A., E. Bergmannova, V. Kudela, and L. Taimr. 1978. The effect of bacterial wilt on the uptake of manganese and zinc in alfalfa. Acta Phytopathol. Acad. Sci. Hungary 13: 121–132.

Lindsay, W. L. 1979. Chemical Equilibrium in Soils. Wiley, New York.

Marschner, H. 1995. Relationship between mineral nutrition and plant diseases and pests, pp. 436–460. *In*: Mineral nutrition of higher plants. 2nd Ed. Academic Press, New York.

Matsubayashi, M., R. Ito, T. Takase, T. Nomoto, and N. Yameda. 1963. Theory and practice of growing rice. Futi, Tokyo.

Mengel, K. and E. A. Kirkby. 1982. Principles of plant nutrition. 3rd ed. International Potash Institute, Bern, Switzerland.

Nogushi, J. and T. Sugawara. 1966. Potassium and japonica rice. International Potash Institute, Bern, Switzerland.

Ou, S. A. 1972. Rice diseases. Commonwealth Mycological Institute, Kew, Surrey, England.

Park, C. S. 1970. The micronutrient problem of Korean Agriculture, pp. 847–862. *In*: Proc. Int. Symp. Commemorating the 30th Anniversary of Korean Liberation. Nat. Acad. Sci. Rep. Korea, Seoul.

Peltonen, J., S. Kittila, P. Peltonen-Sainio, and R. Karjalainen. 1991. Use of foliar applied urea to inhibit the development of *Septoria nodorum* in spring wheat. Crop Protection 10: 260–284.

Perrenoud, S. 1977. Potassium and plant health, pp. 1–118. *In*: International Potash Institute (ed.). Research topics, No. 3. International Potash Institute, Bern, Switzerland.

Piening, L. J. 1989. Fertilizers can reduce plant diseases, pp. 18–20. *In*: PPI (ed.). Better crops with plant food. Summer 1989. Atlanta, Georgia.

Potash and Phosphate Institute (PPI). 1988. Phosphorus nutrition improves plant disease resistance, pp. 22–23. *In*: PPI (ed.). Better crops with plant food. Fall 1988. Atlanta, Georgia.

Schutte, R. H. 1967. The influence of boron and copper deficiency upon infection by *Erysiphe graminis* D. C. the powdery mildew in wheat var. Kenya. Plant Soil 27: 450–452.

Snyder, G. H., D. B. Jones, and G. Gascho. 1986. Silicon fertilization of rice on Everglades Histosols. Soil Sci. Soc. Am. J. 50: 1259–1263.

Sojka, R. E., D. L. Karlen, and W. J. Busscher. 1991. A conservation tillage research update from the Coastal Plain Soil and Water Conservation Research Center of South Carolina: A review of previous research. Soil Tillage Res. 21: 361–376.

Tandon, H. L. S. and G. S. Sekhon. 1989. Potassium research and agricultural production in India. Potash Rev. 1: 1–11.

Taylor, R. G., T. L. Jackson, R. L. Powelson, and N. W. Christensen. 1983. Chloride, nitrogen form, lime, and planting date effects on take-all root of winter wheat. Plant Dis. 67: 1116–1120.

Tisdale, S. L., W. L. Nelson, and J. D. Beaton. 1985. Soil Fertility and Fertilizers. Macmillan, New York.

Usherwood, N. R. 1980. The effects of potassium on plant diseases, pp. 151–164. *In*: Potassium for agriculture: A situation analysis. Potash and Phosphate Institute (ed.). Atlanta, Georgia.

Wallace, A. and C. P. North. 1962. Metal chelates and virus disease, pp. 142–145. *In*: A. Wallace (ed.). A decade of synthetic chelating agents in inorganic plant nutrition. Edward Brothers, Ann Arbor, Michigan.

Weeks, G. 1993. Reduction of heavy metal uptake by peanut grown on flue dust-amended soils. M.S. Thesis, Univ. Georgia, Athens.

Winslow, M. D. 1992. Silicon, disease resistance, and yield of rice genotypes under upland cultural conditions. Crop Sci. 32: 1208–1213.

Yamauchi, M. and M. D. Winslow. 1987. Silica reduces disease on upland rice in high rainfall area. Int. Rice Res. Inst. Newsletter 12: 22–23.

Yamauchi, M. and M. D. Winslow. 1989. Effect of silica and magnesium on yield of upland rice in humid tropics. Plant Soil 113: 265–269.

Zadoks, J. C. and R. D. Schein. 1979. Epidemiology and plant disease management. Oxford University Press, New York.

7

Simulation of Crop Growth and Management

I. INTRODUCTION

Agriculture is dynamic. Research produces a constant flow of new crop cultivars, machinery, pesticides, and fertilizer products. Concerns for environmental quality and soil conservation modify social and governmental acceptance of farming practices. Price structures and government policies change the economically optimum mix of crops and management practices for a particular set of environmental conditions. Weather changes from year to year. How, then, can a farmer design a successful cropping system for a unique and dynamic set of natural resources and economic conditions?

Agricultural research has traditionally relied heavily on field experiments to evaluate management practices. However, they are costly and time consuming, and can sample only a fraction of the management alternatives available to farmers. As a result, new tools are needed to supplement field experiments and to extrapolate their results in time and space.

Recent advances in computer hardware and software have stimulated development and use of natural resource databases, expert systems, and simulation models to manage and predict the behavior of agricultural systems. Such tools capture our knowledge of agricultural resources and processes and help us to supplement and extrapolate results of field experimentation. Natural resource databases give decision makers detailed descriptions of the edaphic and climatic environment. Expert systems utilize heuristic knowledge of human experts, mimic their reasoning processes, and help nonspecialists make more expert decisions.

Simulation models use our understanding of biophysical processes to predict the crop's and the cropping system's response to environment and management. During the past 15 years, simulation models have been developed for a number of agricultural processes, including weather, hydrology, soil erosion, nutrient transformations and movement, and crop phenology, growth, and yields. Models of specific processes have been combined into comprehensive crop growth and cropping system models that predict the effects of environment, crop genotype, and management on crop yields and, in some cases, the soil resource. In many respects, knowledge described in earlier chapters of this book has been incorporated into simulation models. Such models are being used by agricultural researchers, consultants, and decision makers worldwide to predict the site-specific behavior of alternative management strategies and cropping systems.

Simulation models are being used to:

1. Predict crop adaptation to specific environmental conditions
2. Predict crop response to alternative management practices
3. Predict the behavior of complex cropping systems sensitive to biophysical, social, and economic constraints
4. Predict short- and long-term changes in soil fertility and productivity due to management

This chapter is not intended to be a comprehensive review of specific models or processes. Rather, it is a personal view, illustrated with examples, of how simulation models of agricultural processes and systems are being used. Throughout the discussion, however, it is important to recognize that agricultural systems are complex and dynamic. Our understanding of them is imperfect, and our representation of that understanding in simulation models is incomplete. By definition, models are simplifications of reality. We should view them as imperfect but useful tools in our continuing struggle to understand and manipulate agricultural systems.

II. CROP ADAPTATION

Several chapters describe the growth and development of several important field crops. Most of the relationships described are based on research from areas where the crop has been grown for a number of years and is relatively well known. However, agriculture in many areas of the world is based on only a few crops. Increasing populations and changing economic conditions can change the needs of particular regions. And "new" crops may prove to be important components of future agricultural systems.

Consider, for a moment, the importance of soybeans in the United States and Brazil, maize in Europe, and sunflowers in the Soviet Union. At some time in the past, all these were "new" crops in areas where they are now very important.

Simulation models are increasingly being used to evaluate crop adaptation to particular sets of environmental conditions. Nowhere is this trend more evident than in Australia, a country with underutilized agricultural lands and economic dependence on export of agricultural products. Since the mid-1970s, simulation models have been used to match crop requirements to climate and soil characteristics for many regions of Australia.

A. Guayule

Guayule (*Parthenium argentatum*) is a drought-tolerant, rubber-producing shrub native to the deserts of the southwestern United States and northern Mexico. Though experimental work has been conducted in the United States since the early 1900s and commercial production from wild populations occurred in Mexico prior to 1950, little was known of guayule's adaptation to Australian conditions. In the 1940s experimental plantings in South Australia and Canberra resulted in poor plant survival. The high oil prices of the 1970s renewed interest in guayule. This resulted in an agroclimatic analysis (Nix, 1981) of potential sites for guayule research and, ultimately, industrial production. A generalized plant growth simulation model (GROWEST), incorporating weekly water balance, temperature, and light response functions, was used to evaluate the potential of 1000 Australian sites for guayule production. Sites with high probabilities of long periods of soil saturation or unfavorable ratios of radiation to temperature were excluded. Results indicated that, in Australia, guayule research should be located in the eastern wheat belt of Queensland and New South Wales (Nix and McMahon, 1979), and long-term field research has begun in the most promising regions. Nix and McMahon (1979) also conducted a study on jojoba (*Simmondsia chinensis*) adaptation to Australian climates and soils similar to that described for guayule.

B. Sunflower

Sunflower (*Helianthus annuus*) production became economically attractive in Australia in the 1970s; however, at that time little was known about the crop's adaptation to Australian environments. As a result, field experiments were conducted to determine functional relationships needed to simulate sunflower growth, and a simulation model was developed (Anderson et al., 1978a,b; Smith et al., 1978a,b). The model was designed to simulate crop development, growth, and yield from mean weekly temperature, day length, solar radiation, and precipitation.

To evaluate crop adaptation, the model was executed for 455 Australian sites, each with 13 dates of seeding and four levels of antecedent soil water. The model identified sites where frost would limit seeding dates, where disease would probably limit yields, and where stored soil moisture and growing season precipitation would be important factors.

In a subsequent study, historical daily precipitation records (51–97 years) were used with mean weekly temperature, radiation, evaporation, and day length to evaluate risk associated with different seeding dates at 13 locations (Smith et al., 1978b). Results suggested that in the southern areas of the continent early seeding was less risky than late seeding. In the central highlands of Queensland the reverse was true, and in southern Queensland risk was reduced by seeding at either the beginning or end of the summer period.

A similar Queensland study of optimum seeding dates for short- and long-season cultivars (Hammer and McKeon, 1984) was conducted with a different simulation model (Goyne and Hammer, 1982; Hammer and Goyne, 1982; Hammer et al., 1982) and gave similar results.

C. Fresh Potato Production

Consumer demand for fresh produce is often quite stable throughout the year, and large changes in supply can cause ruinous price fluctuations. Australia's potato-growing districts extend over wide ranges of latitude and elevation, and potatoes (*Solanum tuberosum*) are produced during different seasons at different sites (Hackett and Rattigan, 1978). In the late 1970s, a potato simulation model was developed to help identify alternative production areas, to predict the effects of varying planting times to take advantage of favorable climatic and market conditions, to improve the accuracy of crop forecasting, and to help agronomists analyze the performance of experimental and commercial plantings (Hackett et al., 1979a,b; Sands et al., 1979).

The model estimated phenological development of the crop and tuber growth (bulking). The initial version required weekly mean minimum and maximum temperatures throughout the crop, solar radiation for the three weeks following emergence, and an estimate of the maximum soil water deficit experienced during bulking. Later versions simulated the effects of day length, plant population, and crop cultivar on bulking and estimated graded as well as total tuber yields (Sands and Regel, 1983).

The model was used to evaluate several aspects of production, including the possibility of yield improvement in several potato-growing districts, the effects of altering the planting date to take advantage of seasonal price fluctuations, the potential for commercial production in several areas not producing potatoes at the time, and the potential for exporting seed potatoes to lowland tropical regions (Hackett et al., 1979b).

III. CROP MANAGEMENT

The studies cited above illustrate that crop models can be used to identify potential sites for production of specific crops and to evaluate risks associated with climatic factors such as frost and drought stress. In many cases, we know that crops can be grown in an area, but it would be desirable either to reduce the risk of production or, in the case of perishable vegetables, to match production with demand for the product. Simulation models have been used to accomplish both these objectives.

A. Sowing Date and Maturity Group

In the mid-1970s, the semiarid tropical area around Katherine, Australia, had little history of agricultural production. Some 25 years of traditional agronomic experimentation showed that grain sorghum (*Sorghum bicolor*) could be grown there, but yields were low. Researchers suspected that the proper combination of soil type, sowing date, cultivar maturity type, and canopy development rate could result in higher and more stable yields. Angus et al. (1974) used the CROPEVAL method to evaluate alternative grain sorghum cropping systems. Results for a clay loam soil with 100 mm of available water indicated that early-maturing cultivars sown early in the wet season produced the highest yields and had the lowest year-to-year variation. These simulated results were consistent with previous field experimentation in the area and were obtained much more quickly and less expensively. A different management strategy was recommended for deep sandy loam soils with 150 mm of available water. For these soils, late sowing coupled with slow canopy development allowed the crop to conserve water during the vegetative phase and use it more effectively for grain filling later in the season when the radiation-temperature balance was more favorable. This combination of soil and management practices had not been tested in previous field experiments, and the study illustrates how simulation models can be used to identify promising combinations of soils, cultivars, and management strategies for further testing in field experiments.

In a similar study (T. G. Gerik, personal communication), the SORKAM grain sorghum growth model was used to evaluate alternative planting dates and sorghum maturity types in Texas. For locations in west Texas, where rainfall and soil moisture frequently limit growth, early-maturing cultivars sown early (1–15 April) had the highest yield and least variability. In central Texas, medium-maturing cultivars sown near the normal date (15 March) had the highest yield and least variability, while in south and east Texas, late-maturing cultivars sown late (15 May to 15 June) had the highest simulated yields and least variability. However, in south and east Texas, sorghum is frequently sown earlier to avoid pests or other yield-reducing factors. The results of this study illustrate that uncritical acceptance of simulated results can be dangerous. Model users must constantly remind themselves that factors other than those considered by the model may affect the cropping system.

Stapper et al. (1983) conducted similar studies on the risk of wheat (*Triticum aestivum*) production along rainfall gradients in the Middle East. As expected, both measured and simulated results indicated that short-season cultivars often have a relative advantage over longer-season cultivars when drought stress severely limits crop yields.

B. Plant Populations

Plant population or seeding rate is an important management variable. Small populations are often used when anticipated yields are small due to expected drought or inadequate nutrients. In contrast, large plant populations are used when soil water and fertility are expected to be adequate and high yields are anticipated. Crop simulation models are often sensitive to plant population. In most cases, small plant populations limit crop growth by limiting the rate of leaf area development, which affects light interception and dry matter accumulation (Jones and Kiniry, 1986; Kiniry et al., 1992b).

C. Irrigation and Fertilizer Nitrogen

The maize (*Zea mays*) simulation model, CERES-Maize, is one of several cereal models (Jones et al., 1985; Ritchie and Otter, 1985). The model simulates crop phenological development, growth, and economic yield. It operates on a daily time step and requires minimal weather, soil, and genotype-specific inputs. In addition to simulating the effects of genotype, weather, and soil properties on the crop, the model can simulate irrigation and nitrogen fertilization effects.

Researchers and farmers have long used irrigation scheduling models for irrigation water management. Recently, simulation models like CERES-Maize have been used to predict the effects of different irrigation strategies on both crop yields and irrigation water losses. In dry regions, irrigators often need to utilize scarce irrigation water efficiently. In humid regions where the probability of rainfall during the season is high, irrigation strategies should take the probability of rainfall into account. By using simulation models with long-term weather records, researchers can devise strategies that maximize yields while minimizing water losses due to runoff and deep percolation.

Martin et al. (1985) used the CERES-Maize model to compare three alternative irrigation strategies for each of two soils in Michigan. Thirty-year simulations using weather data generated stochastically (Richardson, 1984) were used. Different irrigation amounts (6, 19, and 32 mm) were applied when drought stress reached a predetermined level, and the model predicted total irrigation amounts, crop yields, drainage losses, and irrigation water use efficiencies. Soils differed in their responses to irrigation strategies. For example, small amounts of water could be applied to a soil with high plant-extractable water without significantly reducing crop yields. This strategy used less irrigation water and resulted in less deep drainage than other methods. However, use of the same strategy on a soil with low plant-extractable water reduced crop yields, although it increased irrigation water use efficiency. The study illustrates the important effects of local conditions on irrigation water management, and studies of the type described are valuable sources of data for site-specific economic analyses.

Effective fertilizer management is often complicated by nitrogen's mobility and soil transformations. It is subject to loss via ammonia volatilization, denitrification, immobilization, and leaching. Its availability to crops can be increased by contributions from rainfall, biological fixation, and mineralization.

The CERES-Maize model simulates many of these processes, and it can be used to simulate crop response to fertilizer nitrogen for specific environmental and management conditions. Singh (1985) used the model to simulate corn response to weather, soil, and nitrogen fertilizer rates for 14 field experiments conducted at different times of the year and on either Oxisols or Andepts in Hawaii.

The model was able to simulate effects of soils, seasons, hybrids, and nitrogen rates on corn yield. Within a soil family, the optimum nitrogen fertilizer rate varied among seasons. In cooler months, the nitrogen response curve reached a maximum at lower fertilizer rates because growth was limited by temperature and solar radiation. Grain yields were higher for the Oxisols than for the Andepts because of higher solar radiation and temperatures at the Oxisol sites. Cultivars differed in yield because of variation in the durations and rates of grain filling.

Although the model normally simulated the shape of the fertilizer nitrogen response curve accurately, it consistently overestimated or underestimated yields in about half the experiments. These apparent errors might be explained by lack of accurate initial soil mineral and organic nitrogen data and the inability of the model to simulate insect, weed, disease, and wind damage and other nutrient deficiencies and toxicities.

D. Furrow Diking

For dryland crops, farmers cannot alter the amount or timing of precipitation; however, they can often manage its utilization. Proper tillage and residue management can improve infiltration and minimize evaporation losses from the soil surface. Fallow can be used to increase stored soil water, and furrow dikes (also known as tied ridges, row dams, or basin listing) can trap runoff until it infiltrates.

Yield response to furrow dikes depends on the soil's water-holding and infiltration characteristics, initial soil water content, and within-season precipitation amounts and timing. The benefits of furrow dikes may be small in years with little potential runoff or with adequate rainfall. However, benefits may be large in years when dikes increase infiltration and that water is later used by the crop (Bilbro and Hudspeth, 1977; Gerard et al., 1983, 1984; Lyle and Dixon, 1977). Simulation models recently have been used to supplement the experimental data and to estimate the probability and magnitude of increased yields due to furrow diking. Krishna et al. (1987) used the SORGF model (Arkin et al., 1976) and runoff algorithms from the Erosion-Productivity Impact Calculator (EPIC) (Williams et al., 1984) to simulate grain sorghum yields. The model was first validated using experimental runoff and crop yield data for diked and undiked situations in the few years and sites with adequate data. Effects of furrow diking were then simulated for three sites in central, southwest, and northwest Texas using 20 years of daily weather records. Results indicated that furrow diking would increase mean grain sorghum yields for all three locations, but the amount of increase depended on the year, the site, and whether the dikes were maintained all year or only during the growing season. The central Texas site had the highest annual precipitation (854 mm) and the lowest simulated mean yield response to furrow diking (490 kg ha^{-1}) due to the relatively low probability of severe drought stress. The southwest Texas site (618 mm) had the highest mean yield response (1160 kg ha^{-1}). The northwest Texas site (467 mm) had an intermediate response (720 kg ha^{-1}) because, even though drought stress severely limited yields, furrow dikes trapped relatively little runoff in this dry environment.

Studies of this type, using validated simulation models, can provide farmers with valuable information concerning the benefits of water conservation practices such as furrow diking. A farmer might even base the decision to install dikes on soil hydrologic characteristics and soil water contents at planting.

E. Ratoon Crops

A few crops, like sugarcane, rice, and grain sorghum, can regrow after harvest to produce a second (or ratoon) crop. In the case of sugarcane, many ratoons may be produced before the field must be plowed and the crop replanted.

Jones et al. (1988) developed a version of EPIC, named AUSCANE, that simulates the growth, biomass, and sugar yields of sugarcane. In addition to being sensitive to weather, irrigation, nitrogen, phosphorus, drainage, micronutrient deficiencies, and soil toxicities, AUSCANE simulates both plant and ratoon cane growth.

When rainfall and temperatures are adequate, as in the southeastern United States and some tropical areas, grain sorghum can produce good ratoon yields (Duncan, 1981). However, the practice is less feasible in regions where the probability of adequate rainfall during the ratoon crop is low. Central Texas is an area with erratic late-season rainfall, and few experimental studies of ratoon sorghum have been conducted. Consequently, Stinson et al. (1981) conducted a feasibility study using the SORGF crop model in conjunction with the HYMO hydrologic model (Williams and Hann, 1978). Results were validated by comparing simulated ratoon yields with those obtained by farmers in the area. Subsequent simulations using historical weather records from the area (Stinson et al., 1981) suggested that, though no ratoon yields would be produced in some dry years, yields as high as 3500 kg ha^{-1} were possible. Median ratoon yields were 500 kg ha^{-1}. The low investment required to produce the ratoon crop and its high potential returns have encouraged more farmers to consider ratooning sorghum and have stimulated research to optimize varietal selection and cultural practices (G. F. Arkin, 1986, personal communication). Subsequent field and simulation studies have better defined areas in Texas with good potential for production of ratoon grain sorghum (T. J. Gerik, 1986, personal communication).

F. Weed Competition

Several crop simulation models have been adapted to predict weed growth and/or the dynamic interactions between a crop and a selected weed, usually taking into consideration factors such as population densities, light, water, and nutrients (Debaeke et al., 1993; Graf et al., 1990; Kiniry et al., 1992b; Retta et al., 1996; Wilkerson et al., 1990). Most such weed-crop competition models are specific to a single crop and one or more weeds (Graf et al., 1990; Wilkerson et al., 1990). Others, like ALMANAC, are capable of simulating a number of crop and weed species (Debaeke et al., 1993; Kiniry et al., 1992a). Like EPIC (Williams et al., 1989) from which it was derived, ALMANAC uses a generic crop growth model with a number of crop-specific variables that differentiate the growth and development of different crop and weed species. Water movement and nutrient movement and transformations are simulated in detail using a daily time step. Potential biomass growth (including roots) is calculated as a function of intercepted solar radiation and radiation use efficiency, which varies among species. Light interception by leaf canopies is simulated using Beer's law and leaf area index. Intercepted light is partitioned among competing species by their ratio of leaf area indexes, weighted by their respective extinction coefficients and the height of their canopies. Thus taller species and those with larger extinction coefficients are given preference for light interception. The model also simulates competition for water and nutrients, taking into account their availability, the relative demands of the competing species, and the depth and distribution of their root systems. This approach can produce relatively accurate estimates of the effects of interspecies competition (Debaeke et al., 1993; Kiniry et al., 1992b).

G. Soil Fertility and Plant Nutrition

Nitrogen (N) and phosphorus (P) play key roles in plant nutrition. Nitrogen is the mineral element required in greatest quantity by cereal crops, it is the most widely used fertilizer nutrient, and demand for it is likely to grow in the foreseeable future (Novoa and Loomis, 1981).

Despite our large investment in fertilizer N, the efficiency with which crops utilize it is poor. Power (1981) concluded that plants generally recover from 20 to 90% of the fertilizer N applied. The N which is not recovered by the crop may be lost through volatilization of ammonia, leaching of nitrate, or denitrification. Organic N may also be lost with eroded sediments, and mineral nitrogen can be rendered inaccessible to the plant through immobilization in the soil organic matter or through lack of soil water.

Phosphorus is also an essential element for plant growth, and in many cases low soil P levels limit crop yields. Consequently, P is often added to the soil as mineral fertilizer or manure. Considerable information is available on the complex chemical and physiological transformations of P in the soil and uptake by plants. Like N, P exists in both mineral and organic forms. Both forms can be lost with eroded sediments, and mineral P can be rendered inaccessible to the plant due to immobilization by soil organic matter or sorption on clay particles.

Traditionally, information needed to make fertilizer recommendations has been obtained from field experiments with different amounts of applied nutrients. However, regression models of fertilizer response are static in nature and are unable to account for variability in crop nutrient demand or the supply of water or nutrients.

Computer models that simulate the effects of weather, soil properties, and agronomic practices on nutrient dynamics and crop growth might improve our understanding and management of fertilizer in complex cropping systems. However, the many transformations of soil N and P and their sensitivity to soil and weather conditions make simulation of these two elements complex and difficult.

Many simulation models describe some or all of the processes affecting N transformations (Frissel and van Veen, 1981; Tanji, 1982). Several of the same models are concerned with specific aspects of the N cycle such as ammonia volatilization (Parton et al., 1981), leaching (Addiscott, 1981; Burns, 1980), and denitrification (Smith, 1981; Leffelaar, 1981). Most of the models are concerned primarily with the major soil processes in the N cycle and few consider the N dynamics of a growing crop. Several very comprehensive models containing details linking together all the major soil transformations have also been developed. Among them, those of van Veen (1977) and Molina et al. (1983) provide considerable detail about the functioning of the various transformations at the microbiological level.

Several models have been developed to simulate pollution from organic wastes and from excessive fertilization (Rao et al., 1981; Selim and Iskandar, 1978, 1981; Donigian and Crawford, 1976). Others simulate most of the major soil N transformations but do not consider crop growth and its response to other factors (Watts and Hanks, 1978; Tillotson et al., 1980; Tanji et al., 1981).

Several models integrate simulation of soil N transformations with crop growth and yield. PAPRAN (Seligman and van Keulen, 1981) simulates the effects of N and water on pasture production. The CERES-N models (Jones and Kiniry, 1986; Otter-Nacke et al., 1986; Godwin and Vlek, 1985; Godwin et al., 1984) simulate the effects of weather, crop genotype, irrigation, soil N transformations, and fertilizer N management on corn and wheat growth and yield. The NTRM (Shaffer and Larson, 1987) and EPIC (Williams et al., 1984) models include many important N transformations as components of comprehensive cropping system models.

Thus, models are now available to predict the effects of soil moisture properties, weather, fertilizer products and practices, tillage, residue management, and crop rotations on the more important soil and plant N transformations, soil fertility, and crop growth and yield.

Phosphorus (P) is an essential constituent of all living organisms, and its deficiency limits crop yields. Consequently, P is often added to the soil-plant system to improve and/or increase crop yields. At least three schools of P modeling can be distinguished: 1) relatively simple, user-oriented, predictive models of fertilizer P effectiveness; 2) detailed models that simulate flux of solution P to plant roots, and 3) more comprehensive models of soil P transformations, with or without uptake by the plant.

A simple user-oriented model that predicts the effects of soil P buffering capacity on residual fertilizer was developed by Bennett and Ozanne (1972) and subsequently modified by Helyar and Godden (1976). Another user-oriented model, DECIDE, helps make fertilizer P recommendations based on crop, soil type, past soil fertility, and yield goals.

The second type of model simulates P flux to plant roots (Claassen and Barber, 1976; Cole et al., 1977; Cushman, 1982; Helyar and Munns, 1975; Olsen and Kemper, 1967; Schenk and Barber, 1979a). These models generally follow the work of Nye and Marriott (1969) to describe the movement of P to individual roots by mass flow and diffusion. Absorption kinetics of the roots are assumed to follow Michaelis-Menten kinetics. For example, Claassen and Barber (1976) proposed a model based on theoretical consideration of potassium uptake by plant roots growing in soil. This model was subsequently modified to predict P uptake by corn (Schenk and Barber, 1979a) and soybean (Silberbush and Barber, 1983). Soil parameters required in the model include the effective average diffusion coefficient, initial P concentration in solution, and buffer power. Plant parameters include maximum influx rate, the Michaelis-Menten constant, water influx, root radius, initial root length, and rate of root growth. The model has subsequently been improved (Cushman, 1984); although it does not simulate the cycling of P in inorganic (solution and absorbed) and organic pools, it provides a quantitative assessment of the importance of root morphology (Schenk and Barber 1979a,b), soil temperature (Mackay and Barber, 1984), and fertilizer placement (Anghinoni and Barber, 1980) on P uptake by roots.

A number of more comprehensive soil and plant P models are available. A mechanistic model was proposed by Mansell et al. (1977) to describe transformations and transport of applied P under steady water flow conditions in uniform soils. Inorganic soil P was considered in five phases (Ryden et al., 1977): solution, physically absorbed, chemisorbed, immobilized, and precipitated. However, for practical applications of the model, a method is needed to assign rate constants based on soil properties and to simulate organic P transformations and their effects on inorganic pools.

Cole et al. (1977) constructed a simulation model of the P cycle in semiarid grasslands. Basically, P supply to the plant is assumed to be limited by diffusion. As plant uptake reduces the P concentration at the root surface, a concentration gradient is established in the surrounding water films and phosphate ions flow toward the root. The soluble P pool is replenished from slightly soluble phosphate minerals, phosphate-absorbing surfaces, and organic P mineralization (Cole et al., 1977). This and similar models (Chapin et al., 1978; Harrison, 1978; Mishra et al., 1979) incorporate the effects of moisture, temperature, soil properties, plant phenology, and organic matter decomposition on P flows. Although they require input data that are not readily available, they have revealed gaps in our knowledge of processes and provide valuable direction for future research.

An even more comprehensive soil and plant P model capable of long-term simulations of soil organic, inorganic, and plant processes affecting P fertility was developed as a component of the EPIC model described earlier (Williams et al., 1984). EPIC was designed to simulate the effects of soil, weather, and crop management on soil erosion and erosion's long-term effects on crop productivity (Putman and Dyke, 1987; Putman et al., 1987). Consequently, its soil P model simulates changes in P fertility for a wide variety of soils, climates, crop management practices, and crop rotations (Jones et al., 1984a,b; Sharpley et al., 1984). Solution P lost in runoff is predicted using labile P concentration in the top soil layer, runoff volume, and a partitioning factor. Sediment transport of P is estimated from concentration in the topsoil layer, sediment yield, and an enrichment ratio.

Mineralization from the fresh organic P pool is governed by C: N and C: P ratios, soil water, temperature, and the stage of residue decomposition. Mineralization from the stable organic P pool associated with humus is estimated as a function or organic P weight, labile P concentration, soil water, and temperature. Daily P immobilization is computed by subtracting the amount of P contained in the crop residue from the amount assimilated by the microorganisms.

Mineral P is transferred among three pools: labile, active mineral, and stable mineral. When P fertilizer is applied, a fraction remains in the labile pool (available for plant use). The remainder is quickly sorbed by the active mineral pool. Simultaneously, P flows from the active mineral pool to the labile pool (usually at a much lower rate). Flow between the labile and active mineral pools is governed by temperature, soil water, a P sorption coefficient, and the amount of material in each pool. The P sorption coefficient is a function of soil chemical and physical properties.

Crop P uptake is governed by the demand of the crop and the labile P concentration in the soil. Although the EPIC P model is comprehensive, it is designed to use only readily available soil inputs: soil test P and soil chemical, physical, and taxonomic data from routine soil pedon characterization.

In recent years, models capable of simulating long-term changes in soil fertility have been developed. A good example is CENTURY, a mechanistic simulation model that mimics the effects of cropping systems and fertilizer inputs on soil C, N, and P pools. In a study by Cole et al. (1988), CENTURY accurately predicted the major changes in organic matter and nutrient-supplying capacity of the soil as the result of conversion of grasslands to croplands in the U.S. Great Plains. The model captured the effects of precipitation, temperature, and soil texture on changes in organic matter levels across the region. Results showed that, in contrast to organic matter losses in earlier periods, improved management practices over the last 40 years have stabilized soil organic matter pools and have improved degraded soils. Use of higher-yielding crop varieties, increased fertilizer rates, reduced tillage, and residue management are keys to recent stabilization of organic matter levels.

IV. COMPLEX CROPPING SYSTEMS

Cropping systems are often complex, with several possible crops that can be grown in monoculture or in rotation, dryland or irrigated, with different amounts of fertilizer and crop protection inputs, and using a variety of tillage implements. Only recently have cropping systems models been developed that are capable of simulating most of the variables that farmers must manage. The most widely used cropping systems model is EPIC, the Erosion-Productivity Impact Calculator. This model is capable of simulating weather, hydrology, soil erosion by water and wind, irrigation, fertilization, tillage, nutrient (nitrogen and phosphorus) and pesticide transformations and movement, crop development and growth, economic yields, long-term impacts of crop rotations, erosion control practices, and nutrient depletion or build-up (Williams et al., 1984, 1989). The model has been used in numerous studies to predict the effects of crop rotations (Dyke et al., 1990; Cabelguenne et al., 1988, 1990), impacts of carbon dioxide and climate change (Robertson et al., 1990; Stockle et al., 1992), irrigation (Cabelguenne et al., 1988; Dyke et al., 1990), contamination of surface and ground waters by agricultural nutrients and pesticides (Benson et al., 1990; Lacewell et al., 1993; Williams et al., 1994), effects of weed competition (Debaeke et al., 1993), soil erosion and soil degradation (Benson et al., 1990), and impact of animal manure application (Edwards et al., 1994). In addition, the crop model component of EPIC (Williams et al., 1989) that can simulate multiple crops grown in complex rotations with realistic management practices is well established (Cabelguenne et al., 1988, 1990; Dyke et al., 1990; Kiniry et al., 1992a, 1995, 1996). These capabilities permit EPIC to predict the long-term sustainability of specific complex cropping systems (Jones et al., 1991). For example, Sharpley et al. (1987) used EPIC to simulate the effects of crop rotations, tillage, and fertilization on long-term depletion of organic C, N, and P throughout soil profiles. Benson et al. (1989) use the same approach to predict soil erosion and long-term degradation of the soil's productive potential, taking into account the effects of soil type, climate, and soil conservation practices.

Two studies illustrate how complex, multicrop systems models like EPIC and ALMANAC can be used to predict the performance of alternative cropping systems.

A. Rainfed Rice

In the first study, ALMANAC was calibrated to simulate upland rice growth and yields using data from line-source experiments (Chinchest, 1981; Cruz and O'Toole, 1984) and heuristic information concerning differences between traditional upland, intermediate upland, and high-yielding semidwarf cultivars (O'Toole et al., 1982). The model simulated alternative rainfed rice cropping systems for the drought-prone rainfed rice belt across the Ganga Plains of the Indian states of Bihar and Uttar Pradesh. ALMANAC simulated the effects of cultivar type (traditional upland, intermediate, and semidwarf), maturity group (short, medium, and long season), fertilizer nitrogen amount, runoff control with bunds, soil type, and presence of a legume relay crop on rice yields and yield variability.

Among cultivar types, traditional cultivars had the advantage of great year-to-year stability, but yields were low. High-yielding semidwarf cultivars had the greatest yield potential but the least yield stability. Among maturity groups, long-season (140–day) cultivars had the greatest yield potential but the least stability. Bunding increased average yields, especially when combined with fertilization. Chickpeas grown on stored soil moisture after the rice harvest could furnish the subsequent rice crop with more than 20 kg N ha^{-1} and provide an additional source of protein.

From an agronomic standpoint, the most desirable alternatives appeared to be to use a cultivar with intermediate stature and intermediate (105–day) maturity, construct bunds to retain runoff and reduce drought stress, and apply over 20 kg N ha^{-1} or plant chickpeas after the rice harvest to provide adequate nitrogen nutrition. However, it is clear that the farmer would have to evaluate the increased labor and/or capital required to care for the crop, maintain the bunds, and plant/harvest the chickpeas. Some farmers might not choose to expend additional labor or capital in order to increase crop yields.

B. Dryland Crop Rotations

Lacewell et al. (1988) used the EPIC cropping systems model to evaluate alternative crop rotations for the Texas high plains. They used results of short-term farming system experiments (Keeling, 1986–1987; Hatfield, 1985–1986) to calibrate the model. The model was then used to evaluate the long-term effects of selected dryland cropping systems on crop yield, soil erosion, and economic returns (McGrann et al., 1986).

The cropping systems selected included monocultures and rotations of cotton, grain sorghum, and wheat under conventional and reduced tillage. Additional systems consisted of cotton or grain sorghum followed immediately by winter wheat (to control wind erosion) and terminated with herbicide in later winter or early spring. Twelve cropping systems were simulated for two soil types and for a total of ten 48–year periods. Yields, wind and water erosion, and net returns assuming the producer did or did not participate in the Federal Farm Program were simulated.

Average annual soil erosion rates varied from 3 to 32 t ha^{-1}. As expected, the highest simulated erosion rates were for monoculture cotton and grain sorghum or rotations of the two. The lowest simulated erosion was found for rotations including wheat. Participation in the Federal Farm Program increased net returns, and cotton was found to be an essential part of a profitable dryland farming system in the region. However, wheat should be used in the rotation to decrease soil erosion, and continuous cotton with wheat terminated by herbicide in late winter appeared to offer a viable system.

This study illustrates the usefulness of cropping system models for simulation of complex systems sensitive to biophysical (soil and weather), social (soil erosion, Federal Farm Program), and economic constraints.

V. SUMMARY

Agriculture is a complex and dynamic activity that is sensitive to the natural environment, economics, and government policies. Management of complex agricultural systems is not simple and has traditionally relied on strategies based on field experimentation and expert opinion. However, recent advances in computer hardware and software have stimulated development and use of natural resource databases, expert systems, and simulation models to predict and manage the behavior of agricultural systems.

Simulation models are being used to predict crop adaptation to specific environmental conditions, predict crop response to alternative management practices, predict the behavior of complex cropping systems, and predict short- and long-term changes in soil fertility and productivity due to soil and crop management. Studies of guayule, sunflower, and potato adaptation to diverse environments in Australia illustrate the use of models to match crop requirements with soils and climates. Examples are given to show how simulation models can be used to evaluate alternative sowing dates, crop maturity groups, irrigation and fertilizer strategies, furrow diking, and ratooning grain sorghum.

Models that simulate soil and crop nutrient dynamics can be used to improve our understanding and management of fertilizer and soil fertility. Models of specific N and P transformations as well as comprehensive models including inorganic, organic, and plant components are available. Complexity and structure of the models vary with their intended use, and the most comprehensive models are capable of simulating long-term changes in soil properties and fertility due to weather and management.

Comprehensive simulation models can be used to optimize complex cropping systems. Examples are drawn from studies of rainfed rice in India and dryland crop rotations on the high plains of Texas.

Simulation of agricultural processes and systems has advanced dramatically in the last 15 years. The future promises even more rapid advances and the routine use of models to manage agricultural systems.

REFERENCES

Addiscott, T. M. 1981. Leaching of nitrate in structured soils. *In*: M. J. Frissel and J. A. van Veen (eds.). Simulation of nitrogen behavior of soil-plant systems. PUDOC, Wageningen, Netherlands.

Anderson, W. K., R. C. G. Smith, and J. R. McWilliam. 1978a. A systems approach to the adaptation of sunflower to new environments. I. Phenology and development. Field Crops Res. 1: 141–152.

Anderson, W. K., R. C. G. Smith, and J. R. McWilliam. 1978b. A systems approach to the adaptation of sunflower to new environments. II. Effects of temperature and radiation on growth and yield. Field Crops Res. 1: 153–163.

Anghinoni, I. and S. A. Barber. 1980. Predicting the most efficient phosphorus placement for corn. Soil Sci. Soc. Am. J. 44: 1016–1020.

Angus, J. F., J. J. Basinski, and H. A. Nix. 1974. Weather analysis and its application to production strategy in areas of climatic instability. Report to the IAD/UNDP international expert consultation on the use of improved technology for food production in rainfed areas of tropical Asia. FAO, Rome.

Arkin, G. F., R. L. Vanderlip, and J. T. Ritchie. 1976. A dynamic grain sorghum growth model. Trans. ASAE 19: 622–626.

Bennett, D. and P. G. Ozanne. 1972. Australia, CSIRO Division of Plant Industry, Annual Report.

Benson, V. W.. O. W. Rice, P. T. Dyke, J. R. Williams, and C. A. Jones. 1989. Conservation impacts on crop productivity for the life of a soil. J. Soil Water Conserv. 44: 600–604.

Benson, V. W., W. A. Goldstein, D. L. Young, J. R. Williams, C. A. Jones, and J. R. Kiniry. 1990. Impacts of integrated cropping practices on nitrogen use and movement, pp. 426–428. *In*: P. W. Unger, W. R. Jordan, T. V. Sneed, and R. W. Jensen (eds.) Challenges in dryland agriculture—a global perspective. Texas Agric. Exp. Sta., College Station.

Bilbro, J. D. and E. B. Hudspeth, Jr. 1977. Furrow diking to prevent runoff and increase yields of cotton. Texas Agric. Exp. Stn. PR-3436.

Burns, I. G. 1980. A simple model for predicting the effects of leaching of fertilizer nitrate during the growing season on the nitrogen fertilizer needs of crops. J. Soil Sci. 31: 175–202.

Cabelguenne, M., C. A. Jones, J. R. Marty, and H. Quinones. 1988. Contribution a l'etude des rotations culturales: Tentative d'utilisation d'un modele agronomique. Agronomie 8: 549–556.

Cabelguenne, M., C. A. Jones, J. R. Marty, P. T. Dyke, and J. R. Williams. 1990. Calibration and validation of EPIC for crop rotations in southern France. Agric. Syst. 333: 153–171.

Chapin, F. S., R. J. Barsdate, and D. Barel. 1978. Phosphorus cycling in Alaska coastal tundra: a hypothesis for the regulation of nutrient-cycling. Oikos 31: 181–199.

Chinchest, A. 1981. The effects of water regimes and nitrogen rates on nitrogen uptake and growth of rice varieties. Ph.D. Thesis, Cornell University, Ithaca, New York.

Claassen, N. and S. A. Barber. 1976. Simulation model for nutrient uptake from soil by a growing plant root system. Agron. J. 68: 961–964.

Cole, C. V., I. C. Burke, W. H. Parton, D. S. Schimel, and J. W. B. Stewart. 1988. Analysis of historical changes in soil fertility and organic matter levels of the North American Great Plains, pp. 199–209. *In*: Proc. International Conference on Dryland Farming, Amarillo and Bushland, Texas, 15–19 August.

Cole, C. V., G. S. Innis, and J. W. B. Stewart. 1977. Simulation of phosphorus cycling in semiarid grasslands. Ecology 58: 1–15.

Cruz, R. T. and J. C. O'Toole. 1984. Dryland rice response to an irrigation gradient at flowering stage. Agron. J. 76: 178–183.

Cushman, J. H. 1982. Nutrient transport inside and outside the root rhizosphere: theory. Soil Sci. Soc. Am. J. 46: 704–709.

Cushman, J. H. 1984. Nutrient transport inside and outside the root rhizosphere: generalized model. Soil Sci. 138: 164–171.

Debaeke, P., J. Kiniry, J.-P. Caussanel, and B. Kafiz. 1993. An integrated model of crop-weed interaction: Application to wheat-oats mixtures. Proc. 8th EWRS Symp., "Quantitative approaches in weed and herbicide research and their practical application," Braunschweig, pp. 5–15.

Donigian, A. S. and N. H. Crawford. 1976. Modelling pesticides and nutrients on agricultural lands. EPA-600/2–76–043, U.S. Environmental Protection Agency, Office of Research and Development, Environmental Research Laboratory, Athens, Georgia.

Duncan, R. R. 1981. Ratoon cropping of sorghum for grain in the southern United States. Univ. Georgia College Agric. Exp. Stn. Res. Bull. 269, Athens, Georgia.

Dyke, P. T., C. A. Jones, J. W. Keeling, J. E. Matocha, and J. R. Williams. 1990. Calibration of a farming systems model for the southern Coastal Plain and High Plains of Texas. Texas Agric. Exp. Stn. MP-1696.

Edwards, D. R., V. W. Benson, J. R. Williams, T. C. Daniel, J. Lemunyon, and R. G. Gilbert. 1994. Use of the EPIC model to predict runoff transport of surface-applied inorganic fertilizer and poultry manure constituents. Trans. ASAE 37: 403–409.

Frissel, M. J. and J. A. van Veen. 1981. Simulation of nitrogen behavior of soil-plant systems. PUDOC, Wageningen, Netherlands.

Gerard, C. J., P. D. Sexton, and D. M. Conover. 1984. Effects of furrow diking, subsoiling, and slope position on crop yields. Agron. J. 76: 945–950.

Gerard, C. J., P. D. Sexton, and D. M. Matus. 1983. Furrow diking for cotton production in the rolling plains. Texas Agric. Exp. Stn. PR-4174.

Godwin, D. C., C. A. Jones, J. T. Ritchie, P. L. G. Vlek, and L. G. Youngdahl. 1984. The water and nitrogen components of the CERES models, pp. 95–100. *In*: ICRISAT (International Crops Research Institute for the Semi-Arid Tropics). Proc. Int. Symp. on Minimum Data Sets for Agrotechnology Transfer, 21–26 March, 1983. ICRISAT Center, Patancheru, India.

Godwin, D. C. and P. L. G. Vlek. 1985. Simulation of nitrogen dynamics in wheat cropping systems, pp. 311–332. *In*: W. Day and R. K. Atkin (eds.). Wheat growth and modelling. Plenum, New York.

Goyne, P. J. and G. L. Hammer. 1982. Phenology of sunflower cultivars. II. Controlled environment studies of temperature and photoperiod effects. Aust. J. Agric. Res. 33: 251–261.

Graf, B., A. P. Gutierrez, O. Rakotobe, P. Zahner, and V. Delucchi. 1990. A simulation model for the dynamics of rice growth and development. II. The competition with weeds for nitrogen and light. Agric. Systems 32: 367–392.

Hackett, C. and K. Rattigan. 1978. The location and timing of potato production in Australia. CSIRO Australian Div. Land Use Res. Tech. Paper No. 38.

Hackett, C., P. J. Sands, and H. A. Nix. 1979a. A model of the development and bulking of potatoes. (*Solanum tuberosum* L.). II. Prediction of district commercial yields. Field Crops Res. 2: 333–347.

Hackett, C., P. J. Sands, and H. A. Nix. 1979b. A model of the development and bulking of potatoes. (*Solanum tuberosum* L.). III. Some implications for potato production and research. Field Crops Res. 2: 349–364.

Hammer, G. L. and P. J. Goyne. 1982. Determination of regional strategies for sunflower production. Proc. 10th Int. Sunflower Conf., March, Surfers Paradise, Australia.

Hammer, G. L., P. J. Goyne, and D. R. Woodruff. 1982. Phenology of sunflower cultivars. III. Models for prediction in field environments. Aust. J. Agric. Res. 33: 263–274.

Hammer, G. L. and G. M. McKeon. 1984. Evaluating the effect of climate variability on management of dryland agricultural systems in northeastern Australia. *In*: E. A. Fitzpatrick (ed.). Need for climatic and hydrologic data in agriculture of Southeast Asia, Canberra, 12–15 December 1983. Symposium sponsored by United Nations University.

Harrison, A. F. 1978. Phosphorus cycles of forest and upland grassland systems and some effects of land management practices, pp. 175–195. *In*: Phosphorus in the environment: its chemistry and biochemistry. CIBA Foundation Symp. 57, Elsevier/North-Holland, Amsterdam.

Hatfield, J. L. 1985–1986. Wheat yield data. U.S. Department of Agriculture, Agricultural Research Service, Cropping Systems Research Laboratory, Lubbock, Texas.

Helyar, K. R. and D. P. Godden. 1976. The phosphorus cycle—What are the sensitive areas? pp. 23–30. *In*: G. J. Blair (ed.). Prospects for improving efficiency of phosphorus utilization. Proceedings of a symposium at the University of New England, Armidale, N.S.W., Australia. Reviews in Rural Sci. III.

Helyar, K. R. and D. N. Munns. 1975. Phosphate fluxes in the soil-plant system: a computer simulation. Hilgardia 43: 103–130.

Jones, C. A. and J. R. Kiniry. 1986. CERES-Maize: a simulation model of maize growth and development. Texas A&M University Press, College Station.

Jones, C. A., C. V. Cole, A. N. Sharpley, and J. R. Williams. 1984a. A simplified soil and plant phosphorus model. I. Documentation. Soil Sci. Soc. Am. J. 48: 800–805.

Jones, C. A., A. N. Sharpley, and J. R. Williams. 1984b. A simplified soil and plant phosphorus model. III. Testing. Soil Sci. Soc. Am. J. 48: 810–813.

Jones, C. A., J. R. Williams, A. N. Sharpley, and C. V. Cole. 1985. Testing and nutrient components of the Erosion-Productivity Impact Calculator (EPIC), pp. 211–213. *In*: Proc. Natural Resources Modeling Symposium. ARS-30, U.S. Department of Agriculture, Agricultural Research Service, Washington, DC.

Jones, C. A., M. K. Wegener, J. S. Russell, I. M. McLeod, and J. R. Williams. 1988. AUSCANE—Simulation of Australian sugarcane with EPIC. CSIRO Div. of Tropical Crops and Pastures, Technical Paper 29.

Jones, C. A., P. T. Dyke, J. R. Williams, J. R. Kiniry, V. W. Benson, and R. H. Griggs. 1991. EPIC: An operational model for evaluation of agricultural sustainability. Agric. Systems 37: 341–350.

Keeling, J. W. 1986–1987. Farming systems research data. Texas Agric. Exp. Stn., Lubbock.

Kiniry, J. R., R. Blanchet, J. R. Williams, V. Texier, C. A. Jones, and M. Cabelguenne. 1992a. Sunflower simulation using the EPIC and ALMANAC models. Field Crops Res. 30: 403–423.

Kiniry, J. R., J. R. Williams, P. W. Gassman, and P. Debaeke. 1992b. A general process-oriented model for two competing plant species. Trans. ASAE 35: 801–810.

Kiniry, J. R., D. J. Major, R. C. Izaurralde, J. R. Williams, P. W. Gassman, M. Morrison, R. Bergentine, and R. P. Zentner. 1995. EPIC model parameters for cereal, oilseed, and forage crops in the northern Great Plains region. Can. J. Plant Sci. 75: 679–688.

Kiniry, J. R., M. A. Sanderson, J. R. Williams, C. R. Tischler, M. A. Hussey, W. R. Ocumpaugh, J. C. Read, G. Van Esbroeck, and R. L. Reed. 1996. Simulating alamo switchgrass with the ALMANAC model. Agron. J. 88: 602–613.

Krishna, J. H., G. F. Arkin, J. R. Williams, and J. R. Mulkey. 1987. Simulating furrow-dike impacts on runoff and sorghum yields. Trans. ASAE 30: 143–147.

Lacewell, R. D., M. E. Chowdhury, K. J. Bryant, J. R. Williams, and V. W. Benson. 1993. Estimated effect of alternative production practices on profit and ground water quality: Texas Seymour aquifer. Great Plains Res. 3: 189–213.

Lacewell, R. D., J. G. Lee, C. W. Wendt, R. J. Lascano, and J. W. Keeling. 1988. Implications of alternative dryland crop rotations: Texas High Plains, pp. 635–638. *In*: Proc. Int. Conf. on Dryland Farming, Amarillo and Bushland, Texas, 15–19 August.

Leffelaar, P. A. 1981. A model to simulate partial anaerobiosis. *In*: M. J. Frissel and S. A. van Veen (eds.). Simulation of nitrogen behavior of soil-plant systems. PUDOC, Wageningen, Netherlands.

Lyle, W. M. and O. R. Dixon. 1977. Basin tillage for rainfall retention. Trans. ASAE 20: 1013–1017.

Mackay, A. D. and S. A. Barber. 1984. Soil temperature effects on root growth and phosphorus uptake by corn. Soil Sci. Soc. Am. J. 48: 818–823.

Mansell, R. S., H. M. Selim, and J. G. A. Fiskell. 1977. Simulated transformations and transport of phosphorus in soil. Soil Sci. 124: 102–109.

Martin, E. C., J. T. Ritchie, and T. L. Loudon. 1985. Use of the CERES-Maize model to evaluate irrigation strategies, pp. 342–350. *In*: Advances in evapotranspiration. Proceedings of the national conference, 16–17 December, St. Joseph, Michigan. American Society of Agricultural Engineers.

McGrann, J. M., K. D. Olson, T. A. Powell, and T. R. Nelson. 1986. Microcomputer budget management system user's manual, version 2.6. Department of Agricultural Economics, Texas A&M University, College Station.

Mishra, B., P. K. Khanna, and B. Ulrich. 1979. A simulation model for organic phosphorus transformation in a forest soil ecosystem. Ecol. Model. 6: 31–46.

Molina, J. A. E., C. E. Clapp, M. J. Shaffer, F. W. Chester, and W. E. Larson. 1983. NCSOIL, a model of nitrogen and carbon transformations in soil: description, calibration, and behavior. Soil Sci. Soc. Am. J. 47: 85–91.

Nix, H. A. 1981. Simplified simulation models based on specified minimum data sets: the CROPEVAL concept, pp. 151–169. *In*: A. Berg (ed.). Application of remote sensing to agricultural production forecasting. Balkema, Rotterdam.

Nix, H. A. and J. P. McMahon. 1979. Jojoba in Australia: where? *In*: Proc. First Australian Jojoba Conf., Bathurst, N.S.W., AAAT, Sydney.

Novoa, R. and R. S. Loomis. 1981. Nitrogen and plant production. Plant Soil 58: 177–204.

Nye, P. H. and F. H. C. Marriott. 1969. A theoretical study of the distribution of substances around roots resulting from simultaneous diffusion and mass-flow. Plant Soil 30: 459–472.

Olsen, S. R. and W. D. Kemper. 1967. Movement of nutrients to plant roots. Adv. Agron. 20: 91–151.

O'Toole, J. C., T. T. Chang, and B. Somrith. 1982. Research strategies for improvement of drought resistance in rainfed rices, pp. 201–222. *In*: Research strategies for the future. IRRI, Los Banos, Philippines.

Otter-Nacke, S., D. C. Godwin, and J. T. Ritchie. 1986. Yield model development: Testing and validating the CERES-WHEAT model in diverse environments. Agristars Publ. No. YM-15–00407, JSC 20244, NASA, Houston, Texas.

Parton, W. J., W. D. Gold, F. J. Adamson, S. Torbit, and R. G. Wodmansee. 1981. NH_3 volatilization model, pp. 233–244. *In*: M. J. Frissel and J. A. van Veen (eds.). Simulation of nitrogen behavior of soil-plant systems. PUDOC, Wageningen, Netherlands.

Power, J. F. 1981. Nitrogen in the cultivated ecosystem. *In*: F. E. Clark and T. Rosswell (eds.). Terrestrial nitrogen cycles: processes, ecosystems strategies, and management imports. Ecol. Bull. No. 33, Stockholm.

Putman, J. W. and P. T. Dyke. 1987. The erosion-productivity impact calculator as formulated for the Resource Conservation Act appraisal. Staff report AGES861204, Natural Resource Economics Division, Economic Research Service, U. S. Department of Agriculture, Washington, DC.

Putman, J. W., P. T. Dyke, G. L. Wistrand, and K. F. Alt. 1987. The erosion-productivity index stimulator model. Staff report AGES870602, Natural Resource Economics Division, Economic Research Service, U.S. Department of Agriculture, Washington, DC.

Rao, P. S., J. M. Davidson, and R. E. Jessup. 1981. Simulation of nitrogen behavior in the root zone of cropped land areas receiving organic wastes, pp. 81–95. *In*: M. J. Frissel and J. A. van Veen (eds.). Simulation of nitrogen behavior of soil-plant systems. PUDOC, Wageningen, Netherlands.

Retta, A., R. L. Vanderlip, R. A. Higgins, and L. J. Moshier. 1996. Application of SORKAM to simulate shattercane growth using forage sorghum. Agron. J. 88: 596–601.

Richardson, C. W. 1984. WGEN: a model for generating daily weather variables. ARS-8, U.S. Department of Agriculture, Agricultural Research Service.

Ritchie, J. T. and S. Otter. 1985. Description of performance of CERES-Wheat: a user-oriented wheat yield model, pp. 159–175. *In*: W. D. Willis (ed.). ARS wheat yield project. ARS-38, U.S. Department of Agriculture, Agricultural Research Service.

Robertson, T., C. Rosenzweig, V. W. Benson, and J. R. Williams. 1990. Projected impacts of carbon dioxide and climate change in the Great Plains, pp. 675–677. *In*: P. W. Unger, W. R. Jordan, T. V. Sneed, and R. W. Jensen (eds.). Challenges in dryland agriculture—a global perspective. Texas Agric. Exp. Sta., College Station.

Ryden, J. C., J. R. McLaughlin, and J. K. Syers. 1977. Mechanisms of phosphate sorption of soils and hydrous ferric oxide gel. J. Soil Sci. 28: 72–92.

Sands, P. J., C. Hackett, and H. A. Nix. 1979. A model of the development and bulking of potatoes (*Solanum tuberosum* L.) I. Derivation from well managed field crops. Field Crops Res. 2: 309–331.

Sands, P. J. and P. A. Regel. 1983. A model of the development and bulking of potatoes (*Solanum tuberosum* L.). V. A simple model for predicting graded yields. Fields Crops Res. 6: 25–40.

Schenk, M. K. and S. A. Barber. 1979a. Phosphate uptake by corn as affected by soil characteristics and root morphology. Soil Sci. Soc. Am. J. 43: 880–883.

Schenk, M. K. and S. A. Barber. 1979b. Root characteristics of corn genotypes as related to P uptake. Agron J. 71: 921–924.

Seligman, N. C. and H. van Keulen. 1981. PAPRAN: a simulation model of annual pasture production limited by rainfall and nitrogen, pp. 192–221. *In*: M. J. Frissel and J. A. van Veen (eds.). Simulation of nitrogen behavior of soil-plant systems. PUDOC, Wageningen, Netherlands.

Selim, H. M. and I. K. Iskandar. 1978. A simplified nitrogen model for land treatment of wastewater. *In*: International Symposium on Land Treatment of Wastewater. U.S. Army Cold Regions Research Laboratory, Hanover, New Hampshire.

Selim, H. M. and I. K. Iskandar. 1981. Modeling nitrogen transport and transformations in soils. I. Theoretical considerations. Soil Sci. 131: 133–241.

Shaffer, M. J. and W. E. Larson. 1987. NTRM, a soil-crop simulation model for nitrogen, tillage, and crop residue management. Conservation Res. Rep. 34–1, U.S. Department of Agriculture, Agricultural Research Service.

Sharpley, A. N., C. A. Jones, Carl Gray, and C. V. Cole. 1984. A simplified soil and plant phosphorus model. II. Prediction of labile, organic, and sorbed phosphorus. Soil Sci. Soc. Am. J. 48: 805–809.

Sharpley, A. N., S. J. Smith, J. R. Williams, and C. A. Jones. 1987. Simulation of soil formation and degradation processes, pp. 495–506. *In*: V. Gardiner (ed.). International geomorphology 1986, Part II. Wiley, London.

Silberbush, M. and S. A. Barber. 1983. Prediction of phosphorus and potassium uptake by soybean with a mechanistic mathematical model. Soil Sci. Soc. Am. J. 47: 262–265.

Singh, U. 1985. A crop growth model for predicting corn (*Zea mays* L.) performance in the tropics. Ph.D. Thesis, University of Hawaii.

Smith, R. C. G., W. K. Anderson, and H. C. Harris. 1978a. A systems approach to the adaptation of sunflower to new environments. III. Yield predictions for continental Australia. Field Crops Res. 1: 215–228.

Smith, R. C. G., S. D. English, and H. C. Harris. 1978b. A systems approach to the adaptation of sunflower to new environments. IV. Yield variability and optimum cropping strategies. Field Crops Res. 1: 229–242.

Smith, K. A. 1981. A model of denitrification in aggregated soils. *In*: M. J. Frissel and J. A. van Veen (eds.). Simulation of nitrogen behavior of soil-plant systems. PUDOC, Wageningen, Netherlands.

Stapper, M., H. C. Harris, and R. C. G. Smith. 1983. Risk analysis of wheat yields in relation to cultivar maturity type and climatic variability in semi-arid areas, using a crop growth model. *In*: E. A. Fitzpatrick (ed.). Need for climatic and hydrologic data in agriculture of Southeast Asia. Canberra, 12–15 December 1983. Symposium sponsored by United Nations University.

Stinson, D. L., G. F. Arkin, T. A. Howell, C. W. Richardson, and J. R. Williams. 1981. Modeling grain sorghum ratoon cropping and associated runoff and sediment losses. Trans. ASAE 24: 631–635.

Stockle, C. O., J. R. Williams, N. J. Rosenberg, and C. A. Jones. 1992. A method for estimating the direct and climatic effects of rising atmospheric carbon dioxide on growth and yield of crops: Part I—Modification of the EPIC model for climate change analysis. Agric. Systems 38: 225–238.

Tanji, K. K. 1982. Modelling of the nitrogen cycle, pp. 721–772. *In*: F. J. Stevenson (ed.). Nitrogen in agricultural soils. Monogr. 22, Am. Soc. Agron., Madison, Wisconsin.

Tanji, K. K., M. Mehran, and S. K. Gupta. 1981. Water and nitrogen fluxes in the root zone of irrigated maize, pp. 51–66. *In*: M. J. Frissel and J. A. van Veen (eds.). Simulations of nitrogen behavior in soil-plant systems. PUDOC, Wageningen, Netherlands.

Tillotson, W. R., C. W. Robbins, R. J. Wagenet, and R. J. Hanks. 1980. Soil water, solute and plant growth simulation. Utah Agric. Exp. Stn. Bull. 502, Utah State University.

van Veen, J. A. 1977. The behavior of nitrogen in soils. A computer simulation model. Ph.D. Thesis. Vrije Universiteit, Amsterdam.

Watts, C. C. and R. J. Hanks. 1978. A soil-water-nitrogen model for irrigated corn on sandy soils. Soil Sci. Soc. Am. Proc. 42: 492–499.

Wilkerson, G. G., J. W. Jones, H. D. Coble, and J. L. Gunsolus. 1990. SOYWEED: A simulation model of soybean and common cocklebur growth and competition. Agron. J. 82: 1000–1010.

Williams, J. R. and R. W. Hann. 1978. Optimal operation of large agricultural watersheds with quality constraints. Tech. Rep. No. 96. Texas Water Resources Institute, Texas A&M University, College Station.

Williams, J. R., C. A. Jones, and P. T. Dyke. 1984. A modelling approach to determining the relationship between erosion and soil productivity. Trans. ASAE 27: 129–144.

Williams, J. R., C. P. Jones, J. R. Kiniry, and D. A. Spanel. 1989. The EPIC crop growth model. Trans. ASAE 32: 497–511.

Williams, J. R., J. G. Arnold, C. A. Jones, V. W. Benson, and R. H. Griggs. 1994. Water quality models for developing soil management practices, pp. 349–382. *In*: R. Lal and B. Stewart (eds.) Soil processes and water quality.

8

Wheat and Barley

I. INTRODUCTION

Wheat (*Triticum aestivum* L.) and barley (*Hordeum vulgare* L.) together constitute the world's most important cereal crops. Wheat is the leading cereal in terms of total world production, and barley is the world's fourth most important cereal crop after wheat, rice, and corn (FAO, 1992). These two cereals contribute about 41% of the world production of important cereal crops. Wheat and barley were cultivated in western Asia at least 9000 years ago. All the important early civilizations were based on some kind of cereal– wheat and barley in the Middle East and Mediterranean, rice in southern and eastern Asia, and corn in the New World (Purseglove, 1985). During the last two decades, developing countries have raised wheat production more rapidly than have wheat-producing countries of the developed world. According to a study conducted by Centro Internacional de Mejoramento de Maiz Y Trigo (CIMMYT; 1985), during 1970–72 to 1982–84, developing countries increased wheat production on the order of 54% of total production, whereas in developed countries the increase was 46% of total production. The increase in production was mainly due to use of better adapted high-yielding, semidwarf cultivars and better management practices. From 1900 until 1951, Kansas winter wheat yields averaged only 1079 kg ha^{-1} (Shroyer and Cox, 1993). But over the past 15 years, with the advent of semidwarf cultivars and increased nitrogen use, yields have more than doubled to more than 2300 kg ha^{-1} (Shroyer and Cox, 1993). Cox et al. (1988) reported that breeding efforts alone had increased grain yields of hard red winter wheat in the Great Plains of the United States by about one percent of the baseline yield per year between 1919 and 1987. In reviewing the literature, Austin et al. (1989) concluded that genetic improvement in winter wheat yields across several countries around the world ranged from 0.4 to 0.8% per year.

The leading wheat- and barley-producing countries in the world are China, India, United States, Canada, USSR, Australia, France, United Kingdom, Germany, Argentina, Turkey, and Pakistan. Detailed data related to production in different continents, yield per hectare, and area planted are given by Briggle and Curtis (1987) for wheat and by Poehlman (1985) for barley.

Both of these cereals are important sources of food for human consumption and feed for livestock. On a worldwide scale, wheat contributes approximately 30% of the total cereal production (FAO, 1992), making wheat a major source of minerals for many people. Wheat starch is used for laundering, paper laminating and corrugating, adhesives, textiles, wallpaper, billboard paste, and paper additives (Miller, 1974). Low-grade wheat also is fermented for the production of alcohol. Similarly, use of barley malt in brewing is well known. Several of the cereal grains may be used for malt, but barley, wheat, and rye are unique in the production of α-amylase and β-amylase enzymes, which hydrolyze starch to dextrans and fermentable sugars (Dickson, 1979; Peterson and Foster, 1973).

Grain protein content is one of the most important malt quality traits in barley. Protein content of barley grain malted in the United States should not exceed 135 g kg^{-1} for midwestern six-rowed genotypes and 130 g kg^{-1} for western two-rowed genotypes. Grain protein content that exceeds recommended levels is undesirable for malting because it may increase steep times and may cause uneven water uptake during steeping, uneven germination during malting and increasing malt loss due to abnormal growth, excessive enzymatic activity, low extract yields, excessive nitrogenous compounds in the wort during brewing, and chill-haze formation in beer (Goblirsch et al., 1996).

II. CLIMATE AND SOIL REQUIREMENTS

Wheat and barley are the most important cereals of the temperate regions, but are grown at high altitudes in the tropics and also extend into the tropical lowlands. Wheat production is concentrated between latitudes 30 and 60°N and 27 and 40°S (Nuttonson, 1955), but wheat is also grown beyond these

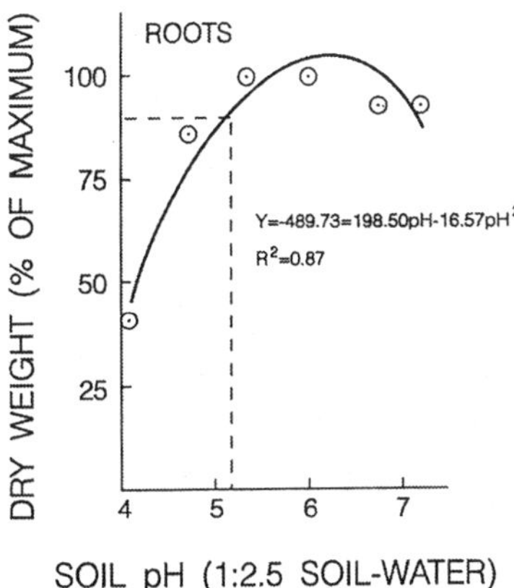

Figure 8.1 Relationship between soil pH and dry weight of roots of wheat in an Oxisol.

limits. Similarly, barley has a broad ecological adaption. It is grown at latitude 64°N in Alaska, latitude 67°N in Finland, and latitude 70°N in Norway. It is the only cereal that matures at these high latitudes (Nuttonson, 1955). The minimum temperature for growth of wheat is about 3–4°C, the optimum temperature is about 25°C, and the maximum is about 30–32°C (Briggle, 1980). Barley also has more or less similar temperature requirements for growth. Both these cereals may be injured by frost during the flowering and early grain-filling periods. Wheat and barley can withstand heat in a dry climate or high humidity in a cool climate, but they perform poorly in a hot, humid climate due to increased infestation of diseases.

The availability of water is a major factor limiting cereal production in most regions of the world. Wheat can be grown in regions where annual precipitation ranges from 250 to 1750 mm; about three-fourths of the land area used for wheat production receives an average of 375–875 mm annually (Briggle and Curtis, 1987). The response of wheat to six levels of soil-water tension was studied under field conditions in an Oxisol of the Brazilian Savannah Region (Guerra, 1995). The crop was irrigated when the soil-water tension, measured with tensiometers or gypsum blocks, reached values of 41, 51, 69, 185, 562, and 993 kPa at a depth of 10 cm throughout the crop cycle. The water was applied to the crop by using a plastic hose connected to a perforated PVC tube to assure uniform water distribution on the experimental plots. The results indicate a reduction in grain yield with the increase in soil-water tension. The highest yield (6952 kg/ha) was obtained with the lowest soil-water tension (41 kPa). The yield components that caused reduc-

Table 8.1 Grain Yield (kg/ha) of Wheat Genotypes under Different Aluminum Saturation

	Aluminum saturation (%)[a]				
Genotype	0	15	30	45	Mean
Trigo BR 23	2928A	2773B	2511B	2262C	2619B
IAPAR 29	3411A	2884B	2311B	1431D	2510B
IAPAR 60	3505A	3219A	2886A	2564B	3043A
OCEPAR 16	3391A	2518B	2254B	2233C	2599B
Trigo BR 35	3505A	3367A	3225A	3023A	3280A
IAPAR 6	3286A	2836B	2813B	2297C	2808B
IAPAR 53	3384A	3252A	2918A	2503B	3014A
Trigo BR 18	3420A	2843B	2454B	2291C	2752B
IAC 5-Maringà	3444A	3179A	3116A	2845A	3146A
Anahuac	3308A	2161B	1521C	1189D	2045C
MEAN	3358a	2903b	2601c	2264d	2782
F[b]	0.44	3.72**	4.64**	9.40**	12.68**
C.V.(%)	12.97	11.47	15.56	14.19	14.04

[a] Means followed by the same capital letters in the columns, and lower case letters in the rows, do not differ significantly among themselves by the Scott-Knott test at 5% probability.
[b] Significant to a 1% (**) probability level.
Source: Costa et al., 1996.

tion in yield were the number of spikes per square meter, number of spikelets per spike, and number of grains per spike. No significant difference was observed for 1,000–grain weight and hectoliter-grain weight. The results indicate that plants define their development and yield as a function of soil-water tension to maintain grain quality.

Barley is often regarded as a drought-resistant crop. The water requirement for production of a unit weight of grain is less than for other cereals (Carlton, 1916), its transpiration rate being the lowest among the small grains (Nuttonson, 1955). Almost two-thirds of the world's barley production is in subhumid or semiarid regions (Poehlman, 1985). Short periods (3–5 days) of very high temperature (33–40°C) can markedly reduce the yield and quality of wheat (Randall and Moss, 1990). The tolerance of wheat yield to very high temperatures is known to vary with genotypes. Genotypic variation of response to high temperature of the order of 20% was recorded for the majority of yield and quality components by Stone and Nicolas (1995). The fact that responses of this magnitude were caused by exposure to high temperatures lasting only

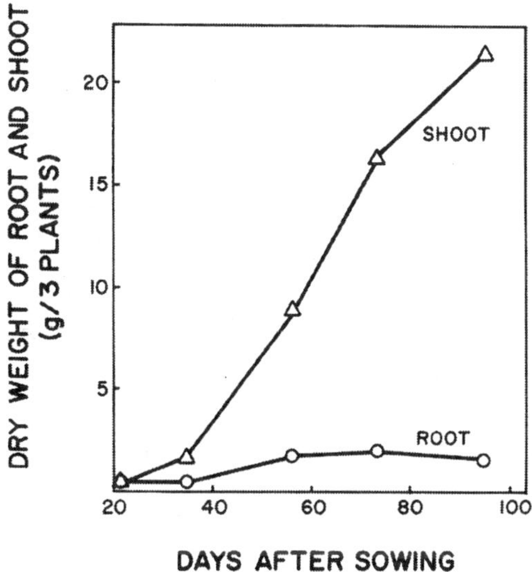

Figure 8.2 Dry weight of root and shoot of wheat plant days after sowing in Oxisol of central Brazil under greenhouse conditions. (From Fageria, 1990.)

5–6% of the grain-filling period demonstrates the extent to which short periods of very high temperature may affect wheat yield and quality (Stone and Nicolas, 1995). Barley is an important crop in arid and semiarid regions of the United States. According to the U.S. Department of Commerce, slightly over 20% of barley is irrigated (Farm and Ranch Irrigation Survey, 1979) and that the average yield of irrigated barley was about 4000 kg ha^{-1} (compared to about 2000 kg ha^{-1} for nonirrigated barley).

In the tropical climates, irrigation is necessary to produce good yields of wheat and barley when precipitation is less than 200 mm during crop growth. Rainfall distribution during crop growth is a critical factor in most production environments. When irrigation is needed, the irrigation should be scheduled so that the available soil water in a 120–cm profile does not fall below 50–60% to minimize water stress at all times (Baldridge et al., 1985).

Wheat and barley can be grown on a variety of soils, but reasonable drainage and good water-holding capacity are preferred. In terms of soil, both these cereals are considered medium acid-tolerant (Adams, 1981; Carver and Ownby, 1995), but according to Doll (1964) the permissible soil pH range for wheat is 5.5–7, whereas for barley the range is 6.5–7.8. Figure 8.1 shows root dry weight of wheat as a function of increasing soil pH in an Oxisol of central Brazil. Root dry weight was increased with increasing soil pH up to a value of about 6, then it was decreased. This means that in Oxisols the pH should be raised up to 6 for getting optimum growth of wheat. Aluminum is an important component of soil acidity, and genotype differences have been reported in relation to Al tolerance in wheat (Table 8.1). According to Bower and Fireman (1957), barley is more tolerant than other cereal crops in alkaline soils and less tolerant to acid soils. A soil pH in the range of 6–8.5 is generally acceptable for barley growth (Poehlman, 1985). Both wheat and barley are also considered to be medium tolerant to soil salinity. Salinity at initial yield decline (threshold) is reported to be about 6 dS m^{-1} for both species (Maas and Hoffman, 1977). However, differences in tolerance to acidity exist among cultivars in both species.

III. GROWTH AND DEVELOPMENT

Plant growth and development are complex processes, depending on many factors both internal and external. They can be considered to begin with germination, followed by a large complex series of morphological and physiological events that are called growth and development (Ting, 1982). These two terms are sometimes confused and used interchangeably, but they are different. Growth is an irreversible increase in size or volume accompanied by the biosynthesis of new protoplasmic constituents. Development is a combination of both growth and cellular differentiation, a higher order of change that involves anatomical and physiological specialization and organization. According to Salisbury and Ross (1985), the process by which cells become specialized is called differentiation, and the processes of growth and differentiation of individual cell into tissues and organs are often called development. Another useful term for this process is morphogenesis.

According to the above definition, growth is a quantitative aspect of development, representing an increase in the number and size of cells, and differentiation is the qualitative aspect of development (Ting, 1982). As a quantitative aspect, growth can be measured in terms of increase in volume or weight of the crop plant or its organ. In field crops, productivity is the main objective of cultivation; in this situation, dry weight of the plant or plant parts is the preferred measurement of growth. Figure 8.2 shows the growth curve of root and shoot dry weight of the wheat plant as a function of time. Root and shoot growth was slow in the beginning. Root growth increased until 60 days of growth and then remained more or less constant. But shoot growth increased significantly after about 35 days of growth and continued to increase until maturity. The slow early growth of the shoot may be attributed to the relatively small number of cells that can divide, the small leaf area available for light interception and photosynthesis, and perhaps the relatively large percentage of photosynthate going to the roots (Brown, 1984).

The root is a very important organ of the plant which provides mechanical support and absorbs nutrients and water. The roots contribute about 15% of the total weight of the wheat plant at 60 days of growth and 3% at maturity (Figure 8.2). The extent of the root system is more important than its weight.

Bauer et al. (1987) studied dry matter distribution in the wheat plant and showed that the leaves constituted 100% of aereal dry matter after emergence, about 50% at the flag-leaf stage, and 20% at anthesis. Dry matter accumulation in winter wheat increased rapidly from the jointing growth stage through grain in the stiff dough stage (Waldren and Flowerday, 1979). Translocation of dry matter from leaves to grain begins at flowering, and translocation from culms and head to grain begin at grain filling.

Plant growth occurs through cell division and cell elongation. Areas where cells are actively dividing are called meristem regions. The plant embryo contains only meristem tissue, so all plant parts such as roots, stems, leaves, and flowers are derived from the meristem (Stoskopf, 1981). A detailed description of growth and development of barley is given by Wych et al. (1985) and of wheat by Simmons (1987).

A. Growth Analysis

Growth analysis is the procedure of analyzing plant growth rate by expressing it as the algebraic product of a series of factors (Hardwick, 1984). The growth analysis formulas, together with necessary and sufficient conditions for their use, have been discussed in several articles (e.g., Emecz, 1962; Radford, 1967; Evans, 1972; Hunt, 1979; Charles-Edwards and Fisher, 1980; Warren, 1981; Wilson, 1981; Jolliffe et al., 1982; Hardwick, 1984; Brown, 1984). Growth is analyzed principally as crop growth rate, relative growth rate, and net assimilation rate. The plant parameters which are commonly measured to calculate growth rates are dry matter and leaf area.

Crop Growth Rate

The dry matter accumulation rate per unit of land area is referred to as crop growth rate (*CGR*), normally expressed as g m^{-2} day^{-1} (Brown, 1984). It can be calculated by:

$$CGR = \frac{W_2 - W_1}{SA\,(t_2 - t_1)}$$

where W_1 and W_2 are crop dry weight at the beginning and end of the interval, t_1 and t_2 are corresponding days, and *SA* is the soil area occupied by the plants at each sampling. Crop growth pattern can be defined accurately by taking plant samples at different time intervals during the growing season. Crop growth rate is normally low in the early growth stage and increases with time, reaching a maximum value around flowering. Crop growth rate studies help in the interpretation of experimental results of different crop cultivars and other management practices and in the evaluation of fertility status of the soils.

Relative Growth Rate

The relative growth rate (*RGR*) of a plant at an instant in time (*t*) is defined as the increase of plant material per unit of material present per unit of time (Radford, 1967). It can be calculated and expressed in g(g dry wt.)$^{-1}$ day^{-1}:

$$RGR = \frac{1}{w}\frac{dw}{dt} = \frac{d}{dt}(\log_e w)$$

where w is the dry weight and dw/dt is the change in dry weight per unit time.

Net Assimilation Rate

The dry matter accumulation per unit of leaf area is termed net assimilation rate (*NAR*) and is expressed as g(m leaf area)$^{-2}$ day^{-1} (Brown, 1984):

$$NAR = \frac{1}{A}\frac{dw}{dt}$$

where A is the leaf area and dw/dt is the change in plant dry matter per unit time. The objective of measuring *NAR* is to determine the efficiency of plant leaves in dry matter production. Net assimilation rate decreases with crop growth due to mutual shading of leaves and reduced photosynthetic efficiency of older leaves.

B. Growth Stages

Starting from germination until harvest, the crop plant develops through different growth stages which are determined by environmental factors, cultivars, and cultural practices. Knowledge of these growth stages is useful in deciding the right time for cultural operations such as top-dressing of N and application of herbicides, insecticides, and fungicides. Further, identification of growth stage is useful in many physiological studies to identify the critical stages in the life cycle of the plant which are sensitive to environmental factors in order to take necessary measures to improve crop yields. In cereals, for example, the reproductive and ripening stages are more sensitive to water deficiency than the vegetative stage (Yoshida, 1972). In wheat the embryo ear has its full complement of spikelets well before the ear emerges from the flag leaf sheath, and any treatment intended to increase spikelet number per ear must consider this (Kirby and Appleyard, 1984).

The most common method for identifying growth stages in cereals is known as the Feekes scale (Table 8.2). The Feekes scale is based on the external appearance of the plant or plant organs. The Feekes scale was improved by Zadoks et al. (1974) and later reproduced by Tottman and Makepeace (1979) and Tottman (1987) with drawings of selected growth stages of wheat, barley, and oats. The whole growth cycle of cereals is divided into 10 principal growth stages, and each principal growth stage is subdivided into secondary growth stages. A detailed description of these principal and secondary growth stages is given in Table 8.3. Some of the growth stages representing wheat and barley growth and development from germination to ripening are present in Figures 8.3 through 8.10.

During the seedling growth stage, only leaves on the main shoot should be counted, excluding tillers and their leaves. A leaf can be described as unfolded when its ligule has emerged from the sheath of the preceding leaf (Tottman, 1987). The inflorescence development of cereals was described in detail by Bonnet (1966), Nerson et al. (1980), Kirby and Appleyard (1984), Reid (1985), and Lersten (1987), and readers may refer to these publications for detailed information.

Waldren and Flowerday (1979) also gave a growth description for winter wheat based on a 10–point scale (Table 8.4). According to these authors, the growth stages described by the Feekes scale are not easily distinguished by farmers because the scale is based on small morphological changes which are not readily apparent, especially at the later stages. Waldren and Flowerday (1979) claim that the growth stages described by them (Table 8.4) are identified by morphological changes which can be easily detected by farmers, students, and others. The stages have less secondary and tertiary division than those of the Feekes scale, and separate distinctly different periods of development, such as heading and flowering, in separate stages.

C. Partitioning of Dry Matter

Partitioning of photosynthetic products to vegetative and reproductive organs is an important factor in improving crop yield. Wheat and barley are determinate plants and divert photosynthetic products to vegetative growth early in the season and to grain later (Brown, 1984). Grain yield improvement in cereals like wheat and barley over the past four decades was achieved through an increase in harvest index and a decrease in plant height (Polisetty, 1993). Harvest index, the ratio of grain to total biological yield, is a measure of the degree to which a crop partitions photosynthetic products into grains (Donald and Hamblin, 1976; Wych et al., 1985). In wheat the harvest index reported in the literature is around 0.34 in traditional old cultivars and about 0.44 in improved new cultivars (CIMMYT, 1972). This means that the recent gains

(text continues on p. 264)

Table 8.2 Growth Stages in Cereals Based on the Feekes Scale

Growth stage		Description	
1		One shoot (number of leaves can be added) = "braiding."	Tillering
2		Beginning of tillering.	
3		Tillers formed, leaves often twisted spirally. In some varieties of winter wheat, plants may be "creeping" or prostate.	
4		Beginning of the erection of the eudostem, leaf sheaths beginning to lengthen.	
5		Pseudostem (formed by sheaths of leaves) strongly erected.	
6		First node of stem visible at base of shoot.	Stem extension
7		Second node of stem formed, next to last leaf just visible.	
8		Last leaf visible, but still rolled up, ear beginning to swell.	
9		Ligule of last leaf just visible.	
10		Sheath of last leaf completely grown out, ear swollen but not yet visible.	
10.1		First ears just visible (awns showing in barley, ear escaping through split of sheath in wheat or oats).	Heading
	10.2	Quarter of heading process completed.	
	10.3	Half of heading process completed.	
	10.4	Three-quarters of heading process completed.	
	10.5	All ears out of sheath.	
	10.5.1	Beginning of flowering (wheat).	Flowering
	10.5.2	Flowering complete to top of ear.	
	10.5.3	Flowering over at base of ear.	
	10.5.4	Flowering over, kernel watery ripe.	
11.1		Milky ripe.	Ripening
11.2		Mealy ripe, contents of kernel soft but dry.	
11.3		Kernel hard (difficult to divide by thumbnail).	
11.4		Ripe for cutting. Straw dead.	

Source: Large, 1954.

Table 8.3 Cereal Growth Stages: Description of the Principal and Secondary Growth Stages

0	Germination
00	Dry seed
01	Start of imbibition (water absorption)
02	—
03	Imbibition complete
04	—
05	Radicle (root) emerged from caryopsis (seed)
06	—
07	Coleoptile (shoot) emerged from caryopsis
08	—
09	Leaf just at coleoptile tip
1	Seedling growth
10	First leaf through coleoptile
11	First leaf unfolded
12	2 leaves unfolded
13	3 leaves unfolded
14	4 leaves unfolded
15	5 leaves unfolded
16	6 leaves unfolded
17	7 leaves unfolded
18	8 leaves unfolded
19	9 or more leaves unfolded
2	Tillering
20	Main shoot only
21	Main shoot and 1 tiller
22	Main shoot and 2 tillers
23	Main shoot and 3 tillers
24	Main shoot and 4 tillers
25	Main shoot and 5 tillers
26	Main shoot and 6 tillers
27	Main shoot and 7 tillers
28	Main shoot and 8 tillers
29	Main shoot and 9 or more tillers
3	Stem elongation
30	Ear at 1 cm (pseudo-stem erect)
31	First node detectable
32	Second node detectable
33	Third node detectable
34	Fourth node detectable
35	Fifth node detectable
36	Sixth node detectable
37	Flag leaf just visible
38	—
39	Flag leaf ligule just visible

Table 8.3 (Continued)

4	Booting
40	—
41	Flag leaf sheath extending
42	—
43	Boots just visibly swollen
44	—
45	Boots swollen
46	—
47	Flag sheath opening
48	—
49	First awns visible
5	Inflorescence (ear/panicle emergence
50	—
51	First spikelet of inflorescence just visible
52	—
53	1/4 of inflorescence emerged
54	—
55	1/2 of inflorescence emerged
56	—
57	3/4 of inflorescence emerged
58	—
59	Emergence of inflorescence completed
6	Anthesis (flowering)
60	—
61	Beginning of anthesis
62	—
63	—
64	—
65	Anthesis halfway
66	—
67	—
68	—
69	Anthesis complete

Table 8.3 (Continued)

7	Milk development
70	—
71	Caryopsis (kernel) water ripe
72	—
73	Early milk
74	—
75	Medium milk
76	—
77	Late milk
78	—
79	—
8	Dough development
80	—
81	—
82	—
83	Early dough
84	—
85	Soft dough
86	—
9	Ripening
90	—
91	Caryopsis hard (difficult to divide)
92	Caryopsis hard (not dented by thumbnail)
93	Caryopsis loosening in daytime
94	Overripe, straw dead and collapsing
95	Seed dormant
96	Viable seed giving 50% germination
97	Seed not dormant
98	Secondary dormancy induced
99	Secondary dormancy lost

Source: Tottman, 1987.

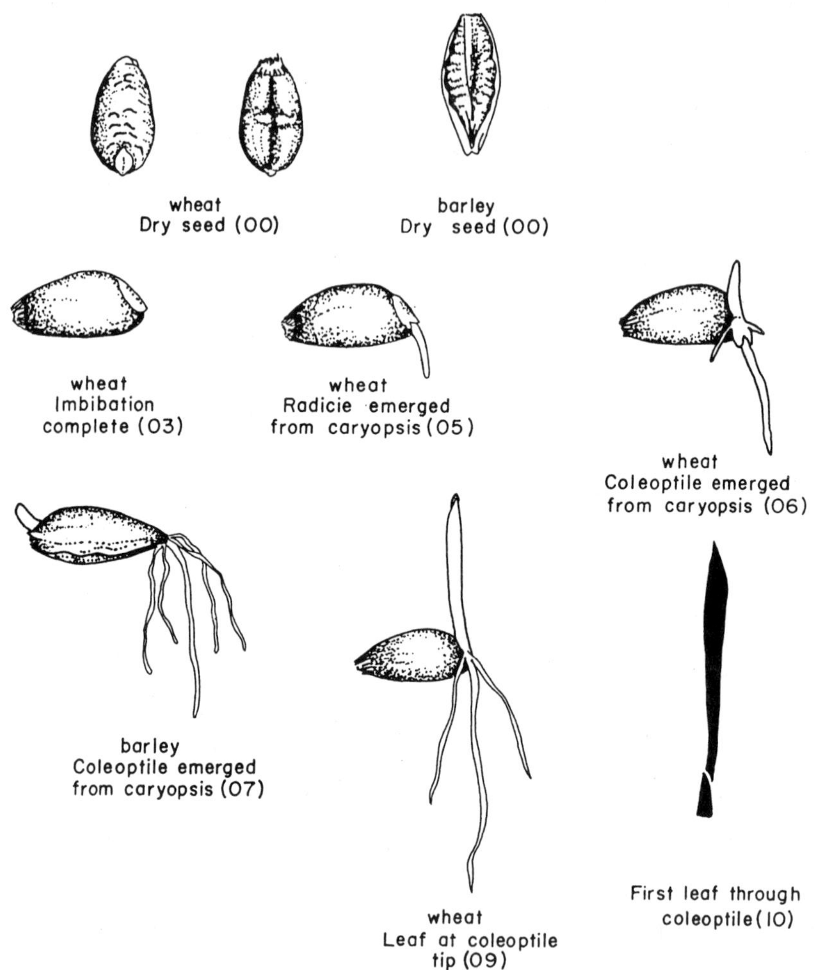

Figure 8.3 Germination and seedling growth in wheat and barley. (From Tottman, 1987.)

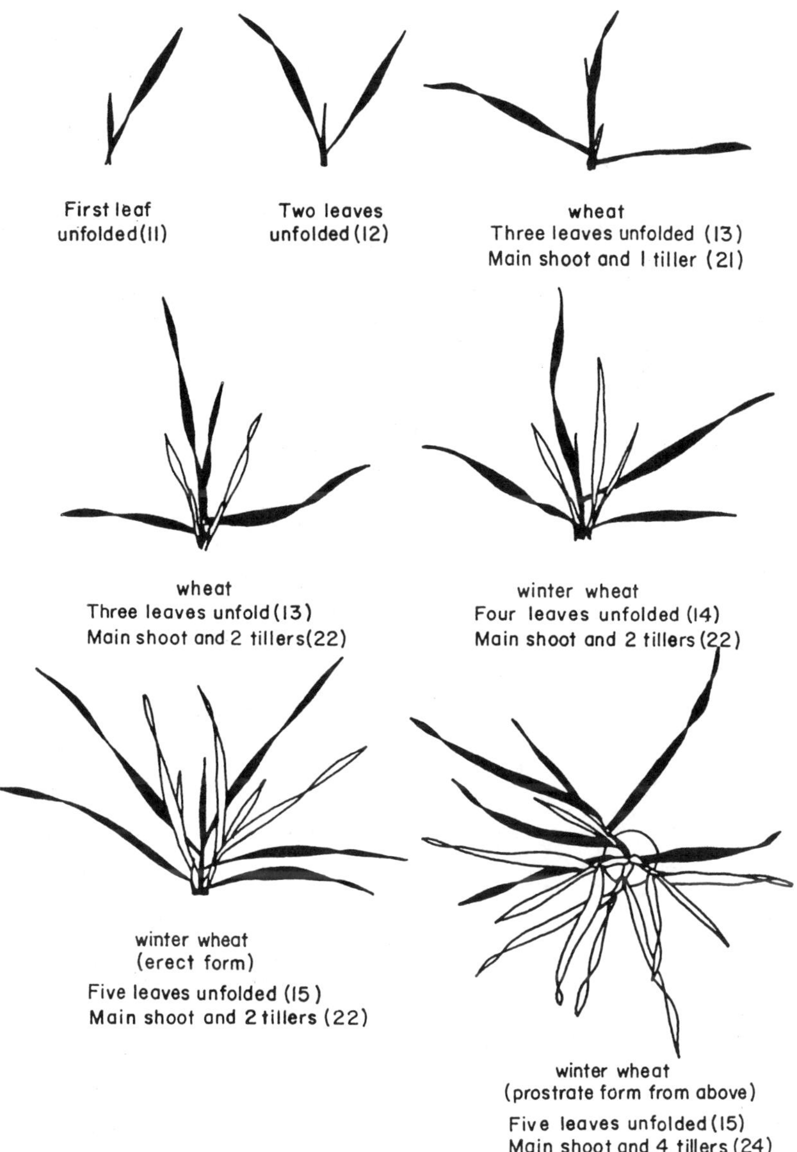

Figure 8.4 Seedling growth and tillering in wheat. (From Tottman, 1987.)

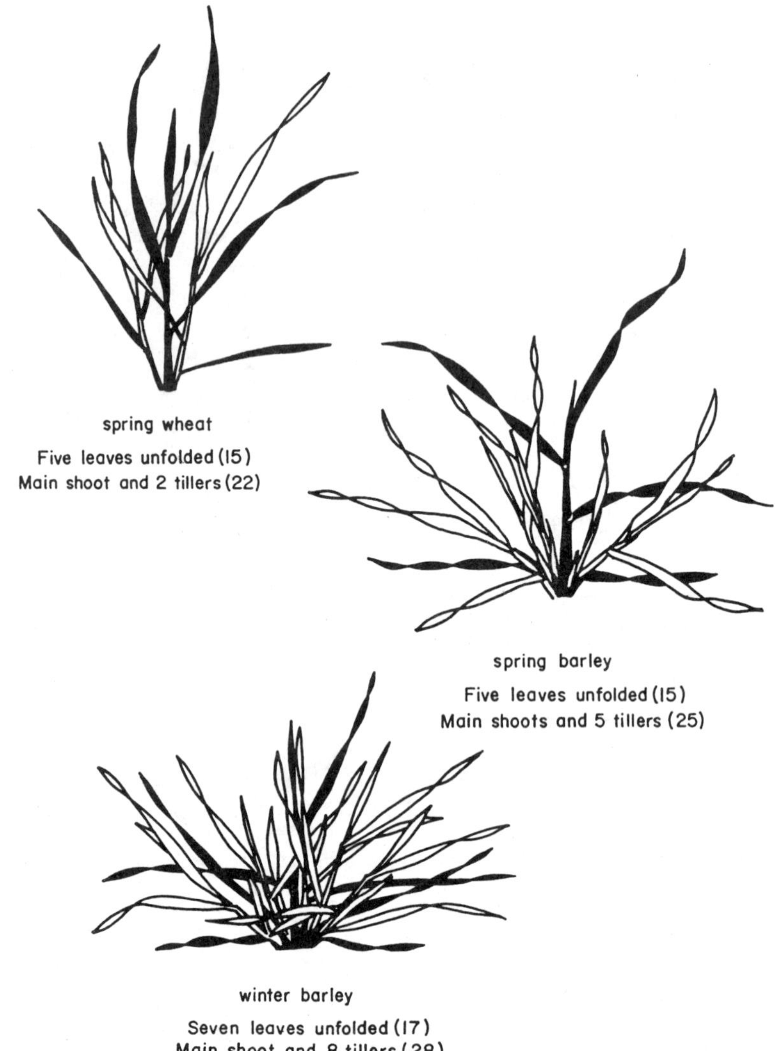

Figure 8.5 Seedling growth and tillering in wheat and barley. (From Tottman, 1987.)

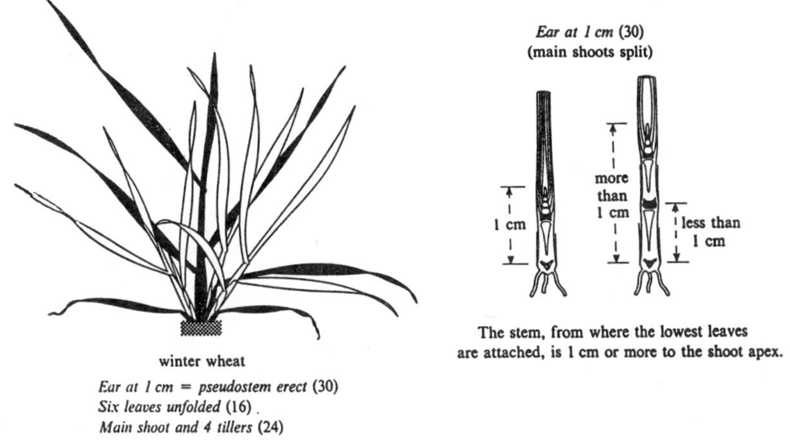

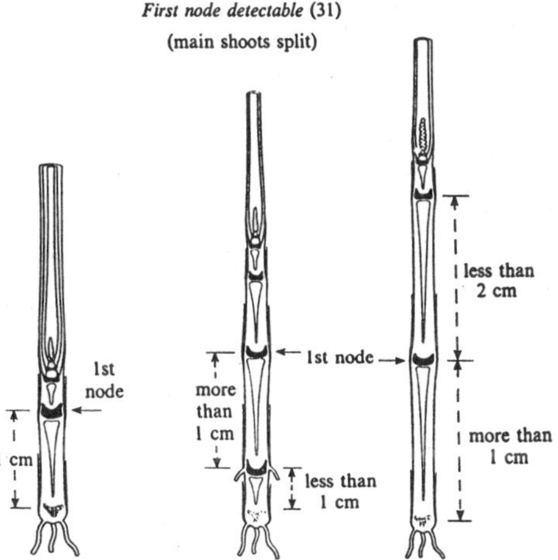

Figure 8.6 Stem elongation in winter wheat. (From Tottman, 1987.)

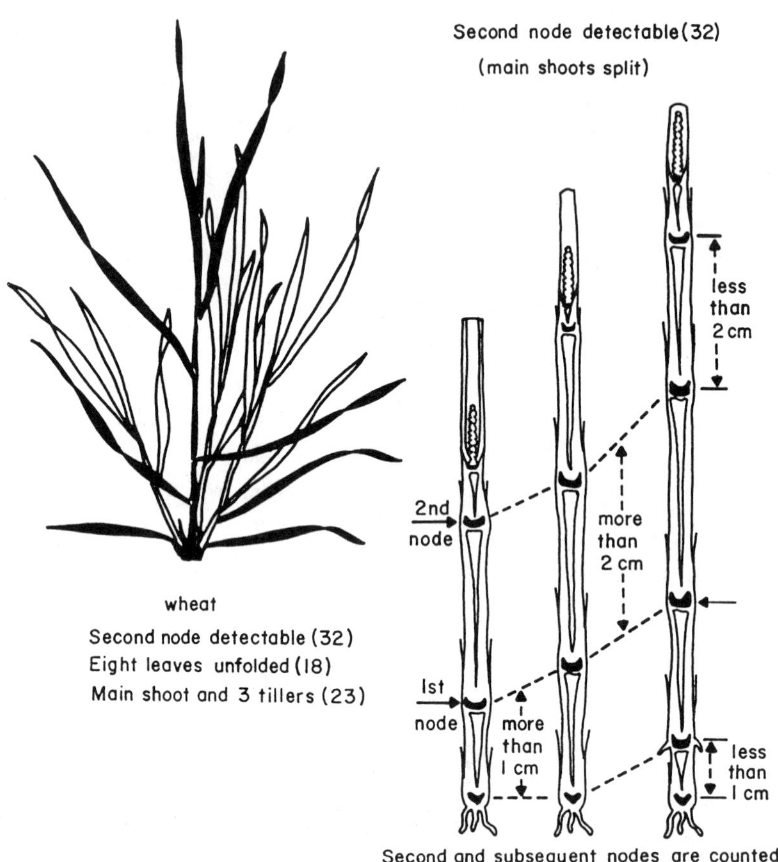

Figure 8.7 Stem elongation in winter wheat. (From Tottman, 1987.)

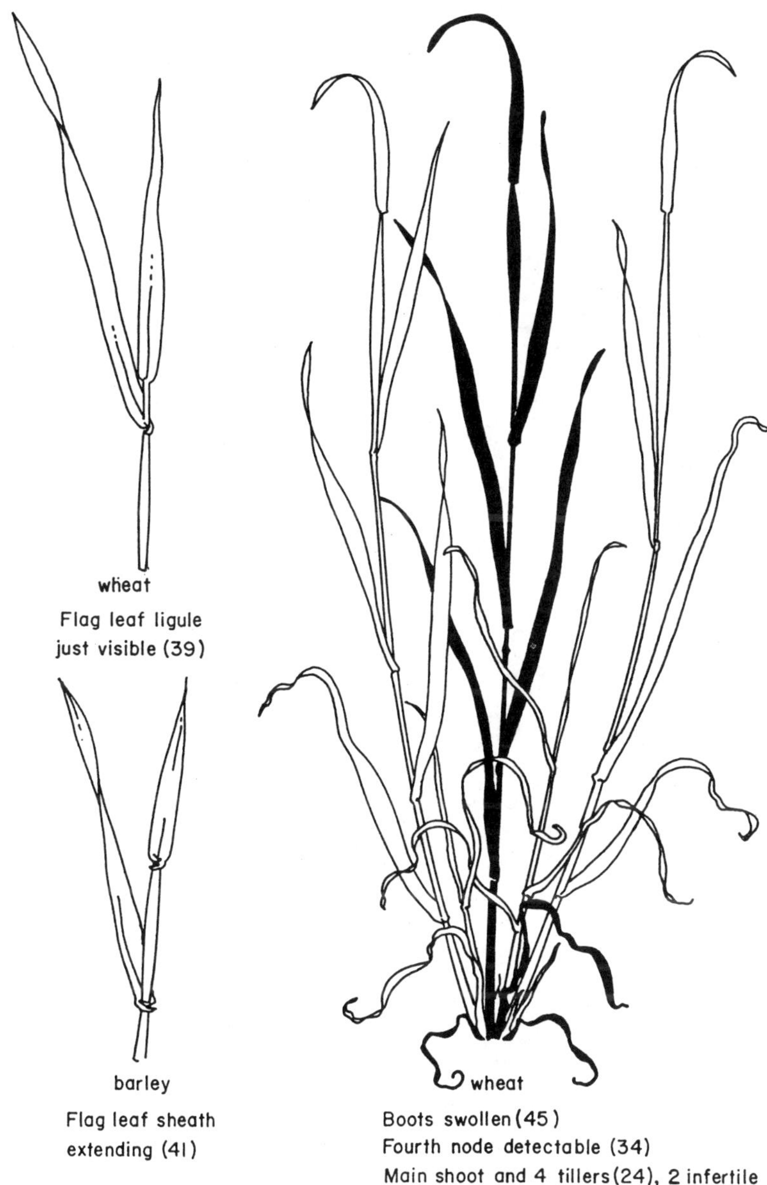

Figure 8.8 Booting growth stage in wheat and barley. (From Tottman, 1987.)

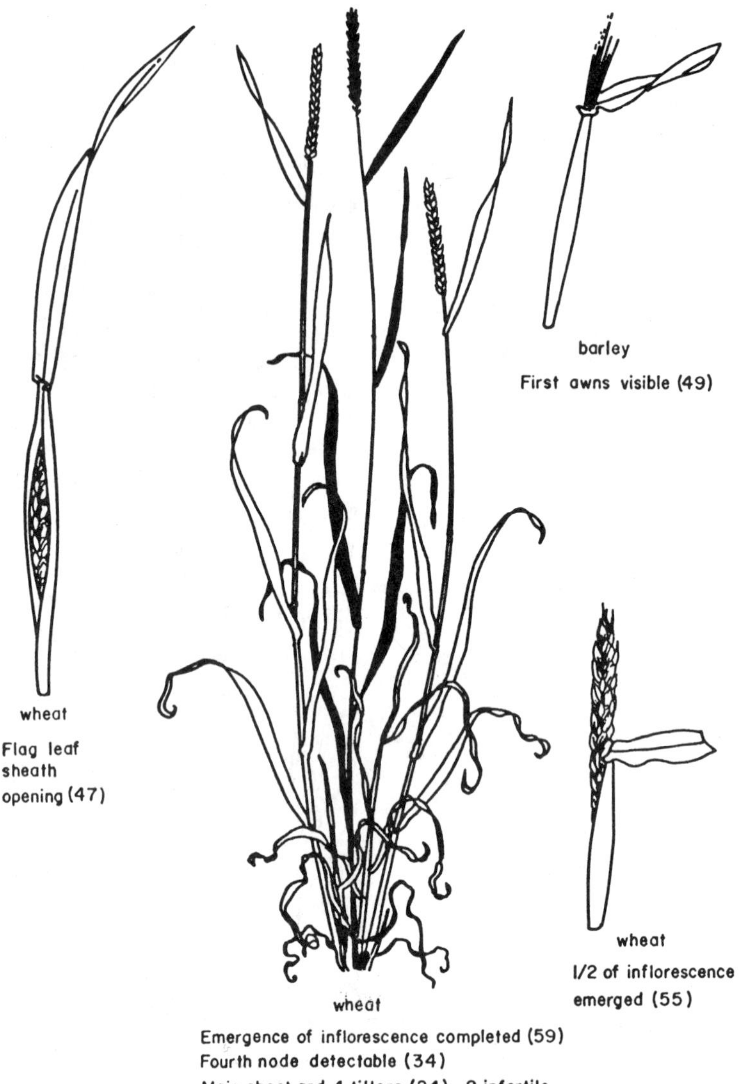

Figure 8.9 Booting and ear emergence in wheat and barley. (From Tottman, 1987.)

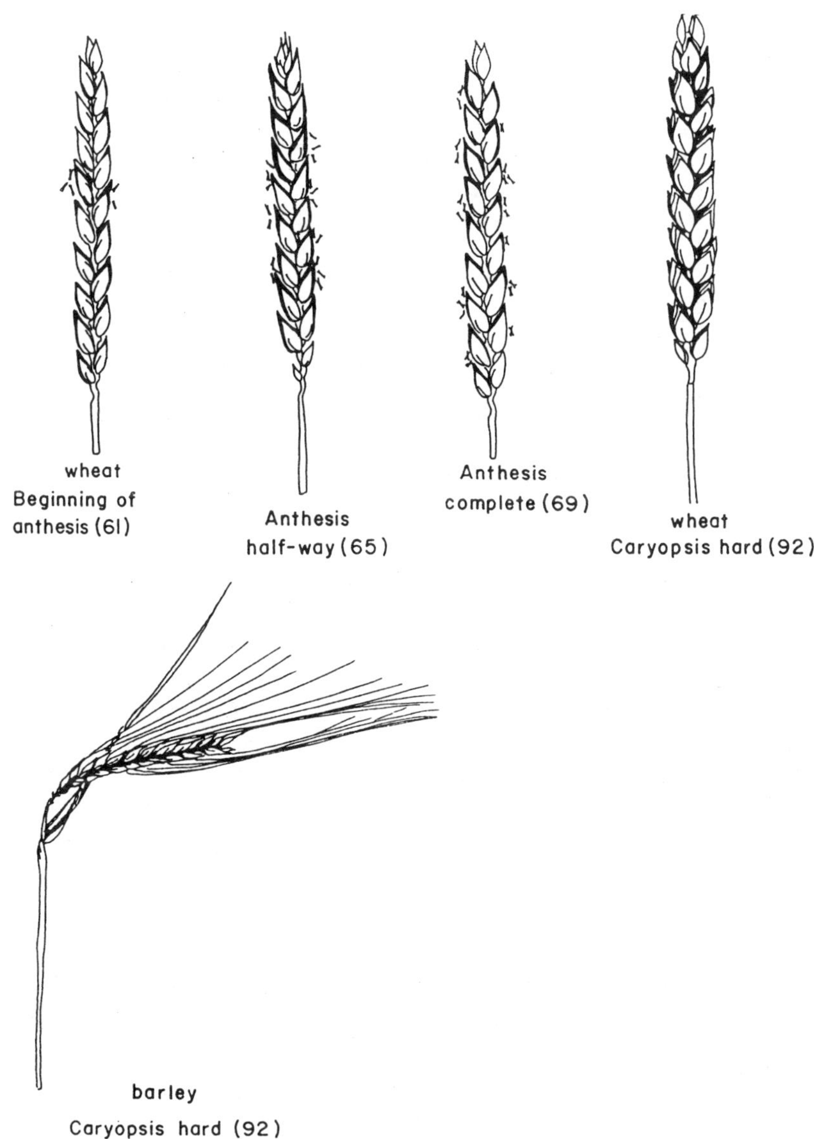

Figure 8.10 Anthesis and ripening growth stages in wheat and barley. (From Tottman, 1987.)

Table 8.4 Growth Stages of Winter Wheat and Their Descriptions

Growth stage	Description
0	Emergence of coleoptile.
1	Crown is visible, tillers develop.
2	Leaf sheaths elongate and form a false stem. Collars visible.
3	Culm elongation. First internode visible (jointing).
4	Tip of flag leaf visible (boot stage).
5	Peduncle elongates. Inflorescence emerges (heading).
6	Flowering (anthesis).
7	Anthesis complete. Grain filling begins. Lower leaves turn color.
8	Ripening. Grain hard but will not crack.
9	Ripening. Grain hard but will not crack. Inflorescence has lost all green color. Uppermost node still green.
10	Maturity. Grain cracks and is easily separated from chaff.

Source: Waldren and Flowerday, 1979.

in wheat yield with modern cultivars can be attributed to improved harvest index. Similarly, there was improvement in harvest index in barley in Britain from 0.33 to 0.50 between 1880 and 1980 (Riggs et al., 1981), and in the United States the harvest index of midwestern malting barley increased from 0.27 to 0.40 between 1920 and 1978 (Wych and Rasmusson, 1983).

Austin et al. (1980) suggested that harvest index of cereals might be increased from the current range of 0.4–0.5 to around 0.6 before diminishing return sets in. This suggests that there is considerable scope for improving wheat and barley yield by improving harvest index.

D. Relationship Between Growth and Yield

Plant growth in cereals is directly related to grain yield. In a broad sense, grain yield is a function of the number of grains per unit area and the weight of individual grains. The grains per unit area are determined by the number of ears per unit area and ear size. After the ear size is formed, grain weight depends on translocation of carbohydrates, principally from the leaves and ears (Carrasco and Thorne, 1979). Yield of grain is directly dependent on the photosynthetic capacity of the crop after anthesis and the sink capacity of the grain. Under optimum conditions, the contribution of preanthesis photosynthate reserves to final grain weight was found to be 5–10% in wheat and 20%

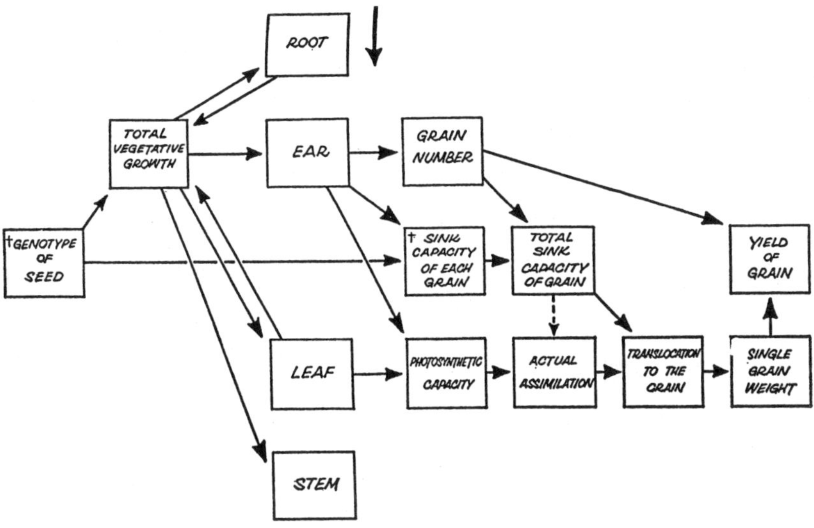

Figure 8.11 Relationships between principal physiological characters and yield in cereals. (From Bingham, 1969.)

in barley (Sharma, 1992). The photosynthetic capacity after anthesis is mainly dependent on leaf area duration, which is a function of leaf area index (*LAI*) at anthesis and leaf longevity (Bingham, 1969). McCaig and DePauw (1995) analyzed wheat trials data from western Canada from 1947 to 1992. Historical data from these tests were analyzed with the objectives of comparing grain-yield-related variables of recently registered cultivars with those of earlier cultivars and determining the yield advances made within the Canada western red spring (CWRS) wheat class. Canadian cultivars increased maximum yield potential approximately 6–9 kg ha^{-1} yr^{-1} during a 90–yr period. Yield potential of sawfly-resistant cultivars has been increasing at a rate of 11 kg ha^{-1} yr^{-1}, although they consistently yielded less than the highest-yielding hollow-stem cultivars. In general, the genetic yield increases resulted from an increase in the number of kernels produced rather than an increase in kernel size. This suggests that bread wheat grown on the prairies has been sink-limited during grain filling. The sink capacity of the grain is a product of the number of grains set and the growth characteristics of the individual grains. Figure 8.11 shows the relationship between yield and yield components; it can be concluded that grain yield in cereals is related to plant development in the vegetative, reproductive, and ripening stages.

IV. NUTRIENT REQUIREMENT

Nutrient requirements of a crop are determined by soil, climate, yield potential of the cultivar, and cropping system. Fertilizer recommendations are usually based on results of field trials in which crop response to various rates of fertilizer application is determined. The response curve then provides, for each trial, the relationship between the amount of fertilizer and crop yield. From this curve the economically optimum application rate of fertilizer needed for maximum yield can be derived (Neeteson and Wadman, 1987). Supply of adequate amounts of nutrients is one of the most important factors in increasing the yield of wheat and barley. It is estimated that in the last decade fertilizers have increased cereal yields in developing countries by 50% (Peter, 1980; Greenwood, 1981). Table 8.5 shows that wheat yields in India, Egypt, and Ethiopia are almost proportional to the rate of fertilizer application. Tables 8.6 and 8.7 show nitrogen, phosphorus, and potassium use for wheat and barley crops in different countries. Yields of these two crops are generally higher where higher levels of N, P, and K nutrients are applied (Martinez, 1990).

Nitrogen fertilizer use efficiency by winter wheat is highest when applications are timed when crop use of N is high. Winter wheat N uptake is most rapid from tillering through booting developmental growth stages with 80% of the total accumulation occurring before grain filling (Knowles et al., 1994).

Preplant versus spring N fertilizer applications have been compared for optimum N use efficiency by winter wheat grown in rainfed regions of the United States. Lutcher and Mahler (1988) found that applications of N fertilizer through the jointing (Feekes 6) growth stage (Large, 1954) gave maximum wheat grain yields. They also found lower wheat grain yields with N applied after booting (Feekes 10). Lutcher and Mahler (1988) spring topdressed 101 kg N/ha as either ammonium nitrate or urea ammonium nitrate (UAN; 32% N solution) on winter wheat from early tillering (Feekes 4) through booting on a silt loam soil in Idaho. They found that maximum grain yields were produced from both N forms applied until jointing; however, a decline in yield resulting from their late spring application was less for UAN than for ammonium nitrate. Figure 8.12 shows a quadratic relationship between rates applied and N uptake by wheat grains; maximum N uptake occurred at 183 kg N ha^{-1} (calculated by quadratic equation).

After nitrogen, phosphorus deficiency is a widespread problem in wheat and barley production in various parts of the world. The phosphorus requirement should be determined on the basis of soil test P and crop yield response. Numerous approaches are taken to interpreting soil tests. Mathematical functions are used to describe the response of a crop to changes in the soil concentration of an extractable immobile nutrient. Crop response may be

Table 8.5 Wheat Yield in Relation to Fertilizer Use in Different Countries

Country	NPK applied (kg ha^{-1})	Grain yield (t ha^{-1})
Egypt	176	3.33
Ethiopia	2	1.07
India	20	1.39

Source: Peter, 1980.

Table 8.6 Nitrogen, Phosphorus, and Potassium Application Rates (kg/ha) for Wheat in Different Countries[a]

Country	N	P	K	Total
Algeria	14	13.11	0	27.11
South Africa	20	19.23	10.79	50.02
Cyprus	60	13.11	0	73.11
Oman	84	18.35	34.86	137.21
Egypt	182	15.73	0	197.73
Zimbabwe	194	6.56	10.79	211.35
Saudi Arabia	306	121.48	0	427.48
Ecuador	7	7.43	0	14.43
Uruguay	39	13.55	0	52.55
Argentina	8	3.06	0	11.06
Brazil	9	22.72	45.96	77.68
Colombia	39	24.47	39.01	102.48
Australia	13	9.61	0	22.61
Turkey	44	11.36	0	55.36
Spain	46	9.61	9.96	65.57
Italy	83	29.72	18.26	130.98
Bulgaria	138	53.31	29.88	221.19
Sweden	124	13.55	19.92	157.47
Germany	140	26.22	49.80	216.02
United Kingdom	186	22.72	42.33	2151.05
France	156	31.90	64.74	252.64
Ireland	153	44.14	176.79	373.93
Denmark	183	17.48	34.86	235.34
Hungary	155	38.46	79.68	273.14
Japan	176	65.98	121.18	363.16
United States	50	9.61	9.96	69.57
Canada	50	17.48	4.15	71.63

[a] Values correspond to the 1986/87 cropping year.
Source: Compiled from Martinez, 1990.

Table 8.7 Nitrogen, Phosphorus, and Potassium Application Rates (kg/ha) for Barley in Different Countries.[a]

Country	N	P	K	Total
South Africa	22	19	11	52
Colombia	21	21	34	76
Ecuador	4	5	1	10
Uruguay	78	37	0	115
Cyprus	45	13	0	58
Saudi Arabia	56	19	0	75
Turkey	26	7	0	33
Hungary	90	29	65	184
Denmark	139	17	39	195
France	138	38	80	256
Federal Republic of Germany	105	26	66	197
Ireland	103	35	119	257
Spain	42	9	10	61
Sweden	73	12	24	109
United Kingdom	99	16	35	150
Australia	9	9	0	18

[a] Values correspond to 1986/87 cropping years.
Source: Martinez, 1990.

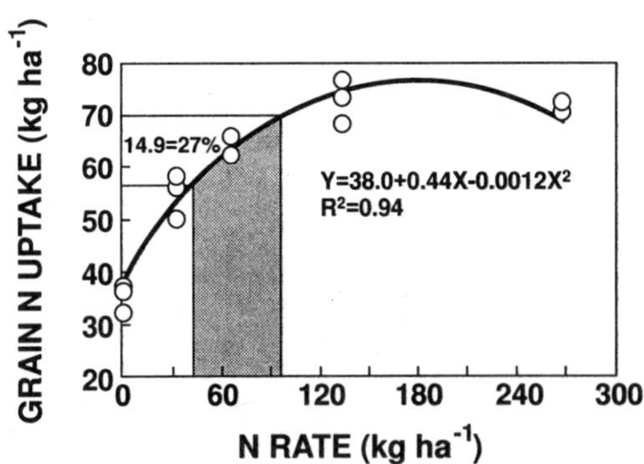

Figure 8.12 Relationship between nitrogen rates and N uptake by wheat grains. (From Raun et al., 1995.)

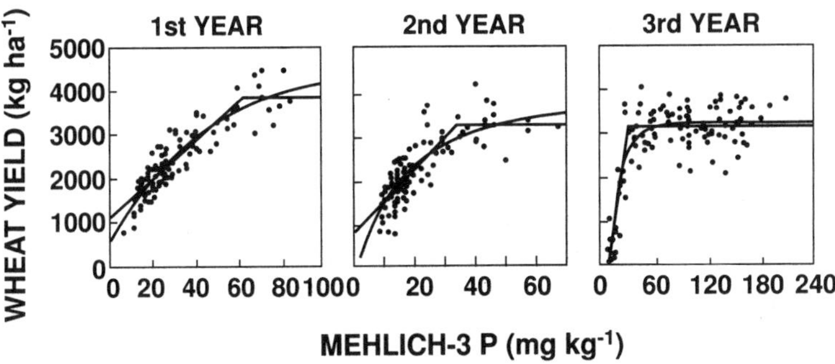

Figure 8.13 Response of wheat to Mehlich-3 extractable P with linear-plateau and exponential prediction functions for three crops. (From Cox, 1992.)

expressed as actual yield or as a percentage of maximum yield, sometimes termed *relative yield* (Cox, 1992).

When the critical level for wheat was evaluated using the exponential function to an economic optimum, the interpretation was much closer to that found with the linear-plateau function (Figure 8.13). The critical level ranged from 31 to 47 mg/kg with a mean of 39. This value is similar to the average critical value for corn (38 mg/kg) and somewhat more than that for soybean (36 mg/kg) using the economic optimum approach. The much wider range in critical levels found when arbitrarily using 95% maximum yield with the exponential function indicates that economic conditions have a major effect on the critical level interpretation and should be considered.

Genotypic differences have been observed for P requirements of a wheat crop. A study conducted in the greenhouse at the National Rice and Bean Research Center, Goiania, Brazil, showed that root length, root dry weight, and P-use efficiency were different among genotypes (Table 8.8). Phosphorus concentration in shoots did not differ significantly among genotypes across the three P levels. Root length varied from 32.7 to 44 cm, root dry weight varied from 0.41 to 0.54 g/pot, P-use efficiency varied from 125 to 188 mg dry matter/mg P absorbed. Based on P-use efficiency, genotypes were classified into four groups (Figure 8.14). This type of grouping was suggested by Fageria and Baligar (1993) for crop genotypes for P-use efficiency. The first group was genotype-efficient and responsive (ER) and which produced above the average yield of all the genotypes and responded well to applied P were classified in this class. The genotypes falling into this group were: BR10, CPAC89128, and NL459.

The second classification was genotype-efficient and nonresponsive (ENR). These genotypes produced more than average yield, but response to P appli-

Table 8.8 Root Length, Root Dry Weight, P Concentration in Shoot, and P Use Efficiency of 15 Wheat Genotypes Across Three P Levels

Genotype	Root length (cm)	Root dry weight (g/pot)	P conc. in shoot (g kg^{-1})	P use efficiency (mg DM/mg P absorbed)
1. Anahuac	39.8abc	0.42	3.33	152ab
2. BR10	40.0abc	0.50	3.11	188a
3. BR26	43.2ab	0.44	3.26	152ab
4. BR33	41.1abc	0.42	3.54	138ab
5. PF87949	38.0abc	0.48	3.36	185a
6. PF87950	37.2abc	0.52	3.26	150ab
7. PF89481	41.7ab	0.42	3.34	168ab
8. PF89490	39.7abc	0.43	3.10	182a
9. CPAC8909	35.7abc	0.41	3.31	183a
10. CPAC8947	35.2bc	0.42	3.27	143ab
11. CPAC89128	32.7c	0.54	3.12	182a
12. CPAC89194	41.3abc	0.44	3.13	125b
13. CPAC89321	41.5ab	0.41	3.30	145ab
14. IAPAR8745	44.0a	0.51	3.21	152ab
15. NL459	42.8ab	0.47	3.85	162ab

Means in the same column followed by the same letter are not significantly different at 5% probability levels by Tukey's test.

cation was lower than the average of all genotypes classified in this group. The genotypes BR33, PF87950, CPAC8947, CPAC89194, CPAC89321, and IAPAR8745 fall into this group. The third type of genotype is known as nonefficient and responsive (NER). The genotypes which produced less than average dry matter yield but responded to P application above the average of 15 genotypes are classified in this group. The genotypes falling into this group were PF87949, PF89481, PF89490, and CPAC8909.

The last group of genotypes is those which produced less than average yield, and response to applied P was also less than average. This type of genotype was classified as nonefficient and nonresponsive (NENR). Only two genotypes fall into this group, Anahuac and BR26. From a practical point of view, the genotypes which fall into the group efficient and responsive are the most desirable. Because these genotypes produced well at a low P level and also responded well to applied P, this means this type of genotype can be utilized under low as well as high technology with reasonably good yield. The second most desirable group is genotypes efficient and nonresponsive.

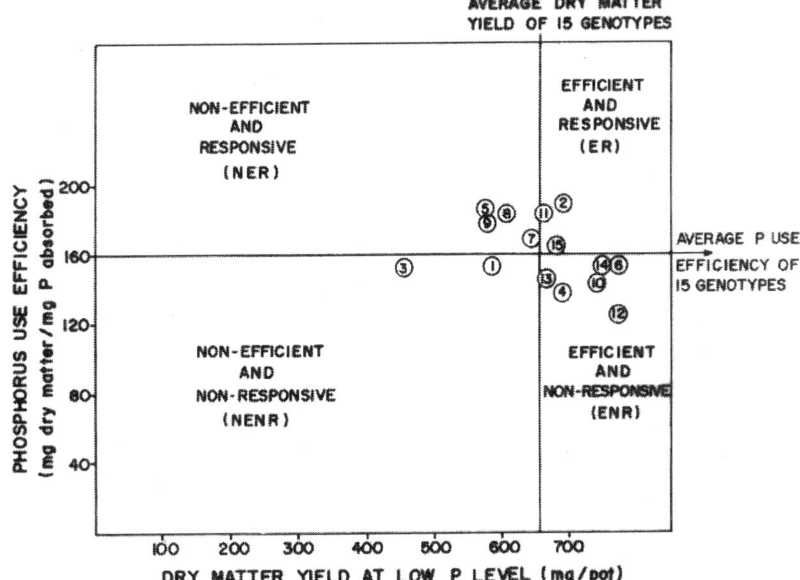

Figure 8.14 Classification of wheat genotypes for P use efficiency. Numbers in the circles correspond to genotypes listed in Table 8.8.

Genotypes of this type can be planted under low P levels with more than average yield.

The genotypes nonefficient and responsive sometimes can be used in breeding programs for their P-responsive characteristics. The most undesirable genotypes are the nonefficient and nonresponsive type.

We compared the growth and P-uptake parameters of efficient and responsive genotypes (most desirable) to nonefficient and nonresponsive (most undesirable) to draw a conclusion regarding the reason for P-use efficiency differences among genotypes. Phosphorus uptake in root and dry weight of roots of efficient and responsive genotypes were superior to nonefficient and nonresponsive genotypes. Other plant characteristics such as root length and P concentration were similar, but all the efficient and nonresponsive genotypes produced higher shoot weight as well as high shoot P uptake at the same P concentration as compared to nonefficient nonresponsive genotypes. This means the greater P efficiency was attributed to greater P-use efficiency than to differences in P concentration. Gardiner and Christensen (1990) also reported similar results for wheat cultivars in relation to P use efficiency.

Table 8.9 Uptake of Macronutrients by Wheat and Barley[a]

Crop		Yield ($t\ ha^{-1}$)	Nutrient ($kg\ ha^{-1}$)					
			N	P	K	Ca	M	S
Wheat	Grain	3	75	15	12	3	9	5
	Straw	5	50	7	80	13	5	9
	Total	8	125	22	92	16	14	14
	ER, grain		40	200	250	1000	333	600
	ER, straw		100	714	63	385	1000	556
Barley	Grain	5.4	123	20	32	—	9	11
	Straw	—	45	7	107	—	10	11
	Total		168	27	139	—	19	22
	ER, grain		44	270	169	—	600	491

[a] Nutrient efficiency ratio (ER) = yield kg^{-1} nutrient uptake.
Nutrient uptake = concentration × dry matter.
Sources: Malavolta, 1979; Munson, 1982.

Data related to nutrient uptake by wheat and barley in tropical and subtropical climates are presented in Table 8.9. When 8 t ha^{-1} dry matter (grain + straw) of wheat was produced, the uptake of N was 125 kg ha^{-1}, of K was 92 kg ha^{-1}, and of P was 22 kg ha^{-1}. A barley grain yield of 5.4 t ha^{-1} resulted in uptake of 168 kg N ha^{-1}, 139 kg K ha^{-1}, and 27 kg P ha^{-1} in both grain and straw combined. These data show that N is required in the greatest quantity, followed by K and P. More N and P were translocated to grain, and most of the K remained in the straw. Halvorson et al. (1987) compiled data from the literature and came to the conclusion that the amount of N needed by a wheat crop (roots, vegetative portion, and grain) ranged from 30 to 50 kg N per ton of grain in various parts of the United States. In the case of barley, approximately 1 kg of N is required for each 34 kg of grain produced (McGeorge, 1953; Jensen and Lund, 1967).

Phosphorus efficiency is low, and only about 15–20% of the applied P is used by the first crop. Phosphorus requirements are estimated for vegetative and grain portions of the wheat plant to be from 6 to 8 kg P per ton of grain (Halvorson et al., 1987). Potassium removal in wheat grain is about 5 kg/ton of grain. The dry matter of cereal crops grown with adequate nutrients seldom contains less than 1.5% N, 0.30% P, and 1.5% K (Greenwood et al., 1980a,b,c). The Food and Agricultural Organization (FAO) carried out more than 100,000 fertilizer trials with the indigenous cereal varieties and cultural practices of 40 developing countries and concluded that usually 10–20 kg of grain was

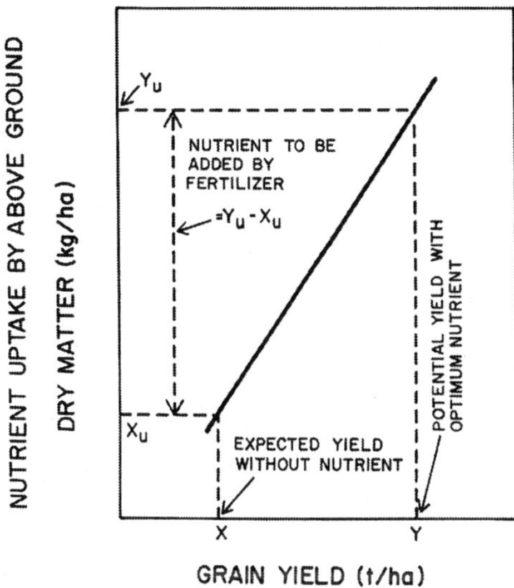

Figure 8.15 Nutrient uptake and yield relationship used to calculate nutrient requirements in crop plants.

produced for every kg of N applied when growth was limited by N. The corresponding values for P and K were 20–25 and 3–5 kg cereal grain produced per 1 kg of nutrient applied, respectively (Peter, 1980; Greenwood, 1981).

The nutrient requirements of a crop can be estimated on the basis of nutrient uptake by the aboveground part using the following formula:

$$N_R = \frac{Y_u - X_u}{E_f}$$

where N_R is the nutrient requirement, Y_u the uptake of nutrient in the aboveground dry matter to attain the desired yield, X_u the uptake of nutrient from the soil without fertilizer addition, and E_f the increase in nutrient uptake of aboveground dry matter per unit of nutrient applied. The nutrient uptake and yield relationship is shown in Figure 8.15.

Micronutrient-deficient soils are widespread; many millions of hectares of arable land worldwide are deficient in one or more micronutrient elements (Welch et al., 1991). For example, Zn deficiency was reported for soils of various characteristics: High and low pH, high and low organic matter, calcareous, sodic, sandy, wetland or ill-drained, limed acid soils, etc. (Takkar and Walker, 1993; Rengel and Graham, 1995).

Large areas of the Australian cereal belt are deficient in micronutrients, notably Zn and Mn, and wheat grown under such deficient conditions produces seed with low Zn or Mn content. When seed with low Mn content was resown in Mn-deficient soil (as is commonly done by Australian farmers), wheat had a poor seedling vigor and low yields at harvest (Marcar and Graham, 1986; Singh and Bharti, 1985). Similar results were obtained with *Hordeum vulgare* where elevated Mn content in seed increased both vegetative and grain yield under Mn-deficient conditions in the field (Longnecker et al., 1991).

A. Nutrient Concentration

The determination of nutrient concentrations in the plant is an important method for assessing plant nutrient status. The principles and relationship between nutrient concentrations and yield are discussed in Chapter 4 (diagnostic techniques for nutritional disorders). Deficient, critical or adequate, and high levels of nutrients in wheat and barley plants are given here to compare plant analytical results. These values are affected by several factors but may serve as general guidelines in the interpretation of plant analysis results. Nutrient concentration data presented in Tables 8.10 and 8.11 show that deficient, adequate, and high values decreased with increased age of the wheat and barley plants. These values were higher when the leaf was analyzed rather than whole plant tops.

B. Nutrient Distribution in Plant Parts

The nutrient distribution in different plant parts of wheat at harvest is presented in Figure 8.16. These values were determined in a greenhouse experiment conducted in an Oxisol of central Brazil. On an average basis, about 4% P was retained in roots, 17% in tops, 67% transported to the grain, and 12% in ear chaff. Potassium distribution in different plant parts was 2% in roots, 72% in tops, 11% in grain, and 15% in chaff; for Ca, 5% was in the roots, 77% in the tops, 6% in the grain, and 12% in ear chaff. The magnesium distribution was 2% in roots, 53% in tops, 35% in grain, and 10% in chaff.

Micronutrient distribution was as follows: 18% Zn was retained in the roots, 22% in tops, 47% in grain, and 13% in ear chaff. Roots retained 15% Cu, tops 14%, grain 29%, and chaff 42%. Distribution of Fe was 73% in roots, 15% in tops, 5% in grains, and 7% in chaff. Distribution of Mn was 6% in roots, 55% in tops, 21% in grain, and 18% in ear chaff. In conclusion, roots retained maximum Fe; tops retained maximum K, Ca, Mg, and Mn; and grain retained maximum P and a considerable amount of Mg and Zn.

According to Waldren and Flowerday (1979), nitrogen uptake in winter wheat was most rapid from stage 2 to stage 4 and 80% of the total accumulation occurring by stage 7 (anthesis complete). Over 70% of the total N

Table 8.10 Approximate Nutrient Concentrations in Mature Wheat Tissue That May Be Classified as Deficient, Adequate, or High

			g kg^{-1}		
Nutrient	Growth stage	Plant part	Deficient	Adequate	High
N	Tillering	Leaf blade	< 38	43–52	> 52
	Shooting	Leaf blade	< 30	36–44	> 44
	Heading	Whole tops	< 15	21–30	> 30
	Flowering	Leaf blade	< 24	27–30	> 30
P	Tillering	Leaf blade	< 3.1	3.5–4.9	> 4.9
	Shooting	Leaf blade	< 2.8	3.2–4.0	> 4.0
	Heading	Whole tops	< 1.5	2.1–5.0	> 5.0
	Flowering	Leaf blade	< 2.2	2.5–3.4	> 3.4
K	Tillering	Leaf blade	< 28	34–42	> 42
	Shooting	Leaf blade	< 26	31–36	> 36
	Heading	Whole tops	< 13	15–30	> 30
	Flowering	Leaf blade	< 20	23–32	> 32
Ca	Heading	Whole tops	< 2.0	2.0–5	> 5
Mg	Heading	Whole tops	< 1.2	1.5–5	> 5
S	Heading	Whole tops	< 1.2	1.5–4	> 4

			mg kg^{-1}		
Nutrient	Growth stage	Plant part	Deficient	Adequate	High
Cu	Heading	Leaf blade	< 1.6	> 2.2	—
Zn	Heading	Whole tops	< 15	15–70	> 70
Mn	Mid-till-S-E[a]	Whole tops	—	11–13	—
Fe	Preheading	Upper leaf blade	—	25–100	—
B	Stem extension	Whole tops	< 5	6–10	> 16
Mo	Stem extension	Leaf blade	< 0.05	0.05–0.1	> 0.1

[a] till = tillering; S-E = stem elongation.
Sources: Ward et al., 1973; Reuter, 1986.

Table 8.11 Approximate Nutrient Concentrations in Mature Barley Tissue That May Be Classified as Deficient, Adequate, or High

			g kg^{-1}		
Nutrient	Growth stage	Plant part	Deficient	Adequate	High
N	Tillering	Leaves	< 39	47–51	> 51
	Shooting	Leaves	< 36	45–47	> 47
	Heading	Whole tops	< 15	20–30	> 30
	Flowering	Leaves	< 26	29–35	> 35
P	Tillering	Leaves	< 4.4	5.0–6.8	> 6.8
	Shooting	Leaves	< 3.7	4.2–4.8	> 4.8
	Heading	Whole tops	< 3.5	2.0–5.0	> 5.0
	Flowering	Leaves	< 2.7	3.1–4.2	> 4.2
K	Tillering	Leaves	< 35	42–47	> 47
	Shooting	Leaves	< 30	35–41	> 41
	Heading	Whole tops	< 13	15–30	> 30
	Flowering	Leaves	< 20	23–28	> 28
Ca	Heading	Whole tops	< 3.0	3–12	> 12
Mg	Heading	Whole tops	< 1.5	1.5–5	> 5
S	Heading	Whole tops	< 1.5	1.5–4	> 4
			mg kg^{-1}		
Nutrient	Growth stage	Plant part	Deficient	Adequate	High
Cu	Stem extension	Whole tops	< 2.3	4.8–6.8	> 6.8
Zn	Heading	Whole tops	< 15	15–70	> 70
Mn	Heading	Whole tops	< 24	25–100	> 100
Fe	Heading	Whole tops	—	50–100	—
B	Stem extension	Whole tops	1.9–3.5	6–10	> 16
Mo	Heading	Whole tops	—	0.3–5.0	—

Sources: Ward et al., 1973; Reuter, 1986.

uptake was translocated to the grain at maturity. Uptake of P and K was most rapid from jointing to anthesis complete growth stages. Seventy-five percent of P uptake was translocated to the grain at maturity, but only 15% of K present in the plant was found in the grain at maturity.

Knowles and Watkins (1931) found most of the N and P taken up by wheat plants was translocated to the grain either directly or by mobilization from other parts, and only a small amount of K was translocated to grain from

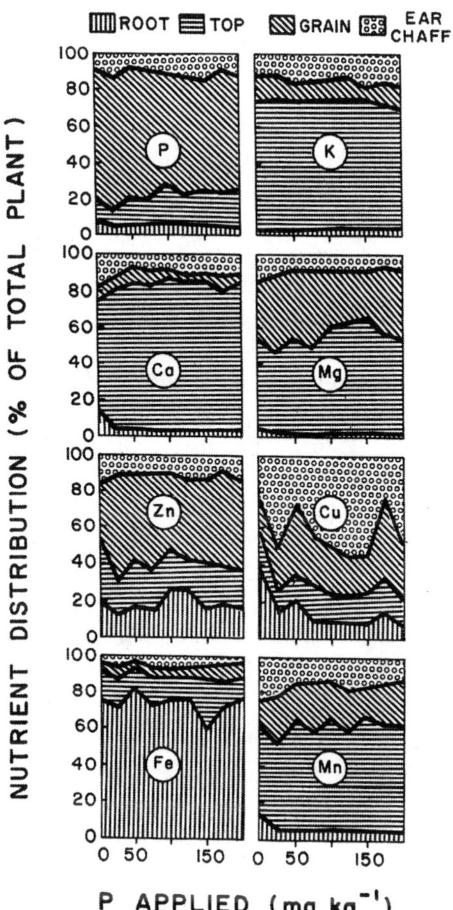

Figure 8.16 Distribution of nutrients in different plant parts of wheat. (From Fageria, 1990.)

other parts of the plant. A similar pattern of N mobilization was shown by McNeal et al. (1968).

V. SUMMARY

Wheat and barley temperate cereals that originated in western Asia are now cultivated in all regions of the world except the lowland humid tropics, where foliar disease, insects, and high mean temperatures preclude successful cultivation. Both crops are adapted to daily mean temperatures in the range of 10–20°C. Winter wheat and barley cultivars are planted in the autumn or winter and usually require vernalization for normal phenological development. Spring cultivars are planted in the spring in temperate climates or in the winter or autumn in lower latitudes where temperatures are too high for vernalization of winter cultivars.

High-yielding semidwarf cultivars with maximum yields exceeding 10 Mg ha^{-1} have contributed to increased yields for farmers in both developed and developing countries. These cultivars normally have greater harvest indices than taller cultivars and are more tolerant to lodging.

Wheat and barley are produced under both irrigated and dryland conditions, and barley is generally recognized as having greater drought and salinity tolerance than wheat. Neither crop is adapted to acid soils, although wheat can be produced at a lower pH than barley, and some genetic tolerance to very acid soils has been identified. If commercial cultivars with tolerance to significant levels of aluminum saturation can be developed, wheat could be expanded in many cool tropical areas that have acid soils.

In most wheat-growing areas, nitrogen is the most limiting mineral nutrient, and 1 metric ton of wheat grain contains about 21 kg N, 4 kg P, and 4 kg K. Barley grain nutrient concentrations are often lower than those in wheat, and 1 metric ton removes approximately 19 kg N, 4 kg P, and 5 kg K. Because of low yields, many dryland wheat- and barley-producing areas as well as in grassland soils with large reserves of organic matter, both soil and plant analyses can be used to diagnose and correct nutrient deficiencies and toxicities.

REFERENCES

Adams, F. 1981. Alleviating chemical toxicities: Liming acid soils, pp. 269–301. *In*: G. F. Arkin and H. M. Taylor (eds.). Modifying the root environment to reduce crop stress. Monogr. 4, Am. Soc. Agric. Eng., St. Joseph, Michigan.

Austin, R. B., J. Bingham, R. D. Blackwell, L. T. Evans, M. A. Ford, C. L. Morgan, and M. Taylor. 1980. Genetic improvements in winter wheat yields since 1900 and associated physiological changes. J. Agric. Sci. 94: 675–689.

Austin, R. B., M. A. Ford, and C. L. Morgan. 1989. Genetic improvement in the yield of winter wheat: A further evaluation. J. Agric. Sci. Camb. 122: 295–301.

Baldridge, D. E., D. E. Brann, A. H. Ferguson, J. L. Henry, and R. K. Thompson. 1985. Cultural practices, pp. 457–482. *In*: D. C. Rasmusson (ed.). Barley. Monogr. 26, Am. Soc. Agron., Madison, Wisconsin.

Bauer, A., A. B. Frank, and A. L. Black. 1987. Aerial parts of hard red spring wheat. I. Dry matter distribution by plant development stage. Agron. J. 79: 845–852.

Bingham, J. 1969. The physiological determinations of grain yield. Agric. Prog. 44: 30–42.

Bonnet, O. T. 1966. Inflorescences of maize, wheat, rye, barley, and oats: Their initiation and development. Agric. Exp. Stn. Bull. 721, University of Illinois College of Agriculture.

Bower, C. A. and M. Fireman. 1957. Saline and alkali soils, pp. 282–290. *In*: Soil. USDA Yearbook of Agriculture.

Briggle, L. W. 1980. Origin and botany of wheat, pp. 6–13. *In*: E. Hafliger (ed.). Wheat. Documenta Ciba-Geigy, Basel.

Briggle, L. W. and B. C. Curtis. 1987. Wheat worldwide, pp. 1–31. *In*: E. G. Heyne (ed.). Wheat and wheat improvement, 2nd Ed. Monogr. 13, Am. Soc. Agron., Madison, Wisconsin.

Brown, R. H. 1984. Growth of the green plant, pp. 153–174. *In*: M. B. Tesar (ed.). Physiological basis of crop growth and development. Am. Soc. Agron., Madison, WI.

Carlton, M. A. 1916. The small grain. Mcmillan, New York.

Carrasco, R. M. and G. N. Thorne. 1979. Physiological factors limiting grain size in wheat. J. Exp. Bot. 30: 669–679.

Carver, B. F. and J. D. Ownby. 1995. Acid soil tolerance in wheat. Adv. Agron. 54: 117–173.

Charles-Edwards, D. A. and M. J. Fisher. 1980. A physiological approach to the analysis of crop growth data. I. Theoretical considerations. Ann. Bot. 46: 413–423.

CIMMYT (Centro Internacional de Mejoramento de Maiz Y Trigo). 1972. Annual Report, Londres, Mexico.

CIMMYT (Centro Internacional de Mejoramento de Maiz Y Trigo). 1985. World wheat facts and trends. Report three: A discussion of selected wheat marketing and pricing issues in developing countries. CIMMYT, Mexico, DF.

Costa, A., L. A.-C. Campos, D. Brunetta, and C. R. Riede. 1996. Reaction of wheat genotypes to soil aluminum saturation. Paper presented at the 4th Int. Symp. Soil-Plant Interactions at Low pH, March 17–24, 1996, Belo-Horizonte, Brazil.

Cox, F. R. 1992. Range in soil phosphorus critical levels with time. Soil Sci. Soc. Am. J. 56: 1504–1509.

Cox, T. S., J. P. Shroyer, L. Ben-Hui, R. G. Sears, and T. J. Martin. 1988. Genetic improvement in agronomic trials of hard red winter wheat cultivars from 1919 to 1987. Crop Sci. 28: 756–680.

Dickson, A. D. 1979. Barley for malting and food, pp. 136–146. *In*: Barley: Botany, culture, winter hardiness, genetics, utilizations, and pests. USDA Handbook of Agriculture, 338.

Doll, E. C. 1964. Lime for Michigan soils. Michigan Agric. Exp. Stn. Bull. 471.

Donald, C. M. and J. Hamblin. 1976. The biological yield and harvest index of cereals as agronomic and plant breeding criteria. Adv. Agron. 28: 361–405.

Emecz, T. I. 1962. Suggested amendments in growth analysis and potentiality assessment in relation to light. Ann. Bot. 26: 517–527.

Evans, G. C. 1972. The quantitative analysis of plant growth. Blackwell, Oxford.

Fageria, N. K. 1990. Response of wheat to phosphorus fertilization on an Oxisol. Pesq. Agropec. Bras. Brasilia 25: 530–537.

Fageria, N. K. and V. C. Baligar. 1993. Screening crop genotypes for mineral stress, pp. 142–159. Proc. Workshop on Adaptation of Plant to Soil Stresses, August 1–4, 1993. Univ. Nebraska, Lincoln. INSTORMIL Publication No. 84–2.

FAO. 1992. Production Yearbook 1991. Vol. 45. Rome.

Goblirsch, C. A., R. D. Horsley, and P. B. Schwarz. 1996. A strategy to breed low-protein barley with acceptable kernel colro and diastatic power. Crop Sci. 36: 41–44.

Farm and Ranch Irrigation Survey, 1979. pp. 80–81. U.S. Government Printing Office, Washington, DC.

Gardiner, D. T. and N. W. Christensen. 1990. Characterization of phosphorus efficiencies of two winter wheat cultivars. Soil Sci. Soc. Am. J. 54: 1337–1340.

Greenwood, D. J. 1981. Fertilizer use and food production: World scene. Fertilizer Res. 2: 33–51.

Greenwood, D. J., T. J. Cleaver, M. K. Turner, J. Hunt, K. B. Niendor, and S. M. H. Loquens. 1980a. Comparisons of the effects of potassium fertilizer on the yield, potassium content and quality of 22 different vegetable and agricultural crops. J. Agric. Sci. 95: 441–456.

Greenwood, D. J., T. J. Cleaver, M. K. Turner, J. Hunt, K. B. Niendor, and S. M. H. Loquens. 1980b. Comparisons of the effects of phosphate fertilizer on the yield, phosphate content and quality of 22 different vegetable and agricultural crops. J. Agric. Sci. 95: 457–469.

Greenwood, D. J., T. J. Cleaver, M. K. Turner, J. Hunt, K. B. Niendor, and S. M. H. Loquens. 1980c. Comparisons of the effects of nitrogen fertilizer on the yield, nitrogen content and quality of 21 different vegetable and agricultural crops. J. Agric. Sci. 95: 471–485.

Guerra, A. F. 1995. Irrigation scheduling on wheat for maximum yield in the Cerrado region. Pesq. Agropec. Bras. 30: 515–521.

Halvorson, A. D., M. M. Alley, and L. S. Murphy. 1987. Nutrient requirements and fertilizer use, pp. 345–383. *In*: E. G. Heyne (ed.). Wheat and wheat improvement, 2nd ed. Monogr. 13, Am. Soc. Agron., Madison, Wisconsin.

Hardwick, R. C. 1984. Some recent developments in growth analysis. A review. Ann. Bot. 54: 807–812.

Hunt, R. 1979. Plant growth analysis: The rationale behind the use of the fitted mathematical function. Ann. Bot. 43: 245–249.

Jensen, L. A. and H. R. Lund. 1967. How cereal crops grow. North Dakota State University Coop. Exp. Serv. Bull. 3.

Jolliffe, P. A., G. E. Eaton, and J. L. Doust. 1982. Sequential analysis of plant growth. New Phytol. 92: 287–296.

Kirby, E. J. M. and M. Appleyard. 1984. Cereal development guide, 2nd ed. Arable Unit, National Agricultural Center, Stoneleigh.

Knowles, R. and J. E. Watkins. 1931. The assimilation and translocation of plant nutrients in wheat during growth. J. Agric. Sci. 21: 612–637.

Knowles, T. C., B. W. Hipp, P. S. Graff, and D. S. Marshall. 1994. Timing and rate of topdress nitrogen for rainfed winter wheat. J. Prod. Agric. 7: 216–220.

Large, E. C. 1954. Growth stages in cereals. Illustrations of the Feekes Scale. Plant Pathol. 3: 128–130.

Lersten, N. R. 1987. Morphology and anatomy of the wheat plant, pp. 33–75. *In*: E. G. Heyne (ed.). Wheat and wheat improvement. 2nd ed. Monogr. 13, Am. Soc. Agron., Madison, Wisconsin.

Longnecker, N. E., N. E. Marcar, and R. D. Graham. 1991. Increased manganese content of barley seeds can increase grain yield in manganese-deficient conditions. Aust. J. Agric. Res. 42: 1065–1074.

Lutcher, L. K. and R. L. Mahler. 1988. Sources and timing of spring top-dress nitrogen on winter wheat in Idaho. Agron. J. 80: 648–654.

Maas, E. V. and G. J. Hoffman. 1977. Crop salt tolerance—current assessment. J. Irrig. Drainage Div. Am. Soc. Civil Eng. 103: 115–134.

Malavolta, E. 1979. Potassium, magnesium and sulphur in Brazilian soils and crops. Tech. Bull. 4, Potash and Phosphate Institute, Piracicaba, Brazil.

Marcar, N. E. and R. D. Graham. 1986. Effect of seed manganese content on the growth of wheat under manganese deficiency. Plant Soil 96: 165–174.

Martinez, A. 1990. Fertilizer use statistics and crop yields. International Fertilizer Development Center, Tech. Bull. T-37, Muscle Shoals, Alabama.

McGeorge, W. T. 1953. Fertilization of field crops in Arizona. Arizona Agric. Exp. Stn. Bull. 247.

McCaig, T. N. and R. M. DePauw. 1995. Breeding hard red spring wheat in western Canada: Historical trends in yield and related variables. Can. J. Plant Sci. 75: 387–393.

McNeal, F. H., G. O. Boatwright, M. A. Berg, and C. A. Watson. 1968. Nitrogen in plant parts of seven spring wheat varieties at successive stages of development. Crop Sci. 8: 535–537.

Miller, D. L. 1974. Industrial uses of wheat and flour, pp. 398–411. *In*: G. E. Inglett (ed.). Wheat: Production and Utilization. AVI Publishing Co., Westport, Connecticut.

Munson, R. D. 1982. Potassium, calcium, and magnesium in the tropics and subtropics. Tech. Bull. T-23, International Fertilizer Development Center, Muscle Shoals, Tennessee.

Neeteson, J. J. and W. P. Wadman. 1987. Assessment of economically optimum application rates of fertilizer N on the basis of response curves. Fertilizer Res. 12: 37–52.

Nerson, H., M. Sibony and M. J. Pinthus. 1980. A scale for the assessment of the developmental stages of the wheat (*Triticum aestivum* L.) spike. Ann. Bot. 45: 203–204.

Nuttonson, M. Y. 1955. Wheat-climatic relationship and the use of phenology in ascertaining the thermal and photothermal requirements of wheat. Am. Inst. Crop Ecology, Washington, DC.

Peter, A. V. 1980. Fertilizer requirements in developing countries. Proc. Fert. Soc. London 188: 1–58.

Peterson, G. A. and A. E. Foster. 1973. Malting barley in the United States. Adv. Agron. 25: 327–378.

Poehlman, J. M. 1985. Adaption and distribution, pp. 1–17. *In*: D. C. Rasmusson (ed.). Barley. Monogr. 26, Am. Soc. Agron., Madison, Wisconsin.

Polisetty, R., R. Chandra, and S. Khetarpal. 1993. Physiological analysis of barley (*Hordeum vulgare* L.) genotypes differing in harvest index. Indian J. Plant Physiol. 36: 90–93.

Purseglove, J. W. 1985. Tropical crops: Monocotyledons. Longman, New York.

Radford, P. J. 1967. Growth analysis formulae—Their use and abuse. Crop Sci. 7: 171–175.

Randall, P. J. and H. J. Moss. 1990. Some effects of temperature regime during grain filling on wheat quality. Aust. J. Agric. Res. 41: 603–617.

Raun, W. R. and G. V. Johnson. 1995. Soil-plant buffering of inorganic nitrogen in continuous winter wheat. Agron. J. 87: 827–834.

Reid, D. A. 1985. Morphology and anatomy of the barley plant, pp. 73–101. *In*: D. C. Rasmusson (ed.). Barley. Monogr. 26, Am. Soc. Agron., Madison, Wisconsin.

Rengel, Z. and R. D. Graham. 1995. Importance of seed Zn content for wheat growth on Zn-deficient soil. I. vegetative growth. Palnt Soil 173: 259–266.

Reuter, D. J. 1986. Temperate and sub-tropical crops, pp. 38–99. In D. J. Reuter and J. B. Robinson (eds.). Plant analysis: An interpretation manual. Inkata Press, Melbourne.

Riggs, T. J., P. R. Danson, and N. D. Start. 1981. Genetic improvement in yield of spring barley and associated changes in plant phenotype, pp. 97–103. *In*: M. Asher (ed.). Barley genetics. IV. Proc. 4th Int. Barley Genetics Symposium, Edinburgh University Press, Edinburgh.

Salisbury, F. B. and C. W. Ross. 1985. Growth and development, pp. 290–308. *In*: Plant Physiol. Wadsworth, Belmont, California.

Sharma, R. C. 1992. Duration of the vegetative and reproductive period in relation to yield performance of spring wheat. Eur. J. Agron. 1: 133–137.

Shroyer, J. P. and T. S. Cox. 1993. Productivity and adaptive capacity of winter wheat land-races and modern cultivars grown under low-fertility conditions. Euphytica 70: 27–33.

Simmons, S. R. 1987. Growth, development, and physiology, pp. 77–113. *In*: E. G. Heyne (ed.). Wheat and wheat improvement, 2nd ed. Am. Soc. Agron., Madison, Wisconsin.

Singh, D. K. and S. Bharti. 1985. Seed manganese content and its relationship with the growth characteristics of wheat cultivars. New Phytol. 101: 387–391.

Stone, P. J. and M. E. Nicolas. 1995. A survey of the effects of high temperature during grain filling on yield and quality of 75 wheat cultivars. Aust. J. Agric. Res. 46: 475–492.

Stoskopf, N. C. 1981. The botany of crop production, pp. 13–38. *In*: Understanding crop production. Reston Publishing Co., Reston, Virginia.

Takkar, P. N. and C. D. Walker. 1993. The distributoin and correction of zinc deficiency, pp, 151–165. *In*: A. D. Robson (ed.) Zinc in Soils and Plants. Kluwer Academic Publishers, Dordrecht.

Ting, I. P. 1982. Growth, growth kinetics and growth movement, pp. 459–480. *In*: Plant physiology. Addison-Wesley, Reading, Massachusetts.

Tottman, D. R. 1987. The decimal code for the growth stages of cereals, with illustrations. Ann. Appl. Biol. 110: 441–454.

Tottman, D. R. and R. J. Makepeace. 1979. An explanation of the decimal code for the growth stages of cereals, with illustrations. Ann. Appl. Biol. 93: 221–234.

Waldren, R. P. and A. D. Flowerday. 1979. Growth stages and distribution of dry matter, N, P, and K in winter wheat. Agron. J. 71: 391–397.

Ward, R. C., D. A. Whitney, and D. G. Westfall. 1973. Plant analysis as an aid in fertilizing small grains, pp. 329–348. *In*: L. M. Walsh and J. D. Beaton (eds.). Soil testing and plant analysis. Soil Sci. Soc. Am., Madison, Wisconsin.

Warren, W. J. 1981. Analysis of growth, photosynthesis and light interception for single stands. Ann. Bot. 31: 41–57.

Welch, R. M., W. H. Allaway, W. A. House, and J. Kubota. 1991. Geographic distribution of trace-element problems, pp. 31–57. *In*: J. J. Mortvedt, F. R. Fox, L. M. Shuman, and R. M. Welch (eds.) Soil Sci. Soc. Am., Madison, Wisconsin.

Wilson, J. W. 1981. Analysis of growth, photosynthesis and light interception for single plants and stands. Ann. Bot. 48: 507–512.

Wych, R. D. and D. C. Rasmusson. 1983. Genetic improvement in malting barley cultivars since 1920. Crop Sci. 23: 1037– 1040.

Wych, R. D., S. R. Simmons, R. L. Warner, and E. J. Kirby. 1985. Physiology and development, pp. 103–125. *In*: D. C. Rasmusson (ed.). Barley. Monogr. 26, Am. Soc. Agron., Madison, Wisconsin.

Yoshida, S. 1972. Physiological aspects of grain yield. Ann. Rev. Plant Physiol. 23: 437–464.

Zadoks, J. C., T. T. Chang, and C. F. Konzak. 1974. A decimal code for the growth stages of cereals. Weed Sci. 14: 415–421.

9

Rice

I. INTRODUCTION

Rice is the most important staple food for more than half of the world's population, including regions of high population density and rapid growth. It provides about 21% of the total caloric intake of the world population. In Asia, where more than 3.1 billion people live, it provides an average 35% of the total calories consumed, ranging up to 80% in Cambodia (IRRI, 1993). It is the most important tropical cereal and, on a world basis, rice production is slightly below that of wheat. The world average yield of rice is about 3500 kg ha^{-1}, more than wheat (2310 kg ha^{-1}) and corn (3200 kg ha^{-1}) (FAO, 1992). Rice is cultivated in 111 countries on all continents (except Antarctica), at latitudes from 36°S in Australia, to 45°N in Japan, 49°N in Czech Republic, and similar latitudes in Hungary and in the Heilongjiang Province of China (Lu and Chang, 1980; McDonald, 1994). Rice production is concentrated in Asia, where more than 90% of the world's supply is produced. China and India are the leading producers as well as consumers of rice. Other major rice-producing countries are Japan, Thailand, Vietnam, and Indonesia.

Oryza sativa L. and *Oryza glaberrima* Steud. are cultivated species of rice. *Oryza sativa* is widely cultivated, but *O. glaberrima* is mainly cultivated in Africa where it is rapidly being replaced by *O. sativa*. The two species show small morphological differences but hybrids between them are always sterile (Chang, 1976). The origin of *O. sativa* is controversial but it is thought to have been domesticated in India or Indochina. *Oryza glaberrima* originated in Africa. A detailed description of the biosystematics and cytogenetics of the genus is given by Chang (1964) and Nayar (1973).

Oryza sativa is further divided into the japonica, indica, and javanica ecological groups. Japonica rice, adapted to cooler areas, is widely grown in temperate countries such as central and northern China, Korea, and Japan, while indica rice is widely grown in tropical regions. Both of these species can be grown in subtropical regions. Javanica rice is the tall, large, and bold grain bulu cultivar of Indonesia, but it has spread to Japan, Taiwan, and the Philippines (Yoshida, 1983).

Unhusked grain, as well as the growing crop, is known as "paddy." Husked or hulled rice, usually termed brown rice, is milled to remove the outer layers, including the aleurone layer and the germ, after which it is polished to produce white rice. Paddy, on milling, gives approximately 20% husk, 50% whole rice, 16% broken rice, and 14% bran and meal (Purseglove, 1985). The endosperm is highly digestible and nutritious and on average contains about 8% protein.

Rice culture is divided into two broad groups, upland and lowland culture. Upland rice refers to rice grown on both flat and sloping fields that are prepared and seeded under dryland conditions and depend on rainfall for moisture. This is also known as dryland or rainfed rice. This type of rice cultivation is most common on small- and medium-size farms in South America, Asia, and Africa. Brazil is the world's largest producer of upland rice (Fageria et al., 1982; IRRI, 1984b). On the other hand, flooded rice is grown on flat land with controlled irrigation. It is also known as irrigated rice, lowland rice, or waterlogged rice. Lowland rice is commonly flooded when seedlings are 25–30 days old. The water level varies from 10 to 15 cm and is maintained until 7–10 days before harvesting. About 77 million ha, 53% of the world's rice area, is cultivated in this manner (IRRI, 1984a). In much of this area, rainfall supplements irrigation water. Irrigated areas with good water control are suitable for growing improved varieties of lowland rice with short stature and lend themselves to improved cultural practices. Perhaps 70–75% of the world's rice production comes from irrigated areas (IRRI, 1984a). A small percentage of rice area in some countries consists of deep water and floating rice culture (IRRI, 1982). A comparison of lowland and upland rice cultures is given in Table 9.1.

Table 9.1 Comparison of Upland and Lowland Rice Cultures

	Lowland	Upland
1.	Cultivated on leveled, bunded, undrained soils.	Cultivated on undulating or leveled naturally drained soils.
2.	Water supply through rainfall or irrigation.	Water supply through rainfall.
3.	Water accumulation in the field during major part of crop growth.	No water accumulation during crop growth.
4.	Reduced root zone during major part of crop growth.	Oxidized root zone during crop growth.
5.	Direct seeding or transplanting.	Direct seeding.
6.	Thin and shallow root system.	Vigorous and deep root system.
7.	High tillering.	Relatively low tillering.
8.	Short and thin leaves.	Long and thick leaves.
9.	Environmental conditions are stable and uniform.	Environmental conditions are unstable and variable.
10.	Incidence of diseases and insects low.	Incidence of diseases and insects high.
11.	Weeds are not a serious problem.	Weeds are a serious problem.
12.	Needs high input.	Needs low input.
13.	High cost of production.	Low cost of production.
14.	Stable and high yield.	Unstable and low yield.

An outstanding breakthrough occurred in rice cultivation with the introduction of high-yielding cultivars. In work carried out at the International Rice Research Institute (IRRI), a cross between the tall tropical cultivar "Peta" from Indonesia and the subtropical semidwarf variety "Dee-geo-woo-gen" from Taiwan produced the semidwarf IR8. IR8 produced a record yield of 11 t ha^{-1} and responded to nitrogen up to 150 kg N ha^{-1} at several locations in tropical Asia during 1966–1968 (Chang, 1976). Dissemination of this improved plant type throughout Latin America was initiated in 1968 by the Colombian-based program of the Centro Internacional de Agricultur Tropical (CIAT), Federacion Nacional de Arroceros (FEDEARROZ), and the Instituto Colombiano Agropecuario (ICA) (Cuevas-Perez et al., 1995). Breeders took advantage of Colombian environmental diversity, developing materials that could be transferred to other countries through trainees who had participated in the evaluation of these materials along with CIAT scientists. This breeding approach increased national rice yields in Colombia from 1.5 t ha^{-1} in 1965 to 4.4 t ha^{-1} in 1975, and in Latin America from 1.8 t ha^{-1} during the period

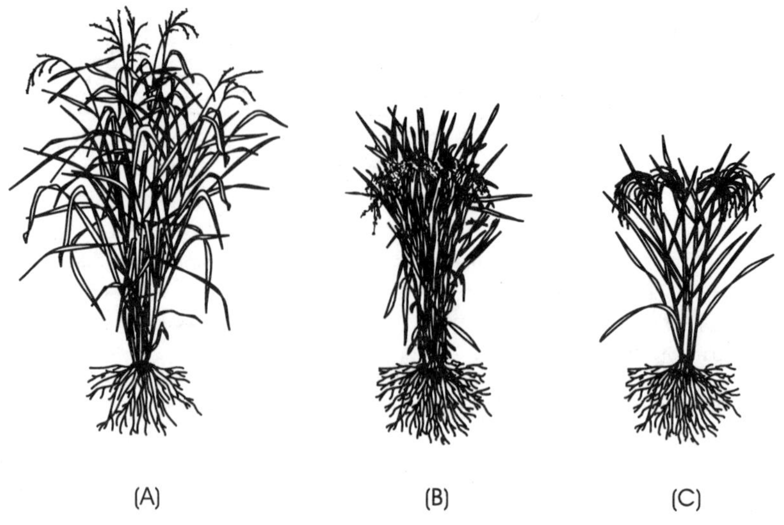

Figure 9.1 (A) Traditional old, (B) modern high yielding, and (C) future ideotype plants of lowland rice. (From Kush, 1995.)

1950–1964 to 2.3 t ha^{-1} in 1974 (Cuevas-Perez et al., 1995). Later estimates indicated that semidwarf cultivars contributed 20% additional rice production in Latin America in 1981 (Cuevas-Perez et al., 1995). Scientists at IRRI and several national breeding programs combined most of the desired features in the improved plant type, including reduced height (about 100 cm), erectness, short, dark-green leaves, high tillering, stiff culms, early maturity, photoperiod insensitivity, nitrogen responsiveness, and high harvest index. The wide adoption of IR8 and other high-yielding cultivars like IR20 and IR22 made it possible for the semidwarf varieties to become major cultivars in Brazil, Colombia, Peru, Ecuador, Cuba, Mexico, Indonesia, Malaysia, Philippines, India, Pakistan, Bangladesh, and South Vietnam. By 1972–1973, semidwarf varieties occupied a large part of the area planted in high-yielding cultivars—about 10% of the world total and 15% in tropical Asia (Chang, 1976). At present, high-yielding semidwarf cultivars predominate in most lowland rice-producing areas. Work is in progress at IRRI and at many national research centers to further improve plant type, grain quality, and pest resistance. Figure 9.1 shows the development of modern high-yielding rice cultivars from traditional old cultivars and ideotype rice plants of the future. At the world level, rice yield is highest in Australia. Rice crop yields in New South Wales are very high by world standards, averaging 8.9 t/ha in 1991–92 (Beecher et al., 1994). These high yields have been attributed to the absence of serious pests and diseases, high solar radiation, high-yielding varieties, and the use of legume-based pasture phases in a traditional rotation (Beecher et al., 1994).

Production, based on population growth rates in countries where rice is the main food crop, indicates that rice production must increase 65% by the year 2020 (IRRI, 1989). This production increase will have to come from the same or an even less amount of land, and water and other production inputs will have to be used more efficiently in the future. Under these situations, a better understanding of growth habits of rice plants, and the effects of environmental factors such as climate, soil, and nutrient requirements should be well understood for improving yield.

II. CLIMATE AND SOIL REQUIREMENTS

Rice is widely distributed throughout the tropical, subtropical, and temperate zones of all continents. Temperature, solar radiation, and rainfall are the important climatic components which affect rice growth and yield. Table 9.2 shows the minimum, optimum, and maximum temperatures for rice growth at different stages. Depending on growth stages, injury to rice may occur when the daily minimum temperature drops below 20°C. Rice is most sensitive to low temperature 14–7 days before heading and at flowering (Satake, 1976; Yoshida, 1983). Low temperatures at these critical stages cause a high percentage of spikelet sterility. Table 9.3 shows the effect of minimum temperatures (15 or 19°C) around flowering (15 days before to 15 days after) on

Table 9.2 Minimum, Optimum, and High Temperatures for Rice Growth at Different Stages

	Critical temperature (°C)		
Growth stage	Low	Optimum	High
Germination	10	20–35	45
Seedling emergence and establishment	12–13	25–30	35
Rooting	16	25–28	35
Leaf elongation	7–12	31	45
Tillering	9–16	25–31	33
Initiation of panicle primordium	15	25–30	35
Panicle differentiation	15–20	25–28	38
Anthesis	22	30–33	35
Ripening	12–18	20–25	30

Source: Yoshida, 1981.

Table 9.3 Effect of Temperature on Grain Yield and Spikelet Sterility in Three Brazilian Upland Rice Cultivars

Cultivar	Minimum[a] (°C)	Maximum[a] (°C)	Grain yield (g/pot)	Grain sterility (%)
IAC 25	15	30	3.45	96
	19	27	26.00	9
IAC 164	15	30	5.56	97
	19	27	29.54	19
IAC 165	15	30	7.37	97
	19	27	30.73	15

[a] Minimum and maximum temperature values were observed about 1 month around flowering.
Source: Fageria, 1984a.

three Brazilian upland rice cultivars in a greenhouse experiment. When the temperature was 15°C for one month around flowering, yield was reduced significantly in all cultivars due to very high grain sterility. Little sterility was observed at 19°C for the same growth stage.

Solar radiation is an important climatic factor that affects temperature, evapotranspiration, and photosynthesis of rice. Solar radiation requirements of rice differ from one growth stage to another (Yoshida and Parao, 1976). Yield is significantly reduced if sufficient solar radiation is not received during the reproductive (panicle initiation to flowering) and ripening (flowering to maturity) growth stages. During reproductive growth, insufficient solar radiation reduces the number of spikelets and consequently grain yield. Similarly, during ripening, the percentage of filled spikelets is also reduced. Inadequate light during the vegetative growth stage affects yield and yield components only slightly (Yoshida and Parao, 1976). Solar radiation of 300 cal cm^{-2} day^{-1} during the reproductive stage allows yields of 5 t ha^{-1} (Yoshida and Parao, 1976). Less solar radiation is required during the ripening growth stage to achieve the same yield level. As such, the effect of solar radiation is generally more pronounced if yield is higher than 5 t ha^{-1} under tropical and subtropical environments.

Rice is a semiaquatic plant that is commonly grown under flooded conditions. However, about half of the rice area in the world does not have sufficient water to maintain flooded conditions, and yield is reduced to some extent by drought, defined here as a period of no rainfall or no irrigation that affects crop growth (Hanson et al., 1990).

Studies of the constraints to rice yield have confirmed that water deficit is the most serious factor limiting production (Widawsky and O'Toole, 1990).

One-half of global rice land, and two-thirds of the rice land in South and Southeast Asia is cultivated under rainfed conditions (Huke, 1982). In addition, much of the rice land classified as irrigated is subject to periodic water deficits (Kush, 1984) due to inequitable water distribution systems and deforestation-related reduction in the water yield of many surface irrigation systems.

Areas designated as "drought prone" in the classification of rainfed rice environments (Garrity et al., 1986) are widespread in Asia, Africa, and Latin America. Studies which evaluated research priorities for genetic improvement of rice implicate water or drought stress as one of the major targets for research (Garrity and O'Toole, 1994).

The most effective method of minimizing the adverse effect of drought is for the crop to grow during the period of high rainfall and high soil water availability (i.e., to escape the drought period). Crop duration is important in determining grain yield because early maturing cultivars often escape a terminal stress while late-maturing cultivars may be affected by it. Timing of drought development in relation to phenology is also important for determination of grain yield. The stage from panicle development to anthesis is most susceptible to water stress in rice (O'Toole, 1982). Boonjung (1993) showed that grain yield decreases at the rate of 2% per day delay as a 15–day stress period (morning leaf water potential less than –1.0 MPa) occurs later during panicle development. Assuming a reduction of 2% grain yield per day with the delay in termination of a 15–day stress, a 20–day difference in flowering time between two cultivars of equal yield potential could cause a grain yield difference of about 40%. Thus it is likely that cultivars with different phenology will react differently to a drought, depending on the timing of stress (Maurya and O'Toole, 1986). Hence, genotypes should be compared for drought resistance/ susceptibility within the same phenology group, or at least genotypic variation in phenology should be corrected in some way before differences in drought resistance are estimated (Garrity and O'Toole, 1994). Alternatively, it is possible to implement a strategy of staggered planting of lines so that they flower at about the same time (Lilley and Fukai, 1994).

Several drought-resistance mechanisms (and the putative traits which contribute to them) have been identified for rice, important among these being drought escape via appropriate phenology, root characteristics, specific dehydration avoidance and tolerance mechanisms, and drought recovery. Some of these mechanisms/traits confer drought resistance, while others show potential to do so in rice. The most important is the appropriate phenology which matches crop growth and development with the water environment. A deep root system with high root length density at depth is useful in extracting water thoroughly in upland conditions, but does not appear to offer much scope for improving drought resistance in rainfed lowland rice where the development of a hard pan may prevent deep root penetration. Under water-limiting environments, genotypes which maintain the highest leaf water potential generally grow best, but it is not known if genotypic variation in leaf water potential is solely caused by root factors. Osmotic adjustment is promising because it can potentially counteract the effects of a rapid decline in leaf water potential, and there is large genetic variation for this trait. There is genotypic variation in expression of green leaf retention which appears to be a useful character for prolonged droughts, but it is affected by plant size which complicates its use as a selection criterion for drought resistance (Fukai and Cooper, 1995).

Despite our increased understanding of stress physiology, the development of drought-resistant cultivars (i.e., cultivars which produce higher yield than others in drought conditions) has been slow in rice and other crops (Fukai and Cooper, 1995). A major reason for the slow progress in developing drought-resistant rice cultivars is the incidence of large genotype by environment (G × E) interactions, which result from a combination of differences in genotypic adaptation and the heterogeneous environments within the target areas. A consequence of G × E interactions is that particular lines do not perform well under all conditions encountered in the target population of environments, complicating selection of new cultivars (Fukai and Cooper, 1995).

As rice is a semiaquatic plant, it can grow successfully in standing water. It can transport oxygen or oxidized compounds from the leaves to the roots and into the rhizosphere. The oxygen in the rice leaves and roots comes from atmospheric oxygen absorbed by the leaves and oxygen released in photosynthesis through the hydrolysis of water. The water requirement of rice, as measured by the transpiration ratio, is similar to that of most major crops and is mostly affected by climate and soil. The average water requirement of irrigated rice at various locations in various countries of Southeast Asia is about 1240 mm per crop (Kung, 1971). Consequently, rice cultivation appears to be limited to areas where the annual rainfall exceeds 1000 mm (Yoshida, 1983).

Evapotranspiration of upland rice increases with increasing plant age, with a maximum in the reproductive stage followed by a decrease during the ripening stage. Water stress during the growth period adversely affects growth, but the magnitude of growth reduction varies from one growth stage to another (Figure 9.2).

The response of rice to water stress at the vegetative stage is primarily reduced height, tillers, and leaf area (IRRI, 1975), while at more sensitive reproductive stages (e.g., flowering) high spikelet sterility results in the greatest reduction in grain yield (Matsushima, 1968; Fageria, 1980a; Cruz and O'Toole, 1984). The most critical water deficit stage is the period from about 10 days before flowering to flowering (Matsushima, 1962). High sterility

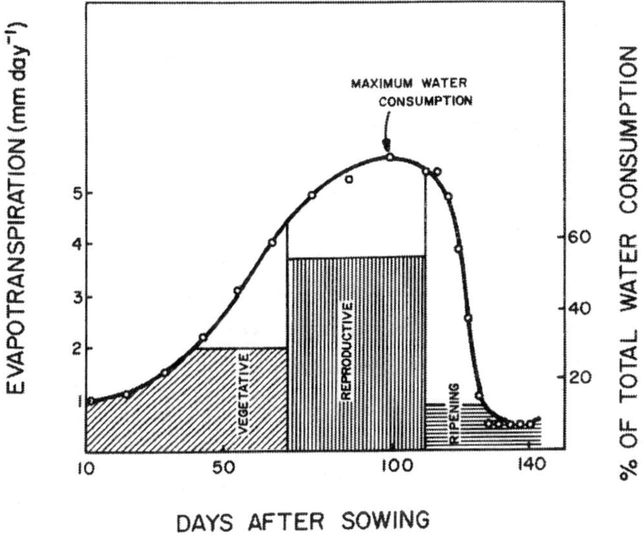

Figure 9.2 Evapotranspiration and water consumption during growing season of upland rice in central Brazil. (From Fageria, 1980a.)

resulting from water stress at this time is irreversible, and adequate water at later stages is totally ineffective in reducing its effects. If water stress occurs during the early vegetative stage, however, plants can recover.

Tomar and Ghildyal (1975) studied the differences in resistance to water transport between plants grown on upland soils and those grown on flooded soils. They concluded that resistance to water transport in the nonflooded rice was nearly twice as high as in the flooded plants. The nonaerenchymatous roots of nonflooded plants had about 17 times more resistance than aerenchymatous roots of flooded rice.

Rice is adapted to a wide range of soil conditions. It can be grown on sandy loams, shallow lateritics, and heavy clays, provided there is adequate water. Heavy soils that have an impervious subsoil layer that limits drainage are usually better suited to lowland rice, as they reduce loss of water and nutrients through percolation. Rice is an acid-tolerant crop with an optimum pH for upland culture around 5 (Fageria and Zimmermann, 1996). In lowland culture, soil pH rises during flooding. Increases from acidic to neutral (or slightly basic) pH are caused by reactions that reduce iron oxides and increase the concentration of carbon dioxide in the soil. Rice can also be grown on alkaline soils. It is considered moderately susceptible to soil salinity, with a threshold salinity about 3 dS m^{-1} (Maas and Hoffman, 1977).

III. GROWTH AND DEVELOPMENT

Rice is an annual grass. Irrigated rice has a seed dormancy period ranging from 1 to 3 months, depending on the cultivar and seed moisture content. The most practical method of breaking seed dormancy is a thermal seed treatment of 50°C for about 4–5 days. Normally, upland and japonica-type cultivars do not have a seed dormancy which significantly affects germination. After dormancy is broken, rice seeds sown in soil germinate in 5–7 days. Germination of rice is hypogeal. When grown in well drained soils, the coleorhiza with the radical is the first organ to emerge, while under inundation the coleoptile precedes the coleorhiza (Purseglove, 1985).

Although rice can germinate and grow at low oxygen concentrations, germination is poor and seedlings have low vigor (Ponnamperuma, 1965). Oxygen can be supplied to germinating seeds under reduced conditions by coating the seeds with strong oxidizing agents such as calcium peroxide (CaO_2) and magnesium peroxide (MgO_2). If seeds are coated with 60% calcium peroxide powder at a level of 35% (by weight), the 1 mg of oxygen generated is sufficient to enable germination in an anoxic environment (Baker and Hatton, 1987; Hatton and Baker, 1987). The decomposition of inorganic Ca and Mg peroxides in a well buffered, neutral environment initially involves a hydrolysis step (Baker and Hatton, 1987):

$$MO_2 + 2H_2O \leftrightarrows M(OH)_2 + H_2O_2$$

The decomposition of H_2O_2 liberates the O_2 according to the following reaction:

$$2H_2O_2 \rightarrow 2H_2O + O_2$$

Some of the hydroxide may subsequently react with atmospheric carbon dioxide with following results:

$$M(OH)_2 + CO_2 \rightarrow MCO_3 + H_2O$$

In a poorly buffered environment, inorganic peroxide decomposition would result in an increased pH and reduced decomposition, while under acid conditions its decomposition would result in the production of metal cation and hydrogen peroxide (Baker and Hatton, 1987):

$$MO_2 + 2H^+ \rightarrow M^{2+} + H_2O_2$$

Baker and Hatton (1987) also reported that coating of seed with calcium peroxide, while having a beneficial effect in providing oxygen to a seed germinating under reduced O_2 conditions, might cause reduction in overall germination under aerobic conditions compared to that of an uncoated seed.

A. Roots

Monocotyledonous plants, such as rice, form a fibrous root system consisting of seminal, nodal, and lateral roots (Yamachi et al., 1987b). The seminal root originates from the embryo, whereas the nodal roots emerge from stems; both produce numerous first-, second-, and third-order lateral roots (Yamachi et al., 1987a). The seminal and nodal roots constitute a framework, whereas lateral roots form the network of finer roots.

Yamachi et al. (1987b) studied the root system of rice under controlled conditions and concluded that most of the nodal roots grew vertically into the soil profile. These roots were mostly distributed just below the shoot bases and produced profusely branching fine and relatively short laterals.

On germination the primary root emerges, followed by two additional seminal roots. Adventitious roots are then produced from the basal nodes of the primary stem and tillers (Purseglove, 1985). All the roots of rice have a definite relationship with the emergence and development of the leaves. When leaf *n* is developing, roots emerge simultaneously at node *n*–3 of the same stem (Fujii, 1961). Root weight increases with age of the plant (Figure 9.3). Root development is affected by environmental factors such as moisture, temperature, aeration, and nutrients, as well as by cultivar (Fageria, 1984a). According to Okajima (1960), the emergence of root is closely correlated with the N concentration of the stem base, since active emergence of roots takes place only when the N concentration is above 1%. The root system of upland rice is more vigorous than that of lowland rice. In the case of lowland rice, roots which arise from the lower nodes spread laterally and do not usually penetrate more than 15 cm in the soil (Adair et al., 1962). Sharma et al. (1987) reported that about 75% of the roots of rainfed lowland rice was found in the top 10 cm of soil at flowering, and root length density at the flowering stage was less in dry seeded than in transplanted rice.

When grown in flooded soil, rice roots develop aerenchyma that permits diffusion of O_2 from the leaves through the roots to the soil. This permits root growth in reduced soils. Quantitative and qualitative estimates of root oxidizing activity indicate that several semi-independent enzyme systems may operate, including both oxidases and dehydrogenases (Yamada and Ota, 1958; Yoshida and Takahashi, 1960; Armstrong, 1969). In wetland species, oxygen diffusion from roots contributes to rhizosphere oxidation and is a source of O_2 used in respiration by rhizosphere microorganisms (Pugh, 1962).

B. Tops

Rice has a determinate growth habit. The culm is more or less erect and cylindrical with solid nodes and hollow internodes. Leaves alternate in two

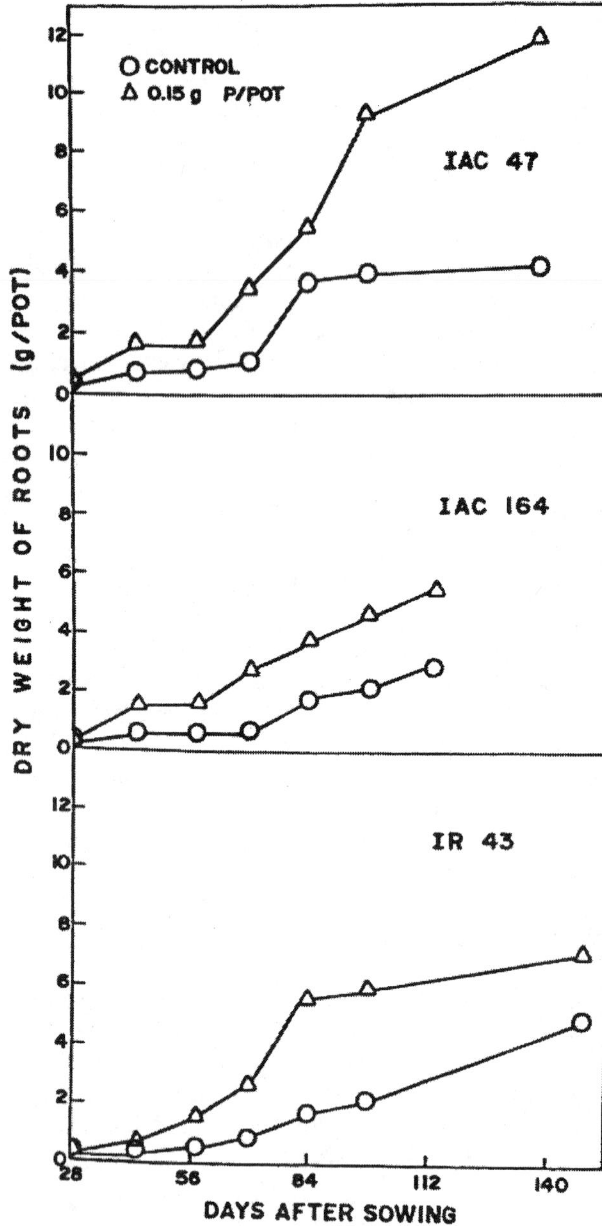

Figure 9.3 Effect of P fertilization on dry weight of roots of three upland rice cultivars during growth cycle in Oxisol of central Brazil.

ranks with a single leaf at each node. The numbers of nodes and leaves are greatest on the main culm and decrease on successive tillers. A terminal panicle varies from compact and erect to open and drooping. The base of the peduncle is enclosed in the sheath of the flag leaf. Each spikelet has one flower and is strongly compressed laterally, articulating below the two small outer glumes. Unlike other cereals, each floret has six (rather than three) stamens and two long styles. A detailed description of rice inflorescence is given by Chang and Bardenas (1965). Depending on cultivar and growing conditions, rice tillers prolifically. Tillering begins at the four- to five-leaf stage (Murata and Matsushima, 1975). According to Katyama (1951), emergence of tillers is closely related to that of leaves. The primary tiller emerges from the axil of leaf *n*–3, when *n* of the main stem elongates. Secondary and tertiary tillers emerge in the same way, and all tillers are synchronized with the development of leaves on the main stem. Nitrogen and phosphorus content of the plant, solar radiation, and temperature are the most important factors affecting tillering in rice. Nitrogen concentration of 3.5% is necessary for active tillering. At 2.5% N tillering stops, and below 1.5% death of tillers takes place (Ishizuka and Tanaka, 1963). Similarly, no tillering takes place when the P concentration of the mother stem is below 0.25% (Honya, 1961). Initial tillers depend on the nutrient supply from the mother stem but become autotrophic when they have three leaves and four or five roots (Ishizuka and Tanaka, 1963). Figures 9.4 and 9.5 show the effects of P fertilization and plant age on tiller of upland and lowland rice grown under Brazilian conditions. The decrease in tiller number after 70–100 days results from death of the last tillers due to their failure to compete for light and nutrients (Ishizuka and Tanaka, 1963). Another explanation is that during the period of growth beginning with panicle development, competition for assimilates exists between the developing panicles and young tillers. Eventually, growth of many young tillers is suppressed, and they may senesce without producing seed (Dofing and Karlsson, 1993).

Tillering is an important growth parameter in transplanted rice, and in some situations heavy tillering is essential. However, when rice is broadcast or aerially sown, there may be potential gains in yield from free tillered or even uniculum rice or higher harvest index sown at heavier seeding rates (Donald and Hamblin, 1983).

Leaf Area Index (*LAI*)

Variations in leaf area and duration are important physiological parameters that determine crop yield. Light interception by the canopy is strongly influenced by leaf size and shape, angle and vertical separation, and horizontal arrangements. Thick, short, and erect leaves are associated with higher photosynthetic rate and thereby higher dry matter production and yield (Yoshida,

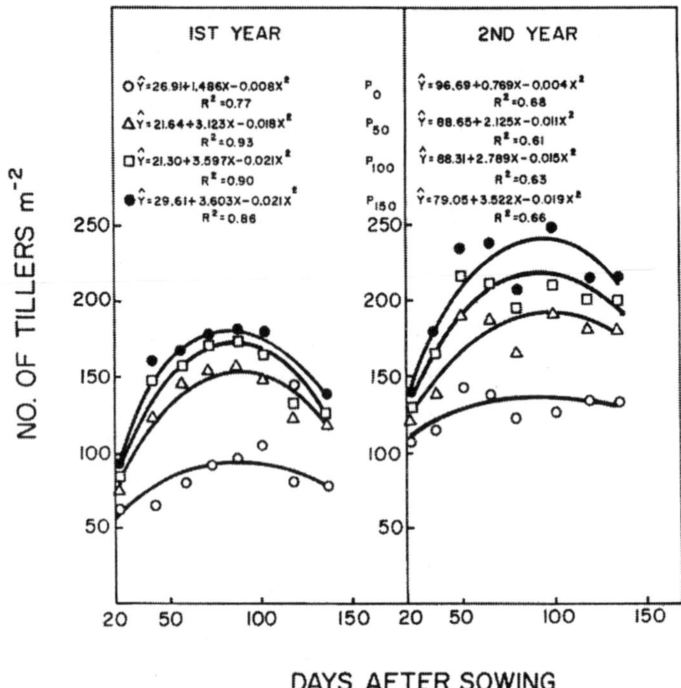

Figure 9.4 Tillering in upland rice cultivar IAC47 affected by P (kg P_2O_5 ha^{-1}) fertilization and period of growth in Oxisol of central Brazil. (From Fageria et al., 1982.)

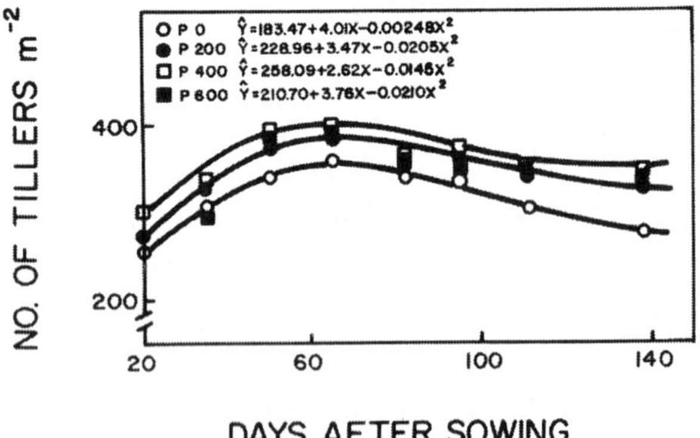

Figure 9.5 Tillering in lowland rice cultivar IAC 435 as influenced by P (kg P_2O_5 ha^{-1}) fertilization and plant age in gley humic Brazilian soil. (From Fageria, 1980b.)

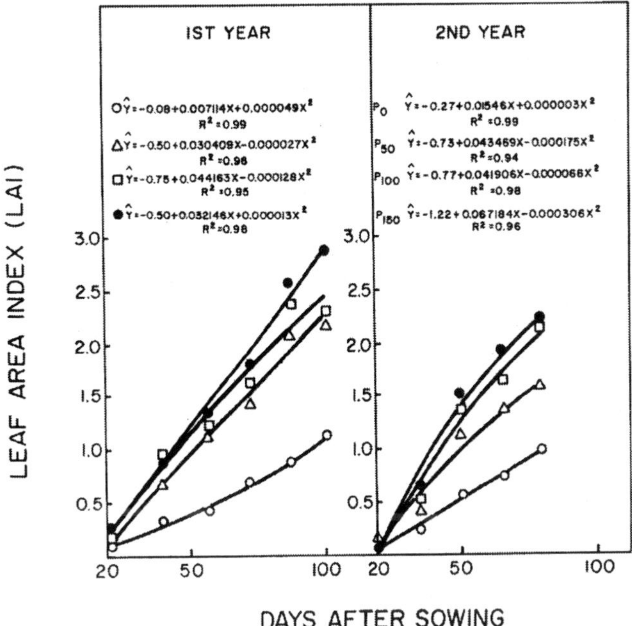

Figure 9.6 Effects of P fertilization on leaf area index of upland rice days after sowing. (From Fageria et al., 1982.)

1972). The *LAI* of rice increases as crop growth advances and reaches a maximum at about heading (Yoshida, 1983). The increase in *LAI* is caused by an increase in tiller number and in size of successive leaves. Among various environmental factors, N and P nutrients have the most marked effect on *LAI* by increasing the number of tillers as well as leaf size. Figures 9.6 and 9.7 show the effects of P fertilization and plant age on *LAI* of upland and lowland cultivars. The *LAI* in upland as well as lowland rice significantly increased with P fertilization and age. The optimum *LAI* for upland rice is about 2–3 at 85–100 days after sowing (Figure 9.6). This can be compared to the optimum *LAI* for lowland rice, which is about 4–7 (Yoshida, 1972), or approximately double that of upland rice (Figures 9.6 and 9.7). Such a low *LAI* of upland rice may be related to differences in plant type. Rice grown under upland conditions is often subjected to moisture stress and in general has fewer tillers and less leaf area than rice grown under lowland conditions (Chang and Vergara, 1975).

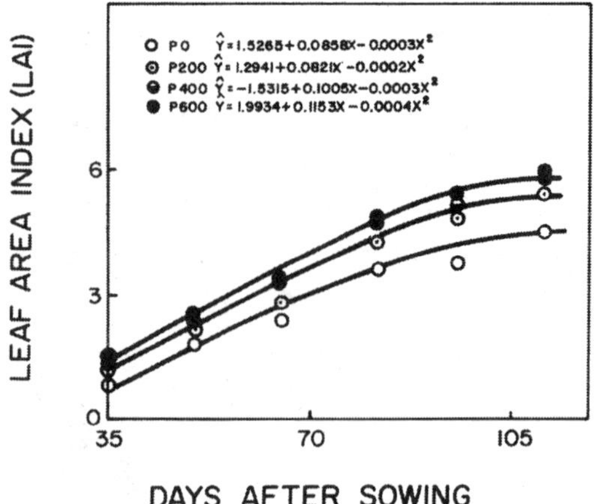

Figure 9.7 Effects of P fertilization on leaf area index of lowland rice days after sowing. (From Fageria, 1980b.)

C. Dry Matter

Rice is a C_3 plant with lower dry matter production than C_4 cereals like corn and sorghum. Like that of all other field crops, dry matter production in rice increases as leaf area and light interception increase and then may decrease as the crop approaches physiological maturity. The rate of production is influenced by environmental condition, plant density, cultivar, and management practices.

Figure 9.8 shows the dry matter production of upland rice at successive stages of growth during 2 years of study under field conditions in Brazil. Although total dry matter production was not the same for two seasons, growth curves were quadratic for both seasons. Plant growth was slow during the first 50–60 days, followed by a sharp increase in dry matter production from 60 to 110 days. This age approximately corresponds to panicle initiation to heading for cultivar IAC 47, which has a growth period of 130–135 days when grown in the central part of Brazil (Fageria et al., 1982). The increase in dry matter at this stage is mainly attributed to an increase in weight of leaves and stems (Ishizuka, 1973). Figure 9.8 also shows that dry matter production increased with higher levels of P in both years. The higher dry matter production with application of P could be related to a higher leaf area index and tillers (Fageria et al., 1982).

Dry matter production at different growth stages of lowland rice is presented in Figure 9.9. The growth curve was exponential quadratic. Plant growth was slower during the first 40–50, days and then there was a sharp

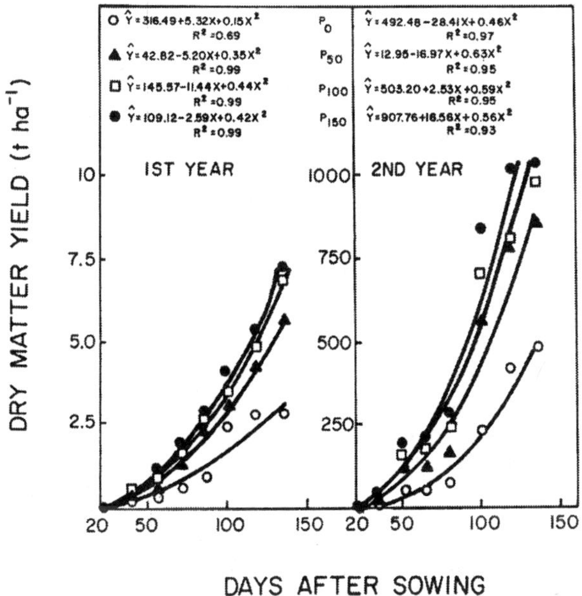

Figure 9.8 Dry matter production by upland rice cultivar IAC 47 as influenced by plant age and phosphorus fertilization (kg P_2O_2 ha^{-1}) in Oxisol of central Brazil. (From Fageria et al., 1982.)

increase in dry matter production from 50 to 91 days. This age approximately corresponded to 13 days before panicle initiation to flowering for cultivar Javae which had a growth period of 120 days in the central part of Brazil. The increase in dry matter at this growth period is mainly attributed to an increase in the weight of leaves and stems (Fageria et al., 1982). From flowering to maturity, dry matter production decreased (Figure 9.9). Dry matter loss from the vegetative tissues from the interval of flowering to maturity was 24%, suggesting active transport of assimilates to the panicles, which resulted in a grain yield of about 6600 kg ha^{-1} (Figure 9.10). The limited amount of assimilate loss compared with the large dry matter gain in the panicles is in agreement with the generally established concept that photosynthesis during grain filing is the major contributor to the grain yield of rice (Guindo et al., 1994).

Crop growth rate (Figure 9.11), calculated from the data of Fageria (1980b), is about 240 g ha^{-1} day^{-1} at 20 days of growth and decreases to about 10 g ha^{-1} day^{-1} at the end of the growth cycle. The decrease in crop growth rate with increasing age is related to a decreased photosynthetic rate with age and higher respiration rate due to increased leaf canopy (Yoshida, 1972).

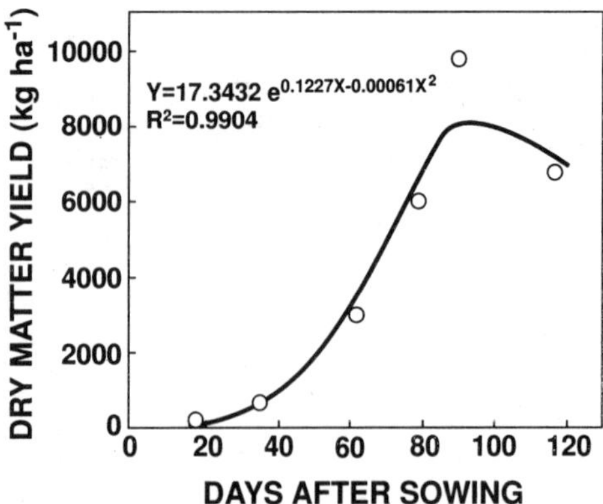

Figure 9.9 Dry matter production by lowland rice cultivar Javae in Inceptisol of Brazil during growth cycle. (From Fageria, 1996.)

The relative importance of number of panicles, spikelet sterility, and 1000–grain weight was evaluated by the order of inclusion into the multiple forward step-wise regression procedure and the additional contribution that each yield component made to the total variation in yield (Fageria, 1996). Panicles per pot was the most important component of yield, accounting for 86.64% of the variation in yield. Spikelet sterility accounted for a 7.14% variation in yield, and 100–grain weight a 2.6% variation. Grain yield was significantly related to panicles per pot and harvest index, but not with the panicle length

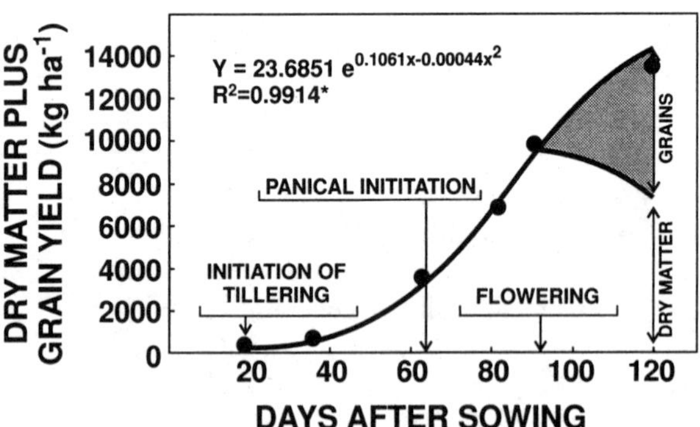

Figure 9.10 Dry matter and grain yield of lowland rice during the growth cycle of the crop. (From Fageria, 1996.)

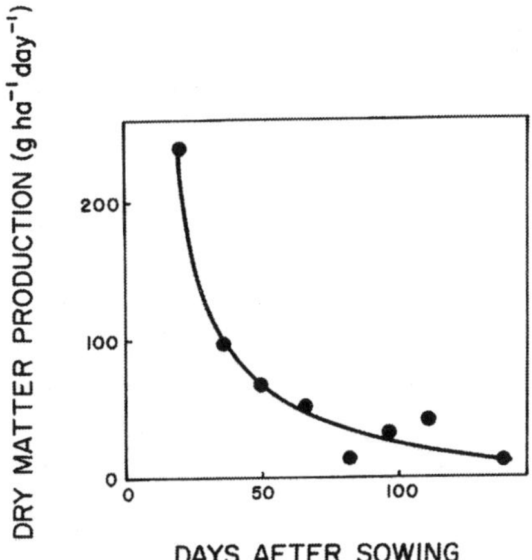

Figure 9.11 Lowland rice crop growth rate under field conditions. (From Fageria, 1980b.)

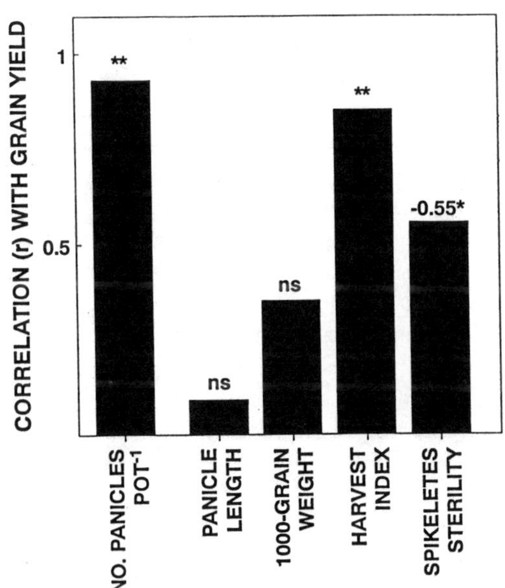

Figure 9.12 Correlation (r) of grain yield components with rice yield. *, ** Significant at 0.05 and 0.01 probability levels, respectively. NS = not significant. (From Fageria, 1996.)

and 1000–grain weight (Figure 9.12). Spikelet sterility was also significantly related to grain yield, but correlation was negative ($r = 0.55^*$).

D. Harvest Index

The biological yield of a cereal crop is the total yield of plant tops and is an indication of the photosynthetic capability of a crop (Yoshida, 1981). Harvest index is the ratio of the yield of grain to the aboveground biological yield (Donald and Hamblin, 1976). Grain yield or economic yield can be increased either by increasing total dry matter production or by increasing harvest index. Harvest index varies with variety and environment, and a negative correlation exists between plant height and harvest index (IRRI, 1977). For example, the harvest index of improved semidwarf indica cultivars is in the range of 0.45–0.55, whereas that of japonica cultivars ranges from 0.40 to 0.49 (Yoshida, 1983). The harvest index of traditional tall rice cultivars is generally less than 0.40. The relationship between total dry matter and harvest index is not very clear. Yoshida (1983) reported that rice grain yield (Y) is related to total dry matter production (W) by:

$$Y = 0.631W^{0.901}$$

This relationship holds true up to 10 t rough rice ha^{-1} (Kiuchi et al., 1966). This suggests that harvest index decreases slowly with increasing total plant dry weight. However, records of high-yielding crops indicate that high rice yields are achieved by increasing total dry weight and by maintaining a high harvest index (Shiroshita et al., 1962).

Some authors have speculated that a further increase in grain yield through breeding can only be accomplished with an increase in total biological yield (Rahman, 1984) and thus total straw yield. The highest harvest index exhibited by California cultivars under direct seeding was 0.59 (Roberts et al., 1993), which is consistent with the maximum harvest indices of 0.60–0.65 reported for high-yielding semidwarf cultivars in transplanted rice (Rahman, 1984). This 0.60–0.65 range is viewed as a theoretical maximum because of the structural difficulty of supporting more than 65% of the total biological yield as grain on less than 35% of the biological yield as straw (Roberts et al., 1993).

IV. YIELD AND YIELD COMPONENTS

Average grain yield varies among continents and countries. A maximum average yield of 6460 kg ha^{-1} has been reported in Australia, followed by North America at 6170 kg ha^{-1}, Europe at 5130 kg ha^{-1}, Asia at 3400 kg ha^{-1}, and Africa at 1730 kg ha^{-1} (FAO, 1989). These average yields include both lowland and upland rice cultures. Normally, average yields are much higher for lowland rice than upland rice. For example, in Brazil the average upland rice yield is about 1.5 t ha^{-1}, whereas the average lowland or irrigated rice yield is about 4.5 t ha^{-1}.

The maximum recorded grain yield is 13.2 t ha^{-1} in Japan, 11 t ha^{-1} in the Philippines, and 17.8 t ha^{-1} in India (Yoshida, 1983). If these yields are compared with a world average of about 3.5 t ha^{-1} (FAO, 1992), it is evident that there is a lot of potential to improve rice yield.

The major yield components in rice are number of panicles per unit area, spikelet number per panicle, spikelet fertility percentage, and grain weight, normally expressed as 1000–grain weight. Taking into consideration these yield components, grain yield can be expressed as follows (Yoshida, 1983):

Grain yield (t ha^{-1}) = panicle number m^{-2} × spikelet number per panicle × % filled spikelets times 1000–grain weight (g) × 10^{-5}

or

= spikelet m^{-2} × % filled spikelets × 1000 grain weight (g) × 10^{-5}

Among these yield components, spikelets m^{-2} is usually the most variable yield component, accounting for about 74% of the variation in yield. Filled spikelet percentage and grain weight together account for 26% of the yield variation (Yoshida and Parao, 1976). Park and Cho (1989) characterized the physiological characters of high-yielding rice varieties as increased number of spikelets per unit area, high photosynthetic efficiency, and high harvest index. Rao (1988) and Venkateswarlu et al. (1990) suggested increasing number of high-density grain (HDG) per panicle as one of the physiological components of high-yielding cultivars. Thus it is evident that efficient sink size and number play a pivotal role in enhancing rice yield potential. Panicle number is determined in the vegetative growth stage, grains per panicle in the reproductive growth stage, and grain weight in the ripening stage (Figure 9.13). Variation among cultivars in the growth cycle is mainly due to variation in the vegetative growth stage. The reproductive and ripening stages are more or less constant in all cultivars. Some factors which affect yield components, and consequently yields, are climatic factors like temperature, solar radiation, water availability, and nutrients, especially N and P. The reduction division growth stage (Figure 9.13) is very sensitive to extreme temperatures, low solar radiation, drought, and N deficiency.

V. NUTRIENT REQUIREMENTS

Nutrient requirements for upland and lowland rice are different due to differences in yield levels and growing conditions. In lowland rice, environmental

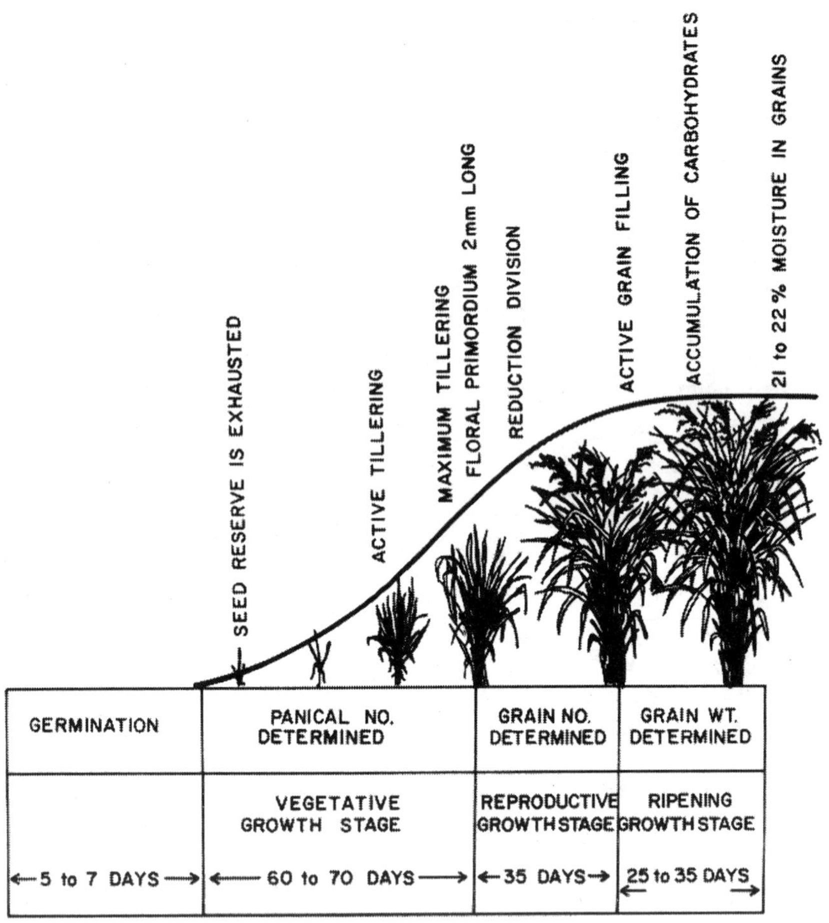

Figure 9.13 Different growth stage and yield component determinations in rice cultivar of 130 and 135 days growth cycle. (From Fageria, 1984a.)

conditions are mostly stable and favorable for plant growth, and high fertilizer application can ensure high yields. However, in the case of upland rice, inadequate water, particularly around flowering, often reduces yield significantly (Fageria, 1980a). Under such circumstances, there is little or no difference in yield between well-fertilized and underfertilized crops. In fact, the higher leaf area and rate of water use of well-fertilized crops can render them more susceptible to the stress of prolonged drought (> 10 days) around flowering (Fageria, 1980a). Therefore, fertilizer recommendations for upland rice should take into account the risk of drought. Another factor that should be taken into consideration is the incidence of blast disease in upland rice. If precautions are not taken to control this disease, panicle neck blast can cause total crop failure. Excess N application sometimes increases blast infection in upland rice (Faria et al., 1982).

A. Upland Rice

Soil Constraints

Rice grown in rainfed, naturally drained soils, without surface water accumulation, normally without phreatic water supply, and normally not bunded, is called upland rice (Garrity, 1984). About half of the world's 135 million ha of rice land is rainfed, and, of this, 19.1 million ha is dryland. As such, dryland accounts for 14% of the total rice area and 29% of the rainfed area (Garrity, 1984). Most upland rice production occurs in South America, Asia, and Africa (Gupta and O'Toole, 1986). Soil fertility is one of the major constraints in upland rice production. Figure 9.14 shows the major soil constraints in the upland rice areas of South and Southeast Asia. Soil acidity, low CEC, and high P fixation capacity are the major soil chemical properties affecting upland rice production. Similar soil constraints are present in South America. For example, the cerrado region is situated in the central part of Brazil, and its total area is about 200×10^6 ha. At present, 3% of the cerrado

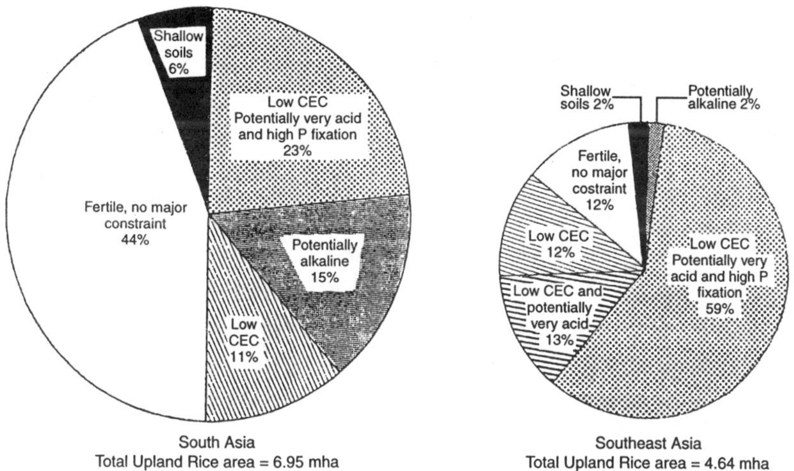

Figure 9.14 Comparative upland rice areas by type of soil constraints based on the fertility capability classification system. (From Garrity and Agustin, 1984.)

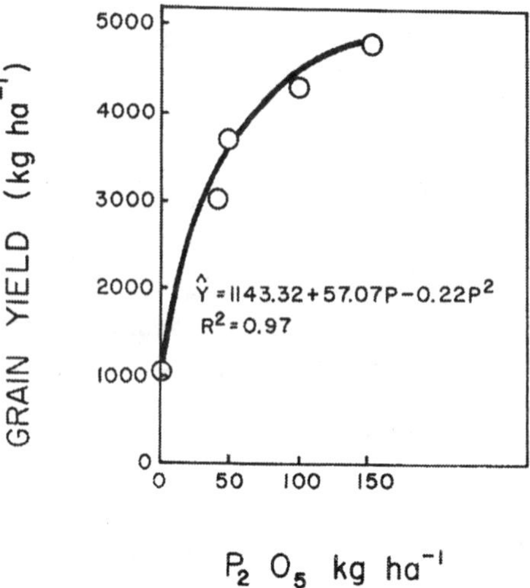

Figure 9.15 Upland rice response to P fertilization in Oxisol of central Brazil. (From Fageria et al., 1982.)

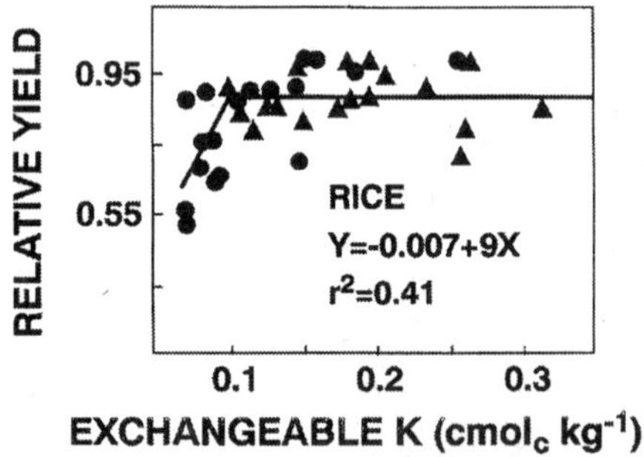

Figure 9.16 Relationship between exchangeable K concentration in the humid tropical Ultisol and relative grain yield of upland rice. (From Cox and Uribe, 1992.)

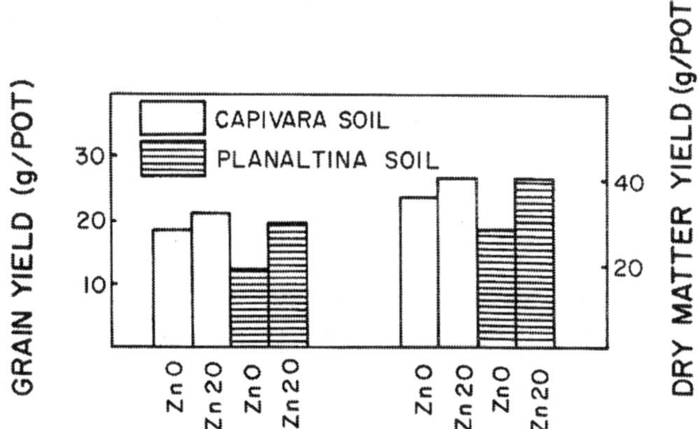

Figure 9.17 Upland rice response to zinc fertilization in an Oxisol. (From Fageria and Zimmermann, 1979.)

is under cultivation, and it is estimated that at least 50×10^6 ha could be used for crop production (Goedert, 1983). Low soil fertility is the main constraint on crop production in cerrado soils. Figure 9.15 shows the response of upland rice cultivar IAC47 to phosphorus in a cerrado Oxisol. Rice yield increased nearly five-fold with the application of 150 kg P_2O_5 ha^{-1} compared to control treatments. For upland rice production in Brazil, P deficiency is the most important yield-limiting factor among all essential plant nutrients. This is due to the low inherent P level of the soil (< 2 mg kg^{-1}) and high P fixation capacities (Fageria and Barbosa Filho, 1987). Deficiencies of N, K, Ca, Mg, and Zn and toxicity of Al have also been reported (Barbosa Filho and Fageria, 1980; Goedert, 1983). Figure 9.16 shows the effect of extractable K concentration on the relative grain yield of upland rice on an Ultisol in the Amazon Basin. The extractable K (modified Olsen) critical level was 0.1 cmol kg^{-1} of soil. Figure 9.17 shows the response of upland crop to Zn application in cerrado soils of Brazil.

Nutrient Concentration and Uptake

Nutrient concentration (content per unit dry weight) and uptake or accumulation (concentration × dry matter) are often used to assess crop nutrient deficiencies and toxicities. Phosphorus is the most yield-limiting nutrient in the major upland rice-producing regions of the world. Critical P concentrations in three upland rice cultivars are presented in Table 9.4. These values were determined under greenhouse conditions at the National Rice and Bean Research Center of EMBRAPA, Brazil. Critical concentrations of other nutrients under field conditions at different growth stages are presented in Table 9.5.

Table 9.4 Critical P Concentration in the Tops of Three Upland Rice Cultivars at Different Growth Stages in Oxisol of Central Brazil

	Cultivar (g kg^{-1})			
Plant age	IAC 47	IAC 164	IR 43	Average
28	2.8–3.7	2.1–2.5	2.8–3.6	2.6–3.3
43	1.8–2.2	2.2–2.5	2.4–3.0	2.1–2.6
57	2.1–2.3	2.1–2.3	2.2–2.4	2.1–2.3
70	1.5–1.8	1.7–1.8	1.6–1.8	1.6–1.8
84	1.3–1.5	1.6–1.8	1.3–1.5	1.4–1.6
98	1.1–1.3	1.3–1.5	1.3–1.5	1.2–1.4

Source: Fageria, 1987.

Table 9.5 Nutrient Concentrations in the Shoot of Upland Rice During the Crop Growth Cycle in an Oxisol of Central Brazil

	Days after sowing						
						130 (PM)	
Nutrient concentration	19 (IT)	43 (AT)	68 (IP)	90 (B)	99 (F)	(Shoot)	(Grains)
N (g kg^{-1})	50	32	20	17	16	12	17
P (g kg^{-1})	4.1	1.7	1.0	0.8	0.8	0.6	1.9
K (g kg^{-1})	35	37	32	26	25	26	3
Ca (g kg^{-1})	3.2	4.0	3.4	2.8	2.8	4.1	0.5
Mg (g kg^{-1})	2.7	2.8	2.6	2.1	2.2	2.8	1.1
Zn (mg kg^{-1})	47	26	24	22	24	24	28
Cu (mg kg^{-1})	21	13	8	6	6	5	10
Mn (mg kg^{-1})	300	240	193	143	147	197	20
Fe (mg kg^{-1})	1283	257	127	73	57	80	30
B (mg kg^{-1})	7	6	6	6	7	6	4

IT = initiation of tillering; AT = active tillering; IP = initiation of panicle; B = booting; F = flowering; and PM = physiological maturity.
Source: Fageria, 1996.

Concentration of almost all nutrients in the shoot decreased with the advancement of age of the plant. This is as expected because with increasing plant age, more dry matter was produced which diluted the concentration of nutrients accumulated. Among macronutrients, concentration of K was higher in the shoot at all growth stages except at 19 days of growth. Among micronutrients, the concentration of iron was maximum, followed by manganese. Concentrations of N, P, Zn, and Cu were higher in the grain as compared to shoot at the physiological maturity growth stage (Table 9.5).

Accumulation of N, P, and K in straw and grains of the five upland rice cultivars most commonly planted in Brazil are presented in Tables 9.6 and 9.7. On a whole-plant basis, rice cultivars accumulated more K than N and more N than P. Grains accumulated more N and P than straw, but K accumulation was higher in straw than in grain. Among micronutrients, the accumulation order was: Mn > Fe > Zn > Cu > B.

Distribution of N, P, and K in the roots, tops, and grains of two upland rice cultivars is presented in Figure 9.18. There was little difference in N, P, and K distribution in two cultivars, except for the K content in the grain, which was substantially higher in IR43. Distribution of N was approximately 15% in the roots, 35% in the straw, and 50% in the grain, whereas P was about 10% in the roots, 25% in the straw, and 65% in the grain. However, for K, 3% was retained in the roots, 76–86% remained in the straw, and 11–21% was exported to the grain. These results clearly show that most K remained in the straw, whereas most N and P were translocated to the grain.

B. Lowland Rice

Soil Constraints

Lowland rice is grown in submerged or waterlogged soils. The most remarkable differences between submerged soils and upland soils are caused by the supply of O_2 to the soils. Under submerged conditions, the supply of O_2 is ordinarily restricted within the surface layer and close proximity of roots. Except for these sites, depletion of O_2 significantly affects the microbial metabolism and N transformation in soil. Patrick (1982) summarized the characteristic features of anaerobic bacterial degradation of organic matter in submerged soils. He indicated that the accumulation rates of inorganic N (ammonium) under anaerobic conditions are faster than would be expected from the C/N ratio of organic matter and slow microbial decomposition due to the low requirement of N in anaerobic metabolism (Nishio et al., 1994).

Ponnamperuma (1972) divided the submerged soils into three groups: 1) waterlogged (gley) soils, 2) marsh soils, and 3) paddy soils. He described waterlogged soils as those saturated with water for a sufficiently long time to give the soil the distinctive gley horizons resulting from oxidation-reduction

Table 9.6 Accumulation of N, P, and K in Five Brazilian Upland Rice Cultivars (kg ha^{-1})

			Straw			Grain			Total		
Cultivar	Straw yield	Grain yield	N	P	K	N	P	K	N	P	K
IAC 25	4073	3256	31.7	1.87	82.2	41.4	5.48	5.88	73	7	88
EEPG369	4636	3576	32.4	1.66	115.9	45.5	6.33	9.22	78	8	125
Jaguary	4451	2427	40.5	2.93	76.6	32.3	4.09	4.38	73	7	80
Batatais	3242	2063	20.1	0.97	55.7	36.7	5.22	6.04	57	6	62
IAC 47	5302	2260	50.3	3.02	67.8	28.5	4.25	3.88	79	7	72
Average	4341	2716	35	1.89	79.4	36.9	5.07	5.88	72	7	85

Source: Ohno and Marur, 1977.

Table 9.7 Dry Matter and Grain Yield and Nutrient Accumulation in Shoot and Grains of Upland Rice During Crop Growth Cycle in an Oxisol of Central Brazil

Dry matter and grain production and nutrient accumulation	Days after sowing							
						130 (PM)		
	19 (IT)	43 (AT)	68 (IP)	90 (B)	99 (F)	(Shoot)	(Grains)	Total[a]
Dry matter (kg ha^{-1})	87	1123	4751	6395	8455	6642	—	
Grains (kg ha^{-1})	—	—	—	—	—	—	4794	11436
N (kg ha^{-1})	4.3	35.8	93.0	105.9	132.1	79.5	79.9	159.4
P (kg ha^{-1})	0.34	1.13	4.75	5.34	6.76	3.98	9.35	13.3
K (kg ha^{-1})	3.1	41.3	170.9	165.9	208.3	174.2	14.8	189
Ca (kg ha^{-1})	0.28	4.43	16.36	17.01	23.73	27.43	2.46	29.89
Mg (kg ha^{-1})	0.23	3.15	12.43	13.01	18.90	18.65	5.41	24.06
Zn (g ha^{-1})	4.14	28.88	112.31	138.57	204.41	159.29	136.23	295.52
Cu (g ha^{-1})	1.85	14.28	38.01	39.02	51.72	34.82	51.04	85.86
Mn (g ha^{-1})	24.46	270.86	915.75	903.33	1247.77	1307.79	395.47	1703.28
Fe (g ha^{-1})	108.68	289.62	588.55	450.26	483.52	522.83	147.61	670.44
B (g ha^{-1})	0.57	7.24	28.51	40.68	60.01	39.85	19.67	59.52

[a] Total of grains + straw at harvest or PM growth stage.

IT = initiation of tillering; AT = active tillering; IP = initiation of panicle; B = booting; F = flowering; and PM = physiological maturity.

Source: Fageria, 1996.

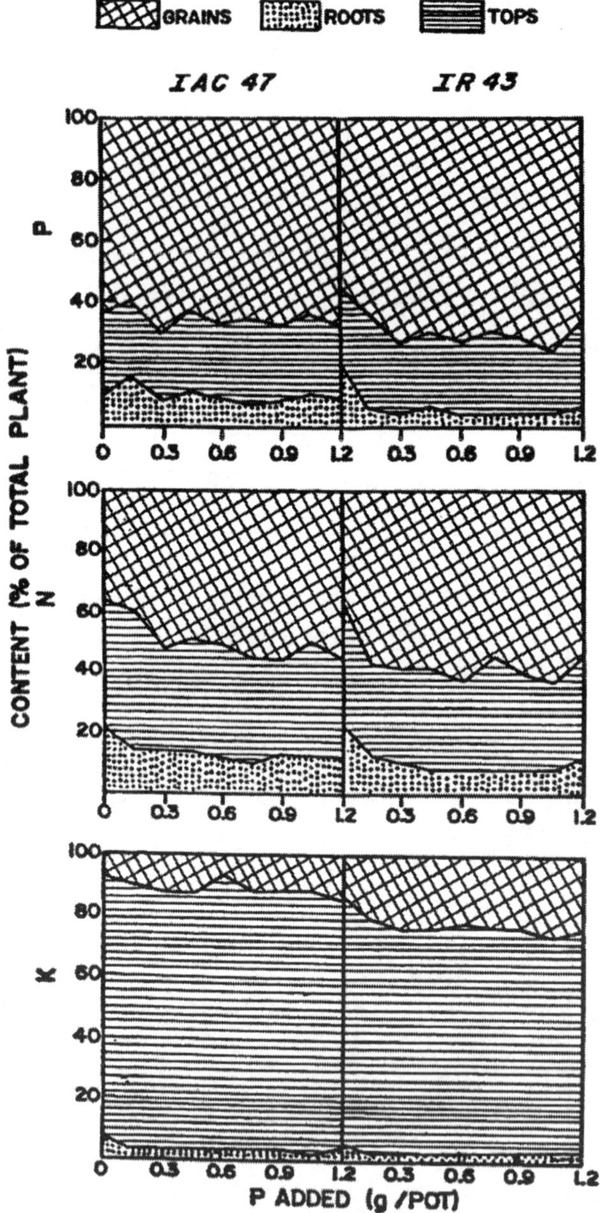

Figure 9.18 Distribution of N, P, and K in roots, tops, and grains of two upland rice cultivars.

processes. Normally, in such soils, there are three distinct horizons. A partially oxidized A horizon is high in organic matter due to reduced rates of oxidation in intermittently saturated soils. The second horizon is a mottled zone in which oxidation and reduction alternate, and iron and manganese are deposited as rusty mottles or streaks if the diffusion of oxygen into the soil aggregate is slow. The third horizon is a permanently reduced zone which is bluish green due to the presence of ferrous compounds. Waterlogged soils occur in almost any climatic zone from the tundra to the desert or humid tropics. Saturation with water may be due to impermeability of the soil material, the presence of an impervious layer, or a high water table (Ponnamperuma, 1972).

Marsh soils may be defined as soils that are more or less permanently saturated or submerged. The important characteristics of these soils are the accumulation of plant residues in the surface horizon and the presence of a permanently reduced G horizon below it. The third group of submerged soils consists of paddy soils that are managed in a special way for flooded rice cultivation. Land for flooded rice cultivation is leveled, bunded, provided with a surface drainage system, puddled (plowing and harrowing saturated soil), flooded to maintain 10–15 cm of standing water during most of the crop growth period, and drained at harvest. Because of the submergence of these soils, certain electrochemical changes take place in them which affect nutrient availability and growth.

Electrochemical Changes in Submerged Soils

Flooding the soil drastically reduces soil oxygen content and causes several electrochemical changes which influence rice growth and yield. Detailed discussions of these changes are given by Ponnamperuma (1972, 1978), Patrick and Mikkelsen (1971), De Datta (1981), Fageria (1984a), and Barbosa Filho (1987). They are:

1. depletion of molecular oxygen
2. decrease in redox potential (Eh)
3. increase in pH of acid soils and decrease in pH of alkaline soils
4. increase in specific conductance and ionic strength of the soil solution
5. reduction of Fe^{3+} to Fe^{2+} and Mn^{4+} to Mn^{2+}
6. reduction of NO_3^- and NO_2^- to N_2 and N_20
7. reduction of SO_4^{2-} to S^{2-}
8. increase in supply and availability of N, P, K, Ca, Mg, Fe, Mn, Mo, and Si
9. decrease in supply and availability of Zn, Cu, and S
10. liberation of CO_2, CH_4, organic acids, and H_2S

The change in soil pH to near neutrality affects growth by influencing the following processes (Ponnamperuma, 1978; De Datta and Mikkelsen, 1985):

1. The adverse effects of low or high pH per se are minimized.
2. The availability of many nutrients is increased.
3. Excess Al^{3+} and Mn^{4+} in acid soils are rendered harmless.
4. Iron toxicity in acid soils is reduced.
5. Organic acids decompose and are not highly ionized.

Management of N Supply in Lowland Rice

Nitrogen is considered to be the major nutrient limiting the yield of wetland rice (De Datta, 1981). Figure 9.19 shows response of lowland rice to N application in a Brazilian Inceptisol. Grain yield response was quadratic, and maximum yield was obtained at about 209 kg N ha^{-1} (calculated by quadratic equation); however, 90% of the grain yield was achieved at 120 kg N ha^{-1}. Nitrogen is one of the most difficult plant nutrients to manage because of the large number of potential transformation pathways. Normal recovery of fertilizer nitrogen applied to wetland rice crop is seldom more than 30–40% and, even with the best agronomic practices, rarely exceeds 60–68% (De Datta et al., 1983). Fageria (1996) determined N-use efficiencies of lowland rice under field conditions in an Inceptisol (Table 9.8); N-use efficiencies were lower at the higher N levels, and apparent recovery efficiency was about 44%. Low efficiency of fertilizer N in lowland rice is related to ammonia volatilization, denitrification, leaching, ammonium fixation, immobilization, and runoff (Savant and De Datta, 1982). Recent nitrogen transformation studies indicate that NH_3 volatilization in lowland rice soils is an important loss mechanism, causing a 5–47% loss of applied fertilizer under field conditions. De Datta (1987) estimated that denitrification losses were between 28 and 33%.

The biological oxidation of NH_4^+-N to NO_3^--N (nitrification) results in the conversion of the relatively immobile cation NH_4^+ into a more mobile anion

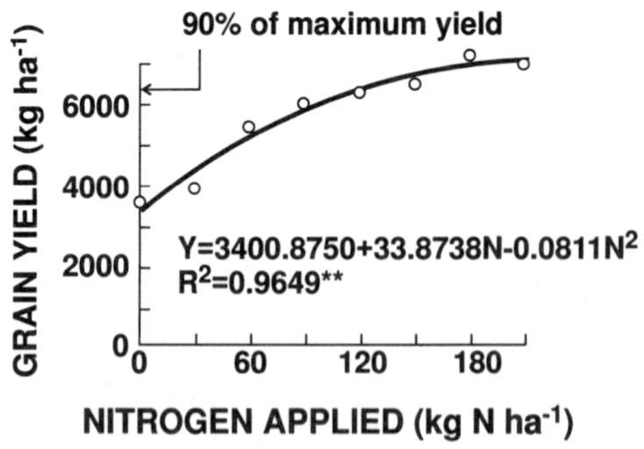

Figure 9.19 Relationship between nitrogen applied and grain yield of lowland rice. (From Fageria, 1996.)

Table 9.8 Nitrogen Use Efficiency in Lowland Rice in an Inceptisol of Central Brazil under Different Nitrogen Rates

N rate (kg ha^{-1})	Agronomic efficiency (kg kg^{-1})	Physiological efficiency (kg kg^{-1})	Agrophysiological efficiency (kg kg^{-1})	Apparent recovery efficiency (%)	Utilization efficiency (kg kg^{-1})
0	—	—	—	—	—
30	31.49	128.39	31.52	58.28	94.03
60	31.07	143.62	62.51	46.63	66.03
90	26.30	180.11	86.89	36.76	59.83
120	22.11	118.99	59.71	40.07	45.19
150	19.07	111.33	55.71	33.93	37.99
180	19.56	94.39	41.96	47.17	44.07
210	15.63	90.98	36.27	43.87	39.62
Average	23.46	123.97	53.54	43.82	55.25

[a] Methods of calculating different efficiencies are given under the section "Nutrient Efficiency."
Source: Fageria, 1996.

(NO_3^-) form, which is susceptible to leaching and denitrification. Submerged soils are an ideal environment for denitrification, since the thin oxidized surface layer promotes nitrification and a deeper reduced zone favors denitrification (Mikkelsen, 1987).

Managing the N supply to rice in flooded soils is an essential precondition for improving N-use efficiency, reducing N losses, and ultimately obtaining high yields. We addressed the possibility of synchronizing N supply with rice crop demands to achieve adequate yield-forming components and, consequently, higher grain yield. The N timing treatments significantly influenced dry matter and grain yield (Figure 9.20). Nitrogen applied in the reproductive growth stage (booting and flowering) did not improve lowland rice grain yield as compared to N applied during early growth stages. Dry matter production exhibited a greater response to late-season application of N than did grain yield. Number of panicles per unit area is the most important yield-contributing trait which can be manipulated significantly with the N fertilization application at an appropriate growth stage during the crop growth cycle. Nitrogen applied late during the reproductive growth stage can be absorbed by the crop, but it is not utilized in grain yield improvement. Although the highest grain yield was obtained with the T_1 treatment (all the N applied at sowing), under field conditions we recommend the T_2 treatment, which produced slightly lower yield (9%) as compared to the T_1 treatment. The reason for this is that denitrification and leaching are the principal sources of N

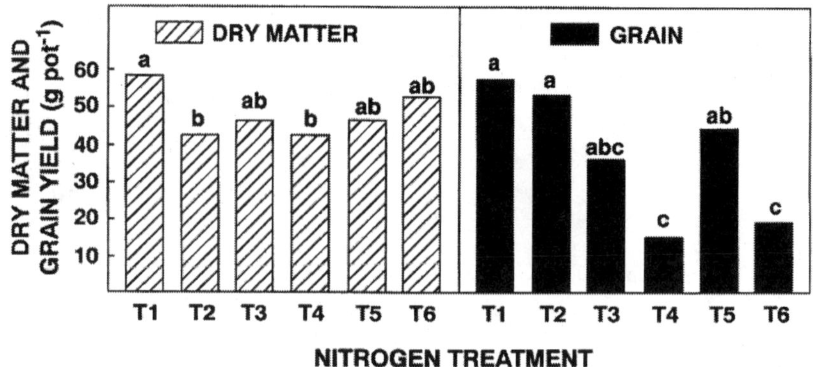

Figure 9.20 Dry matter and grain yield as affected by nitrogen timing treatments. T_1 = all the N applied at the time of sowing; T_2 = one-third of N applied at sowing + one-third applied at active tillering (43 days after sowing) + one-third applied at panicle initiation; T_3 = one-third at sowing + one-third at panicle initiation + one-third at booting; T_4 = one-third at sowing + one-third at panicle initiation + one-third at flowering; T_5 = zero N at sowing + one-half at initiation of tillering + one-half at panicle initiation; and T_6 = zero N at sowing + one-third at initiation of tillering + one-third at booting + one-third at flowering.

losses from soil plant systems under flooded rice cultivation. In the present study, leaching losses did not occur; but, under field conditions, losses of N by leaching will certainly occur.

De Datta (1986) suggested several practices to improve N use efficiency in lowland rice:

1. use of N efficient rice cultivars
2. improved timing of N application
3. deep placement of N fertilizer
4. use of nitrification and urease inhibitors
5. better water management, control of diseases and weeds

Management of P Supply in Lowland Rice

After nitrogen, P may be the second most important nutrient limiting rice yield, especially in acid soils. This means that producing good crop yields on these soils requires good P management. Use of an adequate level of P for maximum yield is possible if P soil test calibration data are available for a given crop and soil type. The amount of nutrient extracted in a soil is of little use until it has been calibrated to crop response in field experiments.

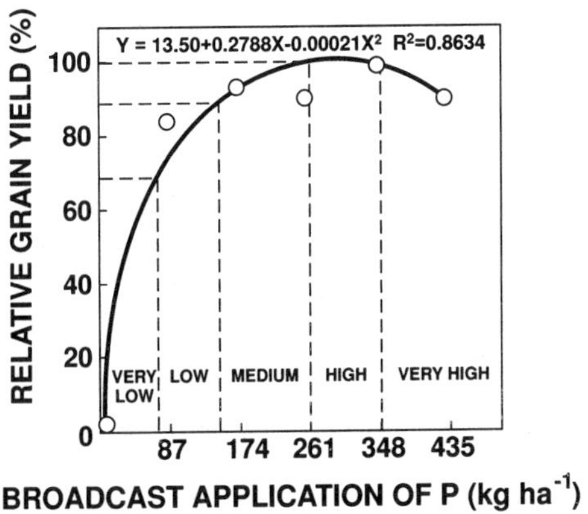

Figure 9.21 The relationship between P applied as broadcast and relative grain yield of lowland rice. The regression equation was calculated using fertilizer application rate expressed as kg P_2O_5 ha^{-1}.(From Fageria, 1996.)

Figure 9.21 shows lowland rice grain yield as a function of phosphorus applied in an Inceptisol. Grain yield increased with P fertilization showing quadratic response. Maximum grain yield calculated on the basis of quadratic regression equation was obtained at about 290 kg P ha^{-1}. Relative grain yield of this experiment was plotted against soil extractable P for classification of soil P test (Figure 9.22).

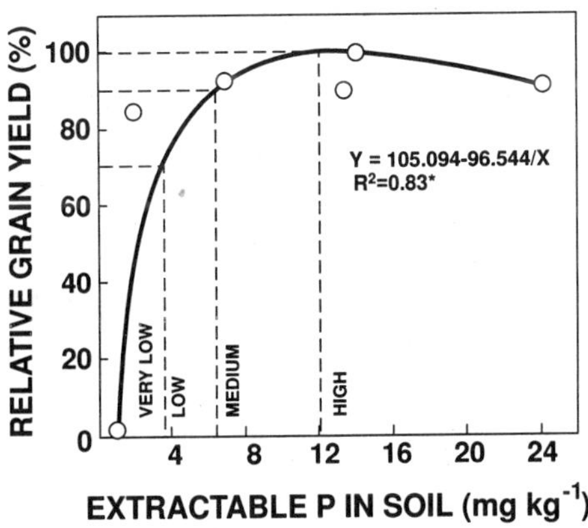

Figure 9.22 The relationship between Mehlich 1 extractable P and relative grain yield of lowland rice. (From Fageria, 1996.)

Table 9.9 Soil P Test Availability Indices and P Fertilizer Recommendations for Lowland Rice in an Inceptisol

Soil P test (mg kg^{-1})	Interpretation	Relative yield (%)	Broadcast P application (kg ha^{-1})	Band P application for maximum yield (kg ha^{-1})
0–3.6	Very low	0–70	100	66
3.6–6.4	Low	70–90	170	66
6.4–12.0	Medium	90–100	275	44
> 12.0	High	> 100	> 275	22

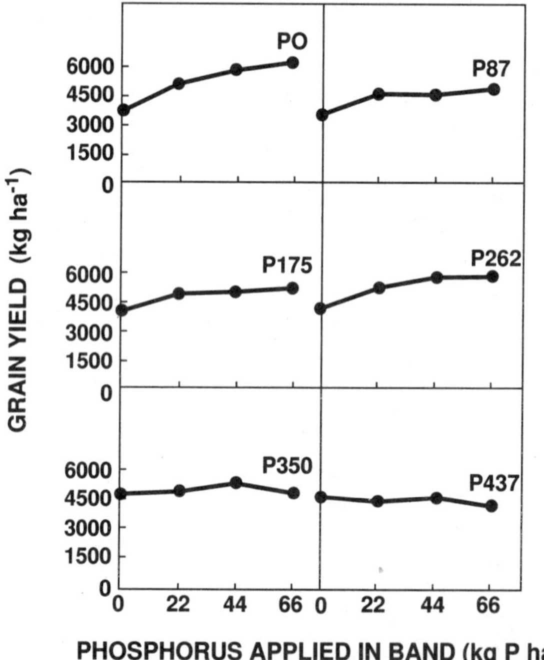

Figure 9.23 The relationship between P applied in band and grain yield of lowland rice. The P values in the figure represent the broadcast application of P in the previous year. (From Fageria, 1996.)

Four categories were established for the P soil test, such as very low (VL), low (L), medium (M), and high (H) in relation to responsive zone, in terms of relative grain yield. The 0–70% relative yield zone is called VL, the 70–90% relative yield zone is called L, the 90–100% relative yield zone is called M, and more than 100% relative yield is called high soil P test. These zones are selected arbitrarily as suggested by Raij (1991) for a soil P calibration study under Brazilian conditions. The soil P test availability indices and P fertilizer recommendations for the soil under investigation (calculated on the basis of Figures 9.21 and 9.22), are presented in Table 9.9.

The P fertilizer recommendation for the very low soil P test was 100 kg P ha^{-1}; for the low P level, it was 170 kg P ha^{-1}; for the medium soil P test, the P required to build up the soil P level was 275 kg P ha^{-1}; and more than 275 kg P ha^{-1} was required for the high soil P test. Once the desired P levels in the soil have been built up by broadcast P application, the next step in managing adequate P levels for maximum yields is to define band P application rates. This can be achieved by applying various P rates on the already built-up P levels and evaluating crop responses. In the present study, these responses were determined (Figure 9.23). Phosphorus rates required for maximum yield were calculated from these response curves. The band P application recommendations were established in this way (Table 9.9). The P rate needed was 60 kg P ha^{-1} in the very low to low level of the soil P test. At the medium P level, it was necessary to apply 44 kg P ha^{-1} for maximum yield.

Nutrient Concentration and Uptake

Plant analysis is an accepted means of predicting fertilizer requirements and of diagnosing nutrient deficiencies based on critical or adequate nutrient values (Sahrawat, 1983). Adequate nutrient concentrations for rice are presented in Tables 9.10 and 9.11, and amounts of nutrients removed by lowland rice under Philippine and Brazilian conditions are presented in Table 9.12 and Figure 9.24. A lowland rice crop producing 9.8 tons of grain ha^{-1} in about 115 days took up 218 kg of N, 31 kg of P, 258 kg of K, 28 kg of Ca, 23 kg of Mg, and 9 kg of S ha^{-1} (Table 9.12).

Accumulation of N, K, Ca, Mg, Fe, Cu, Mn, and Zn increased significantly ($P < 0.01$) with age and followed exponential quadratic responses. Similarly, P and B also increased significantly ($P < 0.01$) with the advancement of the plant age; however, response was quadratic. The accumulation pattern of nutrients was similar to that of dry matter accumulation. The quantity of nutrients accumulated was in the order of K > N > P > Ca > Mg for macronutrients and Mn > Fe > Zn > Cu > B for micronutrients. At the time of harvest, 64% N, 74% P, 15% K, 15% Ca, 58% Mg, 50% Zn, 16% Mn, 94% Cu, 64% Fe, and 47% B were exported to grains. De Datta and Mikkelsen (1985) reported more or less similar quantities of these nutrients exported to

Table 9.10 Adequate Nutrient Concentrations for Rice

Nutrient	Growth stage	Plant part	Adequate concentration
			$g\ kg^{-1}$
N	Heading	Uppermost mature leaves	26–42
P	75 DAS[a]	Whole tops	2.5–4.8
K	75 DAS	Whole tops	15–40
Ca	100 DAS	Whole tops	2.5–4
Mg	100 DAS	Whole tops	1.7–3
S	Tillering	Uppermost mature leaves	2–6
			$mg\ kg^{-1}$
Fe	Tillering	Whole tops	70–300
Zn	Tillering	Whole tops	20–150
Mn	Tillering	Whole tops	30–600
B	Tillering	Uppermost mature leaves	20–100
Cu	Tillering	Uppermost mature leaves	5–20
Mo	Tillering	Uppermost mature leaves	0.5–2

[a] DAS, Days after sowing.
Source: Fageria, 1984a.

Table 9.11 Nutrient Concentrations in the Shoot and Grains of Lowland Rice During the Crop Growth Cycle in an Inceptisol of Central Brazil

Nutrient concentration	Days after sowing						
						141 (PM)	
	23 (IT)	34 (AT)	72 (I)	98 (B)	111 (F)	Shoot	Grains
N (g kg^{-1})	46	24	13	15	10	7	12
P (g kg^{-1})	3.4	2.5	2.4	2.4	2.4	1.6	2.3
K (g kg^{-1})	35	38	29	25	21	26	3.0
Ca (g kg^{-1})	3.2	3.3	2.0	2.4	2.2	3.2	0.3
Mg (g kg^{-1})	2.5	2.2	1.7	2.0	1.9	1.9	1.3
Zn (mg kg^{-1})	45	38	29	32	35	48	28
Cu (mg kg^{-1})	17	11	4	2	2	1	13
Mn (mg kg^{-1})	720	865	718	678	570	815	68
Fe (mg kg^{-1})	535	375	100	173	165	268	48
B (mg kg^{-1})	7	7	7	6	7	8	5

IT = initiation of tillering; AT = active tillering; IP = initiation of panicle; B = booting; F = flowering; and PM = physiological maturity.
Source: Fageria, 1996.

Table 9.12 Nutrient Removal by Rice (IR 36)[a] Yielding 9.8 Tons of Rough Rice ha^{-1}[b] at Harvest, Calauan, Laguna Province, Philippines, 1983, Dry Season

Nutrient element	Mineral concentration ($g\ kg^{-1} \times 10^{-1}$)		Amount of mineral removed				
			At harvest ($kg\ ha^{-1}$)			Product ($kg\ ton^{-1}$)	
	Straw	Grain	Straw	Grain	Total	Straw	Grain
N	0.90	1.46	75	143	218	9.0	14.6
P	0.06	0.26	5.0	25.5	30.5	0.6	2.6
K	2.80	0.27	232	26	258	28	3
Ca	0.32	0.01	27	1.0	28	3.2	0.1
Mg	0.16	0.10	13	10	23	1.6	1.0
S	0.04	0.06	3.3	5.9	9.2	0.4	0.6
Fe	0.018	0.02	1.7	2.0	3.7	0.2	0.20
Mn	0.037	0.006	3.1	0.6	3.7	0.4	0.06
Zn	0.002	0.002	0.2	0.20	0.40	0.02	0.02
Cu	0.0002	0.0025	0.02	0.24	0.26	0.002	0.024
B	0.019	0.0016	0.16	0.16	0.32	0.019	0.016
Si	10.6	2.1	897	206	1086	106	21
Cl	0.65	0.42	54	41	5	6.5	4.2

[a] Fertilizer treatment of 174-17-33 kg of NPK.
[b] Straw yield = 8.3 t ha^{-1}.
Source: De Datta and Mikkelsen, 1985.

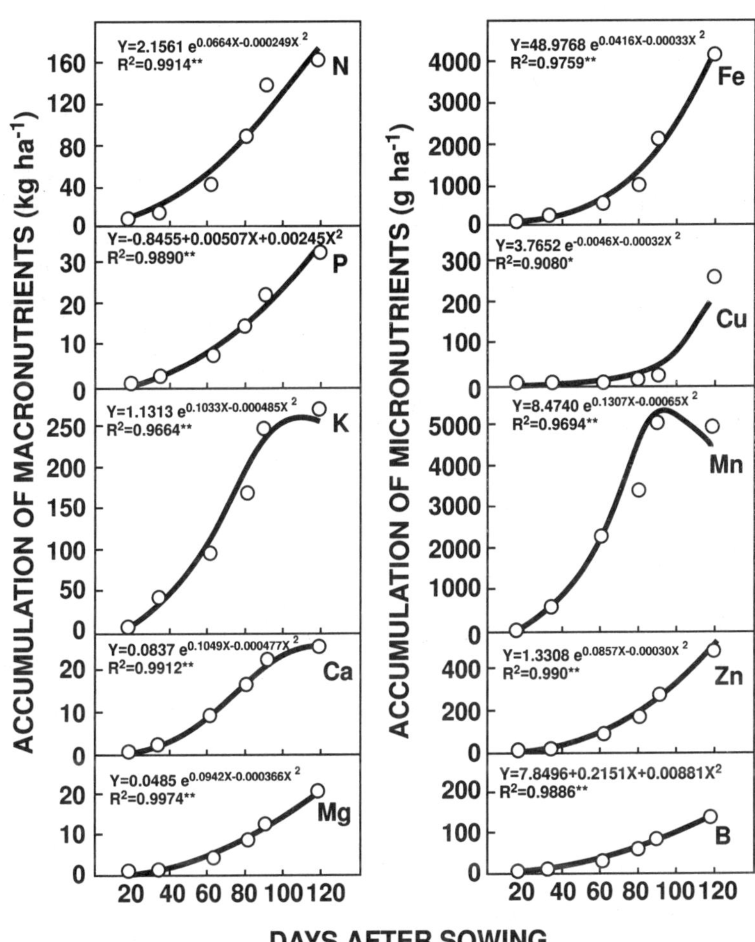

Figure 9.24 Nutrient accumulation in lowland rice crop during the growth cycle. (From Fageria, 1996.)

lowland rice grains. These and other nutrients removed by the crop must be replenished to sustain high production.

C. Nutrient Use Efficiency

Nutrient use efficiency can be defined in several ways. Normally, agronomists define it as grain yield per unit of fertilizer application, whereas physiologists define it as the quantity of dry matter produced per unit weight of nutrient absorbed. The latter definition is also known as nutrient efficiency ratio (ER). A more general definition of nutrient use efficiency is the ability to produce high yields in soils with limited availability of the nutrient being studied (Graham, 1984). For practical purposes, nutrient efficiency is best defined as the increase in yield of the harvested fraction of the crop per unit of nutrient supplied by fertilizer (Cooke, 1987). The highest efficiency is usually obtained with the first increment of nutrient, with additional increments providing smaller increases. It is often important to provide sufficient nutrients for near-maximum yields because high yields minimize the effects of fixed plus variable costs on the costs of production per unit of yield. This often increases profitability to the farmer and lowers costs to the consumer (Cooke, 1987).

It is clear from the above discussion that nutrient use efficiency has been defined in several ways in the literature. This sometimes creates confusion in the analysis and interpretation of experimental results. We define and provide methods of calculation of nutrient use efficiencies here.

1. Agronomic efficiency (*AE*): The agronomic efficiency is defined as the economic production obtained per unit of nutrient applied. It can be calculated by:

$$AE\ (\text{kg kg}^{-1}) = \frac{G_f - G_u}{N_a}$$

where G_f is the grain yield of the fertilized crop (kg), G_u is the grain yield of the unfertilized crop (kg), and N_a is the quantity of nutrient applied (kg).

2. Physiological efficiency (*PE*): The physiological efficiency is defined as the biological production obtained per unit of nutrient absorbed. It can be calculated by:

$$PE\ (\text{kg kg}^{-1}) = \frac{Y_f - Y_u}{N_f - N_u}$$

where Y_f is the total dry matter yield of the fertilized crop (kg), Y_u is the total dry matter yield of the unfertilized crop (kg), N_f is the nutrient uptake of the fertilized crop, and N_u is the nutrient uptake of the unfertilized crop.

3. Apparent recovery efficiency (*ARE*): The apparent recovery efficiency is defined as the quantity of nutrient absorbed per unit of nutrient applied. It can be calculated by:

$$ARE\,(\%) = \frac{N_f - N_u}{N_a} \times 100$$

4. Agrophysiological efficiency (*APE*): Agrophysiological efficiency is defined as the economic production (grain yield in case of annual crops) obtained per unit of nutrient absorbed. It can be calculated by:

$$APE\,(\text{kg kg}^{-1}) = \frac{G_f - G_u}{N_{tf} - N_{tu}}$$

where N_{tf} is the nutrient uptake by straw and grains in the fertilized plot (kg) and N_{tu} is the nutrient uptake by straw and grains in the un fertilized plot (kg).

5. Utilization efficiency (*UE*): Utilization efficiency is the product of physiological and apparent recovery efficiency. It can be calculated by:

$$UE\,(\text{kg kg}^{-1}) = PE \times ARE$$

We conducted a field experiment at the National Rice and Bean Research Center (CNPAF), Goiania-Goias, Brazil, on an Oxisol to determine N, P, and K use efficiencies in upland rice genotypes. The soil of the experimental site was a dark red Latosol in the Brazilian soil classification system and Typic Acrustox (Oxisol) in United States soil taxonomy. Results of N, P, and K use efficiencies calculated by using the previously mentioned formulas are presented in Table 9.13. Efficiencies varied from genotypes to genotypes. Across all the genotypes, physiological efficiency was highest and agronomic efficiency was lowest for N, P, and K. Among three nutrients, phosphorus produced highest grain yield or dry matter per unit applied or absorbed by plants. As far as apparent recovery efficiency is concerned, 39% of the applied N, 14% of the applied P, and 70% of the applied K were recovered by the plants. In the case of K, some genotypes showed recovery efficiency more than 100%. This may be related to a "priming" effect of K fertilizer application on indigenous soil K.

There are several mechanisms and processes that contribute to nutrient use efficiency (Graham, 1984; Lauchli, 1987):

1. root geometry
2. nutrient solubility in the rhizosphere
3. high rates of uptake from low concentrations of the nutrient in the soil solution
4. partitioning of nutrients in the plant
5. utilization efficiency or low functional nutrient requirement

Table 9.13 Nitrogen, Phosphorus, and Potassium Use Efficiencies in Upland Rice Genotypes

Genotypes	Agronomic efficiency (kg/kg)	Physiological efficiency (kg/kg)	Agrophysiolo-ical efficiency (kg/kg)	Apparent recovery efficiency (%)	Utilization efficiency (kg/kg)
			Nitrogen		
CNA6187	11.1	108.0	33.6	24.4	26.3
CNA7127	13.4	67.4	30.9	40.6	27.4
CNA7645	12.6	71.4	28.8	27.0	19.3
CNA6724-1	11.4	44.9	30.0	55.0	24.7
CNA7911	9.0	439.3	44.5	12.0	52.7
CNA7864	8.1	89.1	95.0	16.2	14.4
CNA7875	10.0	70.2	29.0	34.4	24.1
CNA7690	14.9	55.2	46.4	31.6	17.4
Rio Paranaiba	22.8	41.4	41.0	53.8	22.3
CNA6843-1	22.3	88.3	36.4	60.7	53.6
CNA6975-2	10.9	152.9	26.2	50.7	77.5
CNA7460	10.3	87.3	44.2	45.5	39.7
L141	16.9	55.8	55.2	51.9	28.9
Average	13.4	105.5	41.6	38.8	32.9
			Phosphorus		
CNA6187	21.1	1425.9	313.7	5.0	71.3
CNA7127	25.6	1177.9	331.4	8.3	97.8
CNA7645	29.9	138.7	145.6	26.3	36.5
CNA6724-1	21.8	293.7	131.7	16.1	47.3
CNA7911	17.1	1771.6	275.4	16.4	290.5
CNA7864	15.4	1122.3	114.7	11.0	123.5
CNA7875	19.2	492.0	198.2	10.1	49.7
CNA7690	28.4	337.7	242.1	10.7	36.1
Rio Paranaiba	43.7	263.9	307.3	18.2	48.0
CNA6843-1	42.5	774.5	291.2	12.5	96.8
CNA6975-2	42.6	514.3	209.3	17.9	92.1
CNA7460	19.8	1387.6	331.2	8.4	116.6
L141	32.3	462.1	170.0	20.6	95.2
Average	27.6	781.7	235.6	13.9	92.4
			Potassium		
CNA6187	16.6	71.1	26.4	59.9	42.6
CNA7127	20.2	52.1	23.8	79.1	41.2
CNA7645	18.9	50.3	27.4	43.2	21.7
CNA6724-1	17.2	37.3	33.2	78.8	29.4
CNA7911	13.6	75.7	26.8	57.5	43.5
CNA7864	12.3	68.8	38.0	33.1	22.8
CNA7875	15.1	62.2	31.0	65.2	40.6
CNA7690	22.4	56.6	76.6	32.0	19.1
Rio Paranaiba	34.4	63.5	57.4	70.9	45.0
CNA6843-1	33.5	62.9	34.8	96.1	60.4
CNA6975-2	33.6	59.1	39.7	106.4	62.9
CNA7460	15.5	67.1	29.5	119.9	80.4
L141	25.4	44.3	55.9	69.4	30.7
Average	21.4	59.3	38.5	70.1	41.5

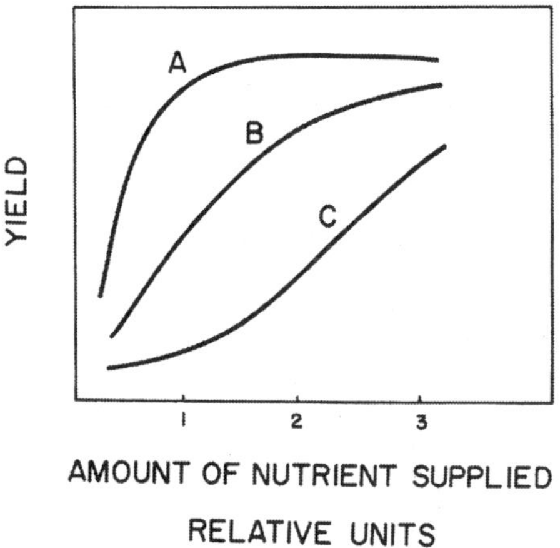

Figure 9.25 Hypothetical yield responses of three genotypes differing in efficiency of utilization of the limiting nutrient as a function of its concentration in the root environment. (A) nutrient efficient, (B) intermediate, and (C) inefficient genotype. (From R. D. Graham. 1984. Reprinted with permission.)

Figure 9.25 shows a hypothetical yield response of three plant genotypes that differ in nutrient use efficiency. Genotype A is very efficient, genotype B is intermediate in nutrient use efficiency, and genotype C is most inefficient. Genotype A produced the highest yield at low nutrient levels as well as at the highest nutrient levels. The most inefficient genotype (C) produced very low yield at low nutrient levels and responded very poorly as nutrient levels were increased in the growth medium.

D. Genetic Potential for Nutrient Use Efficiency

Because of the increasing cost of fertilizers or amendments, genotypic differences in use of nutrients and tolerance to elemental and/or nutrient toxicities deserve special attention in modern agriculture to improve or stabilize crop production. Genetic variability of rice cultivars in nutrient use efficiency and tolerance to Al, Fe, and salinity have been widely reported in the literature (Ponnamperuma, 1977a; Nelson, 1983; Fageria, 1984b; Broadbent et al., 1987).

Nitrogen

The genetics of N efficiency are not well understood. This is largely because this element participates in every plant process, and it is difficult to separate

Table 9.14 Upland Rice Cultivar Response to Nitrogen Fertilization in Oxisol of Central Brazil

Cultivar/line	Grain yield ($kg\ ha^{-1}$) at 15 $kg\ N\ ha^{-1}$	Grain yield ($kg\ ha^{-1}$) at 50 $kg\ N\ ha^{-1}$	N use efficiency (kg grain kg^{-1} applied N)
IAC 114	2319	2800	14
CNA 790124	3208	4075	25
CNA 800160	2929	3142	6
IAC 136	2198	2278	2
IAC 165	1232	1578	10
IR 20	1040	1330	8
CN 770538	896	1687	23
IR 144	1405	1483	2
IRAT 134	959	1073	3

Source: Fageria, 1989.

Table 9.15 Lowland Rice Cultivar Response to Nitrogen Fertilization in Humic-Gley Hydromorphic Soil of Brazil

Cultivar/line	Grain yield ($kg\ ha^{-1}$) at 30 $kg\ N\ ha^{-1}$	Grain yield ($kg\ ha^{-1}$) at 80 $kg\ N\ ha^{-1}$	N use efficiency (kg grain kg^{-1} applied N)
CICA 8	7646	8230	12
BR51-46-1-Cl	7347	8550	24
IAC 435	8526	10265	35
IR 841-63-5-1-9-33	8517	10123	32
IR 841	7373	9183	36
CICA 9	5737	7000	25
IAC 120	3897	5473	32
IR 22	4440	4903	9
Bluebelle	5098	6191	22

Source: Fageria and Barbosa Filho, 1982a.

efficiency from yield attributes in which this nutrient plays a part (Graham, 1984). However, varietal differences in N utilization have been observed in upland as well as lowland rice cultivars (Tables 9.14 and 9.15).

Phosphorus

Phosphorus deficiency has been identified as one of the major limiting factors for rice production, mostly in highly weathered soils such as Oxisols and Ultisols (Sanchez and Salinas, 1981; Fageria et al., 1982). High P fertilization is necessary to obtain good yields on these soils. However, farmers are facing difficulties due to increasing costs of fertilizers, especially in developing countries. An integrated fertilizer-plant breeding approach seems likely to give more economically viable and practical results in the future. The possibility of exploiting genotypic differences in absorption and utilization of P to improve efficiency of P fertilizer use or to obtain higher productivity on P-deficient soils has received considerable attention (Fageria and Barbosa Filho, 1981a,b, 1982b; Fageria et al., 1988a). Table 9.16 shows the results related to dry matter, grain yield, dry matter efficiency index (DMEI), and grain efficiency index in 25 upland rice cultivars at low and high P levels. Rice cultivars responded to P fertilization, but the response varied with cultivar.

Potassium

Potassium deficiencies in upland, as well as lowland, rice occur less frequently than N and P deficiencies. However, with the intensive use of soil and high-yielding cultivars, soil reserves of K will not be sufficient to maintain productivity for a long time. Under these circumstances, use of K-efficient cultivars can be a complementary solution to overcome K deficiency. Figure 9.26 shows the yield differences in lowland and upland rice cultivars under different K levels in a humic-gley hydromorphic (lowland) and an Oxisol (upland).

Calcium, Magnesium, and Sulfur

Reports of calcium, magnesium, and sulfur deficiencies in lowland rice are infrequent. However, deficiencies of these elements in upland rice in tropical soils are quite common (Fageria, 1984a; Fageria and Morais, 1987; Islam et al., 1987; Malavolta et al., 1987). Fageria (1984b) and Fageria and Morais (1987) reported different responses of upland rice cultivars to calcium and magnesium in Oxisols of central Brazil.

Aluminum

Aluminum toxicity is one of the important growth-limiting factors in upland rice production on acid soils. Although rice is an acid-tolerant crop, yield is often reduced by high aluminum saturation in Oxisols and Ultisols. A combination of liming to reduce plow layer Al and selection of cultivars that tolerate high subsoil Al is a possible solution. Genetic variability of rice

Table 9.16 Dry Matter (DM) at Flowering and Grain Yield (GY) of 25 Rice Cultivars at Low (1.5 mg kg^{-1}) and High (5 mg kg^{-1}) P Levels in Oxisol of Central Brazil[a]

Cultivar/line	Physiological maturity (days)	t ha^{-1}					
		DM		GY			
		Low P	High P	Low P	High P	DMEI	GYEI
CNA095-BM30-BM27P-9	108	1.74o	2.19i	0.98i-k	1.78b	0.17m	0.61c-f
CNA095-BM30-BM29P-2	108	2.59n	3.52h	1.73e	2.56b	0.39lm	1.55a-e
CNA511-16-B-5	108	2.48n	3.55h	1.99cd	1.98b	0.38lm	1.38a-f
CNA511-16-B-3	108	3.21m	3.60h	1.96d	2.13b	0.49kl	1.46a-f
CNA511-16-B-6	108	3.31ln	3.82gh	2.13bc	2.13b	0.55j-l	1.59a-d
CNA095-BM30-BM9-4	108	2.98m	3.33h	2.50ab	2.89ab	0.43kl	2.28a
CNA515-11-B-2	120	3.15m	4.69ef	1.38f	1.73b	0.64i-k	0.83b-f
CNA515-11-B-5	120	4.00ij	4.38fg	1.15gh	1.71b	0.76h-j	0.69b-f
IRAT144	120	2.58n	3.44h	2.35a	2.33b	0.39lm	1.92ab
CNA104-B-68-B-2	129	4.32hi	5.60cd	1.82e	1.86b	1.05fg	1.12a-f
CNA108-B-28-8-2B-2	129	4.04ij	5.18de	1.02h-j	1.30b	0.91gh	0.46d-f
CNA444-BM38-7-B-5	129	5.96d	5.53cd	0.87j-l	0.83b	1.43de	0.25f
CNA444-BM38-1-B-2	129	5.11g	6.83b	1.14gh	1.23b	1.51d	0.49d-f
CN449-BM15-3-B-5	129	5.44fg	8.00a	2.18b	2.41b	1.88c	1.83a-c
CNA449-BM15-1-B-5	129	5.79de	8.38a	1.38f	1.45b	2.10b	0.70b-f
CNA449-BM15-1-B-2	129	7.25b	8.16a	1.46f	1.66b	2.56a	0.85b-f
CNA449-BM15-1-B-4	129	4.37h	5.69cd	0.83kl	0.98b	1.07fg	0.29ef
CNA449-BM15-3-B-4	129	6.52c	8.33a	2.37a	2.33b	2.35a	1.93ab
CNA511-12-B-5	129	3.79jk	3.53h	1.22g	0.97a	0.58j-l	2.24a
CNA515-3-1	129	3.65k	5.42d	0.81l	0.89b	0.86g-i	0.25f
CNA104-B-18-P-1-L	129	3.63kl	6.17c	1.96d	2.07b	0.97gh	1.42a-f
IAC 47	129	5.52ef	5.74cd	1.03hi	0.89b	1.37de	0.32d-f
IR3646-8-1-2	142	9.12a	1.11j	2.10b-d	2.02b	0.44kl	1.48a-f
IR5716-18-1	142	5.15g	5.46d	2.36a	2.73ab	1.22ef	2.25a
Salumpikit	142	6.48c	8.55a	1.75e	1.76b	2.41a	1.07a-f

[a] Means in the same column followed by the same letter are not significantly different at P = 0.05 by Duncan's multiple range test.

Dry matter or grain yield efficiency index (DMEI or GYEI) =

$$\frac{\text{Yield at low P level}}{\text{Experimental mean yield at low P}} \times \frac{\text{Yield at high P level}}{\text{Experimental mean yield at high P}}$$

Source: Fageria et al., 1988a.

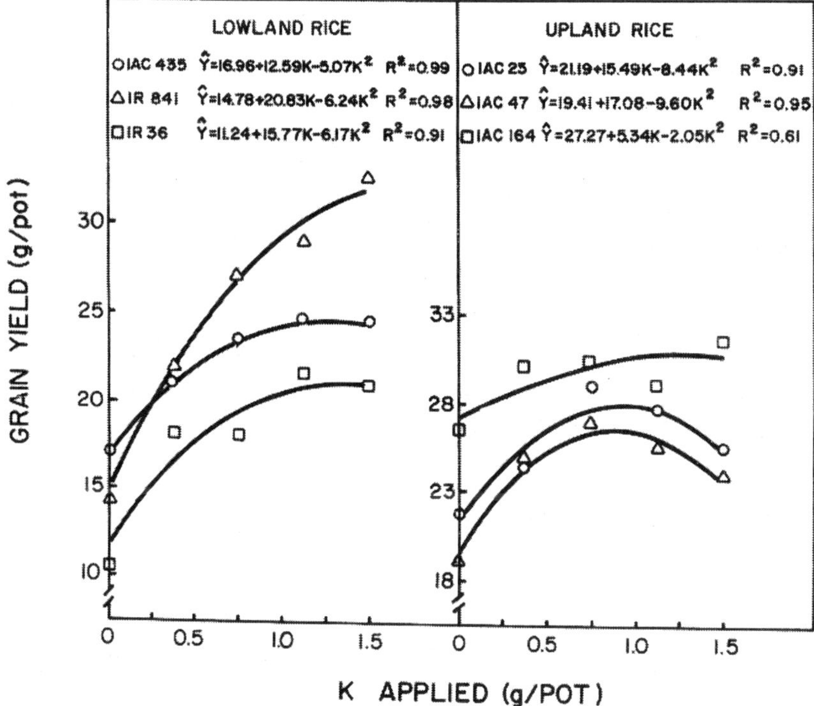

Figure 9.26 Grain yield of lowland and upland rice cultivars as influenced by K fertilization.

cultivars to Al toxicity has been widely reported in the literature (Fageria, 1982, 1985a; Fageria and Carvalho, 1982; Fageria and Barbosa Filho, 1983; Furlani and Hanna, 1984; Chaudhary et al., 1987; Fageria et al., 1988b).

Iron

Iron toxicity in lowland rice has been reported in Southeast Asia, Africa, and South America (Howeler, 1973; Virmani, 1977; Ottow et al., 1982; Fageria and Rabelo, 1987). The solubility of iron in soils is controlled by $Fe(OH)_3$ (soil) in well-oxidized soils, by $Fe_3(OH)_8$ (ferrosic hydroxide) in moderately oxidized soils, and by $FeCO_3$ (siderite) in highly reduced soils (Lindsay and Schwab, 1982). The Fe^{3+} hydrolysis species, $Fe(OH)_2^+$ and $Fe(OH)_3^0$, are the major solution species of inorganic Fe, but their concentrations are too low to supply available iron to plants. Iron is absorbed by plants as Fe^{2+} and must be in the general range $> 10^{-7.7}$ M to avoid iron deficiency. The redox potential of the soil-root environment must be < 12 to supply adequate Fe^{2+} for plants (Lindsay and Schwab, 1982).

The concentration of water-soluble iron, which on submergence rarely exceeds 0.1 mg kg^{-1}, may rise to a peak value of 600 mg kg^{-1} within 2 weeks after submergence and then gradually decrease to values ranging from 20 to 100 mg kg^{-1} (Ponnamperuma, 1977b).

In acid sulfate soils containing high levels of reactive iron oxides, concentrations of water-soluble iron can be as high as 5000 mg kg^{-1} (Ponnamperuma, 1977b). Toxic levels of Fe for rice have been reported to be 300–500 mg kg^{-1} in the soil solution (Tanaka et al., 1966) and 300–600 mg kg^{-1} in rice leaves (Tanaka et al., 1966; Fageria et al., 1981a). Low temperature and the presence of $CaCO_3$ and nitrates tend to retard the decrease in soil Eh and the initial increase in water-soluble iron, but this effect may disappear with time (Savant and McClellan, 1987). Five to fifty percent of the free iron oxides in a rice soil may be reduced within a few weeks of submergence (Ponnamperuma, 1972). According to Ponnamperuma (1978), the increase in the concentration of water-soluble Fe^{2+} after submergence can be described for most of the mineral soils by the equation

$$Eh = 1.06 - 0.059 \log Fe^{2+} - 0.177\, pH$$

Effective measures to ameliorate iron toxicity include periodic surface drainage, liming, and good fertilizer management. If iron toxicity is not severe, the use of tolerant rice cultivars may be a more desirable alternative than these measures (Fageria et al., 1984; Fageria and Rabelo, 1987) (Table 9.17). Because rice roots exude oxygen, ferrous iron is oxidized at the root surface and deposited there as ferric compounds, allowing little iron to enter the plant. Thus, the oxidizing power of the roots is correlated with susceptibility to iron toxicity (Tanaka et al., 1966). Tanaka and Tadano (1972) reported that the iron-excluding power of the normal rice with a 300 mg kg^{-1} Fe solution was 87%. This means that 87% of iron which reached the roots with water was excluded. Respiratory inhibitors such as H_2S, butyric acid, and CH_4 decreased iron exclusion, especially at high concentrations (Tanaka and Tadano, 1972). The oxidizing power of rice roots may be composed of two distinctly different processes: enzymatic oxidation on the root surface and release of molecular oxygen into the rhizosphere. The nature of these two processes and their relative importance in protection against toxic substances may be quite different (Ando et al., 1983).

Salinity

Salinity is an important growth-limiting factor for rice production in arid and semiarid regions of the world. In the arid and semiarid climates, potential evaporation is higher than precipitation. As a result, salts are not leached from

Table 9.17 Shoot Dry Matter Efficiency Index and Classification of Rice Cultiars for Iron Toxicity

Cultivars	Iron concentrations 890 μm	Iron concentrations 1780 μm
P3304F4-27	1.29(T)	1.05(T)
P3293F4-96-6	1.13(T)	0.79(MT)
CNA3814	1.38(T)	1.22(T)
CNA3882	0.71(MT)	0.92(MT)
P2915F4-2-1	0.67(MT)	0.75(MT)
P3038F4-61	1.48(T)	1.59(T)
CNA0007	0.67(MT)	0.69(MT)
CNA3084F4-59	0.89(MT)	0.93(MT)
CNA1051	1.21(T)	1.31(T)
CNA3739	0.87(MT)	0.95(MT)
CNA3762	0.94(MT)	0.95(MT)
CNA5193	0.85(MT)	0.88(MT)

Dry matter efficiency index =

$$\frac{\text{Yield at high Fe level}}{\text{Exp. mean yield at high Fe level}} \times \frac{\text{yield at optimum Fe level}}{\text{Exp. mean yield at optimum Fe level}}$$

Optimum Fe level was 90 μm; T = tolerant; MT = moderately tolerant.
Source: Fageria et al, 1990.

the soil, and they accumulate to levels detrimental to crop growth. Soils are also salinized by tides in coastal areas. Salts generally originate from native soil and irrigation water. Grain yield in rice is more sensitive to salinity than the later stages of vegetative growth (Pearson, 1961), although very young seedlings are particularly sensitive to salinity (Flowers and Yeo, 1981). Thus, salinity imposed at different stages during the life cycle might be expected to have different effects on yield. When the application of salinity coincides with the stage of panicle initiation, there is a particularly marked effect on the reproductive growth of rice (Khatun and Flowers, 1994).

Khatun et al. (1995) assessed the effects of salinity of rice genotypes. Salinity delayed flowering and reduced the number of productive tillers, the number of fertile florets per panicle, the weight per grain, and the grain yield; effects on grain yield were very much more severe than on vegetative growth. Panicle length was also reduced, as was the number of primary branches in a panicle; again, there was genotypic variation in the response of these characters to salinity.

Table 9.18 Salt Tolerance of Important Field Crops

Crop	dS m^{-1} Salinity at initial yield decrease	Salinity at which 25% yield decrease	Salinity at which 25% yield decrease	Classification[a]
Rice	3	6	8	MS
Wheat	6	10	14	MT
Corn	2	6	7	MS
Sorghum	—	9	12	MS
Barley	8	16	18	T
Soybean	5	7	9	MT
Common bean	1	2	3	S
Alfalfa	2	5	8	MS
Cotton	7.7	12	16	T
Potato	1.7	4	6	MS
Sugarcane	1.7	5	8.5	MS

[a] T, tolerant; MT, moderately tolerant; MS, moderately susceptible; and S, susceptible.
Sources: Compiled from Bernstein, 1964; Maas and Hoffman, 1977.

Table 9.18 shows the relative tolerance of important field crops to salinity. Among these crops, barley and cotton are most tolerant to soil salinity and common bean (*Phaseolus vulgaris* L.) is most susceptible to salinity stress. Rice is moderately susceptible to soil salinity, but differences exist among cultivars (Fageria et al., 1981b; Yeo and Flowers, 1982; Fageria, 1985b).

The resistance of salinity by rice is not determined by a single heritable character, but is a complex whole-plant phenomenon involving many interacting processes. Consequently, it has been suggested that the resistance to saline conditions can be increased beyond the existing phenotypic range by selecting individual physiological traits that contribute to resistance and combining them in a breeding program (Yeo and Flowers, 1986; Yeo et al., 1990). While some of the characters determining resistance at the seedling stage have been described (Yeo et al., 1990), rather little is known about those characters that might be important for grain yield under saline conditions. More is known about seedlings than mature plants, primarily because results can be obtained more quickly and easily with small vegetative plants than with large flowering plants. This is especially so for characters where there is great plant-to-plant variation within a supposedly homozygous population (Yeo and Flowers, 1983), requiring the use of large numbers of replicates in an experiment. Experiments on seedlings have been justified, since without seedling establishment there would be no plants to flower.

VI. SUMMARY

After wheat, rice is the most important food crop worldwide. It is a very versatile tropical or warm-season temperate crop that is grown on all continents except Antarctica, but over 90% of world production is in Asia. It is grown from 50°N to 35°S latitude and from sea level to over 3000 m in the Himalayas.

There are two broad types of rice culture, upland and lowland. Upland is defined as that grown in rainfed, naturally well-drained soils without surface water accumulation, normally without a phreatic water supply, and normally without bunds. Brazil is the largest producer of upland rice. Lowland rice is grown on somewhat more than 50% of the world's rice-producing area. It is normally cultivated in bunded paddies with some type of controlled irrigation. Lowland rice soils are normally flooded to a depth of 10–15 cm a few weeks after seeding or transplanting and are drained shortly before harvest. Some varieties of floating rice can grow in waters as deep as 4 m.

Record lowland rice yields are over 10 Mg ha^{-1}, but farmers' yields are usually much lower. In similar environments, upland rice yields are always lower than lowland rice yields, primarily due to the crop's ability to grow and absorb nutrients at optimal rates in flooded soils. In addition, upland rice is often subjected to diseases and infestation, being especially sensitive to blast and drought stress. Development of semidwarf cultivars in the 1960s dramatically increased world rice yields, which now average about 3.5 Mg ha^{-1} worldwide.

Rice can be grown on a wide variety of soils, but slowly permeable clay soils are usually more suitable than soils with little water-holding capacity that drain rapidly. Around flowering, rice is particularly susceptible to drought stress, nitrogen deficiency, blast disease, and low solar radiation. Flooding paddy soils causes a number of electrochemical changes in the soil that, in general, benefit the crop. Many nutrients become more easily available to the crop, and most nutrient toxicities and deficiencies associated with extreme soil pH are eliminated. In lowland soils the most limiting nutrient is normally nitrogen, and fertilizer nitrogen losses are usually high due to ammonia volatilization, denitrification, and leaching. In upland soils, phosphorus is often the most limiting nutrient, especially in acid soils. Both plant and soil analyses are used to detect and correct nutrient deficiencies and toxicities. One metric ton of rice contains approximately 13 kg N, 4 kg P, and 7 kg K. Since straw is often removed or burned, larger amounts of nutrients, especially potassium, are lost. Thus, fertilizers or manures are normally required to maintain soil fertility.

REFERENCES

Adair, C. R., M. D. Miller, and H. M. Beachell. 1962. Rice improvement in the United States. Adv. Agron. 14: 61–108.

Ando, T., S. Yoshida, and I. Nishiyama. 1983. Nature of the oxidizing power of rice roots. Plant Soil 72: 57–71.

Armstrong, W. 1969. Rhizosphere oxidation in rice: An analysis of intervarietal differences in oxygen flux from the roots. Physiol. Plant. 22: 296–303.

Baker, A. M. and W. Hatton. 1987. Calcium peroxide as a seed coating material for padi rice. Plant Soil 99: 357–363.

Barbosa Filho, M. P. 1987. Fertilization and nutrition of rice. Upland and irrigated. Brazilian Potassium and Phosphate Assoc. Tech. Bull. 9, Piracicaba, Brazil.

Barbosa Filho, M. P. and N. K. Fageria. 1980. Occurrence, diagnosis and correction of zinc deficiency in upland rice. EMBRAPA-CNPAF, Tech. Circ. 4, Goiania, Brazil.

Beecher, H. G., J. A. Thompson, P. E. Bacon, and D. P. Hunan. 1994. Effect of cropping sequences on soil nitrogen levels, rice growth, and grain yields. Aust. J. Exp. Agric. 34: 977–986.

Bernstein, L. 1964. Salt tolerance of plants. USDA Inform. Bull. 283.

Boonjung, H. 1993. Modelling growth and yield of upland rice under water limitation conditions. Ph.D. Thesis, The University of Queensland.

Broadbent, F. E., S. K. De Datta, and E. V. Laureles. 1987. Measurement of nitrogen utilization efficiency in rice genotypes. Agron. J. 79: 782–391.

Chang, T. T. 1964. Present knowledge of rice genetics and cytogenetics. Int. Rice Res. Inst. (IRRI) Tech Bull. 1.

Chang, T. T. 1976. Rice, pp. 98–104. *In*: N. W. Simmonds (ed.). Evolution of crop plants. Longman, London.

Chang, T. T. and E. A. Bardenas. 1965. The morphology and varietal characteristics of the rice plant. IRRI Tech. Bull. 4.

Chang, T. T. and B. S. Vergara. 1975. Varietal diversity and morphoagronomic characteristics of upland rice, pp. 72–90. *In*: IRRI (ed.). Major research in upland rice. International Rice Research Institute, Los Banos, Philippines.

Chaudhary, M. A., S. Yoshida, and B. S. Vergara. 1987. Induced mutations for aluminum tolerance after *N*-methyl-*N*-nitrosourea treatment of fertilized egg cells in rice (*Oryza sativa* L.). Environ. Exp. Bot. 27: 37–43.

Cooke, G. W. 1987. Maximizing fertilizer efficiency by overcoming constraints to crop growth. J. Plant Nutr. 10: 1357–1369.

Cox, F. R. and E. Uribe. 1992. Management and dynamics of potassium in a humid tropical Ultisol under a rice-cowpea rotation. Agron. J. 84: 655–660.

Cruz, R. T. and J. C. O'Toole. 1984. Dryland rice response to an irrigation gradient at flowering stage. Agron. J. 76: 178–183.

Cuevas-Perez, F. E., L. E. Berrio, D. I. Gonzalez, F. Correa-Victoria, and E. Tulande. 1995. Genetic improvement in yield of semidwarf rice cultivars in Colombia. Crop Sci. 35: 725–729.

De Datta, S. K. 1981. Principles and practices of rice production. Wiley, New York.

De Datta, S. K. 1986. Improving nitrogen fertilizer efficiency in lowland rice in tropical Asia, pp. 171–186. *In*: S. K. De Datta and W. H. Patrick, Jr. (eds.) Nitrogen economy of flooded rice soils. Martinus Nijhoff, Dordrecht, Netherlands.

De Datta, S. K. 1987. Nitrogen transformation processes in relation to improved cultural practices for lowland rice. Plant Soil 100: 47–69.

De Datta, S. K. and D. S. Mikkelsen. 1985. Potassium nutrition of rice, pp. 665–699. *In*: R. D. Munson (ed.). Potassium in agriculture. Am. Soc. Agron., Madison, Wisconsin.

De Datta, S. K., I. R. P. Fillery, and E. T. Craswell. 1983. Results from recent studies on nitrogen fertilizer efficiency in wetland rice. Outlook Agric. 12: 125–134.

Dofing, S. M. and M. G. Karlsson. 1993. Growth and development of uniculum and conventional tillering barley lines. Agron. J. 85: 58–61.

Donald, C. M. and J. Hamblin. 1976. The biological yield and harvest index of cereals as agronomic and plant breeding criteria. Adv. Agron. 28: 361–405.

Donald, C. M. and J. Hamblin. 1983. The convergent evolution of annual seed crops in agriculture. Adv. Agron. 32: 97–143.

Fageria, N. K. 1980a. Upland rice response to phosphate fertilization as affected by water deficiency in cerrado soils. Pesq. Agropec. Bras., Brasilia 15: 259–265.

Fageria, N. K. 1980b. Influence of phosphorus application on growth yield and nutrient uptake by irrigated rice. R. Bras. Ci. Solo. 4: 26–31.

Fageria, N. K. 1982. Differential aluminum tolerance of rice cultivars in nutrient solution. Pesq. Agropec. Bras., Brasilia 17: 1–9.

Fageria, N. K. 1984a. Fertilization and mineral nutrition of rice. EMBRAPA-CNPAF/ Editora Campus, Rio de Janeiro.

Fageria, N. K. 1984b. Response of rice cultivars to liming in cerrado soil. Pesq. Agropec. Bras., Brasilia 19: 883–889.

Fageria, N. K. 1985a. Influence of aluminum in nutrient solutions on chemical composition in two rice cultivars at different growth stages. Plant Soil 85: 423–429.

Fageria, N. K. 1985b. Salt tolerance of rice cultivars. Plant Soil 88: 237–243.

Fageria, N. K. 1987. Variation in the critical level of phosphorus in the rice plants at different growth stages. R. Bras. Ci. Solo. 11: 77–80.

Fageria, N. K. 1989. Tropical soils and physiological aspects of crop yield. EMBRAPA-CNPAF, Brasilia, Brazil.

Fageria, N. K. 1996. Annual report of the project. The study of liming and fertilization for rice and common bean in Cerrado region. National Rice and Bean Research Center, Goiania, Brazil.

Fageria, N. K., V. C. Baligar, and R. J. Wright. 1988b. Aluminum toxicity in crop plants. J. Plant Nutr. 11: 303–319.

Fageria, N. K. and M. P. Barbosa Filho. 1981a. Screening rice cultivars for efficient phosphorus absorption. Pesq. Agropec. Bras., Brasilia 16: 777–782.

Fageria, N. K. and M. B. Barbosa Filho. 1981b. Screening rice cultivars for higher efficiency of phosphorus absorption. Pesq. Agropec. Bras., Brasilia 16: 777–782.

Fageria, N. K. and M. P. Barbosa Filho. 1982a. Preliminary evaluation of irrigated rice cultivars for higher efficiency of nitrogen utilization. Pesq. Agropec. Bras., Brasilia 17: 1709–1712.

Fageria, N. K. and M. P. Barbosa Filho. 1982b. Evaluation of rice cultivars for low levels of soil phosphorus. R. Bras. Ci. Solo. 6: 142–151.

Fageria, N. K. and M. P. Barbosa Filho. 1983. Upland rice varietal reactions to aluminum toxicity on an Oxisol in central Brazil. IRRI News Letter 8: 18–19.

Fageria, N. K. and M. P. Barbosa Filho. 1987. Phosphorus fixation in Oxisol of central Brazil. Fertilizers Agric. 94: 33–37.

Fageria, N. K., M. P. Barbosa Filho, and J. R. P. Carvalho. 1981a. Influence of iron on growth and absorption of P, K, Ca and Mg by rice plant in nutrient solution. Pesq. Agropec. Bras., Brasilia 16: 483–488.

Fageria, N. K., M. P. Barbosa Filho, and H. R. Gheyi. 1981b. Screening rice cultivars for salinity tolerance. Pesq. Agropec. Bras., Brasilia 16: 677–681.

Fageria, N. K., M. P. Barbosa Filho, and J. R. P. Carvalho. 1982. Response of upland rice to phosphorus fertilization on an Oxisol of central Brazil. Agron. J. 74: 51–56.

Fageria, N. K., M. P. Barbosa Filho, J. R. P. Carvalho, P. H. N. Rangel, and V. A. Cutrim. 1984. Preliminary screening of rice cultivars for tolerance to iron toxicity. Pesq. Agropec. Bras., Brasilia 19: 1271–1278.

Fageria, N. K. and J. R. P. Carvalho. 1982. Influence of aluminum in nutrient solutions on chemical composition in upland rice cultivars. Plant Soil 69: 31–44.

Fageria, N. K. and O. P. Morais. 1987. Evaluation of rice cultivars for utilization of calcium and magnesium in the cerrado soils. Pesq. Agropec. Bras., Brasilia 22: 667–672.

Fageria, N. K., O. P. Morais, V. C. Baligar, and R. J. Wright. 1988a. Response of rice cultivars to phosphorus supply on an Oxisol. Fertilizer Res. 16: 195–206.

Fageria, N. K. and N. A. Rabelo. 1987. Tolerance of rice cultivars to iron toxicity. J. Plant Nutr. 10: 653–661.

Fageria, N. K., R. J. Wright, and V. C. Baligar. 1990. Iron tolerance of rice cultivars, pp. 259–282. *In*: N. El Bassam, M. Dambroth, and B. C. Loughman (eds.) Genetic aspects of plant nutrition. Kluwer Academic Publishers, Dordrecht, The Netherlands.

Fageria, N. K. and F. J. P. Zimmermann. 1979. Interaction among P, Zn, and lime in upland rice. R. Bras. Ci. Solo. 3: 88–92.

Fageria, N. K. and F. J. P. Zimmermann. 1996. Influence of pH on growth and nutrient uptake by crop species in an Oxisol. Paper presented at 4th Int. Symp. on Plant-Soil Interactions at Low pH, March 17–24, 1996, Belo Horizonte, Brazil.

FAO (Food and Agriculture Organization) of United Nations. 1992. Production yearbook, Vol. 45, Rome.

Faria, J. C., A. S. Prabhu, and F. J. P. Zimmermann. 1982. Effects of nitrogen fertilization and spraying fungicides on blast and upland rice productivity. Pesq. Agropec. Bras., Brasilia 17: 847–852.

Flowers, T. J. and A. R. Yeo. 1981. Variability of sodium chloride resistance within rice genotypes. New Phytol. 88: 363–373.

Fujii, Y. 1961. Studies on the regular growth of the roots in the rice plants and wheats. Bull. Fac. Agr., Saga Univ. 12: 1–117.

Fukai, S. and M. Cooper. 1995. Development of drought-resistant cultivars using physiomorphological traits in rice. Field Crops Res. 40: 67–86.

Furlani, P. R. and L. G. Hanna. 1984. Screening of rice and corn plants for tolerance to excess aluminum in nutrient solutions. R. Bras. Ci. Solo. 8: 205–208.

Garrity, D. P. 1984. Asian upland rice environments, pp. 161–183. *In*: International workshop on upland rice. IRRI, Los Banos, Philippines.

Garrity, D. P. and P. C. Agustin. 1984. A classification of Asian upland rice growing environments. Paper presented at the workshop on characterization and classification of upland rice environments, 1–3 August. CNPAF-EMBRAPA, Goiania, Brazil.

Garrity, D. P., L. R. Oldemann, R. A. Morris, and D. Lenka. 1986. Rainfed lowland rice ecosystems: Characterization and distribution, pp. 3–23. *In*: Progress in rainfed lowland rice. International Rice Research Institute, Manila, Philippines.

Garrity, D. P. and J. C. O'Toole. 1994. Screening rice for drought resistance at the reproductive phase. Field Crops Res. 39: 99–110.

Goedert, W. J. 1983. Management of the cerrado soils of Brazil: A review. J. Soil Sci. 34: 405–428.

Graham, R. D. 1984. Breeding for nutritional characteristics in cereals, pp. 57–102. *In*: P. B. Tinker and A. Lauchli (eds.). Advances in plant nutrition, Vol. 1. Praeger, New York.

Guindo, D., B. R. Wells, and R. J. Norman. 1994. Cultivar and nitrogen rate influence on nitrogen uptake and partitioning in rice. Soil Sci. Soc. Am. J. 58: 840–845.

Gupta, P. C. and J. C. O'Toole. 1986. Upland rice: A global perspective. IRRI, Los Banos, Philippines.

Hanson, A. D., W. J. Peacock, L. T. Evans, C. J. Arntzen, and G. S. Kush. 1990. Drought resistance in rice. Nature 234: 2.

Hatton, W. and A. M. Baker. 1987. Calcium peroxide as a seed coating material for Padi rice. Plant Soil 99: 365–377.

Honya, K. 1961. Studies on the improvement of rice plant cultivation in volcanic ash paddy field in Tohoku district. Tohoku Agric. Exp. Stn. Bull. 21: 1–143.

Howeler, R. H. 1973. Iron-induced oranging disease of rice in relation to physio-chemical changes in a flooded Oxisol. Soil Sci. Soc. Am. Proc. 37: 898–903.

Huke, R. E. 1982. Rice area by type of culture: South, Southeast, and East Asia. International Rice Research Institute, Manila, Philippines. 32 p.

IRRI (International Rice Research Institute). 1975. Major research in upland rice. IRRI, Los Banos, Philippines.

IRRI (International Rice Research Institute). 1977. Annual report for 1976. IRRI, Los Banos, Philippines.

IRRI (International Rice Research Institute). 1982. A plan for IRRI's third decade. IRRI, Los Banos, Philippines.

IRRI (International Rice Research Institute). 1984a. Terminology for rice growing environments. IRRI, Los Banos, Philippines.

IRRI (International Rice Research Institute). 1984b. Upland rice in Asia, pp. 45–68. *In*: IRRI (ed.). An overview of upland rice research. Proceedings of the 1982 Bouake, Ivory Coast, upland rice workshop. IRRI, Los Banos, Philippines.

IRRI (International Rice Research Institute). 1989. IRRI toward 200 and beyond. IRRI, Los Banos, Philippines.

IRRI (International Rice Research Institute). 1993. Rice facts. IRRI, Los Banos, Philippines.

Ishizuka, Y. 1973. Physiology of the rice plant. Tech. Bull. 13, Food and Fertilizer Technology Center, Taipai, Taiwan.

Ishizuka, Y. and A. Tanaka. 1963. Studies on nutrio-physiology of the rice plant. Yokendo, Tokyo.

Islam, A. K. M. A., C. J. Asher, and D. G. Edwards. 1987. Response of plants to calcium concentration in flowing solution culture with chloride or sulphate as the counter-ion. Plant Soil 98: 377–395.

Katyama, K. 1951. Studies on the tillering of rice, wheat and barley. Yokendo, Tokyo.

Khatun, S. and T. J. Flowers. 1994. Effects of salinity on seed set in rice. Plant Cell Environ. 18: 61–67.

Khatun, S., C. A. Rizzo, and T. J. Flowers. 1995. Genotypic variation in the effect of salinity on fertility in rice. Plant Soil 173: 239–250.

Kiuchi, T., T. Yoshida, and M. Kono. 1966. Relationship between rice yield and leafiness. Nogyo Gijitsu 21: 551–554.

Kush, G. S. 1984. Terminology for rice growing environments. International Rice Research Institute, Los Banos, Philippines. pp. 5–10.

Kush, G. S. 1995. Increased genetic potential of rice yield: Methods and perspectives, pp. 13–29. *In*: B. S. Pinheiro and E. P. Guimarais (eds.) Rice in Latin America: Perspectives to increase production and yield potential. EMBRAPA-CNPAF, Goiania, Brazil, Document No. 60.

Kung, P. 1971. Irrigation agronomy in monsoon Asia. FAO, Rome.

Lauchli, A. 1987. Soil science in the next twenty-five years. Does biotechnology play a role? Soil Sci. Soc. Am. J. 51: 1405–1408.

Lilley, J. M. and Fukai, S. 1994. Effect of timing and severity of water deficit on four diverse rice cultivars. III. Phenological development, crop growth and grain yield. Field Crops Res. 37: 225–234.

Lindsay, W. L. and A. P. Schwab. 1982. The chemistry of iron in soils and its availability to plants. J. Plant Nutr. 5: 821–840.

Lu, J. and T. T. Chang. 1980. Rice in its temporal and spatial perspectives, pp. 1–74. *In*: B. S. Luh (ed.). Rice production and utilization. AVI, Davis, California.

Maas, E. V. and G. J. Hoffman. 1977. Crop salt tolerance current assessment. J. Irrig. Drainage Div. Am. Soc. Civil Eng. 103: 115–134.

Malavolta, E., G. C. Vitti, C. A. Rosolem, N. K. Fageria, and P. T. G. Guimaraes. 1987. Sulphur responses of Brazilian crops. J. Plant Nutr. 10: 2153–2158.

Matsushima, S. 1962. Some experiments in soil water plant relationship in rice. Div. Agric. Bull. 12, Ministry of Agriculture and Cooperative Federation of Malaya.

Matsushima, S. 1968. Water and physiology of Indica rice. Proc. Crop Sci. Soc. Japan Special Issue: 102–109.

Maurya, D. M. and J. C. O'Toole. 1986. Screening upland rice for drought tolerance. *In*: Progress in upland rice research; proceedings of the 1985 Jakarta Conference. IRRI, Los Banos, Philippines.

McDonald, D. J. 1994. Temperate rice technology for the 21st century: An Australian example. Aust. J. Exp. Agric. 34: 877–888.

Mikkelsen, D. S. 1987. Nitrogen budgets in flooded soils used for rice production. Plant Soil 100: 71–97.

Murata, Y. and S. Matsushima. 1975. Rice, pp. 73–99. *In*: L. T. Evans (ed.). Crop physiology. Cambridge University Press, London.

Nayar, N. M. 1973. Origin and cytogenetics of rice. Genetics 17: 153–292.

Nelson, L. E. 1983. Tolerance of 20 rice cultivars to excess Al and Mn. Agron. J. 75: 134–138.

Nishio, T., H. Sekiya, and K. Kogano. 1994. Transformation of fertilizer nitrogen in soil-plant system with special reference to comparison between submerged conditions and upland field conditions. Soil Sci. Plant Nutr. 40: 1–8.

Ohno, Y. and C. J. Marur. 1977. Physiological analysis of factors limiting growth and yield of upland rice. Annual report of ecophysiological study of rice. IAPAR, Londrina, Brazil.

Okajima, H. 1960. Studies on the physiological function of the root system in the rice plant, viewed from the nitrogen nutrition. Tohoku Univ. Inst. Agr. Res. Bull. 12: 1–146.

O'Toole, J. C. 1982. Adaptation of rice to drought-prone environments, pp. 195–213. *In*: Drought resistance in crops with emphasis on rice. IRRI, Los Banos, Philippines.

Ottow, J. C. G., G. Benchiser, I. Watanable, and S. Santiago. 1982. Multiple nutritional soil stress as the prerequisite for iron toxicity of wetland rice (*Oryza sativa* L.). Trop. Agric. (Trinidad) 60: 102–106.

Park, S. H. and S. Y. Cho. 1989. High yielding rice cultivars in South Korea, p. 75. *In*: IRRI (ed.) Progress in irrigated rice research. IRRI, Los Banos, Philippines.

Patrick, W. H. 1982. Nitrogen transformation in submerged soils, pp. 449–466. *In*: F. J. Stevenson (ed.) Nitrogen in agricultural soils. Am. Soc. Agron., Madison, Wisconsin.

Patrick, W. H., Jr. and D. S. Mikkelsen. 1971. Plant nutrient behavior in flooded soil, pp. 187–215. *In*: R. A. Olsen (ed.). Fertilizer technology and use, 2nd ed. Soil Sci. Soc. Am., Madison, Wisconsin.

Pearson, G. A. 1961. The salt tolerance of rice. Int. Rice Comm. Newsletter 10: 1–4.

Ponnamperuma, F. N. 1965. Dynamic aspects of flooded soils, pp. 295–328. *In*: IRRI (ed.). The mineral nutrition of the rice plant. Johns Hopkins Press, Baltimore.

Ponnamperuma, F. N. 1972. The chemistry of submerged soils. Adv. Agron. 24: 29–96.

Ponnamperuma, F. N. 1977a. Screening rice for tolerance to mineral stresses. IRRI Research Paper, Series No. 6, IRRI, Los Banos, Philippines.

Ponnamperuma, F. N. 1977b. Behavior of minor elements in paddy soils. IRRI Research Paper, Series No. 8, IRRI, Los Banos, Philippines.

Ponnamperuma, F. N. 1978. Electrochemical changes in submerged soils and the growth of rice, pp. 421–441. *In*: IRRI (ed.). Soils and rice. IRRI, Los Banos, Philippines.

Pugh, G. J. F. 1962. Studies on fungi in coastal soils. II. Fungal ecology in a developing salt marsh. Trans. Br. Physiol. Soc. 45: 560–566.

Purseglove, J. W. 1985. Tropical crops: Monocotyledons. Longman, London.

Rahman, M. S. 1984. Breaking the yield barriers in cereals with special reference to rice. J. Aust. Inst. Agric. Sci. 504: 228–232.

Raij, B. V. 1991. Fertilidade do solo e adubacão. Editora Agronomica Ceres, São Paulo, Brazil.

Rao, P. S. 1988. Production trends of high density grain as influenced by nitrogen, season, crop canopy and duration of lowland irrigated paddy. Oryza 25: 47–51.

Roberts, S. R., J. E. Hill, D. M. Brandon, B. C. Miller, S. C. Scardaci, C. M. Wick, and J. F. Williams. 1993. Biological yield and harvest index in rice: Nitrogen response of tall and semidwarf cultivars. J. Prod. Agric. 6: 585–588.

Sahrawat, K. L. 1983. Nitrogen availability indexes for submerged rice soils. Adv. Agron. 36: 415–451.

Sanchez, P. A. and J. G. Salinas. 1981. Low-input technology for managing Oxisols and Ultisols in tropical America. Adv. Agron. 34: 279–398.

Satake, T. 1976. Sterility-type cool injury in paddy rice plants, pp. 281– 300. *In*: IRRI (ed.). Climate and rice. IRRI, Los Banos, Philippines.

Savant, N. K. and S. K. De Datta. 1982. Nitrogen transformation in wetland rice soils. Adv. Agron. 35: 241–302.

Savant, N. K. and G. H. McClellan. 1987. Do iron oxide systems influence soil properties and nitrogen transformations in soils under wetland rice-based cropping systems? Commun. Soil Sci. Plant Anal. 18: 83–113.

Sharma, P. K, S. K. De Datta, and C. A Redulla. 1987. Root growth and yield response of rainfed lowland rice to planting method. Exp. Agric. 23: 305–313.

Shiroshita, T., K. Ishii, J. Kaneko, and S. Kitajima. 1962. Studies on the high productivity of paddy rice by increasing the manurial effects. J. Cent. Agric. Exp. Stn. 1: 47–108.

Tanaka, A., R. Loe, and S. A. Navaroso. 1966. Some mechanisms involved in the development of iron toxicity symptoms in the rice plant. Soil Sci. Plant Nutr. 12: 158–164.

Tanaka, A. and T. Tadano. 1972. Potassium in relation to iron toxicity of the rice plant. Potash. Rev. 9: 1–12.

Tomar, V. S. and B. P. Ghildyal. 1975. Resistance to water transport in rice plants. Agron. J. 67: 269–272.

Venkateswarlu, B., B. S. Versiara, and G. Patola. 1990. Occurrence of different grades of grain during maturation of rice panicle, pp. 87–92. *In*: Proceedings of the International Congress of Plant Physiology, New Delhi, Feb. 15–20, 1988.

Virmani, S. S. 1977. Varietal tolerance of rice to iron toxicity in Liberia. Int. Rice Res. Newsletter 2: 4.

Widawsky, D. A. and J. C. O'Toole. 1990. Prioritizing the rice biotechnology research agenda for eastern India. The Rockefeller Foundation, Grondal Press, New Delhi. 86 pp.

Yamachi, A., Y. Kono, and J. Tatsumi. 1987a. Quantitative analysis on root system structures of upland rice and maize. Japan J. Crop Sci. 56: 608–617.

Yamachi, A., Y. Kono, and J. Tatsumi. 1987b. Comparison of root system structures of 13 species of cereals. Japan J. Crop Sci. 56: 618–631.

Yamada, N. and Y. Ota. 1958. Studies on the respiration of crops: Enzymatic oxidation of ferrous iron by rice roots. Proc. Crop Sci. Soc. Japan 26: 205–210.

Yeo, A. R. and T. J. Flowers. 1982. Accumulation and localisation of sodium ions within the shoots of rice (*Oryza sativa*) genotypes differing in salinity resistance. Physiol. Plant. 56: 343–348.

Yeo, A. R. and T. J. Flowers. 1983. Varietal differences in the toxicity of sodium ions in rice leaves. Physiol. Plant. 59: 189–195.

Yeo, A. R. and T. J. Flowers. 1986. Salinity resistance in rice (*Oryza sativa* L.) and a pyramiding approach to breeding genotypes for saline soils. Aust. J. Plant Physiol. 13: 161–173.

Yeo, A. R., M. E. Yeo, S. A. Flowers, and T. J. Flowers. 1990. Screening of rice (*Oryza sativa* L.) genotypes for physiological characters contributing to salinity resistance and their relationship to overall performance. Theor. Appl. Genet. 79: 377–384.

Yoshida, S. 1972. Physiological aspects of grain yield. Annu. Rev. Plant Physiol. 23: 437–464.

Yoshida, S. 1981. Fundamentals of rice crop science. IRRI, Los Banos, Philippines.

Yoshida, S. 1983. Rice, pp. 103–127. *In*: IRRI (ed.). Potential productivity of field crops under different environments. IRRI, Los Banos, Philippines.

Yoshida, S. and F. T. Parao. 1976. Climatic influence on yield and yield components of lowland rice in the tropics, pp. 471–494. *In*: IRRI (ed.). Climate and rice. IRRI, Los Banos, Philippines.

Yoshida, T. and Takahashi, J. 1960. Studies on the metabolism in roots of lowland rice. Respiratory and enzymatic gradient in rice root. J. Soil Manure Japan 31: 423–426.

10

Corn

I. INTRODUCTION

Corn (*Zea mays* L.), known as maize, is the world's third most important cereal after wheat and rice. Corn is grown primarily for grain and secondarily for fodder and raw material for industrial processes. The grain is used for both human and animal consumption. The vegetative parts of the plant are cut green and either dried or made into silage for animal feed. The domestication and selection of corn probably began in central or southwestern Mexico about 7000 years ago (Goodman, 1988). The date of origin for cultivated maize in the highlands of central Mexico is unknown, but Palomero Toluqeuño was considered an ancient indigenous race by Wellhausen et al. (1952). Kato (1984) proposed a multicenter domestication of maize, with two of the four centers in central Mexico. Probably highland maize came from the higher altitude centers. Excavations at Teotihuacán, in the Valley of Mexico, have uncovered ears with the characteristics of Cónico (Wellhausen et al., 1952), indicating that maize similar to the current types was important for the ancient civilizations of highland Mexico and has been cultivated at altitudes above 2000 m for several millennia (Eagles and Lothrop, 1994). Distinguishable groups of cultivars arose in Mexico and Central America, in the northeastern United States, on the northern coast of South America, in the Andes, and in central Brazil. The Spanish and Portuguese quickly distributed corn throughout the world in the sixteenth century (Jones, 1985). It is grown in more countries than any other cereal and has produced the largest grain yield of any cereal.

The chief corn-producing countries are the United States, USSR, Romania, Yugoslavia, Hungary, Italy, China, Brazil, Mexico, South Africa, Argentina, India, and Indonesia. The main reason for its wide distribution is that corn has many assets. These include its high yield per unit of labor and per unit of area. It is a compact and easily transportable source of nutrition. Its husks give protection from birds and rain. It can be harvested over a long period, stores well, and can even be left dried in the field until harvesting is convenient. It provides numerous useful food products, and it is frequently preferred to sorghum and the millets (Jones, 1985). The chemical composition of the grain on a dry weight basis is about 77% starch, 2% sugar, 9% protein, 5% fat, 5% pentosan, and 2% ash (Purseglove, 1985).

II. CLIMATE AND SOIL REQUIREMENTS

Corn, with its large number of cultivars of different maturity periods, can be grown over a wide environmental range. It is essentially a crop of warm countries with adequate moisture. The bulk of the crop is grown in the warmer parts of the temperate regions and in the humid subtropics. It is mainly grown from 50°N to 40°S and from sea level to 4000 m in the Andes and Mexico. The water balance of a crop is determined by evapotranspiration, rainfall, and soil characteristics. Reported values for seasonal evapotranspiration of corn vary widely (440–1000 mm) and are influenced by available water and local environmental parameters (Musick and Dusek, 1980; Eck, 1984). Water deficit reduces crop yields. Rainfall during the corn-growing period should be in the range of 460–600 mm, and in the tropics corn does best with 600–900 mm of rain during the growing season.

Deficiency of water during any growth stage of corn can reduce grain yield. However, the magnitude of the reduction depends on the growth stage of the crop at the time of stress, the severity and duration of the stress, and the susceptibility of the genotype to stress (Lorens et al., 1987). Water stress during vegetative development reduces expansion of leaves, stems, and roots and ultimately affects the development of reproductive organs and potential grain yield (Denmead and Shaw, 1960). Corn is most sensitive to drought stress during pollination, when delayed emergence of silks may reduce fertilization and subsequent grain yield as a result of fewer seed numbers (Herrero and Johnson, 1981). Drought stress as late as 2–3 weeks following 50% silking

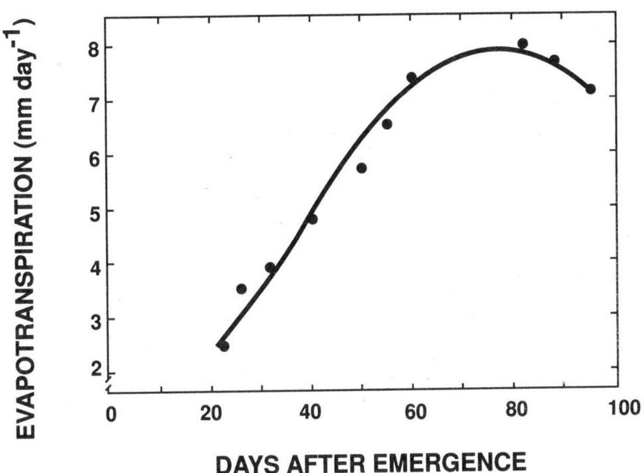

Figure 10.1 Relationship between corn plant age and evapotranspiration. (From Oliveira et al., 1993.)

may also reduce seed number (Frey, 1981). Drought during the linear growth phase of kernel development primarily affects mean kernel weight by reducing assimilate production or duration of grain fill, or by a combination of both factors (Jones and Simmons, 1983). Kernels at the tip of the ear often develop poorly when water stress occurs during the seed-filling period (Tollenaar and Daynard, 1978).

In conclusion, maize grain yield is particularly sensitive to water deficits that coincide with the tasselling–silking period and approximately two weeks after silking (Otegui et al., 1995). The number of kernels per plant is defined during this period. Oliveira et al. (1993) studied evapotranspiration (ET) and water extraction from different soil depths on a red-yellow podzolic soil, Barreiros-Bahia, Brazil. The accumulated ET during the 95-day crop growth was 455 mm. The period of crop maximum demand 8.02 mm per day took place at 81 days' growth (Figure 10.1). Water extraction by 0–20 cm, 20–40 cm, 40–60 cm, and 60–80 cm soil depth was 36%, 39%, 22%, and 3%, respectively.

Soil temperatures of 26–30°C are optimum for both germination and early seedling growth. Emergence is normally reduced below 13°C and fails below 10°C (Riley, 1981). Optimum temperature at tasseling is 21–30°C. High temperature promotes respiration. Chang (1981) reported that the average respiration loss is about 25% of the photosynthetic rate in the temperate zone and about 35% in the tropics. However, an important effect of temperature is that high temperature, particularly at night, shortens the grain-filling period, thereby reducing the yield (Jones et al., 1981; Wilson et al., 1973). High temperature increases the rate of grain filling but greatly reduces the duration of grain-filling period, whereas low temperatures cause an inverse response (Jones et al., 1981). Hunter et al. (1977) suggested that the grain yield of maize, like that of small grains, is higher at lower temperatures because of an increase in the length of the grain-filling period and greater partitioning of postanthesis dry matter to grain. A detailed description of climatic requirements of corn is given by Shaw (1977).

Corn can be grown on a wide variety of soils but performs best on well-drained deep loams and silt loams containing adequate organic matter and available nutrients. Corn can be grown on soils with a pH from 5.0 to 8.0, but corn is moderately sensitive to salinity, and 90% relative yield was obtained at an electrical conductivity of about 1.8 dS m^{-1} (Jones, 1985). Soil salinity can have a marked influence on uptake of a number of nutrients, but decreased dry matter production probably most often results from decreased soil water and increased toxicity of sodium chloride and sulfate in the soil solution (Larson and Hanway, 1977).

III. GROWTH AND DEVELOPMENT

Crop growth and development involve complex physiological and biochemical processes which are influenced by the crop's environment in ways that are still inadequately understood (Daughtry et al., 1984). Soil moisture, temperature, availability of nutrients, soil, aeration, genotype, and cultural practices are the principal environmental variables which influence development of crops. Knowledge of growth sequences during the season is required for better management practices.

A. Germination

The germination of the corn seed is similar to that of many grasses except for scale differences resulting from the relatively large endosperm and embryo (Duncan, 1975). Corn seeds germinate immediately after maturity, even while attached to the plant. With a favorable environment, the radicle emerges 2–3 days after sowing, and the plumule, enclosed in the coleoptile, breaks through the seed coat 1–2 days later. Other seminal roots, usually three in number, soon follow. The tubular white mesocotyl elongates to within 3 cm of the soil surface; its length depends on the depth of planting and in most cultivars its maximum length is 12.5–15 cm. If the seed is planted deeper than this or growing conditions are unfavorable, elongation stops and the seedling fails to emerge (Purseglove, 1985). Adequate moisture and soil temperatures near 30°C are ideal for germination and plant emergence. The morphology of a corn seedling is shown in Figure 10.2.

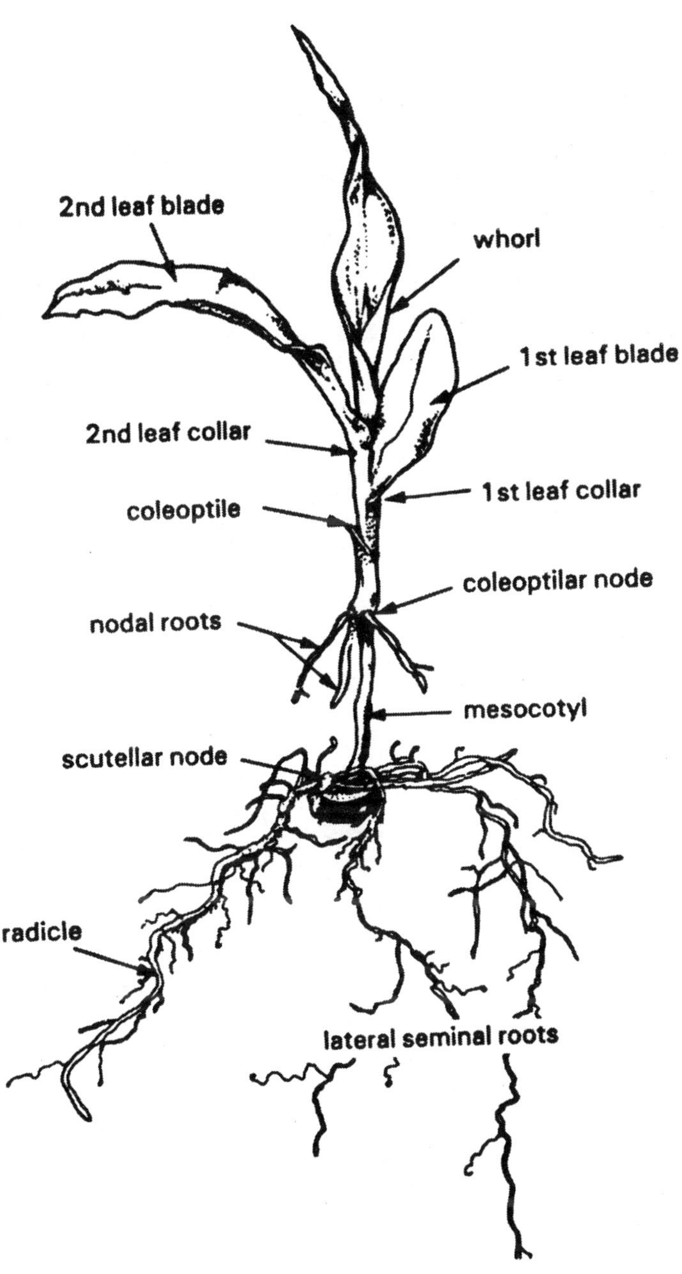

Figure 10.2 The morphology of a corn seedling. (From Stevens et al., 1986.)

B. Roots

The radicle grows out to produce the first seminal root, after which three or more seminal roots grow out sideways from the embryo. They supply most of the nutrition during the first 2 weeks after germination and remain functional for some time (Purseglove, 1985). The seminal roots quickly lose their importance, and the young plant is supported and nourished by the permanent root system which begins to develop from the crown. The main root system continues to grow downward and to branch, and additional roots are produced in successive whorls from stem nodes above the crown. According to Anderson (1987), the most rapid root development in corn occurs in the 8 weeks after planting. The depth of extension of roots in deep soils is a linear function of time until tasseling. From tasseling to the start of grain filling, brace roots develop (Larson and Hanway, 1977). During the rapid grain-filling storage, total root length and root dry weight do not increase and may decrease before grain matures (Mengel and Barber, 1974). Root system growth and configuration respond to soil water, temperature, air, nutrient, toxic chemical, and soil resistance. Management practices can affect root system development by modifying these factors.

C. Stem and Leaves

The corn stem consists of leaf blades, leaf sheaths, nodes, and internodes. Leaves are borne alternately on either side of the stem at the nodes. Each corn leaf consists of a thin, flat blade with a definite midrib and a thicker, more rigid sheath with a smaller midrib. Leaf inclination varies considerably among genotypes, from almost horizontal to almost vertical in one mutant with no collar (Duncan, 1975).

D. Inflorescence

The corn plant is monoecious and diclinous, with male and female inflorescences borne in separate inflorescences on the same plant. The male inflorescence is called the tassel, and the female inflorescence is called the ear. The ear is actually a modified spike produced from a short lateral branch in the axil of one of the largest foliage leaves, about halfway down the stem (Purseglove, 1985). Thus, unlike other major cereals, corn produces its economic yield (grain) on a lateral shoot. Because of this separation of ear and tassel, plus the protandry of flowering, corn is primarily a cross-pollinated species.

The pistil of the female flower, known as the silk, develops from the growing point of the flower. It elongates through the length of the husks propelled by the growth of an intercalary main stem located at its base. Each silk continues to grow until it is pollinated and fertilization takes place (Duncan, 1975).

Table 10.1 Definitions of Growth Stages for Maize

Growth stage	Maize
0	Emergence, coleoptile visible at soil surface
1	Collar of 4th leaf visible
2	Collar of 8th leaf visible
3	Collar of 12th leaf visible
4	Collar of 16th leaf visible. Tips of many tassels visible
5	75% of plants have silks visible
6	Kernels in "blister" stage
7	Very late "roasting ear" stage
8	Early dent stage
9	Full dent stage
10	Physiological maturity

Source: Hanway, 1963.

E. Growth Stages

Several methods have been proposed to describe corn growth stages. Larson and Hanway (1977) described five periods with unique characteristics: 1) planting to emergence, 2) emergence to tasseling, 3) tasseling to silking, 4) silking to physiological maturity, and 5) dry-down period. The classical numerical system proposed by Hanway (1963) has 10 growth stages (Table 10.1). A more recent, more detailed approach proposed by Stevens et al. (1986) is given in Tables 10.2 and 10.3.

F. Dry Matter Production

Corn was one of the first species shown to use the C_4 photosynthetic pathway; thus its photosynthetic metabolism differs from that of rice and wheat, the two other major cereals (Hatch and Slack, 1970). The discovery of the C_4 photosynthetic pathway explained corn's high leaf photosynthetic rates, low CO_2 compensation points, and absence of photosynthetic high saturation up to full sunlight (Figure 10.3). The photosynthetic rate of corn, sorghum, sugarcane, and bermudagrass can reach 60 mg CO_2 dm^{-2} h^{-1}, almost double that of soybean, cotton, and alfalfa, which use the C_3 photosynthetic pathway (Figure 10.3).

Dry matter accumulation for corn grown under good management in South Carolina was described by Karlen et al. (1988) and is shown in Figure 10.4.

Table 10.2 A Field Guide to Pretassel Phenology of Corn

Analog	Stage
0.00–0.90	*Planting to emergence.* By definition, planting was analogous with stage 0.0, emergence of the coleoptile from the soil with stage 0.5, and exposure of the ligule of the first leaf with stage 1.0 (Fig. 10.2). Substages 0.0 to 0.5 were identified according to the terminal location of the coleoptile relative to the surface of the soil. The scutellar node and mesocotyl or scutellar internode were assigned a value of 0.25, while the coleoptilar node and internode were assigned a value of 0.5.
1.00–10.0	*Leaf and tassel emergence.* A leaf was considered fully emerged when the collar and auricles were completely visible. Nodes, internodes, and leaves were numbered sequentially from the first leaf (Fig. 10.2). For example, a complete description of a corn plant in which the collar and auricles of the sixth leaf were located halfway along the sheath of the fifth leaf, the most recent fully emerged leaf (i.e., collar and auricles were visible above the auricles of the preceding leaf), would be: phase pretassel, stage five-leaf, substage 0.5. Substages were, therefore, determined from the position of the collar and auricles of the emerging leaf (indicator leaf plus one) relative to the total length of the sheath of the most recently emerged leaf. As plants developed and lower leaves were lost, elongation of the fourth, fifth, sixth, and seventh internodes provided a basis for determining leaf stage and substage. Leaf number was determined by counting internodes, starting at the first internode, to equal or exceed 2.0 cm in length. This was internode 5.0 in field corn and 6.0 in popcorn.
10.00	*Tassel emergence (complete).* Documented as though it were an additional leaf according to the relative position of basal tassel branches with respect to the collar and auricles of the last (flag) leaf. The tassel was considered to be fully emerged when the lower branches were entirely visible beyond the collar and auricles of the flag leaf.

Source: Stevens et al., 1986.

Table 10.3 A Field Guide to Post-Tassel Phenology of Corn

Analog	Stage
10.25 10.375 (mid) 10.50 (late)	*Anthesis early*. Documented according to the proportion of a fertile tassel bearing fully emerged anthers actively shedding pollen. Characteristically, pollen shed began at the midpoint of the central axis and progressed acropetally (upward), then basipetally (downward), followed by the lateral branches. The adaxial floret of each spikelet shed pollen prior to the abaxial floret. The development of male and female reproductive organs was not always synchronized on the same plant; however, under normal conditions with commercial material, silking coincided with 75 to 100% of the tassel actively shedding pollen. Three substages, early (25%), mid (50%), and late (75%), were recognized as equal divisions.
10.50 10.625 (mid) 10.75 (late)	*Fertilization (early)*. Documented according to the proportion of silks on the primary ear that had stopped elongating and had begun to turn brown. Three substages, early (25%), mid (50%), and late (75%), were recognized as equal divisions.
10.50 10.825 (mid) 11.00 (late)	*Brown silk (early)*. Documented according to the proportion of silks on the primary ear that had turned dark brown. By this stage pollen shed and fertilization had been completed and the endosperm was a beginning to develop rapidly. Three substages, early (25%), mid (50%), and late (75%), were recognized as equal divisions.
11.00 11.25 (mid) 11.50 (late)	*Blister (early)*. Kernels appeared as white translucent grains. The endosperm and abundant inner fluid were clear in color, and the embryo was visible on dissection. Three substages, early, mid, and late, were recognized as equal divisions.
11.50	*Milk (early)*. Association with increasing viscosity of the endosperm to resemble a milklike substance. Three substages, early, mid, and late, were recognized as equal divisions.

Table 10.3 (Continued)

Analog	Stage
12.00 12.25 (mid) 12.50 (late)	*Soft-dough (early).* Increasing deposits of starch within the endosperm of dent corn caused it to develop a heavy doughlike consistency, and foliar detail was evident macroscopically in the embryo. In popcorn, deposits of vitreous or dense horny endosperm gave the impression of a small lens or incomplete cap to the kernel. A reliable indicator of this stage in popcorn involved crushing kernels to locate vitreous starch deposits. Three substages, early, mid, and late, were recognized as equal divisions.
12.50	*Twenty-five percent solids.* Kernels within the basal third of the dent corn ear by this stage had begun to form dents. In popcorn, glazing or capping of kernels was usually evident near the butt of the ear.
13.00	*Fifty percent solids.* Apparent in both dent and popcorn when the endosperm within a majority of kernels displayed definite signs of hardening to a point adjacent to the distal location of the embryo. By this stage, the husk covering the ear had begun to dry rapidly. Denting or capping (glazing) of the kernels (depending on grain type) was usually evident by this stage under normal climatic conditions, unless the crop was planted abnormally late in the growing season.
13.50	*Seventy-five percent solids.* By this stage, with normal climatic conditions and planting dates, a distinct brown coloration, often more prevalent in popcorn, had begun to develop within the placental regions of kernels. The embryo was noted to be close to fully developed.
14.00	*One-hundred percent solids.* With normal climatic conditions and planting dates and in the absence of additional stress factors, this stage was determined as a point of maximum grain dry matter accumulation and assumed to coincide generally with physiological maturity.

Source: Stevens et al., 1986.

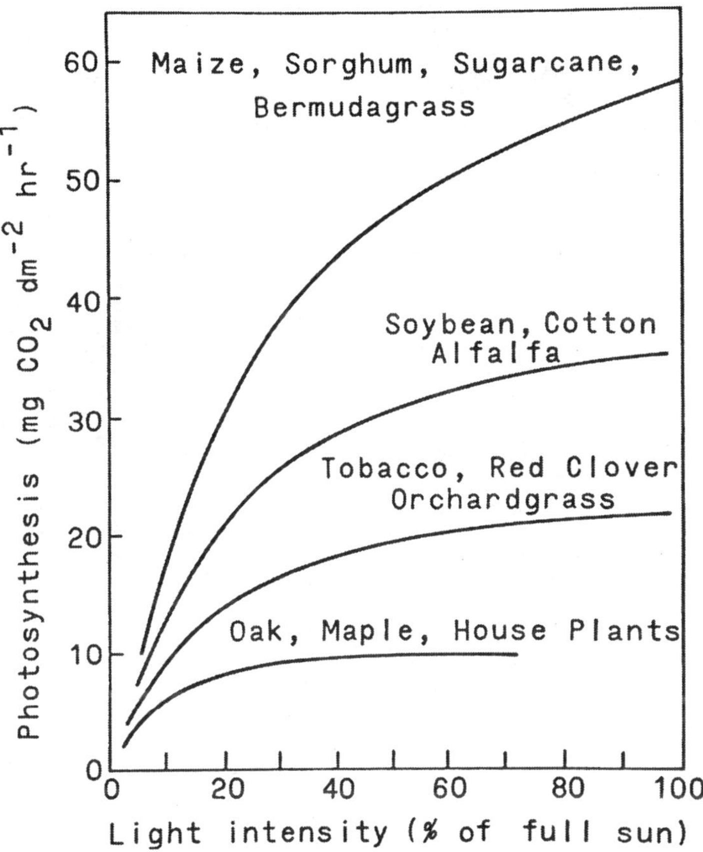

Figure 10.3 Effect of light intensity on average leaf photosynthetic rates of various plant species. (From Mock and Pearce, 1975.)

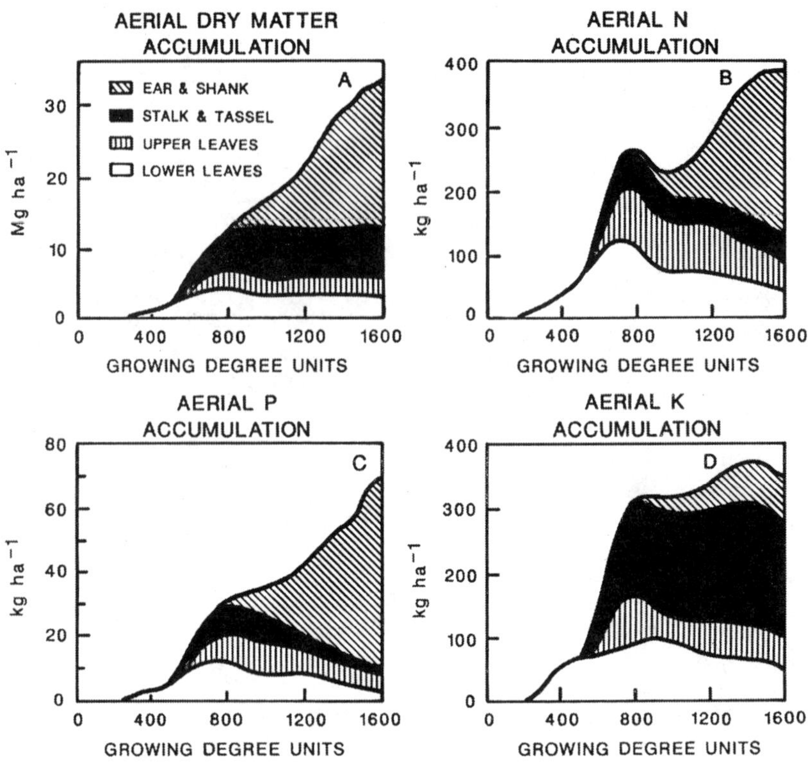

Figure 10.4 Aerial dry matter, N, P, and K accumulation in different parts of corn plant yielding 19.3 Mg ha^{-1} of grain. (From Karlen et al., 1988.)

Total shoot dry matter at physiological maturity was 31.8 Mg ha^{-1}, including 16.3 Mg ha^{-1} of grain dry matter. This resulted in a grain/stover ratio of 0.51, excluding root biomass. Leaves accounted for 18% of the aerial biomass at physiological maturity, with 2.8 mg ha^{-1} of leaves retained both below and above the ear (Figure 10.4). Maximum accumulation in the stalk and tassel was 6.5 Mg ha^{-1} or 20% of the aerial biomass. Nongrain portions of the ear and shank, which were calculated by subtracting grain yield from the ear and shank biomass at physiological maturity, accounted for 3.4 Mg ha^{-1} or approximately 11% of the aerial dry matter. Hanway (1962a) reported a similar dry matter accumulation pattern in corn under field conditions.

According to Karlen et al. (1988), daily dry matter growth rates show two distinct peaks. The first occurs during vegetative growth when the potential ovule number is being established. During this period, maximum rates of dry matter accumulation for lower leaves, upper leaves, and stalk and tassel fractions were approximately 100, 200, and 300 kg ha^{-1} day^{-1}, respectively. The second peak, which occurs during grain fill, shows a peak growth rate of approximately 450 kg ha^{-1} day^{-1}, which goes directly into the ear and shank.

G. Leaf Area Index

Green leaf area, usually expressed as a nondimensional green leaf area index (*LAI*), is a major factor determining solar radiation interception, canopy photosynthesis, and, therefore, yield. Duncan (1975) stated that interception of light is the primary function of corn canopies, but the intercepted light should be used efficiently, and this is a function of leaf angle. He pointed out that between *LAI* values of 3 and 4, horizontal leaves intercept about 90% of the light, while leaves 15° from vertical intercept only 75–80%. Hoyt and Bradfield (1962) reported that the net assimilation rate (rate of growth per unit leaf area) of corn is constant when *LAI* is less than 2.7 but declines rapidly as *LAI* increases above that value. Eik and Hanway (1966) also reported a linear relationship between corn grain yield and *LAI* up to an *LAI* of 3.3 at midsilk. Leaf growth continues until silking, or soon after (Hanway, 1962a), but there is no evidence that leaf growth influences ear growth during the critical growth period. Certainly, there is considerable evidence that grain yields and grain number increase with *LAI* up to values ranging from approximately 3–5 for both the U.S. corn belt (Eik and Hanway, 1966) and tropical conditions (Yamaguchi, 1974).

Development of a corn canopy capable of intercepting almost all photosynthetically active radiation by silking requires a plant density of 5–10 plants m^{-2}, depending on the cultivar and row spacing.

IV. YIELD AND YIELD COMPONENTS

Average grain yields of corn vary substantially among the temperate, subtropical, and tropical regions. In the early 1970s, it was estimated that average corn yields in the temperate, subtropical, and tropical regions were 3.5, 1.8, and 1 t ha^{-1}, respectively (Goldsworthy, 1974).

In the temperate region, a maximum yield of 22 t ha^{-1} has been reported in Michigan, and yields of 10 t ha^{-1} at the commercial level are common (Fischer and Palmer, 1983). In the tropics, most of the high yields reported are confined to intermediate- or high-altitude areas having long rainy seasons. Yields of 12 t ha^{-1} have been reported from experiments in Salisbury, Zimbabwe (latitude 18°S, altitude 1500 m), and yields of 10 t ha^{-1} have been reported from Kitale, Kenya (latitude 2°N, altitude 1890 m) (Fischer and Palmer, 1983). Average yields in several well-managed, irrigated field experiments conducted on Oxisols in Hawaii, Brazil, and Puerto Rico were 7.6–9.5 t ha^{-1} with maximum yields of almost 12 t ha^{-1} (Tsuji, 1985). In the lowland tropics, experimental yields typically range from 5 to 8 t ha^{-1} with good management practices. Differences among experimental yields probably reflect the effects of temperature, solar radiation, rainfall, and agronomic practices in the regions. Low grain yields of most tropical corn cultivars have also been attributed to poor partitioning of total dry matter to the grain (Goldsworthy, 1974; Yamaguchi, 1974).

The yield of corn is the product of kernel number per unit area and kernel weight. Of these, grain weight is the more stable, and large differences in yield are usually the result of fluctuations in grain number. The variation in grain yield of 12 tropical populations, grown at 15 sites in Mexico to provide different environments and planting dates, was linearly related to change in grain number ($Y = 482.4 + 13.6X$, $r^2 = 0.70$) and rate of grain growth ($Y = 310 + 959X$, $r^2 = 0.36$) but not to duration of grain growth (Fischer and Palmer, 1983). Grain number per unit area depends on events before and around flowering. Deficiency of N, moisture stress, and inadequate radiation at that time significantly reduce grain number.

Large differences exist between potential experimental yields and yields obtained by farmers. The difference is particularly great in developing countries, where corn is grown by subsistence farmers without access to improved management techniques. There are many reasons for low yields, including drought and nutrient stresses, inadequate pest control, and use of poorly adapted cultivars with low yield potential. Dramatic yield increases in developing countries will probably require simultaneous improvement of crop management and germplasm. Fischer and Palmer (1984) suggested the following measures to improve corn germplasm for tropical environments:

1. There is variation in the rate and percentage of germination of seeds at both warm and cool soil temperatures, which suggests that it should be possible to develop, through recurrent selection, genotypes that will germinate and grow at a wide range of temperatures (Hardacre and Eagles, 1980).
2. Increasing the duration of growth and the fraction of the total duration devoted to grain filling may provide other opportunities to increase yield.
3. The rate of dry matter production during ear development is critical for yield determination. It is often possible to increase growth rates by improving light interception with higher plant populations. The rate of growth per unit of intercepted light might also be increased by selection for high rates of photosynthesis or by modification of the canopy structure through changes in leaf angle, leaf size, and the vertical distribution of leaves.

4. Selection for shorter plants, with uniform plant height, fewer and narrower leaves, and a small tassel has resulted in a shorter interval between pollen shed and silking and fewer barren plants at dense plant populations.
5. Tolerance to high plant populations is also associated with production of more than one ear at lower plant populations.

Maize yields in the United States have steadily increased since 1930, and genetic improvement has been credited with over 50% of the increase (Russell, 1991). Publicly and privately funded breeding programs have contributed to this improvement of maize hybrids: publicly funded breeding programs emphasized genetic improvement of populations, whereas privately funded breeding programs emphasized inbred development within F_2 and backcross populations. Long-term population improvement programs provide improved germplasm sources and inbreds for breeding programs that emphasize short-term breeding objectives (Fountain and Hallauer, 1996).

V. NUTRIENT REQUIREMENTS

Achieving high corn yields requires an adequate supply and balance of essential nutrients in the growth medium. Additions of plant nutrients as fertilizers to corn can be efficient only if they are absorbed and used for the production of increased yield (Barber and Olson, 1968). Further, the quantity of nutrients required by a crop depends on soil, climate, yield level, cultivar planted, and management practices. Fertilizer needs can be assessed by soil and plant analysis as well as by visual deficiency symptoms. A detailed description of these nutritional diagnostic techniques is given in Chapter 4. No doubt nutrient accumulation and plant concentrations vary among agroecosystems, but these values can serve as guides for assessing the nutritional requirements and status of the corn crop. Actual fertilizer recommendations should be made on the basis of experimental results obtained for different nutrients and specific agroclimatic regions. One approach to reduce the impact of nutrient deficiency on maize production may be to select cultivars which are superior in the utilization of available nutrients, either due to enhanced uptake capacity or because of more efficient use of absorbed nutrient in grain production (Lafitte and Edmeades, 1994).

Nitrogen is one of the most important nutrients limiting maize yield in various parts of the world. Accurate fertilizer N recommendations for corn production are important for maximizing productivity and profit while minimizing environmental impact of fertilizer use. Nitrogen fertilizer requirements depend on many factors, including yield goal, inorganic soil N, potential N mineralization, soil type, and numerous environmental factors (Schlegel and Havlin, 1995).

Maximizing the quantity of fertilizer N recovered by the crop, or minimizing the quantity of residual fertilizer N after harvest, generally will reduce potential groundwater contamination (Keeney, 1987). Apparent fertilizer N recovery (AFNR) generally ranges between 30 and 70% (Legg and Meisinger, 1982) and depends on many factors, including location or soil type, inorganic soil N content, N rate, and other N or water management factors (Walters and Malzer, 1990). Generally, AFNR decreases with increasing N application rate (Oberle and Keeney, 1990b). In their study, AFNR was higher in coarse-textured soils than in fine-textured soils because of greater N mineralization in the fine-textured soils. Use of an adequate N level is one way to get maximum yield and at the same time improve N recovery efficiency.

Fertilizer N rates for maximum or optimum yield also vary widely between locations (or soils) and years (Onken et al., 1985). In long-term studies with corn, the N rate required for maximum yield varied from 174 to 241 kg N/ha on several rainfed silt loam Wisconsin soils, compared with 185–258 kg N/ha on irrigated sandy loam and loamy sand soils (Oberle and Keeney, 1990a). Fertilizer N recommendations for irrigated corn in western and south central Nebraska varied between 56 and 184 kg N/ha and depended on soil profile and nitrate-N content of irrigation water (Ferguson et al., 1991).

Decisions concerning optimum rates of fertilization directly or indirectly involve fitting some type of model to yield data collected when several rates of fertilizer are applied. Although several different models are commonly used to describe crop yield response to fertilizers, it is seldom explained why one model is selected over others. Cerrato and Blackmer (1990) compare and evaluate several models (linear-plus-plateau, quadratic-plus-plateau, quadratic, exponential, and square root) commonly used for describing the response of corn (*Zea mays* L.) to N fertilizer (Figure 10.5). The evaluation involved 12 site-years of data, each having 10 rates of N applied preplanting. All models fit the data equally well when evaluated by using the r^2 statistic. All models indicated similar maximum yields, but there were marked discrepancies among models when predicting economic optimum rates of fertilization. Mean (across all site-years) economic optimum rates of fertilization as indicated by the various models ranged from 128 to 379 kg N ha^{-1} at a common fertilizer-to-corn price ratio. Statistical analyses indicated that the most commonly used model, the quadratic model, did not give a valid description of the yield responses and tended to indicate optimal rates of fertilization that were too high. The quadratic-plus-plateau model best described the yield responses observed in this study. The results clearly show that, especially amid increasing concerns about the economic and environmental effects of over-fertilization, the reason for selecting one model over others deserves more attention than it has received in the past (Cerrato and Blackmer, 1990).

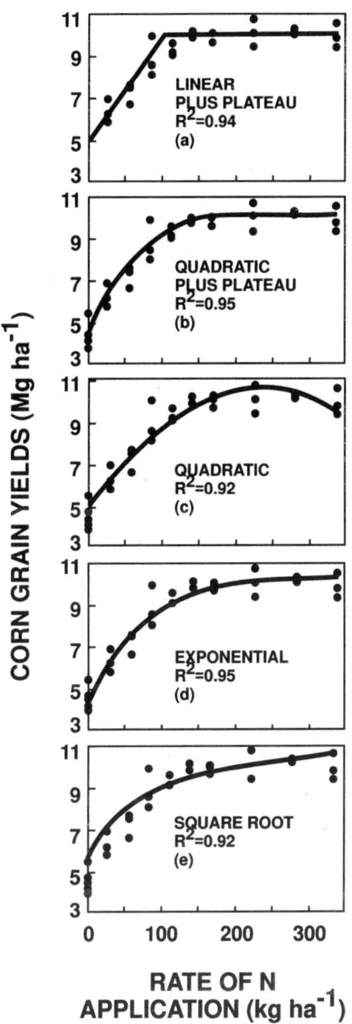

Figure 10.5 Relationship between rate of N applied and corn grain yield as a function of five models. (From Cerrato and Blackmer, 1992.)

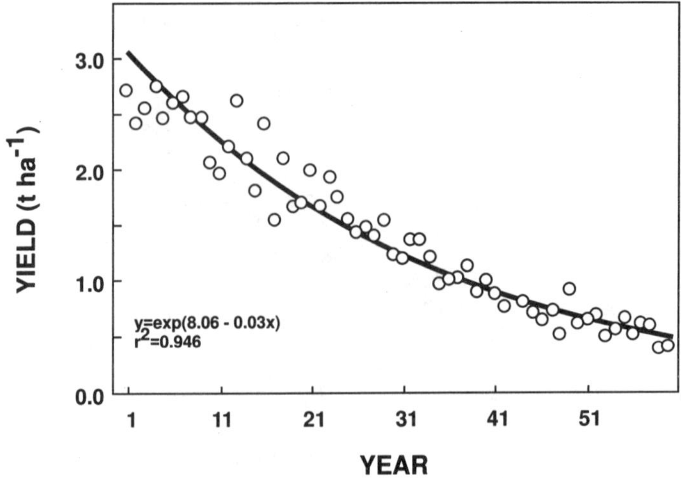

Figure 10.6 Corn grain yield as a function of years of cultivation in monoculture in an Oxisol of central Brazil. (From Thornton et al., 1995.)

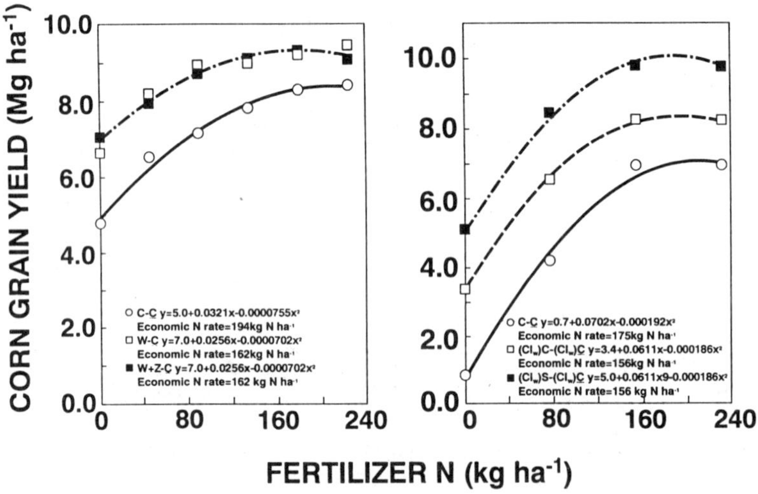

Figure 10.7 Fertilizer N response of continuous corn (C – C), corn following wheat (W – C), and corn following wheat interseeded with alfalfa (W + A – C). On the right side of the figure, continuous corn with winter crop of crimson clover [(CLW)C – (CLW)C] or corn in a soybean-corn rotation with winter crops of crimson clover each year [(CLW)S – (CLW)C]. From Lory et al., 1995.)

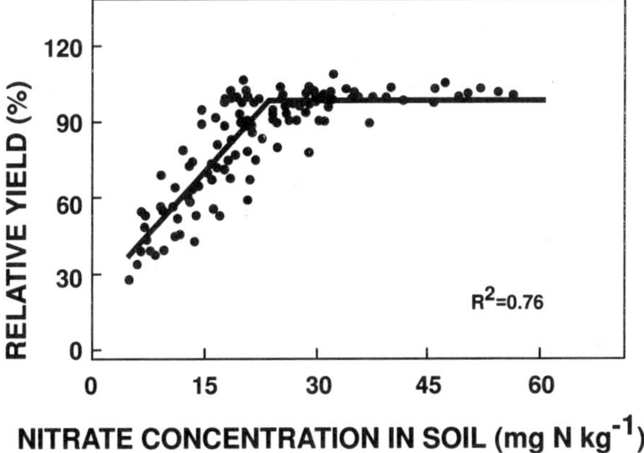

Figure 10.8 Relationship between relative yield and NO3 concentration in soil. (From Binford et al., 1992.)

Adequate nitrogen rates also depend on whether the crop is grown as a monoculture or in rotation. Figure 10.6 shows that corn grain yield decreased when corn in monoculture was grown for 60 years in central Brazil (Thornton et al., 1995).

Cereals and other nonlegumes typically require less fertilizer N when grown following a legume. Nitrogen credits for a previous legume crop often are used to reduce fertilizer N recommendations in combination with other site-specific information (Lory et al., 1995). Figure 10.7 shows that there was a positive N rotation effect associated with corn following wheat or corn following wheat plus and alfalfa. That is, for corn grown in rotation, the economic N rate was smaller and maximum yield was greater than for continuous corn. The N credit estimate based on the different method was 36 kg N ha^{-1} when corn was grown in rotation, as compared to corn grown in monoculture.

Binford et al. (1992) reported that soil tests for evaluating the N status of cornfields show promise as a good tool for improving N management during corn production. Soil NO_3 concentration was a good predictor of final grain yield (Figure 10.8). It explained 76% of the variability in relative yield. The relationship with relative yield is especially important because relative yields provide a useful index of N sufficiency as evaluated by the corn plants. That is, relative yields significantly less than 100% indicate that additions of more fertilizer are likely to increase yields (Binford et al., 1992).

Phosphorus deficiency is a principal yield-limiting factor for annual crop production in acid soils of temperate as well as tropical regions. Economic

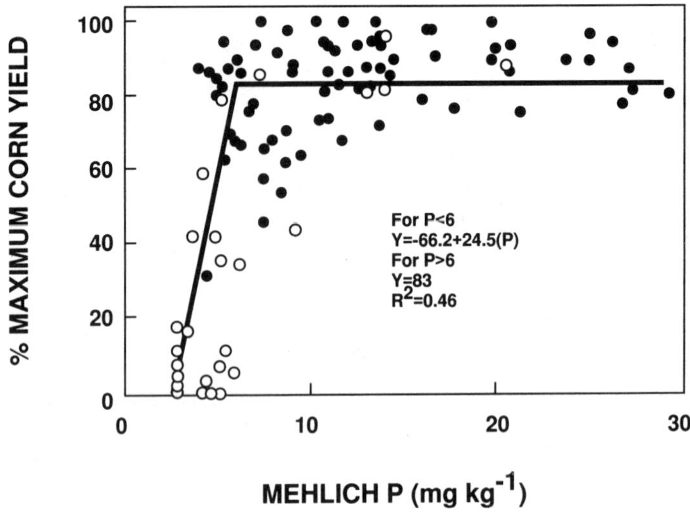

Figure 10.9 Relationship between Mehlich 1 extracting soil phosphorus and yield of corn. (From Smyth and Cravo, 1990.)

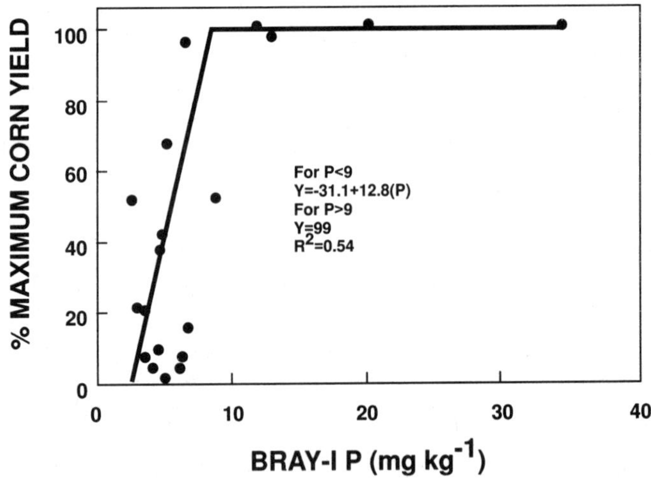

Figure 10.10 Relationship between Bray 1 extracting soil phosphorus and yield of corn. (From Smyth and Cravo, 1990.)

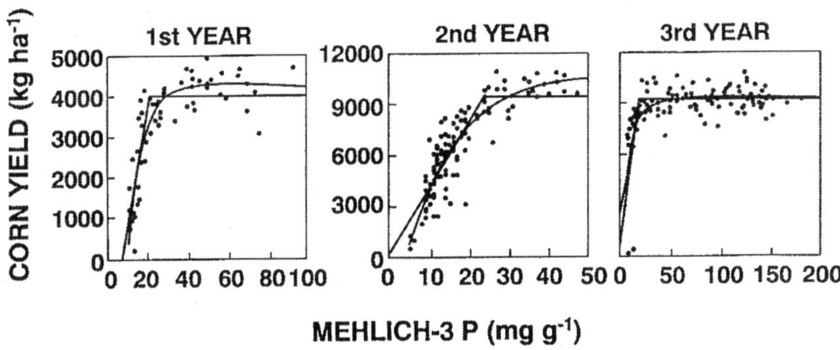

Figure 10.11 Response of corn to Mehlich-3 P with linear-plateau and exponential prediction functions for three corn crops. (From Cox, 1992.)

considerations often require judicious management of P fertilizer inputs to such soils. Fertilizer P recommendations depend on a knowledge of: 1) the existing level of available soil P, 2) the optimum level of soil P for the crop to be grown, and 3) the level of fertilizer which must be added to raise available soil P to the optimum level. Figures 10.9 and 10.10 show the critical P level for corn by Mehlich 1 and Bray 1 extractant in an Oxisol soil of Brazilian Amazon. Maximum corn yield was achieved at about 6 mg P kg^{-1} by Mehlich 1 and 9 mg kg^{-1} by Bray 1 extracting solution. Cox (1992) determined critical P levels for corn in a Typic Umbraquult soil using Mehlich 3 extracting solution (Figure 10.11). He estimated critical P levels by the linear-plateau and exponential models. For the linear-plateau function, the critical P level ranged from 18 to 33 mg kg^{-1}, with an average of 39. This means that soil P test levels for maximum corn yields varied from soil to soil, from extracting solution used, and also the statistical model used for the interpretation of the data.

In addition to the use of adequate P levels, the use of phosphorus-efficient genotypes to increase and/or stabilize crop production has become increasingly attractive in recent years due to the high cost of fertilizer and to pollution problems. Results of a study conducted at the National Rice and Bean Research Center, Goiania, Brazil, showed that corn genotypes differ significantly in their P requirements (Table 10.4).

Phosphorus uptake under three P levels differed significantly among genotypes (Table 10.4). At the low P level, it ranged from 0.87 to 1.63 mg Pot^{-1}; at the medium P level, from 4.93 to 9.79 mg Pot^{-1}; and at the high P level, from 9.18 to 14.05 mg Pot^{-1}. Low and high values of P uptake in shoot

Table 10.4 Phosphorus Uptake in the Shoot and P-Use Efficiency of Nine Corn Genotypes Under Different P Levels

Genotype	P uptake in shoot (mg/pot)				P use efficiency (mg/mg) across medium and high P levels
	Low P	Medium P	High P	Average	
1. AG519	1.11bc	7.41ab	11.29ab	6.60b	140ab
2. BR107	0.99c	5.02b	11.37ab	5.79b	123b
3. BR112	1.63a	9.79a	14.05a	8.48a	145ab
4. BR126	0.87c	7.31ab	9.93ab	6.03b	195ab
5. BR451	1.10bc	7.38ab	12.51ab	6.99ab	128ab
6. C701	0.90c	6.18ab	10.59ab	5.88b	130ab
7. C805	1.20abc	7.94ab	12.99ab	7.38ab	140ab
8. Dina 170	0.80c	4.93b	12.01ab	5.90b	197a
9. Sintetico 6	1.53ab	6.77ab	9.18b	5.82b	153ab
F-test		**	**	*	**
CV %		15	18	14	18

*, ** Significant at 0.05 and 0.01 probability levels, respectively. NS = not significant. Means in the same column followed by the same letter are not significantly different at 0.05 probability level by Tukey's test.

$$\text{P-use efficiency} = \frac{D_{MH}}{P_{MH}} - \frac{D_L}{P_L}$$

where:

D_{MH} = Dry matter yield of root and shoot across medium and high P level
D_L = Dry matter yield of root and shoot at low P level
P_{MH} = P accumulation in root and shoot across medium and high P level
P_L = P accumulation in root and shoot at low P level

are related to dry matter production. Phosphorus use efficiency also differed significantly among genotypes across P levels (Table 10.4). Genotype Diana 170 had the highest P-use efficiency, and genotype BR107 had the lowest. Based on dry matter production (root plus shoot) at the low P level and phosphorus-use efficiency, the genotypes were classified into four groups (Figure 10.12), according to the methodology proposed by Fageria and Baligar (1993). These groups were as follows:

1. *Efficient and responsive*: Genotypes which produced dry matter yield higher than the average of 9 genotypes at the low P level and responded well with the addition of P (average P-use efficiency, higher than the

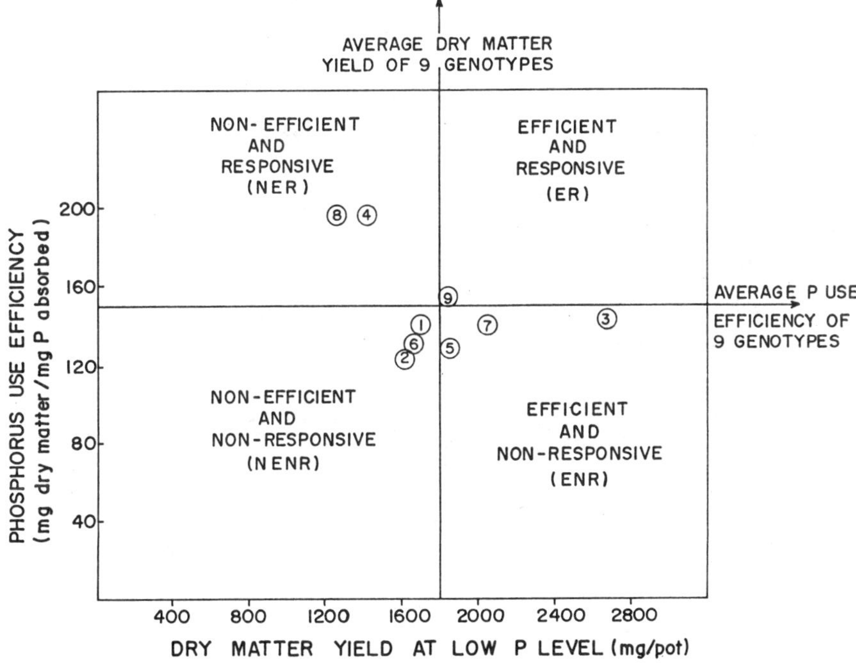

Figure 10.12 Classification of nine corn genotypes for phosphorus use efficiency. Values in the circles refer to genotype numbers, which are numbered in Table 10.4.

average of 9 genotypes, were classified as efficient and responsive. Genotype Sintetico 6 falls into this category.

2. *Efficient and nonresponsive*: In this group are genotypes which produced higher than average dry matter yield, but P-use efficiency was lower than the average of nine genotypes. Genotypes BR112, BR451, and C805 fall into this category.
3. *Nonefficient and nonresponsive*: In this group are genotypes which produced less than average dry matter yield, but P-use efficiency was higher than average. In this group fall genotypes BR126 and Dina 170.
4. *Nonefficient and nonresponsive*: In this group are genotypes which produced lower than average dry matter yield as well as lower than average P-use efficiency. In this group fall the genotypes AG519, BR107, and C701.

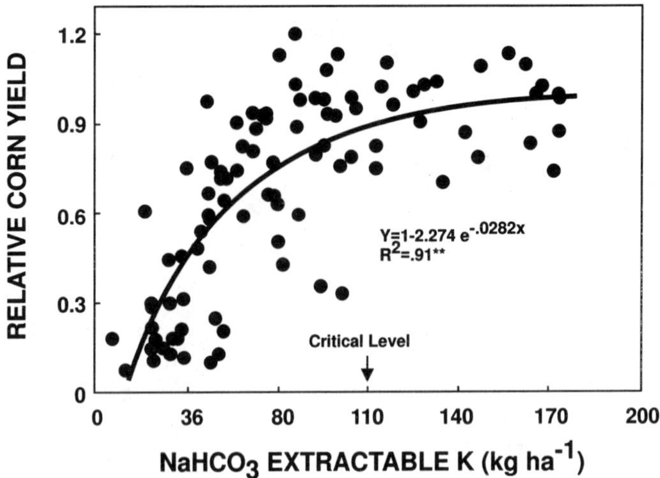

Figure 10.13 Relationship between $NaHCO_3$ extractable K and relative corn yield on the loamy soil. (From Cox and Uribe, 1992.)

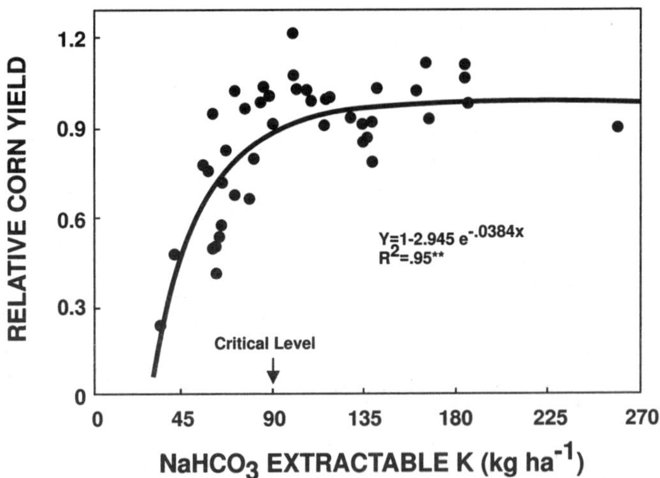

Figure 10.14 Relationship between $NaHCO_3$ extractable K and relative corn yield on the sandy loam soil. (From Cox and Uribe, 1992.)

From a practical point of view, genotypes which fall under the group efficient and responsive are the best ones. These genotypes can produce well under a low P level and respond well with P application. The second group, which can be used under low technology or under a low P level and can produce well, are the efficient and nonresponsive.

Deficiency of K is not as common as nitrogen and phosphorus in corn-producing areas in various parts of the world. However, if this nutrient is extracted in maximum amounts, deficiency can occur in intensive cultivation. Results in Figure 10.13 and 10.14 showed response of corn to extractable K in a humid tropical Ultisol. Critical exchangeable K levels for corn were 110 kg ha^{-1} (55 mg K kg^{-1}) on the loam (Figure 10.13) and 90 kg ha^{-1} (45 mg K kg^{-1}) on the sandy loam Ultisols (Figure 10.14).

A. Nutrient Uptake

Nutrient uptake (concentration × dry matter) is directly related to dry matter production. Hanway (1962b) and Sayre (1955) conducted classical studies on uptake of N, P, and K by corn crops. More recently, Karlen et al. (1988) conducted more detailed studies of uptake and distribution of N, P, K, Ca, Mg, S, B, Cu, Mn, Fe, and Zn in different plant parts as a function of growing degree units (GDU). These authors calculated accumulated growing degree units with the help of the following equation and used them as a time scale for nutrient accumulation.

$$GDU = \frac{T_{\min} + T_{\max}}{2} - 10^{\circ}C$$

where $T_{\min}$ is the minimum daily temperature or 10°C, whichever is larger, and $T_{\max}$ is the maximum daily temperature or 30C°, whichever is smaller. Accumulation of nutrients in different plant parts of corn is presented in Figures 10.4, 10.15, and 10.16. Total shoot N accumulation was approximately 386 kg ha^{-1} in 32 Mg ha^{-1} dry matter at physiological maturity. Maximum N accumulations in lower leaves (leaves below the ear), upper leaves (leaves above the ear), stem and tassel, and ear and shank were approximately 122, 81, 56, and 255 kg ha^{-1}, respectively. Two distinct features of tissue N contents (Figure 10.4) are the gradual decline in vegetative N during reproductive growth stages and the apparent net loss of aerial N shortly after tassel emergence (900 GDU). The former presumably reflects translocation from vegetative plant parts to the developing grain. The latter may result from volatilization losses because there is no major sink for N-rich compounds between tasseling and the beginning of grain filling. Another mechanism restricting N uptake at this time may be feedback inhibition (Karlen et al., 1988).

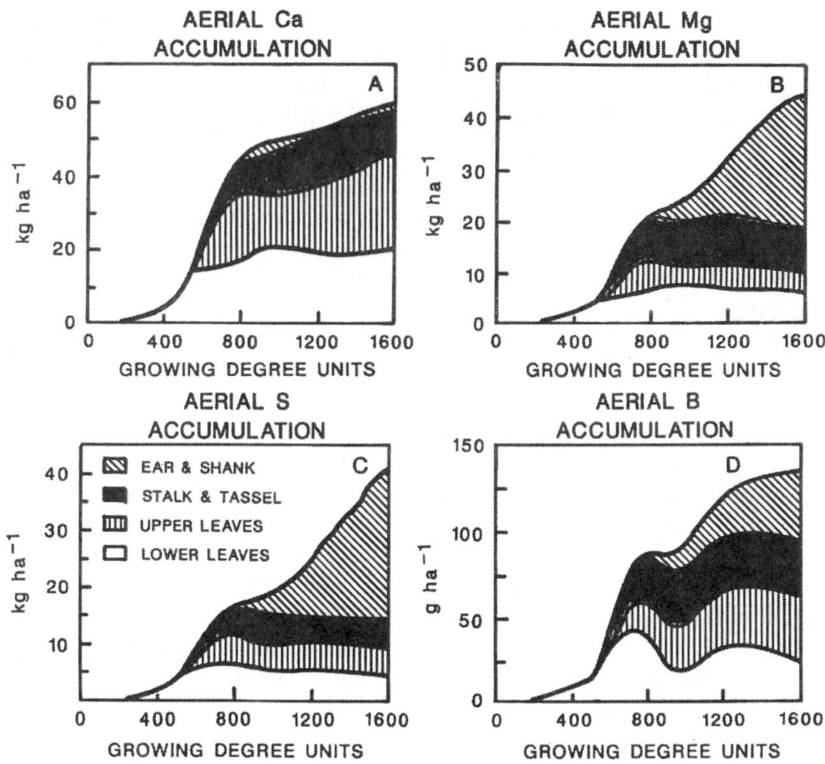

Figure 10.15 Accumulation of Ca, Mg, S, and B in different parts of corn plant for corn yielding 19.3 Mg ha^{-1} of grain. (From Karlen et al., 1988.)

At physiological maturity, total shoot P accumulation was approximately 70 kg ha^{-1}; peak accumulations in lower leaves, upper leaves, stem and tassel, and ear and shank fractions were approximately 13, 10, 9, and 59 kg ha^{-1}, respectively (Figure 10.4). Phosphorus accumulated steadily until maturity, and during grain fill there was considerable translocation of P from vegetative parts to the grain.

With regard to K, 86% was accumulated by silking, and only 19% of the K was contained in the ear and shank portion. Thus, most of the K absorbed remained in the stover and was recycled through crop residues for future crop production.

Accumulation of Ca, Mg, and S continued throughout the growing season. There was essentially no translocation of Ca to the ear during grain fill, but a small amount of Mg was translocated from leaf and stalk fractions to the

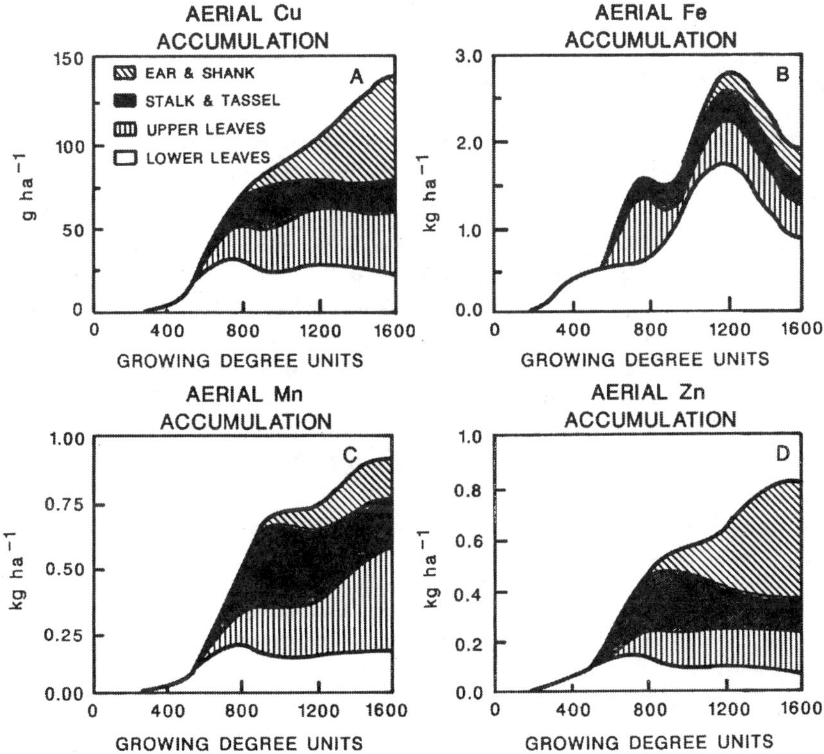

Figure 10.16 Accumulation of Cu, Fe, Mn, and Zn in different parts of corn plant for corn yielding 19.3 Mg ha^{-1} of grain. (From Karlen et al., 1988.)

ear and shank during reproductive growth. In the case of S, approximately 2 kg ha^{-1} was translocated from the lower leaves during early grain fill, but there was essentially no translocation of S from upper leaves or the stalk and tassel fraction.

Plant Boron (B) content declined during early grain fill in all plant parts; this suggests that uptake slowed, since B does not readily translocate from older tissues to meristematic regions. Increased accumulation during the latter stages of grain fill and nearly uniform distribution at physiological maturity suggest that fertilization programs must provide adequate amounts of B throughout the season (Karlen et al., 1988). There was very little translocation of Cu among the plant fractions, and at physiological maturity the distribution was 16, 26, 14, and 44% in the lower leaves, upper leaves, stem and tassel, and ear and shank, respectively.

Accumulation of Fe showed two distinct peaks, one near silking and the other approximately halfway through the grain-fill period (Figure 10.16). Iron accumulation was 45, 21, 16, and 18% in lower leaves, upper leaves, stem and tassel, and ear and shank fractions, respectively.

Total manganese accumulation was approximately 0.9 kg ha^{-1}. Plant Mn content increased throughout the growing season, although more than 70% was accumulated by silking. Approximately 18, 44, 21, and 17% was located in the lower leaves, upper leaves, stem and tassel, and ear and shank fractions, respectively, at physiological maturity.

Zinc accumulation (Figure 10.16) totaled approximately 0.8 kg ha^{-1} at physiological maturity, and distribution was 8, 20, 16, and 56% in lower leaves, upper leaves, stem and tassel, and ear and shank, respectively.

In conclusion, the results of Karlen et al. (1988) showed that the total accumulation at physiological maturity was approximately 31800, 386, 70, 370, 59, 44, 40, 0.13, 0.14, 1.9, 0.9, and 0.8 kg ha^{-1} for dry matter, N, P, K, Ca, Mg, S, B, Cu, Fe, Mn, and Zn, respectively. Amounts of accumulation measured in this study can provide general guidelines for very high corn yields when more economic practices are used.

B. Nutrient Concentration

As corn plants age, the ratio of cytoplasm to structural tissues decreases. This normally causes a gradual decrease in shoot N and P concentrations (Hanway, 1962b). In addition, nutrient deficiencies cause concentrations to decrease. Jones (1983) summarized the effects of both growth stage (Hanway, 1963) and N and P deficiencies on maize shoot N and P concentrations. These data suggest that adequate whole-shoot N concentrations decline from about 4 to 5% in the seedling stage to slightly over 1% at maturity (Figure 10.17). However, N-deficient shoots may have N concentrations as low as 0.4–0.5% (Figure 10.18). Similarly, adequate shoot P concentrations decline from about 0.6% in the seedling to about 0.2% after silking (Figures 10.19 and 10.20).

The grain is an important sink for N. At harvest, grain N concentration ranges from about 1–2%, while total shoot N concentration varies from about 0.6 to 1.6% (Figure 10.21). Thus, while corn grain normally accounts for 40–60% of shoot dry matter, it contains about 50–80% of the nitrogen in the shoot.

Relative yield increases as the N concentration increases in the plant (Figure 10.22). Yield tended to plateau at the higher N concentration in the plant tissue. Figure 10.23 shows a good relationship between soil NO_3 concentration and relative concentration of N in the plant tissue. This relationship suggests that corn plants tend to have a maximum concentration of N that depends upon soil and environmental conditions. Maximum plant N concentration occurred at about 20 mg N kg^{-1} of soil, and this value may be considered as a critical level for corn (Binford et al., 1992).

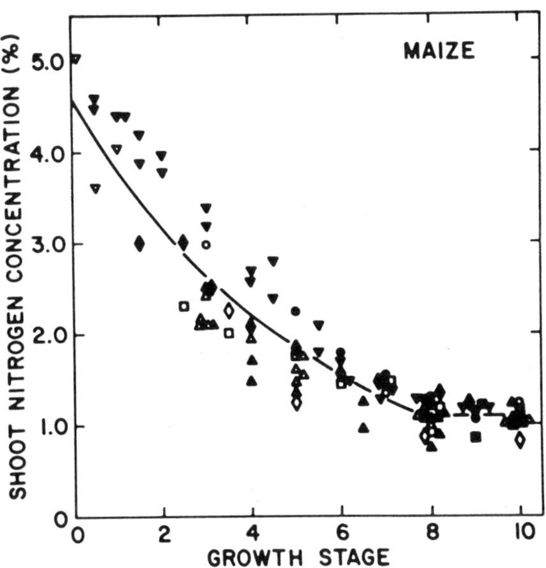

Figure 10.17 Relationship between growth stage and corn shoot N concentration from experiments with near-optimum N nutrition. (From Jones, 1983.)

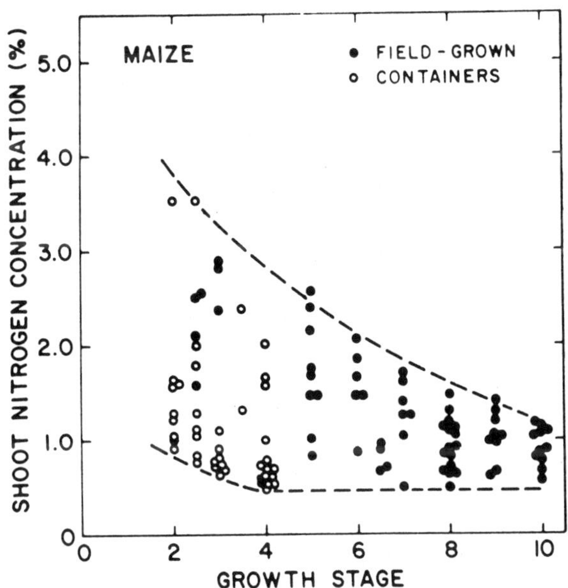

Figure 10.18 Relationship between growth stage and corn shoot N concentration from experiments with varying levels of N nutrition. (From Jones, 1983.)

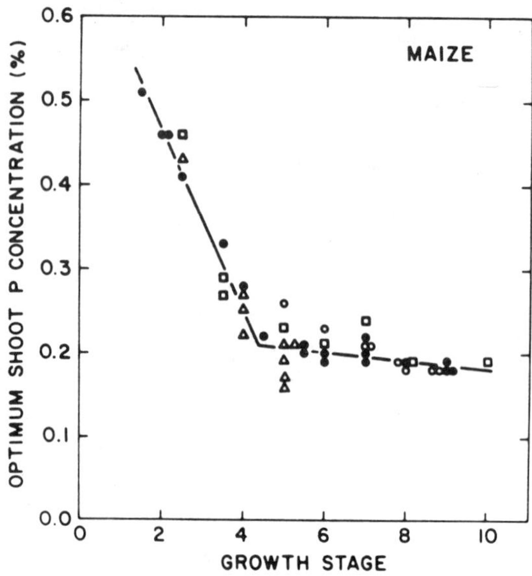

Figure 10.19 Relationship between growth stage and corn shoot P concentration in treatments of field experiments with near-optimum P nutrition. (From Jones, 1983.)

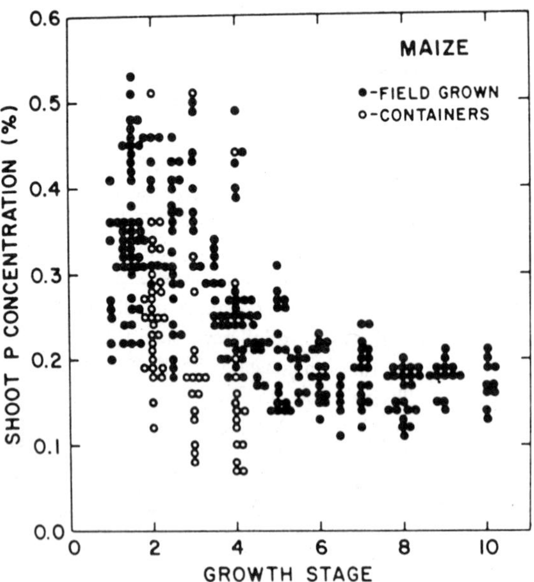

Figure 10.20 Relationship between growth stage and corn shoot P concentration from treatments with varying levels of P. (From Jones, 1983.)

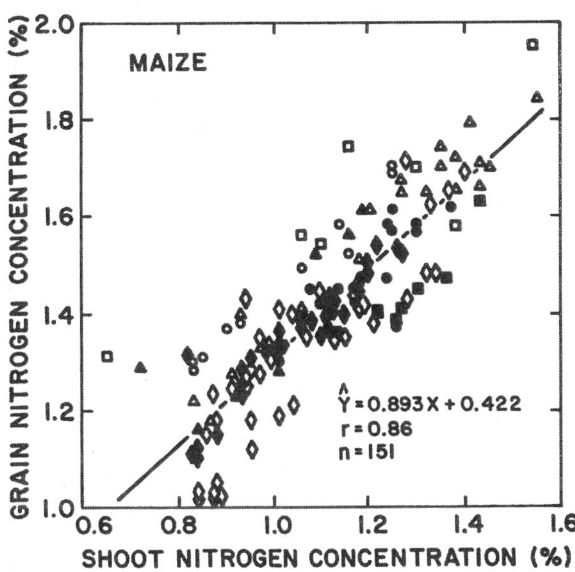

Figure 10.21 Relationship between corn shoot N concentration and grain N concentration. (From Jones, 1983.)

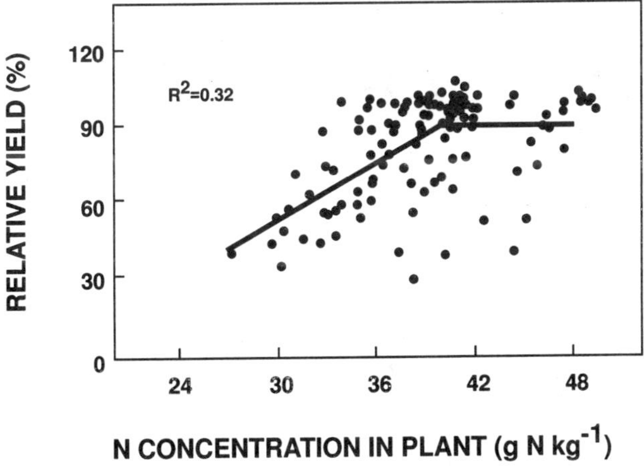

Figure 10.22 Relationship between relative yield and N concentration in corn plant tissue. (From Binford et al., 1992.)

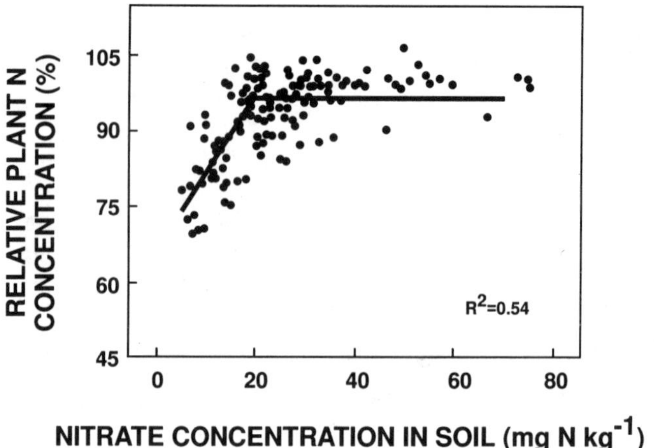

Figure 10.23 Relationship between nitrate concentration in soil and relative plant N concentration in the corn tissue. Relative concentrations are N concentrations in plants expressed as percentages of the mean N concentrations in plants from plots receiving the two highest rates of N fertilizer. (From Binford et al., 1992.)

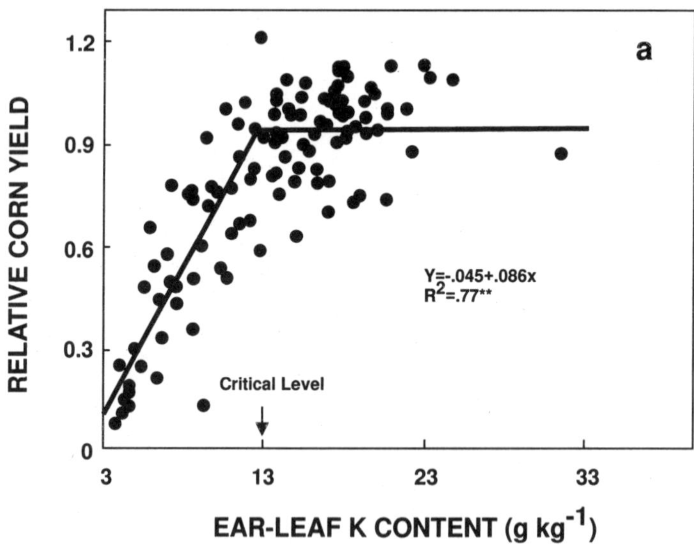

Figure 10.24 Relationship between K concentration of the corn leaf at flowering and relative yield. Equations apply up to the critical level. (From Cox and Uribe, 1992.)

Similarly, a relationship between K concentration in ear leaf of corn and relative grain yield is shown in Figure 10.24. The critical levels of K in plant tissue at flowering was 13 g kg^{-1}.

Plant analysis has been used in various ways to diagnose plant nutrient adequacy and estimate fertilizer needs. The nutrient concentration of a plant is affected not only by the supply of that nutrient but also by supplies of other nutrients and by environmental factors. Table 10.5 provides values of adequate nutrient concentrations in corn plants at different growth stages. These values can be used as a guideline for identifying nutrient deficiencies or sufficiency levels in corn plants.

VI. SUMMARY

Corn is the third most important cereal worldwide, after wheat and rice. It is a warm-season temperate or tropical crop grown for grain, fodder, and raw materials for industrial processes. Corn's main advantages include its high yields per unit of labor and per unit of land area. It is a compact, easily transportable source of nutrition that is relatively tolerant of pests and is easily stored. Corn is grown in a wide range of climates, from the tropics to 50° latitude and from sea level to 4000 m elevation. It tolerates a wide range of soil conditions and is grown under both irrigated and dryland conditions.

A crop with the C_4 photosynthetic pathway, corn has high potential growth rates. Maximum grain yields are over 12 Mg ha^{-1} in most parts of the world where corn is grown, and yields over 20 Mg ha^{-1} have been recorded under the most favorable conditions. Yields of the best farmers can approach experimental yields in developed countries, but average yields are usually much lower due to climatic, nutrient, and biological stresses.

Corn is often grown in rotation with soybean, cotton, or other dicotyledonous crops to facilitate chemical weed control and because of their complementary labor and machinery requirements. In temperate climates it is normally planted earlier and harvested later than soybeans. In many tropical areas, it is planted in association with grain legumes like common bean, and it provides physical support for climbing cultivars. It is less drought tolerant than grain sorghum and pearl millet, but it requires less water than upland rice.

Because of its high potential grain production, corn nutrient requirements can be great. Approximately 16 kg N, 3 kg P, and 3 kg K are removed from the field in each metric ton of grain. Both plant and soil analyses are used to diagnose and correct nutrient deficiencies and toxicities. Because of the large amounts of nutrients removed in the grain, fertilizers and manures are almost always needed to maintain soil fertility.

Table 10.5 Adequate Levels of Nutrients in Corn Plants[a]

Nutrient	Growth stage	Plant part	Adequate concentration g kg^{-1}
N	30 to 45 DAE	Whole tops	35.0–50.0
	Before tasseling	LB below whorl	30.0–35.0
	Silking	BOAC	> 32.0
	Silking	Ear LB	28.0–35.0
P	35 to 45 DAE	Whole tops	4.0–8.0
	Before tasseling	LB below	2.5–4.5
	Silking	BOBC	> 2.9
	Silking	Ear LB	2.5–4.0
K	30 to 45 DAE	Whole tops	30.0–50.0
	Before tasseling	LB below	20.0–25.0
	Silking	BOBC	> 18.0
	Silking	Ear LB	17.0–30.0
Ca	30 to 45 DAE	Whole tops	9.0–16.0
	Before tasseling	LB below	2.5–5.0
	Silking	Ear LB	2.1–5.0
Mg	30 to 45 DAE	Whole tops	3.0–8.0
	Before tasseling	LB below	1.3–3.0
	Silking	Ear LB	2.1–4.0
S	30 to 45 DAE	Whole tops	2.0–3.0
	Silking	Ear LB	1.0–2.4

Table 10.5 (Continued)

Nutrient	Growth stage	Plant part	Adequate concentration
			mg kg^{-1}
Zn	30 to 45 DAE	Whole tops	20–50
	Before tasseling	LB below whorl	15–60
	Silking	Ear LB	20–70
Cu	30 to 45 DAE	Whole tops	7–20
	Before tasseling	LB below whorl	3–15
	Silking	Ear LB	6–20
Mn	30 to 45 DAE	Whole tops	50–160
	Before tasseling	LB below	20–300
	Silking	Ear LB	20–150
Fe	30 to 45 DAE	Whole tops	50–300
	Before tasseling	LB below	30–200
	Silking	Ear LB	21–250
B	30 to 45 DAE	Whole tops	7–25
	Before tasseling	LB below whorl	4–25
	Silking	Ear LB	6–20
Mo	< 30 cm tall	Whole tops	0.1–10
	Before tasseling	LB below whorl	0.1–3
	Silking	Ear LB	> 0.2

[a] DAE, days after emergence; LB, leaf blade; BOAC, blade opposite and above cob; and BOBC, blade opposite and below cob.

Sources: Compiled from Escano et al., 1981; Jones, 1967; Melsted et al., 1969; Neubert et al., 1969; and Reuter, 1986.

REFERENCES

Anderson, E. A. 1987. Corn root growth and distribution as influenced by tillage and nitrogen fertilization. Agron. J. 79: 544–549.

Barber, S. A. and R. A. Olson. 1968. Fertilizer use on corn, pp. 163–168. *In*: R. C. Dinauer (ed.). Changing patterns in fertilizer use. Soil Sci. Soc. Am., Madison, Wisconsin.

Binford, G. D., A. M. Blackmer, and M. E. Cerrato. 1992. Nitrogen concentration of young corn plants as an indicator of nitrogen availability. Agron. J. 84: 219–223.

Cerrato, M. E. and A. M. Blackmer. 1990. Comparison for describing corn yield response to nitrogen fertilizer. Agron. J. 82: 138–143.

Chang, J. H. 1981. Corn yield in relation to photoperiod, night temperature, and solar radiation. Agric. Meteorol. 24: 253–262.

Cox, F. R. 1992. Range in soil phosphorus critical levels with time. Soil Sci. Soc. Am. J. 56: 1504–1509.

Cox, F. R. and E. Uribe. 1992. Potassium in two humid tropical Ultisols under a corn and soybean cropping system. I. Management. Agron. J. 84: 480–484.

Daughtry, C. S. T., J. C. Cochran, and S. E. Hollinger. 1984. Estimating silking and maturity dates of corn for large areas. Agron. J. 76: 415–420.

Denmead, O. T. and R. H. Shaw. 1960. The effects of soil moisture stress at different stages on development and yield of corn. Agron. J. 52: 272–274.

Duncan, W. G. 1975. Maize, pp. 23–50. *In*: L. T. Evans (ed.). Crop physiology. Cambridge University Press, London.

Eagles, H. A. and J. E. Lothrop. 1994. Highland maize from central Mexico—its origin, characteristics, and use in breeding programs. Crop Sci. 34: 11–19.

Eck, H. V. 1984. Irrigated corn yield response to nitrogen and water. Agron. J. 76: 421–428.

Eik, K. and J. J. Hanway. 1966. Leaf area in relation to yield of corn. Agron. J. 58: 16–18.

Escano, C. R., C. A. Jones, and G. Uehara. 1981. Nutrient diagnosis in corn grown on Hydric Dystrandepts: I. Optimum tissue nutrient concentrations. Soil Sci. Soc. Am. J. 45: 1135–1139.

Fageria, N. K. and V. C. Baligar. 1993. Screening crop genotypes for mineral stresses, pp. 152–159. *In*: Proc. workshop on adaptation of plants to soil stresses. University of Nebraska, Lincoln, INTSORMIL Publ. No. 94–2.

Ferguson, R. B., C. A. Shapiro, G. W. Hergert, W. L. Kranz, N. L. Locke, and D. H. Krull. 1991. Nitrogen and irrigation management practices to minimize nitrate leaching from irrigated corn. J. Prod. Agric. 4: 186–192.

Fischer, K. S. and A. F. E. Palmer. 1983. Maize, pp. 155–180. *In*: IRRI (ed.). Potential productivity of field crops under different environments. IRRI, Los Banos, Philippines.

Fischer, K. S. and A. F. E. Palmer. 1984. Tropical maize, pp. 213–248. *In*: P. R. Goldsworthy and N. M. Fisher (eds.). The physiology of tropical field crops. Wiley, New York.

Fountain, M. O. and A. R. Hallauer. 1996. Genetic variation within maize breeding populations. Crop Sci. 36: 26–32.

Frey, N. M. 1981. Dry matter accumulation in kernels of maize. Crop Sci. 21: 118–122.

Goldsworthy, P. R. 1974. Maize physiology. *In*: Proc. worldwide maize improvement in the 70's and the role of CIMMYT. CIMMYT, Mexico.

Goodman, M. M. 1988. The history and evaluation of maize. CRC Critical Rev. Plant Sci. 7: 197–220.

Hanway, J. J. 1962a. Corn growth and composition in relation to soil fertility: I. Growth of different plant parts and relation between leaf weight and grain yield. Agron. J. 54: 145–148.

Hanway, J. J. 1962b. Corn growth and composition in relation to soil fertility: II. Uptake of N, P, and K and their distribution in different plant parts during the growing season. Agron. J. 54: 217–222.

Hanway, J. J. 1963. Growth stages of corn (*Zea mays* L.). Agron. J. 55: 487–492.

Hardacre, A. K. and H. A. Eagles. 1980. Comparisons among populations of maize at 13°C. Crop Sci. 20: 780–784.

Hatch, M. D. and C. R. Slack. 1970. Photosynthetic CO_2 fixation pathways. Annu. Rev. Plant Physiol. 21: 141–162.

Herrero, M. P. and R. R. Johnson. 1981. Drought stress and its effects on maize reproductive systems. Crop Sci. 21: 105–110.

Hoyt, P. and R. Bradfield. 1962. Effect of varying leaf area by partial defoliation and plant density on dry matter production in corn. Agron. J. 54: 523–525.

Hunter, R. B., M. Tollenaar, and C. M. Breuer. 1977. Effect of photoperiod and temperature on vegetative and reproductive growth of a maize (*Zea mays* L.) hybrid. Can. J. Plant Sci. 57: 1127–1133.

Jones, C. A. 1983. A survey of the variability in tissue nitrogen and phosphorus concentrations in maize and grain sorghum. Field Crops Res. 6: 133–147.

Jones, C. A. 1985. C_4 grasses and cereals: Growth, development, and stress response. Wiley, New York.

Jones, J. B., Jr. 1967. Interpretation of plant analysis for several crops, pp. 40–58. *In*: L. M. Walsh and J. B. Beaton (eds.). Soil testing and plant analysis, Part 2. SSSA Special Publ., Ser. No. 2. Soil Sci. Soc. Am., Madison, Wisconsin.

Jones, R. J. and S. R. Simmons. 1983. Effect of altered source-sink ratio on growth of maize kernels. Crop Sci. 23: 129–134.

Jones, R. J., B. G. Gengenback, and V. B. Cardwell. 1981. Temperature effects on in vitro kernel development of maize. Agron. J. 21: 761–766.

Karlen, D. L., R. L. Flannery, and E. J. Sadler. 1988. Aerial accumulation and partitioning of nutrients by corn. Agron. J. 80: 232–242.

Kato, Y. A. T. 1984. Chromosome morphology and the origin of maize and its races. Evol. Biol. 17: 219–255.

Keeney, D. R. 1987. Nitrate in ground water—agricultural contribution and control, pp. 329–351. *In*: Proc. conf. on agricultural impacts on ground water, Omaha, Nebraska, 11–13 Aug. 1986. National Water Well Assoc., Dublin, Ohio.

Lafitte, H. R. and G. O. Edmeades. 1994. Improvement for tolerance to low soil nitrogen in tropical maize. I. Selection criteria. Field Crops Res. 39: 1–14.

Larson, W. E. and J. J. Hanway. 1977. Corn production, pp. 625–669. *In*: G. F. Sprague (ed.). Corn and corn improvement. Monogr. 18, Am. Soc. Agron., Madison, Wisconsin.

Legg, J. O. and J. J. Meisinger. 1982. Soil nitrogen budgets, pp. 503–566. *In*: F. J. Stevenson (ed.) Nitrogen in agricultural soils. Agron. Monogr. 22, Am. Soc. Agron., Madison, Wisconsin.

Lorens, G. F., J. M. Bennett, and L. B. Loggale. 1987. Differences in drought resistance between two corn hybrids. II. Component analysis and growth rates. Agron. J. 79: 808–813.

Lory, J. A., M. P. Russelle, and T. A. Peterson. 1995. A comparison of two nitrogen credit methods: Traditional versus different. Agron. J. 87: 648–651.

Melsted, S. W., H. L. Motto, and T. R. Peck. 1969. Critical plant nutrition composition values useful in interpreting plant analysis data. Agron. J. 61: 17–20.

Mengel, D. B. and S. A. Barber. 1974. Development and distribution of the corn root system under field conditions. Agron. J. 66: 341–344.

Mock, J. J. and R. B. Pearce. 1975. An ideotype of maize. Euphytica 24: 613–623.

Musick, J. T. and D. A. Dusek. 1980. Irrigated corn yield response to water. Trans. ASAE 23: 92–103.

Neubert, P., W. Wrazidlo, N. P. Vielmeyer, I. Hundt, F. Gullmick, and W. Bergman. 1969. Tabellen zur Pflanzenanelze—erste orientierrende Ubersicht. Institut für Pflanzenernährung Jena, Berlin.

Oberle, S. L. and D. R. Keeney. 1990a. Soil type, precipitation, and fertilizer N effects on corn yields. J. Prod. Agric. 3: 522–527.

Oberle, S. L. and D. R. Keeney. 1990b. Factors influencing corn fertilizer N requirements in the northern U.S. corn belt. J. Prod. Agric. 3: 527–534.

Oliveira, F. A., J. J. S. Silva, and T. G. S. Campos. 1993. Evapotranspiration and root development of irrigated corn. Pesq. Agropec. Bras., Brasilia 28: 1407–1415.

Onken, A. B., R. L. Matheson, and D. M. Nesmith. 1985. Fertilizer nitrogen and residual nitrate-nitrogen effects on irrigated corn yield. Soil Sci. Soc. Am. J. 49: 134–139.

Otegui, M. E., F. H. Andrade, and E. E. Suero. 1995. Growth, water use, and kernel abortion of maize suspected to drought at silking. Field Crops Res. 40: 87–94.

Purseglove, J. W. 1985. Tropical crops: Monocotyledons. Longman, New York.

Reuter, D. J. 1986. Temperate and sub-tropical crops, pp. 38–99. *In*: D. J. Reuter and J. B. Robinson (eds.). Plant analysis: An interpretation manual. Inkata Press, Melbourne.

Riley, G. J. P. 1981. Effect of high temperature on the germination of maize (*Zea mays* L.). Planta 151: 68–74.

Russell, W. A. 1991. Genetic improvement of maize yields. Adv. Agron. 46: 245–298.

Sayre, J. D. 1955. Mineral nutrition of corn, pp. 296–314. *In*: G. F. Sprague (ed.). Corn and corn improvement. Academic Press, New York.

Schlegel, A. J. and J. L. Havlin. 1995. Corn response to long term nitrogen and phosphorus fertilization. J. Prod. Agric. 8: 181–185.

Shaw, R. H. 1977. Climatic requirement, pp. 531–523. *In*: G. F. Sprague (ed.). Corn and corn improvement. Monogr. 18, Am. Soc. Agron., Madison, Wisconsin.

Smyth, T. J. and M. S. Cravo. 1990. Critical phosphorous levels for corn and cowpea in a Brazilian Amazon Oxisol. Agron. J. 82: 309–312.

Stevens, E. J., S. J. Stevens, A. D. Flowerday, C. O. Gardner, and K. M. Eskridge. 1986. Developmental morphology of dent corn and popcorn with respect to growth staging and crop growth models. Agron. J. 78: 867–874.

Thornton, P. K., G. Hoogenboom, P. W. Wilkens, and W. T. Bowen. 1995. A computer program to analyze multiple-season crop model outputs. Agron. J. 87: 131–136.

Tollenaar, M. and T. B. Daynard. 1978. Kernel growth at two positions on the ear of maize (*Zea mays*). Can. J. Plant Sci. 58: 189–197.

Tsuji, G. Y. 1985. Agroenvironments for maize production, pp. 74–91. *In*: J. A. Silva (ed.). Soil-bases agrotechnology transfer. Benchmark Soils Project, Department of Agronomy and Soil Science, Hawaii Institute of Tropical Agriculture and Human Resources. College of Tropical Agriculture and Human Resources, University of Hawaii.

Walters, D. T. and G. L. Malzer. 1990. Nitrogen management and nitrification inhibitor effects on nitrogen-15 urea: I. Yield and fertilizer use efficiency. Soil Sci. Soc. Am. J. 54: 115–122.

Wellhausen, E. J., L. M. Roberts, and E. Harnandez. 1952. Races of maize in Mexico. Bussey Inst., Harvard University, Cambridge, Massachusetts.

Wilson, J. H., M. S. J. Cloves, and J. C. S. Allison. 1973. Growth and yield of maize at different altitudes in Rhodesia. Ann. Appl. Biol. 73: 77–84.

Yamaguchi, J. 1974. Varietal traits limiting the grain yield of tropical maize. IV. Plant traits and productivity of tropical varieties. Soil Sci. Plant Nutr. 20: 287–304.

11

Sorghum

I. INTRODUCTION

Sorghum *(Sorghum bicolor L.* Moench) is ranked fifth among cereals behind wheat, rice, corn, and barley in worldwide production and area planted (FAO, 1983). It is the major cereal of rainfed agriculture in the semiarid tropics (SAT). Over 55% of the global sorghum production is in the SAT; and of the total SAT production, Asia and Africa contribute about 65%, of which 34% is harvested in India (Sahrawat et al., 1996). Cultivated sorghum originated in northeast Africa, where the greatest diversity of types exists. The most likely area is that now occupied by Ethiopia and part of Sudan, from which sorghum spread to West Africa (Doggett, 1970). There is evidence of sorghum in Assyria by 700 B.C. and in India and Europe by A.D. 1 (Eastin, 1983). Cultivated sorghums were first introduced to America and Australia about 100 years ago. Domestication and cultivation of sorghum has spread throughout the world, and today it is grown on about 48 million hectares (FAO, 1982). The major sorghum production areas today include the Great Plains of North America, sub-Saharan Africa, northeastern China, the Deccan Plateau of central India, Argentina, Nigeria, Egypt, and Mexico. The USSR, France, and Spain are the leading sorghum-producing countries in Europe.

Sorghum is the basic cereal food in parts of Asia and Africa, while in the United States and Europe it serves mainly as feed for poultry and livestock. Sorghum stems and foliage are often used as animal fodder, and in some areas the stems are used as building material and fuel. Some sorghums have sweet, juicy stems that contain up to 10% sucrose and are chewed or used to produce syrup. Sorghum is also widely used for brewing beer, particularly in Africa, and it is among the most widely adapted of the warm-season cereals with potential for biomass and fuel production. High-energy sorghum consists of hybrids of grain and sweet sorghum types that are currently being developed for both grain and biomass production (Hons et al., 1986). They give slightly lower grain yields than conventional grain sorghums but produce large amounts of stover with high carbohydrate concentrations.

In comparing plant biomass systems for energy production, sweet sorghum is a leading contender because of its C_4 plant characteristics, with a high photosynthetic rate, large biomass yield, high percentage of easily fermentable sugars and combustible organics (fiber), tolerance to water stress, and low fertilizer requirements (Shih et al., 1981). Sweet sorghum has great potential as a field crop for sugar and ethanol production because it is adapted to a wide array of climatic and edaphic environments, whereas sugarcane can be grown only in tropical and subtropical climates. In addition, fiber from the leaves and stalks can provide the fuel required to process the extracted juice. The net energy ratio for sweet sorghum has been estimated by Sheehan et al. (1978) to exceed 1.0; that is more energy can be recovered as ethanol than is used to grow and process the crop.

II. CLIMATE AND SOIL REQUIREMENTS

Sorghum is adapted to tropical, subtropical, and temperate climates. The optimum temperature for photosynthesis is 30–36°C (Vong and Murata, 1977). It does not tolerate frost, and most sorghum production is concentrated between latitudes 40°N and 40°S (Purseglove, 1985). Wardlaw and Bagnall (1981) found a decrease in the flow of ^{14}C-labeled photosynthates through the phloem in sorghum leaf tissue exposed to air temperatures below 10°C. Lang and Minchin (1986) reported reduced translocation in sorghum leaves after exposure to air temperatures of 3°C. These reductions in translocation were attributed to changes in membrane properties that altered the symplast/apoplast ratio of solute movement through leaf tissues. Kendal et al. (1985) found a consistent relationship between cell membrane alterations and cellular stress after freezing in wheat crown tissue.

Air temperatures of –2°C or lower reduced test weight of grain from field-grown plants, whereas only exposure to –4°C reduced test weight of grain from greenhouse-grown plants. Exposure to –2°C for 4 h reduced caryopsis weights 81% at 200, 57% at 300, 25% at 450, and 3% at 600 growing degree days (GDD, base temperature 5.7°C) after anthesis (maturity ≈ 850 GDD) (Staggenborg and Vanderlip, 1996). In temperate climates sorghum is often planted in rotation with other crops and tolerates somewhat drier conditions. It also tolerates temporary waterlogging and can be grown on cracking clay soils with poor internal drainage.

Time from emergence to anthesis is affected by both photoperiod and temperature. Sorghum is a typical short-day plant; its sensitivity to photoperiod is controlled by four genes (Quinby, 1973). One of the most important achievements of grain sorghum breeders has been to reduce photoperiod sensitivity in cultivars and thereby expand adaptation to temperate regions with long days during the growing season (Jones, 1985).

Sorghum is a drought-resistant crop, but cultivars differ in their reactions to drought. Drought resistance is related to morphological and physiological properties, including: 1) slow shoot growth rate until the root system is well developed, 2) great root weight and volume and high root/shoot ratios in resistant cultivars (Nour and Weibel, 1978), 3) a larger adventitious root system and lower leaf area than corn, 4) ability to reduce leaf osmotic potential and maintained turgor during stress (Ackerson et al., 1980), 5) the ability to maintain relatively high leaf water potential under conditions of increasing soil moisture stress (Blum, 1974a,b), and 6) the ability to produce large amounts of epicuticular wax and roll leaves in time of drought to reduce water loss (Blum, 1975).

Garrity et al. (1984) studied the stomatal behavior across growth stages of grain sorghum and found that stomatal resistance was sensitive to small reductions in leaf water potential during the vegetative period. During the reproductive stage, stomates became nearly insensitive to leaf water potential in plants irrigated weekly. Ackerson et al. (1980) observed that stomatal response to increasing water stress was altered after flowering in some sorghum hybrids. They suggested that sorghum regulates water loss by reducing evapotranspiration through increases of stomatal resistance during early periods of growth and that it has the ability to adapt physiologically to water stress through osmotic adjustment during latter growth stages. Although sorghum is able to resist drought, the crop responds well to plentiful water during booting and heading, the growth stages most sensitive to drought (Salter and Goode, 1967).

Sorghum can grow on a wide range of soils. It is better adapted than pearl millet to the deep cracking clays and the black cotton soils of the tropics. At the other extreme, sorghum can be productive on light, sandy soils and can grow with a wide range of soil pH from 5.0 to 8.5 (Doggett, 1970).

Grain sorghum is considered to be moderately tolerant to salinity (Lall and Sakhare, 1970; Maliwal, 1967). Bresler et al. (1982) reported that a 10% reduction in grain yield occurs when the electrical conductivity of the saturation extract is 4.8 dS m^{-1}, and 12 dS m^{-1} causes a 50% reduction. In more recent work, however, Francois et al. (1984) reported that relative grain yields of two cultivars, Double TX and NK-265, were unaffected up to a soil salinity of 6.8 dS m^{-1}. Each unit increase in salinity above 6.8 m^{-1} reduced yield by 10%. Genotypic variation in sorghum salinity tolerance has been reported (Heilman, 1973; Ratandilok, 1978; Taylor et al., 1975) and, as with most other cereals, germination and vegetative growth are less sensitive than grain yield (Francois et al., 1984). In comparison with other cereals, grain sorghum is more sensitive to salinity than barley but less sensitive than maize (Bresler et al., 1982).

III. GROWTH AND DEVELOPMENT

To manage the sorghum crop for maximum productivity, it is important to understand how the plant grows and develops. A brief description of structure and growth and development of the sorghum plant is given in this section. For more detail, readers may refer to Freeman (1970), Doggett (1970), Vanderlip (1972), Vanderlip and Reeves (1972), and Peacock and Wilson (1984).

A. Germination

Germination of the seed is the starting point of growth and development for annual crops. It is defined as the emergence of the radicle from the seed coat. Germination includes uptake of water, called imbibition, mobilization of stored food reserves within the seed, and resumption of growth and development of the embryo to form the shoot and root structures of the seedling (Fisher, 1984). Adequate moisture and suitable temperature are the two environmental requirements for germination. Seeds of tropical crops such as sorghum may not germinate satisfactorily at temperatures below 20°C but will generally germinate well at temperatures as high as 40°C. The optimum range of soil temperature for sorghum seed germination is 21–35°C (Kanemasu et al., 1975; Aisien and Ghosh, 1978). The lethal temperature for germination is from 40 to 48°C (Kailasanathan et al., 1976), and there is genetic variation in germination at high temperatures (Wilson et al., 1982).

Fawusi and Agboola (1980) reported that optimum germination of sorghum seeds occurs between 25 and 50% of field capacity in a sandy loam soil. This ability to germinate at relatively low soil water contents is consistent with results of Evans and Stickler (1961), who found near-optimum germination in mannitol solutions at osmotic potentials as low as –0.5 MPa.

Grain sorghum may exhibit seed dormancy for the first month after harvest under some conditions. Goodsell (1957) recorded that seeds are often dormant soon after harvest, especially when harvested early and dried rapidly. Scarification, immersion for 1 minute in water at 24°C, or immersion for 4 minutes at 21°C, was effective in breaking dormancy (Doggett, 1970). Genetic differences in seed dormancy have been reported (Gritton and Atkins, 1963; Parvatikar et al., 1975).

B. Roots

Since roots are the plant organs responsible for water and nutrient accumulation, it is important to understand their structure and development. Like other grasses, grain sorghum has two distinct types of root systems. The seminal roots develop from the embryo below the scutellar node, while the adventitious roots are produced from the lower stem node or crown of the plant near the soil surface (Zartman and Woyewodzic, 1979; Jordan et al., 1979). Gould (1968) stated that the total number of seminal grass roots varies from one to seven. Sieglinger (1920) determined that the radicle is the only seminal (temporary) root in sorghum, but it develops several branches. The nodal or adventitious roots develop to become the bulk of the sorghum root system. As the adventitious root system develops, there is a concomitant diminution of seminal roots as they abscise. Hackett (1973) reported the seminal root diameter of grain sorghum to be 0.2 mm and the adventitious root diameter to be 0.35 mm.

According to Kaigama et al. (1977) and Myers (1980), maximum root weight of sorghum occurs at about anthesis, and roots can extend to a depth of more than 1.5 m at a rate of 2–5 cm day^{-1} (Kaigama et al., 1977; Nakayama and van Bavel, 1963). Lavy and Eastin (1969) reported lateral extension of over 2 m from the crown and maximum root activity in the top 15 cm. Saint-Clair (1977) found 84% of the roots in the top 25 cm, and Mayaki et al. (1976) reported 80% in the top 30 cm. Bloodworth et al. (1958) found that, on a weight basis, 70% of the sorghum roots were in the 0–7.5 cm depth, 14% in the 7.5–15 cm depth, and 98% in the top 91 cm of the soil profile.

Significant genotypic variation has been found in sorghum root development, suggesting that sorghum could be improved by selection for more extensive root development at depth (Nour and Weibel, 1978). McClure and Harvey (1962) found that the hybrid grain sorghum root system develops much more extensively and rapidly than the parent lines after panicle exertion. Improved soil management practices coupled with selection of hybrids for deep rooting may be an effective way to increase water use efficiency and nutrient uptake under rainfed conditions (Peacock and Wilson, 1984).

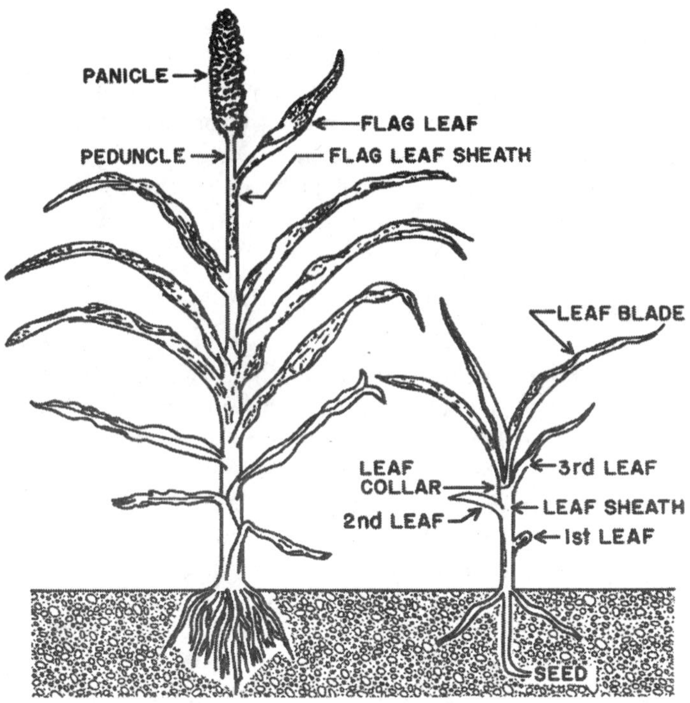

Figure 11.1 Mature and young sorghum plants. (From Vanderlip, 1972.)

C. Stem, Leaves, and Tillers

Sorghum stems are solid, usually erect, dry or juicy, starchy or sweet, and 0.5–3 cm in diameter at base. Plant height depends on number of nodes and length of internodes (Purseglove, 1985). In photoperiod-sensitive cultivars, long photoperiods delay initiation of the panicle and result in tall plants with numerous internodes.

Grain sorghum exhibits a relatively simple leaf area display. As in all grasses, leaves arise on alternate sides of the stem at the nodes and are composed of a sheath and a leaf blade (Maas et al., 1987). Grain sorghum is determinate with respect to the number of leaves per stalk, and initiation of new leaves is terminated by floral initiation. Under normal field conditions, most commercial varieties of grain sorghum in the United States exhibit a unimodal distribution of leaf sizes in which individual leaf areas increase from the ground upward to a maximum value and then decrease to the top of the stalk (Maas et al., 1987). Some tall, tropically adapted varieties and plants experiencing stresses exhibit a bimodal distribution of individual leaf areas (Quinby, 1974). Figure 11.1 shows the major parts of young and mature sorghum plants.

Grass tillers are axillary shoots that grow, produce adventitious roots, and may become completely independent of the stem from which they arise. Grain sorghum cultivars vary in their tendency to produce tillers (Escalada and Plucknett, 1975; Isbell and Morgan, 1982), and environmental conditions can stimulate or retard their production. For example, defoliation or death of the main stem reduces apical dominance and permits rapid tiller production.

D. Inflorescence

The inflorescence is a compact to open panicle with primary branches arising from a central rachis. These give rise to secondary and sometimes tertiary branches, which carry the racemes of spikelets (Doggett, 1970). The racemes bear spikelets in pairs; one of each pair is sessile and fertile, and the other is pedicled and male or sterile. The sessile spikelet has two glumes, which enclose two florets. The upper floret is perfect; the lower is sterile and consists of a lemma which partially enfolds the fertile floret (Doggett, 1970). Anthesis begins near the tip of the panicle 0–3 days after emergence from the boot. Flowering proceeds basipetally for 4–7 days. Sorghum is generally a selfpollinated plant, but some cross-pollination always occurs.

The onset of the reproductive phase commences with the initiation of the panicle, which usually occurs between 30–40 days after emergence but may vary according to genotype and environmental conditions (Peacock and Wilson, 1984). The grain attains maximum dry weight 25–55 days after blooming. The air-dried whole grain contains 8–16% water, 8–15% protein, 2–5% fat, 68–74% carbohydrates, 1–3% fiber, and 1.5–2.0% ash (Purseglove, 1985).

E. Growth Stages

Correct identification of growth stages is very important for tissue sampling for nutrient analysis, top dressing of nitrogen, pest control (e.g., diseases, and insects), irrigation, and harvesting. Table 11.1 describes the different growth stages of sorghum plants. Although cultivar, temperature, and (sometimes) photoperiod affect the duration of growth stages, the same general pattern is found in early, medium, and late maturity hybrids (Vanderlip and Reeves, 1972).

F. Dry Matter Production

Sorghum is a C_4 crop with a high CO_2 assimilation capacity and potentially high rates of dry matter production. Leaf photosynthetic rates of over 72 mg

Table 11.1 Growth Stages of Sorghum

Growth stage	Approximate days after emergence	Identification characteristic
0	0	Emergence. Coleoptile visible at soil surface.
1	10	Collar of third leaf visible.
2	20	Collar of fifth leaf visible.
3	30	Growing point differentiation. Approximately 8-leaf stage by previous criteria.
4	40	Final leaf visible in whorl.
5	50	Boot. Panicle extended into flag leaf sheath.
6	60	Half-bloom. Half of plants at some stage of bloom.
7	70	Soft dough.
8	85	Hard dough.
9	95	Physiological maturity. Maximum dry matter accumulation.

Source: Vanderlip, 1972.

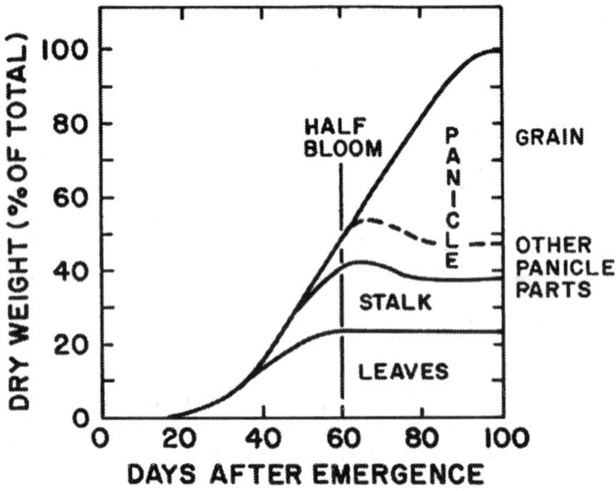

Figure 11.2 Dry weight of different plant parts of grain sorghum as a function of plant age. (From Vanderlip, 1972.)

CO_2 dm^{-1} ha^{-1} have been reported under field conditions (Rawson et al., 1978; Eastin, 1983). Typical patterns of dry weight development and partitioning among different plant parts are shown in Figure 11.2. In the first 30–35 days of plant growth, shoot dry matter consists primarily of leaves. The culm or stalk then begins rapid growth, and both leaf and stalk weights increase until they reach their maximum values at about 60–65 days, respectively. Panicle weight increases rapidly from 50–60 days, and after pollination grain weight increases rapidly.

Eck and Musick (1979) found that the dry matter accumulation rate of grain sorghum was nearly constant from about 40 days after planting until near physiological maturity (115 days). About 59% of the total dry matter had accumulated by flowering. Leaf blades reached their maximum weight 7 days after the boot stage and maintained that weight until near maturity. Stalk weights reached their maximum 7 days after half bloom and declined from then until maturity. At maturity, stems, leaves, and panicles contained 18, 24, and 58% of the total dry matter, respectively. Roy and Wright (1973) reported that the contributions of stems, leaves, and panicles to the total dry matter yield were 32, 18, and 50%, respectively, for a well-fertilized ($N_{120}P_{26}$ kg ha^{-1}) treatment, but they were 41, 23, and 36% for an unfertilized treatment.

IV. LEAF AREA INDEX

Leaf area index (*LAI*) is often used as an indicator of plant growth and for evaluating assimilation and transpiration rates in plant physiological studies. It is also frequently used in agronomic studies to model yield and to make crop production decisions (Hodges and Kanemasu, 1977). Grain yield of annual crops is usually related to duration of leaf area, and establishment of a high leaf area index as early as possible is important in order to obtain maximum yield. Large plant populations, narrow row spacing, and large seeds can contribute to early leaf area development (Doggett, 1970).

In cereals most of the carbohydrate in the grain results from photosynthesis after heading, though reallocation of assimilates from the stem often accounts for 10–12% of total grain weight (Chamberlin and Wilson, 1982; Fischer and Wilson, 1971). Because of the importance of postanthesis photosynthesis, the longer leaf area is retained after heading, the more assimilates are available for grain growth. This leaf area duration may be affected by mineral nutrition, especially N. Nitrogen deficiency and retranslocation of N from the leaves to the grain seed leaf senescence and reduce the duration of leaf area after heading (Yoshida, 1972). Maximum leaf area index in sorghum is usually achieved just before anthesis, and an *LAI* of approximately 5 is found in productive commercial fields in the united states. Fields in drier areas can have *LAI* values of 2–4 (Eastin, 1983).

Both “senescent” and “nonsenescent” genotypes of sorghum are known. Leaves of senescent genotypes senesce soon after physiological maturity of the grain. In contrast, leaves of nonsenescent genotypes remain green and contribute to accumulation of starch and sucrose in the culms (McBee et al., 1983).

V. YIELD AND YIELD COMPONENTS

The highest grain yields reported for sorghum are 16.5 t ha^{-1} (Pickett and Fredericks, 1959) and 14.25 t ha^{-1} (Fischer and Wilson, 1975). The harvest index for the latter was 0.45, and total aboveground dry matter was 31.7 t ha^{-1}. Average worldwide yields are about 1.3 t ha^{-1}, ranging from as low as 0.66 t ha^{-1} in parts of Africa to as high as 4 t ha^{-1} in Latin America (Peacock and Wilson, 1984).

Sorghum grain yield can be expressed by (Quinby, 1973):

$$\text{Yield} = P \times H \times S \times W$$

where

P = number of plants
H = panicles or heads per plant
S = seeds per head
W = weight per seed

High yield can be obtained only if all yield components are at optimum levels. This requires adequate plant population, adequate water and nutrients, optimum crop management, and a cultivar with high yield potential. Hybrid vigor (heterosis) is great in commercial grain sorghum hybrids. A greater number of seeds per plant has been recognized as the yield component that contributes most to heterosis (Quinby, 1973).

VI. NUTRIENT REQUIREMENTS

Nutrient requirements vary according to soil, climate, and cultivar. It is not the purpose of this discussion to give detailed fertilizer recommendations for sorghum, but rather to develop a conceptual framework concerning nutrient requirements of the crop. The best way to determine specific rates of nutrients for a given crop is to determine the yield potential and the amount of mineral nutrients that can be furnished by the soil. Soil tests, yield response curves, and visual assessment of deficiency and/or toxicity symptoms are the important methods for diagnosing nutritional disorders. These techniques were discussed in detail in Chapter 4.

Development of better adapted, higher-yielding cultivars has increased the yield potential of sorghum. This increased yield potential requires more plant nutrients. Consequently, fertilizer application to sorghum has increased tremendously. Nitrogen, phosphorus, and potassium are the essential elements required in relatively large quantities. Nutrient inputs are one of the important components of improved farming systems within the assured rainfall area (> 800 mm annual rainfall) of the Indian SAT. Deficiencies of N and P are common for crops such as sorghum (Katyal and Das, 1993).

Under rainfed cropping in India, it is generally understood that in a soil, if the 0.5 M $NaHCO_3$ extractable P (Olsen P) is less than 5 mg kg^{-1} soil, a response to applied P is likely (El-Swaify et al., 1985). However, recent research at the International Crops Research Institute for the Semi-Arid Tropics (ICRISAT) showed that this critical limit is unlikely to hold for grain sorghum grown in Vertisols. The sorghum crop responded little to applied P unless the extractable Olsen P was less than 2.5 mg kg^{-1} soil. In contrast, substantial responses to fertilizer P were obtained in nearby Alfisols when the Olsen P was greater than 2.5 but less than 5 mg kg^{-1} soil (Sahrawat, 1988). In the Vertisol, 90% relative grain yield of sorghum was obtained at 2.8 mg kg^{-1} Olsen extractable P while in the Alfisol, 90% relative grain yield was achieved at 5.0 mg P kg^{-1} soil. These results suggest that a single critical limit of available P does not hold true for grain sorghum in the two soils types under similar agroclimatic conditions and that the critical limit is lower for the clayey Vertisol than the sandy Alfisol (Sahrawat et al., 1996).

In the United States, most grain sorghum is grown on soils well supplied with calcium and magnesium (Tucker and Bennett, 1968). However, at a soil pH of less than 5.5, aluminum toxicity and Ca and Mg deficiencies can reduce yields. Abruna et al. (1982) reported that in Puerto Rico grain sorghum yields were reduced when Al saturation exceeded 10%, and no yield was obtained when toxicity exceeded 80%. Other cases of Al toxicity have been reported in soils of the semiarid and subhumid zones of West Africa (Doumbia et al., 1993). Dolomitic lime is often applied to acid soils to reduce aluminum toxicity and to supply adequate calcium and magnesium. However, use of acid-tolerant genotypes can be a complementary solution, and conventional breeding systems have been used for improvement of acid soil tolerance (Duncan, 1987; Gourley et al., 1990). Recently, biotechnology has been utilized to produce somaclonal variation in sorghum. Somaclonal variation in sorghum for agronomic, physiological, and morphological traits has been reported (Miller et al., 1992). Tissue culture regenerated lines of sorghum demonstrated improved acid soil tolerance under field conditions and also improved root development at low pH (4.2 to 4.6) during the seedling stage (Miller et al., 1992). Baligar et al. (1989,1993) studied growth and nutrient uptake behavior of Al-sensitive and Al-tolerant sorghum genotypes under field

Table 11.2 Average Response of Growth Traits and Nutrient Concentrations for Al-Sensitive and Al-Tolerant Sorghum Genotypes

Plant parameter	Al-sensitive	Al-tolerant
Shoot dry weight g plant^{-1}	0.09	0.34
Root dry weight g plant^{-1}	0.06	0.14
Shoot:root ratio	1.52	2.49
Al tolerance index for shoot (%)[a]	22.00	61.00
l tolerance index for root (%)[a]	26.00	58.00
N concentration (mg g^{-1})	29.1	33.1
P concentration (mg g^{-1})	1.3	1.6
K concentration (mg g^{-1})	16.5	28.7
Ca concentration (mg g^{-1})	3.1	4.0
Mg concentration (mg g^{-1})	1.4	1.7
Zn concentration (μg g^{-1})	65.6	138.7
Fe concentration (μg g^{-1})	209.6	263.3

[a] Al tolerance index = (Growth with Al / Growth without Al) × 100.
Source: Baligar et al., 1989, 1993.

conditions. Aluminum-tolerant entries had higher shoot and root weight, shoot: root ratio, tolerance index, and nutrient concentrations as compared to Al-sensitive entries (Table 11.2).

Genetic diversity for N use efficiency (NUE) has been demonstrated in sorghum, with some of the most efficient types being cultivars that evolved from low-fertility environments (Gardner et al., 1994). Therefore, exploiting genotypic differences in N demand and efficiency have been proposed as possible alternatives for reducing the cost and reliance upon fertilizer N. However, little is known about the combination of morphological, anatomical, and physiological factors that contribute to improved NUE. Landrace cultivars that have adapted to low N environments may possess different stress-coping mechanisms than do domesticated cultivars developed in contemporary breeding programs (Pearson, 1985). Physiological processes, which are related to N stress tolerance, frequently relate to leaf area and performance in terms of gas exchange rates and stomatal conductance from a given supply of leaf N (Pavlik, 1983). Leaf morphological and anatomical features can also influence these physiological processes and contribute to NUE (Pavlik, 1983; Longstreth and Nobel, 1980). Leaf size (Bhagsari and Brown, 1986), leaf thickness (Alagarswamy et al., 1988) and internal leaf anatomy (Nobel et al., 1975) have been associated with photosynthetic N efficiency.

In areas where soil pH and calcium content are high, the availability of certain micronutrients may be restricted. Nutrients in this category are iron, zinc, manganese, and copper. Of these, iron deficiency is the most common; therefore, more studies have been conducted on iron chlorosis in sorghum than on other micronutrients (Tucker and Bennett, 1968). Iron deficiency on calcareous soils causes dramatic but usually localized decreases in grain sorghum growth. These areas of iron-deficient plants are visible from a distance because of dramatic decreases in growth and severe chlorosis. Symptoms are usually most severe on the most recently expanded leaves, but all leaves may be affected in severe cases. Iron applied in foliar sprays is rapidly incorporated into proteins, but its translocation to younger leaves is usually inadequate to prevent their chlorosis. Thus, multiple foliar applications are often needed. For example, Withee and Carlson (1959) reported that three applications were needed for maximum grain sorghum yields.

In general, grain sorghum and corn are somewhat more susceptible to iron deficiency than most dicots (Brown, 1978). However, significant genotypic differences in sensitivity to iron deficiency have been observed in grain sorghum (Brown and Jones, 1975; Williams et al., 1982). This is probably due to the inability of sensitive genotypes to excrete H^+ and organic reducing agents from the roots.

Most of the tropical soils where sorghum is grown are subject to deterioration with agricultural use, mainly because of a decrease in organic matter content. Management practices that control the amounts of organic matter are therefore an important aspect of soil management for sorghum production.

VII. NUTRIENT UPTAKE

One general guideline for nutrient needs is the nutrient removal by the crop. Figures 11.3, 11.4, and 11.5 show N, P, and K uptake and distribution among various sorghum plant parts during growth. Large quantities of N and P and some potassium are translocated from the other plant parts to the grain as it develops. Unless adequate nutrients are available during grain filling, this translocation may cause deficiencies in the leaves and premature leaf loss, which reduce leaf area duration and may decrease yields. Thus an adequate supply of nutrients at all stages of development of the plant is necessary for maximum yields (Vanderlip, 1972).

Figures 11.3, 11.4, and 11.5 show that a large portion of the N and P but only a small portion of potassium is removed in the grain. Amounts removed depend on nutrient concentrations and total grain production. According to Vanderlip (1972), a grain crop of 8500 kg ha^{-1} contains (in the total aboveground plant) 207 kg of N, 39 kg of P, and 241 kg of K. If the entire plant is harvested for silage or other forms of feed, much more potassium is

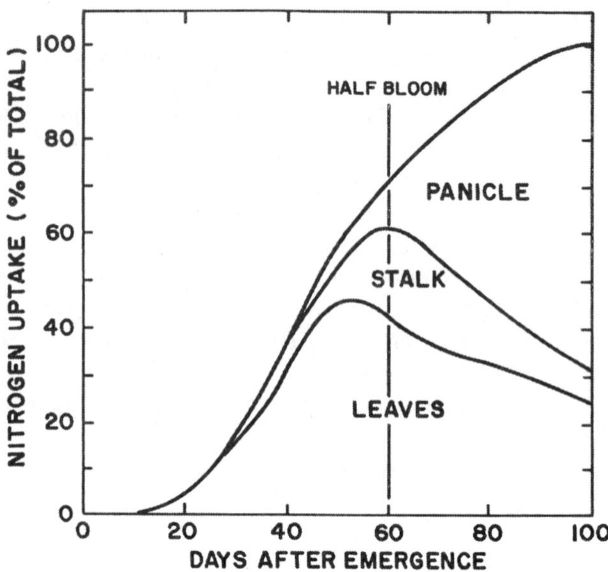

Figure 11.3 Nitrogen uptake in different parts of sorghum plant during crop growth. (From Vanderlip, 1972.)

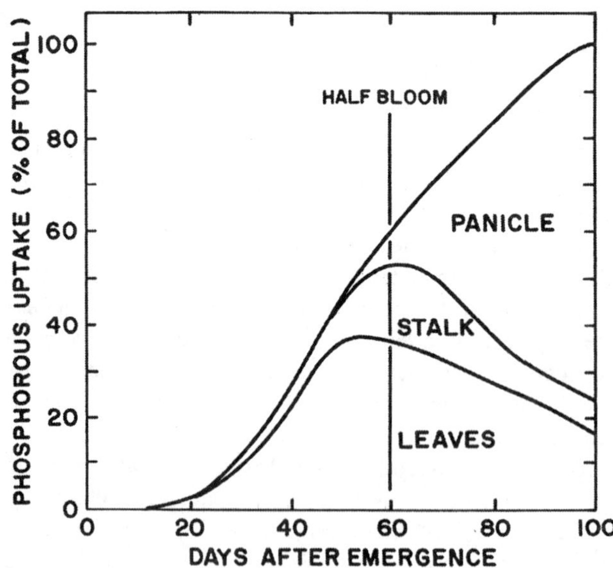

Figure 11.4 Phosphorus uptake in different parts of sorghum plant during crop growth. (From Vanderlip, 1972.)

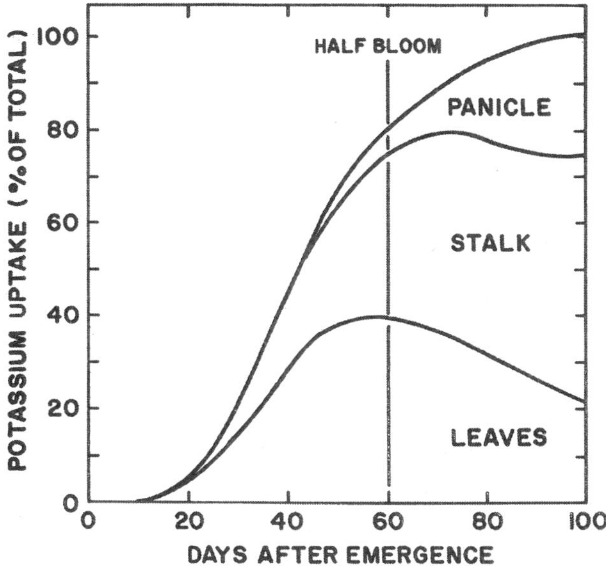

Figure 11.5 Potassium uptake in different parts of sorghum plant during crop growth. (From Vanderlip, 1972.)

removed because most of it is in the vegetative part of the plant. Jones (1983) reported that in a number of experiments the mean N concentration in the grain was 1.67% with a range of 1.02–3.20%. Mean N concentration of the stover was 0.80% with a range of 0.36–1.26%.

Pal et al. (1982) reviewed the 25 years of research in India on the mineral nutrition and fertilizer response of grain sorghum. Accumulations of N, P, and K by grain sorghum were characterized. Usually N and P accumulated slowly compared with the rapid accumulation of K in the early stage of crop growth. In later stages, uptake of K decreased relative to that of N and P. Most of the K remained in the stalk and leaves, while most N and P accumulated in the panicle. Fertilizer responses to N and P were observed throughout India. Improved varieties and hybrids of sorghum responded to N rates ranging from 60 to 150 kg N ha^{-1}, whereas a response to P application was observed up to 40 kg P ha^{-1}. Response to K was inconsistent, depending on the K-supplying power of soils. A balanced fertilizer schedule consisting of 120 kg N ha^{-1}, 20 kg P ha^{-1}, 33 kg K ha^{-1}, and 15–25 kg $ZnSO_4$ ha^{-1} was recommended for improved productivity of grain sorghum.

Knowledge of tissue nutrient concentrations necessary for maximum growth rates is needed to diagnose nutrient deficiencies. It is also required as an input

Table 11.3 Correlation Between Plant Nutrient Concentrations and Grain Yield at Various Growth Stages[a]

Element	Seedling (stage 2)	Early vegetative (stage 3)	Late vegetative (stage 4–5)	Bloom (stage 6)	Fruiting (stage 7–8)
N	+	+	+	+	+V
P	+	+	+	+	+V
K	+	+	+	—	0
Ca		=	=	—	—
Mg	—	—	—	=	–0
B	0+	0	0	0	–0
Cu	0	0	0	0–	0
Fe	—	0–	0–	0	0
Mn	V	V	V	V	—
Zn	V	V	V	0–	0V
Al	0–	0–	0	0	0–

[a] Degree and direction indicated: ++, good, positive correlation; +, fair, positive correlation; 0, no correlation; —, fair, negative correlation; =, good, negative correlation; V, correlation variable or dependent on other conditions.
Source: Lockman, 1972c.

for plant growth models that incorporate nutrition. Lockman (1972a,b,c) provided a useful description of adequate nutritional ranges for several nutrients in grain sorghum tissues at different growth stages. These ranges were summarized from a large body of data and were thus fairly wide. Jones (1983) subsequently attempted to define the variability in N and P concentrations in sorghum. He attempted to normalize and thus reduce variability by expressing the temporal changes as a function of a numerical growth stage system described by Vanderlip and Reeves (1972). Lockman (1972c) correlated plant nutrient concentrations and grain yields at various growth stages of sorghum, as shown in Table 11.3. Nitrogen and phosphorous concentrations were well correlated with yields at all growth stages. Potassium concentrations were correlated with grain yield only for seedling and vegetative samples. Calcium concentrations were only moderately correlated with yield, generally in a negative manner. Magnesium concentrations in grain sorghum plant samples were poorly correlated with yield. Boron, copper, iron, manganese, and aluminum concentrations were not well correlated with yields. Zinc levels in grain sorghum plant samples showed curvilinear correlations with grain yield.

Deficient, low, adequate, and high values for grain sorghum nutrient concentrations at different growth stages are presented in Table 11.4. These values can be used as guidelines in nutritional diagnosis of the crop. Many breeding

Table 11.4 Nutrient Concentrations in Sorghum Plants

Nutrient	Growth stage	Plant part[a]	Deficient ($g\ kg^{-1}$)	Low ($g\ kg^{-1}$)	Adequate ($g\ kg^{-1}$)	High ($g\ kg^{-1}$)
N	Seedling	Whole tops	< 35	30–40	35–51	> 51
	Early vegetative	Whole tops	10 →	30	30–40	> 40
	Vegetative	YMB	—	< 32	32–42	> 42
	Bloom	3BBP	< 25	25–32	33–40	> 40
P	Seedling	Whole tops	< 2.5	2.5–3.0	3.0–6.0	< 6.0
	Early vegetative	Whole tops	1.0 →	1.0–2.0	2.1–5.0	< 5.0
	Vegetative	YMB	< 1.3	1.3–2.5	2.0–6.0	—
	Bloom	3BBP	< 1.3	1.3–1.5	1.5–2.5	< 2.5
K	Seedling	Whole tops	< 25	25–30	30–45	> 45
	Early vegetative	Whole tops	15 →	16–25	25–40	> 40
	Vegetative	YMB	< 15	15–20	20–30	> 30
	Bloom	3BBP	—	—	10–15	—
Ca	Seedling	Whole tops	—	< 12	9–13	> 13
	Early vegetative	Whole tops	2.4	3.0–10.0	10–15	> 15
	Vegetative	YMB	—	—	1.5–9.0	> 9
	Bloom	3BBP	—	< 2.0	2.0–6.0	> 6
Mg	Seedling	Whole tops	—	3.0–3.4	3.5–5.0	> 5
	Early vegetative	Whole tops	2.0	2.0–3.0	4.0–8.0	> 8
	Vegetative	YMB	—	—	2.0–5.0	> 5
	Bloom	3BBP	—	1.0–2.0	2.0–5.0	> 5
S	Vegetative	Whole tops	< 0.3	2.0–2.5	2.5–3	> 3
Fe	Seedling	Whole tops	—	—	160–250	> 300
	Early vegetative	Whole tops	< 30	—	90–120	—
	Vegetative	YMB	—	—	55–200	> 200
	Bloom	3BBP	—	—	65–100	—
Mn	Seedling	Whole tops	—	—	40–150	> 150
	Early vegetative	Whole tops	< 15	11–14	40–70	> 70
	Vegetative	YMB	—	< 10	6–100	> 100
	Bloom	3BBP	—	< 10	8–190	> 190
Zn	Seedling	Whole tops	—	< 30	30–60	> 60
	Early vegetative	Whole tops	< 16	16–20	20–50	> 50
	Vegetative	YMB	—	< 20	20–40	—
	Bloom	3BBP	—	< 15	15–30	—

Table 11.4 (Continued)

Nutrient	Growth stage	Plant part[a]	Deficient ($mg\ kg^{-1}$)	Low ($mg\ kg^{-1}$)	Adequate ($mg\ kg^{-1}$)	High ($mg\ kg^{-1}$)
Cu	Seedling	Whole tops	—	—	8–15	> 15
	Early vegetative	Whole tops	—	< 3	3–14	—
	Vegetative	YMB	—	—	2–15	> 15
	Bloom	3BBP	—	—	2–7	> 10
B	Seedling	Whole tops	—	< 4	4–13	—
	Early vegetative	Whole tops	< 3	3–10	10–15	> 25
	Vegetative	YMB	—	—	1–10	—
	Bloom	3BBP	—	—	1–10	—

[a] YMB, youngest (uppermost) mature leaf blade; 3BBP, third blade below panicle.
Sources: Compiled from Jones and Eck, 1973; Lockman, 1972c; Reuter, 1986.

programs with forage sorghum have emphasized agronomic performance (adaptation, pest resistance, and forage yield and quality), paying little attention to the genotypic differences in mineral concentrations (Kidambi et al., 1993). Mineral concentrations may vary due to genotype, stage of plant development, plant part sampled, time of sampling, and soil fertility. Mineral imbalances or shortfalls may be corrected in the short term by mineral supplementation and in the long term through plant breeding and genotype selection.

VIII. SUMMARY

Sorghum is a warm temperate and tropical cereal with the C_4 photosynthetic pathway and is one of the five major crops of the world. It originated in northeastern Africa in prehistory and is used as a source of food, feed, and industrial raw material. Sorghum has been cultivated for food since ancient times in arid and semiarid regions of Africa and India, where productivity and the area cultivated have remained relatively stable in the recent past. Traditional sorghum cultivars used in these countries are tall, sensitive to photoperiod, and have a low harvest index due to excessive production of stem tissue. In contrast, productivity and area cultivated have increased spectacularly in countries like the United States and Argentina, where hybrid vigor has been exploited through the use of cytoplasmic male sterility. Other factors responsible for yield increases include the development of short cultivars with increased harvest index (0.4–0.5), decreased photoperiod sensitivity, and increased disease and insect resistance, as well as improved management. Potential sorghum grain yields are over 14 Mg ha^{-1}, but average farmers' yields worldwide are only about 1.3 Mg ha^{-1}.

Sorghum can be grown on a wide range of soils with pH values ranging from 5 to 8.5. It is normally more tolerant of drought stress than corn, and it is often grown in areas that are too dry for consistent corn production. Nitrogen is usually the most limiting plant nutrient, and 1 metric ton of sorghum grain contains about 18 kg N, 3 kg P, and 4 kg K. Both soil and plant analysis can be used for diagnosis and correction of nutrient deficiencies and toxicities.

REFERENCES

Abruna, F., J. Rodriguez, and S. Silva. 1982. Crop response to soil acidity factors in Ultisols and Oxisols in Puerto Rico. VI. Grain sorghum. J. Agric. Univ. Puerto Rico 61: 28–38.

Ackerson, R. C., D. R. Krieg, and F. J. M. Sung. 1980. Leaf conductance and osmoregulation on field-grown sorghum genotypes. Crop Sci. 20: 12–14.

Aisien, A. O. and B. P. Ghosh. 1978. Preliminary studies of the germinating behaviour of Guinea corn (Sorghum vulgare). J. Sci. Food Agric. 29: 850–852.

Alagarswamy, G., J. C. Gardner, J. W. Maranville, and R. B. Clark. 1988. Measurement of instantaneous nitrogen use efficiency among pearl millet genotypes. Crop Sci. 28: 681–685.

Baligar, V. C. et al. 1989. Aluminum effects on growth, grain yield, and nutrient use efficiency ratios in sorghum genotypes. Plant Soil 116: 257–264.

Baligar, V. C., R. E. Schaffert, H. L. Santos, G. V. E. Pitta, and A. F. C. Bahia Filho. 1993b. Soil aluminum effects on uptake, influx, and transport of nutrients in sorghum genotypes. Plant Soil 150: 271–277.

Bhagsari, A. S. and R. H. Brown. 1986. Leaf photosynthesis and its correlation with leaf area. Crop Sci. 26: 127–132.

Bloodworth, M. E., C. A. Burleson, and W. R. Cowley. 1958. Root distribution of some irrigated crops using undisrupted soil cores. Agron. J. 50: 317–320.

Blum, A. 1974a. Genotypic responses in sorghum to drought stress. I. Response to soil moisture stress. Crop Sci. 14: 361–364.

Blum, A. 1974b. Genotypic responses in sorghum to drought stress. II. Leaf tissue water relations. Crop Sci. 14: 691–692.

Blum, A. 1975. Effect of the *Bm* gene on epicuticular wax and the water relations of *Sorghum bicolor* (L.) Moench. Israel J. Bot. 24: 50.

Bresler, E., B. L. McNeal, and D. L. Carter. 1982. Saline and sodic soils. Springer-Verlag, Berlin.

Brown, J. C. 1978. Mechanisms of iron uptake by plants. Plant Cell Environ. 1: 249–257.

Brown, J. C. and W. E. Jones. 1975. Phosphorus efficiency as related to iron inefficiency in sorghum. Agron. J. 67: 468–472.

Chamberlin, R. J. and G. L. Wilson. 1982. Development of yield in two grain sorghum hybrids. I. Dry weight and carbon-14 studies. Aust. J. Agric. Res. 33: 1009–1018.

Doggett, H. 1970. Sorghum. Longman, London.

Doumbia, M. D., L. R. Hossner, and A. B. Onken. 1993. Variable sorghum growth in acid soils of subhumid West Africa. Arid Soil Res. and Rehab. 7: 335–346.

Duncan, R. R. 1987. Sorghum genotype comparisons under variable acid soil stress. J. Plant Nutr. 10: 1079–1088.

Eastin, J. D. 1983. Sorghum, pp. 181–204. *In*: IRRI (ed.). Potential productivity of field crops under different environments. IRRI, Los Banos, Philippines.

Eck, M. V. and J. T. Musick. 1979. Plant water stress effects on irrigated grain sorghum. II. Effects on nutrients in plant tissues. Agron. J. 19: 532–538.

El-Swaify, S. A., P. Pathak, T. J. Rego, and S. Singh. 1985. Soil management for optimized productivity under rainfed conditions in the semiarid tropics. Adv. Soil Sci. 1: 1–64.

Escalada, R. G. and D. L. Plucknett. 1975. Ratoon cropping systems of sorghum. I. Origin time of appearance, and fate of tillers. Agron. J. 67: 473–478.

Evans, W. F. and F. C. Stickler. 1961. Grain sorghum seed germination under moisture and temperature stresses. Agron. J. 53: 369–372.

FAO (Food and Agricultural Organization). 1982. Production year book 1981. FAO, Rome.

FAO (Food and Agricultural Organization). 1983. Monthly bulletin of statistics, Vol. 6, No. 6. FAO, Rome.

Fawusi, M. O. A. and A. A. Agboola. 1980. Soil moisture requirements for germination of sorghum, millet, tomato, and celosia. Agron. J. 72: 353–357.

Fischer, K. S. and G. L. Wilson. 1971. Studies of grain production in *Sorghum vulgare*. I. The contribution of pre-flowering photosynthesis to grain yield. Aust. J. Agric. Res. 22: 33–37.

Fischer, K. S. and G. L. Wilson. 1975. Studies of grain production in *Sorghum bicolor* L. Moench. I. Effect of planting density on growth and yield. Aust. J. Agric. Res. 26: 31–41.

Fisher, N. M. 1984. Crop growth and development: The vegetative phase, pp. 119–261. *In*: P. R. Goldsworthy and N. M. Fisher (eds.). The physiology of tropical field crops. Wiley, New York.

Francois, L. E., T. Donouan, and E. V. Maas. 1984. Salinity effects on seed yield, growth, and germination of grain sorghum. Agron. J. 76: 741–744.

Freeman, J. E. 1970. Development and structure of the sorghum plant and its fruit, pp. 28–72. *In*: J. S. Wall and W. M. Ross (eds.). Sorghum production and utilization. AVI Publishing Co., Westport, Connecticut.

Gardner, J. C., J. W. Maranville, and E. T. Paparozzi. 1994. Nitrogen use efficiency among diverse sorghum cultivars. Crop Sci. 34: 728–733.

Garrity, D. P., C. Y. Sullivan, and D. G. Watts. 1984. Changes in grains sorghum stomatal and photosynthetic response to moisture stress across growth stages. Crop Sci. 24: 441–446.

Goodsell, S. F. 1957. Germination of dormant sorghum seed. Agron. J. 49: 387.

Gould, F. W. 1968. Grass systematics. McGraw-Hill, New York.

Gourley, L. M., S. A. Rogers, C. Ruiz-Gomez, and R. B. Clark. 1990. Genetic aspects of aluminum tolerance in sorghum. Plant Soil 123: 211–216.

Gritton, E. T. and R. E. Atkins. 1963. Germination of sorghum seed as affected by dormancy. Agron. J. 55: 169–174.

Hackett, C. 1973. A growth analysis of the young sorghum root system. Aust. J. Biol. Sci. 26: 1211–1214.

Heilman, M. D. 1973. Salinity and iron effects on nutrient uptake by sorghum (*Sorghum bicolor*, L. Moench). Ph.D. dissertation, Texas A&M University, College Station.

Hodges, T. and E. T. Kanemasu. 1977. Modeling dry matter production of winter wheat. Agron. J. 69: 974–978.

Hons, F. M., R. F. Moresco, R. P. Wiedenfeld, and J. H. Cothren. 1986. Applied nitrogen and phosphorus effects on yield and nutrient uptake by high-energy sorghum produced for grain and biomass. Agron. J. 78: 1063–1078.

Isbell, V. R. and P. W. Morgan. 1982. Manipulation of apical dominance in sorghum with growth regulators. Crop Sci. 22: 30–35.

Jones, C. A. 1983. A survey of the variability in tissue nitrogen and phosphorus concentrations in maize and grain sorghum. Field Crops Res. 6: 133–142.

Jones, C. A. 1985. C_4 grasses and cereals: Growth, development, and stress response. Wiley, New York.

Jones, J. B., Jr. and H. V. Eck. 1973. Plant analysis as an aid in fertilizing corn and grain sorghum, pp. 349–364. *In*: L. M. Walsh and J. D. Beaton (eds.). Soil testing and plant analysis. Soil Sci. Soc. Am., Madison, Wisconsin.

Jordan, W. R., M. McCrary, and F. R. Miller. 1979. Compensatory growth in the crown root system of sorghum. Agron J. 71: 803–806.

Kaigama, B. K., I. D. Teare, L. R. Stone, and W.L. Powers. 1977. Root and top growth of irrigated and nonirrigated grain sorghum. Crop Sci. 17: 555–559.

Kailasanathan, K., G. G. S. N. Rao, and S. K. Sinha. 1976. Effect of temperature on the partitioning of seed reserves in cowpea and sorghum. Indian J. Plant Physiol. 13: 171–173.

Kanemasu, E. T., D. L. Bark, and E. Chin Choy. 1975. Effect of soil temperature on sorghum emergence. Plant Soil 43: 411–417.

Katyal, J. C. and S. K. Das. 1993. Fertilizer management in grain sorghum, pp. 61–78. *In*: H. L. S. Tondon (ed.) Fertilizer management in food crops. Fertilizer Development and Consultation Organization, New Delhi, India.

Kendal, E. J., B. D. McKersie, and R. H. Stinson. 1985. Phase properties of membranes after freezing in winter wheat. Can. J. Bot. 63: 2274–2277.

Kidambi, S. P., A. G. Matches, T. P. Karnezos, and J. W. Keeling. 1993. Mineral concentrations in forage sorghum grown under two harvest management systems. Agron. J. 85: 826–833.

Lang, A. and P. E. H. Minchin. 1966. Chilling effects on translocation. J. Exp. Bot. 37: 389–398.

Lall, S. B. and R. S. Sakhare. 1970. Salt tolerance in jowar. Botanique 1: 23–28.

Lavy, T. L. and J. D. Eastin. 1969. Effect of soil depth and plant age on phosphorus uptake by corn and sorghum. Agron. J. 61: 677–680.

Lockman, R. B. 1972a. Mineral comparison of grain sorghum plant samples. I. Comparative analysis with corn at various stages of growth and under different environments. Commun. Soil Sci. Plant Anal. 3: 271–281.

Lockman, R. B. 1972b. Mineral composition of grain sorghum plant samples. II. As affected by soil acidity, soil fertility, stage of growth, variety, and climate factors. Commun. Soil Sci. Plant Anal. 3: 283–293.

Lockman, R. B. 1972c. Mineral composition of grain sorghum plant samples. III. Suggested nutrient sufficiency limits at various stages of growth. Commun. Soil Sci. Plant Anal. 3: 295–303.

Longstreth, D. J. and P. S. Nobel. 1980. Nutrient influences on leaf photosynthesis. Plant Physiol. 65: 541–543.

Maas, S. J., G. F. Arkin, and W. D. Rosenthal. 1987. Relationships between the areas of successive leaves on grain sorghum. Agron. J. 79: 739–745.

Maliwal, G. L. 1967. Salt tolerance studies on some varieties of jowar (*Sorghum vulgare*), mung (*Phaseolus aureus*), and tobacco (*Nicotiana tabacum*) at germination stage. Indian J. Plant Physiol. 10: 95–104.

Mayaki, W. C., L. R. Stone, and I. D. Teare. 1976. Irrigated and nonirrigated soybean, corn, and grain sorghum root systems. Agron. J. 68: 532–534.

McBee, G. G., R. M. Wascon III, F. R. Miller, and R. A. Creelman. 1983. Effect of senescence and nonsenescence on carbohydrates in sorghum during late kernel maturity states. Crop Sci. 23: 372–376.

McClure, J. W. and C. Harvey. 1962. Use of radiophosphorus in measuring root growth of sorghums. Agron. J. 54: 427–439.

Miller, D. R., R. M. Waskom, R. R. Duncan, P. L. Chapman, M. A. Brick, G. E. Hanning, D. A. Timm, and M. W. Nabors. 1992. Acid soil stress tolerance in tissue culture-derived sorghum lines. Crop Sci. 32: 324–327.

Myers, R. J. K. 1980. The root system of a grain sorghum crop. Field Crops Res. 3: 53–64.

Nakayama, F. S. and C. H. van Bavel. 1963. Root activity distribution pattern of sorghum and soil moisture conditions. Agron. J. 55: 271– 274.

Nobel, P. S., L. J. Zaragoza, and W. K. Smith. 1975. Relation between mesophyll surface area, photosynthetic rate, and illumination level during development of leaves of *Plectranthus parviflorus* Henckel. Plant Physiol. 55: 1067–1070.

Nour, A. M. and D. E. Weibel. 1978. Evaluation of root characteristics in grain sorghum. Agron. J. 70: 217–218.

Pal, W. R., U. C. Upadhay, S. P. Singh, and N. K. Umrani. 1982. Mineral nutrition and fertilization response of grain sorghum in India—A review over the last 25 years. Fertilizer Res. 3: 141–159.

Parvatikar, S. R., T. G. Prasad, and D. G. Mestri. 1975. Seed dormancy in sorghum. Curr. Res. 4: 35.

Pavlik, B. M. 1983. Nutrient and productivity relations of the dune grasses *Ammophilia arenaria* and *Elymus mollis*. I. Blade photosynthesis and nitrogen use efficiency in the laboratory and field. Oecologia 57: 227–232.

Peacock, J. M. and G. L. Wilson. 1984. Sorghum, pp. 249–279. *In*: P. R. Goldsworthy and N. M. Fisher (eds.). The physiology of tropical field crops. Wiley, New York.

Pearson, C. J. 1985. Editorial: Research and development for yield of pearl millet. Field Crops Res. 11: 113–121.

Pickett, R. and E. E. Fredericks. 1959. Report Purdue Univ. Agric. Exp. Stn. 2(3): 5–8.

Purseglove, J. W. 1985. Tropical crops: Monocotyledons. Longman, New York.

Quinby, J. R. 1973. The genetic control of flowering and growth in sorghum. Adv. Agron. 25: 125–162.

Quinby, J. R. 1974. Sorghum improvement and the genetics of growth. Texas A&M University Press, College Station.

Ratandilok, N. K. 1978. Salt tolerance in grain sorghum. Ph.D. dissertation. University of Arizona, Tucson.

Rawson, H. M., N. C. Turner, and J. E. Begg. 1978. Agronomic and physiological response of soybean and sorghum crops to water deficits. IV. Photosynthesis, transpiration and water use efficiency of leaves. Aust. J. Plant Physiol. 5: 195–209.

Reuter, D. J. 1986. Temperate and sub-tropical crops, pp. 38–99. *In*: D. J. Reuter and J. B. Robinson (eds.). Plant analysis: An interpretation manual. Inkata Press, Melbourne.

Roy, R. N. and B. C. Wright. 1973. Sorghum growth and nutrient uptake in relation to soil fertility: I. Dry matter accumulation patterns and N content of grain. Agron. J. 65: 709–711.

Sahrawat, K. L. 1988. Overview of research on phosphorus in Vertisols, pp. 4–8. *In*: Phosphorus in Indian Vertisols. Summary proceedings of a workshop, 23–26 Aug. 1988, International Crops Research Institute for the Semiarid Tropics (IRISAT), Patancheru, AP, India.

Sahrawat, K. L., G. Pardhasaradhi, T. J. Rego, and M. H. Rahman. 1996. Relationship between extracted phosphorus and sorghum yield in a Vertisol and an Alfisol under rainfed cropping. Fert. Res. 44: 23–26.

Saint-Clair, P. M. 1977. Root growth of cultivars of grain sorghum *Sorghum bicolor* (L.) Moench. Nat. Can. 104: 537–541.

Salter, P. J. and J. E. Goode. 1967. Crop response to water at different stages of growth. Commonwealth Agricultural Bureau, Farnham Royal, Bucks, England.

Sheehan, G. J., P. F. Greenfield, and D. J. Nicklin. 1978. Energy economics, ethanol—A literature review. Alcohol Fuels Conference, Sydney, Australia.

Shih, S. F., G. J. Gascho, and G. S. Rahi. 1981. Modeling biomass production of sweet sorghum. Agron. J. 73: 1027–1032.

Sieglinger, J. B. 1920. Temporary roots of the sorghum. J. Am. Soc. Agron. 12: 143–145.

Staggenborg, S. A. and R. L. Vanderlip. 1996. Sorghum grain yield reductions caused by duration and timing of freezing temperatures. Agron. J. 88: 473–477.

Taylor, R. M., E. F. Young, Jr., and R. L. Rivera. 1975. Salt tolerance in cultivars of grain sorghum. Crop Sci. 15: 734–735.

Tucker, B. B. and W. F. Bennett. 1968. Fertilizer use on grain sorghum, pp. 189–220. *In*: R. C. Dinauer (ed.). Changing patterns in fertilizer use. Soil Sci. Soc. Am., Madison, Wisconsin.

Vanderlip, R. L. 1972. How a sorghum plant develops. Cooperative Extension Service, Kansas State University, Manhattan.

Vanderlip, R. L. and H. E. Reeves. 1972. Growth stages of sorghum (*Sorghum bicolor* L. Moench). Agron. J. 64: 13–16.

Vong, N. Q. and Y. Murata. 1977. Studies on the physiological characteristics of C_3 and C_4 crop species. I. The effects of air temperature on the apparent photosynthesis, dark respiration, and nutrient absorption of some crops. Jpn. J. Crop Sci. 46: 45–52.

Wardlaw, I. F. and D. Bagnall. 1971. Phloem transport and the regulation of growth of [*Sorghum bicolor* (L.) Moench] at low temperatures. Plant Physiol. 68: 411–414.

Williams, E. P., R. B. Clark, Y. Yusuf, W. M. Rose, and J. W. Maranville. 1982. Variability of sorghum genotypes to tolerate iron deficiency. J. Plant Nutr. 5: 553–567.

Wilson, G. L., P. S. Raju, and J. M. Peacock. 1982. Effect of soil temperature on sorghum seedling emergence. Indian J. Agric. Res. 52: 848–851.

Withee, L. V. and E. N. Carlson. 1959. Foliar and soil applications of iron components to control iron chlorosis of grain sorghum. Agron. J. 51: 474–476.

Yoshida, S. 1972. Physiological aspects of grain yield. Annu. Rev. Plant Physiol. 23: 437–464.

Zartman, R. E. and R. T. Woyewodzic. 1979. Root distribution patterns of two hybrid grain sorghum under field conditions. Agron. J. 71: 325–328.

12

Soybeans

I. INTRODUCTION

The soybean is the most important grain legume crop in the world in terms of its use in human foods and livestock feeds. The soybean belongs to the family Leguminosae, subfamily Papilionoideae, and the genus *Glycine* L. The cultivated species is *Glycine max* (L.) Merrill. Soybean apparently originated in China and was introduced into Europe in the early 1700s and into North America in the early 1800s (Whigham, 1983). Important soybean-producing countries are the United States, Brazil, China, and Argentina. The four major soybean-producing countries together account for 90–95% of the world production (Smith and Huyser, 1987). Since 1970, soybean production has been at least double that of any other oilseed crop. The contribution of soybean to world oilseed production increased from 32% in 1965 to over 50% in the 1980s (Smith and Huyser, 1987). Several important factors contributed to the rapid increase in soybean production, including a steady expansion in the market for soybean oil and meal in various parts of the world. The world average soybean yield is 1.9 t ha^{-1}. The United States has the highest average yield at 2.2 t ha^{-1}, followed by Argentina and Brazil (USDA, 1980). With improved management practices and high-yielding cultivars, yields as high as 7.4 t ha^{-1} have been recorded (Gabel, 1979).

Soybean seeds contain approximately 21% oil and 40% protein on a dry weight basis (Johnson and Bernard, 1962) and provide a valuable food for human consumption. Soybean oil is used for human food, various pharmaceuticals and medicines, and in manufacturing disinfectants, printing inks, and soaps. Various soybean food products are used for human consumption, including fermented food; soy beverage, flour, and whole-bean confectionery products, as well as textural vegetable protein used as simulated meat, fruit, and nut products (Whigham, 1983). After processing, the by-products of the seeds of soybean provide a valuable, protein-rich feed supplement for livestock.

II. CLIMATE AND SOIL REQUIREMENTS

Soybean is a warm-season crop, but cultivation now extends from the tropics to 52°N. The major commercial production of soybean is between 25 and 45° latitude and at altitudes of less than 1000 m. The general climatic requirements are approximately those of corn, and the greatest development in the United States has been in the corn belt (Purseglove, 1987). Soybean is a photoperiod-sensitive short-day plant that has been successfully adapted to 13 maturity groups corresponding to narrow zones of latitude. Cultivars are assigned to maturity groups according to their photoperiod requirements for maturity (Fehr, 1987). Five major genes that affect time of flowering and maturity have been identified (Wilcox et al., 1995). Although expression of these genes is influenced by interactions with latitude, temperature, and photoperiod, environmental effects are thought to be small compared with genotypic effects (McBlain et al., 1987; Wilcox et al., 1995).

Soybean is temperature sensitive and usually is grown in environments with temperatures between 10 and 40°C during the growing season (Whigham, 1983). Controlled environment studies using season-long temperature treatments have shown that final soybean seed yield increased as temperature increased between 18/12 (day/night) and 26/20°C, but yield decreased when plants were grown at temperatures greater than 26/20° (Sionit et al., 1987). Night temperature increases from 10 to 24°C increased soybean seed yield (Seddigh and Jolliff, 1984a,b). Raising temperatures from 29/20 to 34/20°C during seed fill decreased soybean seed yield (Dornbos and Mullen, 1991).

Day/night temperatures of 30/20, 30/30/ 35/20, and 35/30°C were imposed during flowering and pod set (R1–R5), seed fill and maturation (R5–R8), and during the entire reproductive period (R1–R8) (Gibson and Mullen, 1996). Increases in day temperature resulted in decreased seed formation when plants were exposed during flowering and pod set and decreased growth when

Table 12.1 Yield of Selected Soybean Varieties Grown in Soils with 67% and 7% Al saturation near Yurimaguas, Peru

Variety	Yield (t/ha) unlimed soil	Yield (t/ha) limed soil	Relative grain yield
Hardee	1.23	2.13	58
SJ-2	1.20	2.07	58
Mineira	0.93	1.70	55
Jupiter	0.93	2.23	42
Improved Pelican	0.76	2.20	35

Source: Nicholaides and Piha, 1987.

exposed during flowering and pot set or seed fill. Seed growth reductions in plants exposed to the high day temperature were accompanied by decreased photosynthetic rates. The largest yield reduction in this study was 27% and occurred when 35°C occurred for 10 h per day from flowering to maturity. No significant losses in yield occurred at high night temperature at any reproductive growth phase. The only significant interaction between day and night temperature for the yield components was for seed weight per plant during flowering and pod set. Night temperature stress did not occur at 30°C, and a night temperature of 20°C did not reduce the yield loss from daytime high temperature stress. This study suggests that soybean seed yield reductions from high temperatures are primarily a response to day temperature, and moderate to high night temperatures have a small effect on soybean seed yield components (Gibson and Mullen, 1996).

Brown (1960) reported that the maximum rate of development between planting and flowering occurred at 30°C. Soybean plants are the most sensitive of all crop plants to light duration (photoperiod) and are sensitive to light quantity. They are short-day plants, but cultivars differ markedly with respect to the minimum dark period required to induce flowering (Chapman and Carter, 1976). Soybeans can be grown on a wide range of well-drained soils but thrive best on clay loam soils. The crop is better adapted for production on clay than either corn or cotton. The crop is also suited for production on muck. There are significant differences in seed yield among genotypes (Reddy and Dunn, 1987), but the optimum soil pH reported for soybean production is in the range of 6–6.5 (Carter and Hartwig, 1962; McLean and Brown, 1984). Soil Al concentration is an important component of soil acidity, and varietal differences in Al tolerance have been reported (Table 12.1). Soybean is rated as a moderately salt-tolerant crop and the reported salinity threshold is about 5 dS m^{-1} (Maas, 1986).

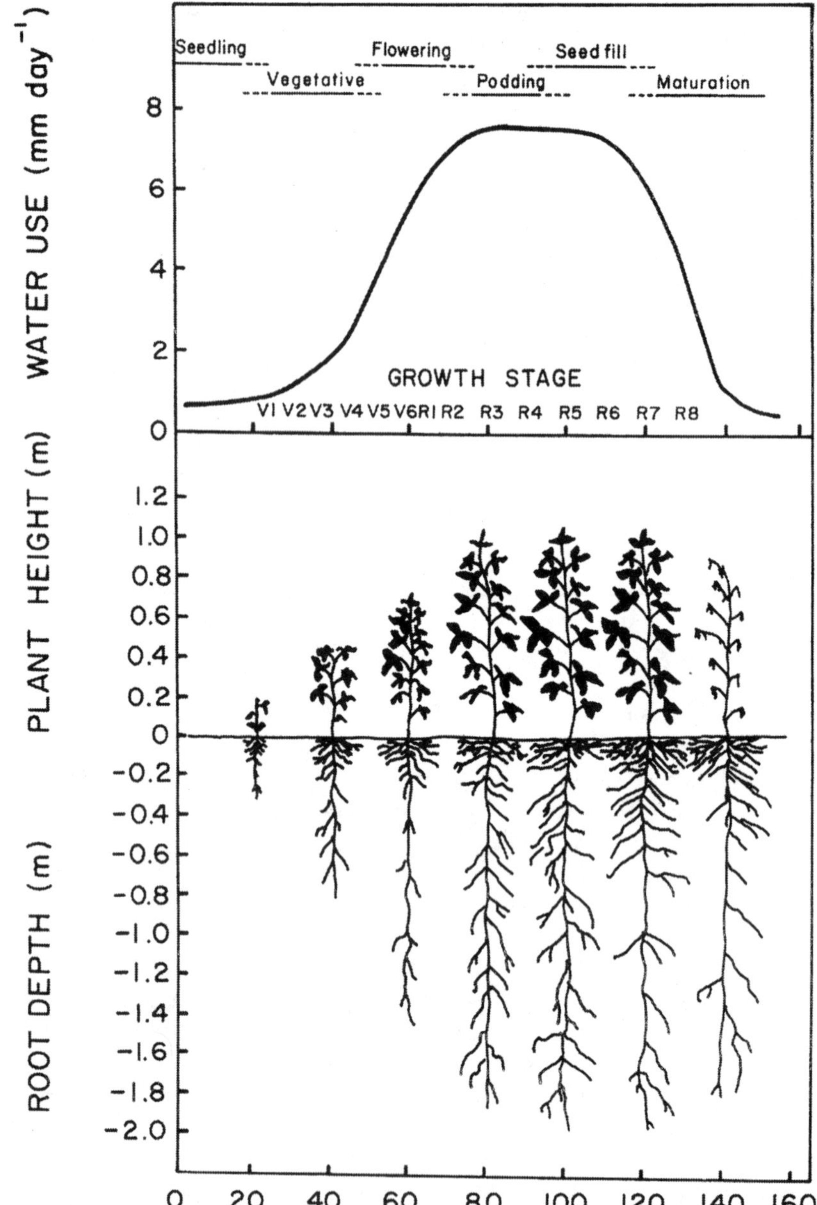

Figure 12.1 Seasonal water use and growth patterns of soybean. (From Van Doren and Reicosky, 1987.)

The water requirements of soybean vary with soil, climatic conditions, growth duration, and yield level of cultivar. Total water requirement for soybeans grown in the midwestern United States has been reported to be in the range of 330–766 mm (Kanemasu et al., 1976; Musick et al., 1976). Water use for soybeans can vary from 450 to 825 mm where the growing season ranges from 100 days at low altitude up to 190 days in higher altitudes (Doorenbos and Pruit, 1977). Total water requirement for soybean in New South Wales (Australia) ranged from 451 to 748 mm per growing season (Mason et al., 1981). The water use pattern during crop growth in the midwestern United States is shown in Figure 12.1. A description of growth stages V_1 to R_8 corresponding to those described by Fehr et al. (1971) is presented in Section III.D. The water requirement for soybeans was low during seedling and maturity growth stages and maximum during the flowering to seed-filling growth stages (Figure 12.1). This pattern of water uptake for soybeans can be applied universally, but the magnitude may vary according to local conditions. The water extraction pattern is related to root distribution (Figure 12.1). Hiler et al. (1974) developed a water stress index for soybeans during different growth stages based on yield reduction. To avoid stress for soybeans, irrigation is needed when the soil water depletion reaches 80% in the vegetative stage, 45% in early to peak flowering, 30% in late flowering to early pod development, and 80% in late pod to maturity. According to Brady et al. (1974), the best yields and most efficient water use are generally obtained when the available soil water in the root zone is not depleted by more than 50–60%.

Research on soybean irrigated by providing either permanent water surrounding raised beds or continuously running water between beds has shown that such regimes can increase yields by 10–20% compared to conventional irrigation practices (Nathanson et al., 1984; Troedson et al., 1986). Termed "saturated soil culture" (SSC), this practice also increased N-fixation and delayed maturity (Nathanson et al., 1984; Troedson et al., 1989). Studies of water relations, photosynthesis, and nitrogen supply further suggested that the effects reflected improved crop water status (Troedson et al., 1989). In one study, no effect of SSC on seed yield was found, but N-fixation did increase (Wang et al., 1993). Comparison of two soybean lines differing in growth duration indicated that SSC was only effective for the line with a longer duration (Nathanson et al., 1984).

Cox and Jolliff (1986) reported leaf area index (*LAI*) and net assimilation rate (*NAR*) of soybean were reduced in the presence of soil water deficits. Pod number was the yield component most sensitive to soil water deficits in soybean. Soybean yield is most sensitive to water stress during the pod-filling period (Momen et al., 1979). Soybean has extracted soil water to a soil depth of 1.5 m (Reicosky and Deaton, 1979), which suggests some drought tolerance for the crop under limited precipitation.

III. GROWTH AND DEVELOPMENT

The growth and development of crop plants comprises several sequential changes from germination to maturity. Information on germination and vegetative and reproductive development of the soybean plant provides a better understanding of soybean production physiology and results in better crop management and higher production.

A. Germination and Seedling Growth

Germination can be defined as the emergence and development from the seed embryo of the essential structures which are indicative of the ability to produce a normal plant under favorable conditions (USDA, 1952). Germination involves mobilization and utilization of food and energy reserves, in contrast to seed development, in which there is a net accumulation of energy materials (Howell, 1960). Soybean germination is epigeal, and under favorable environmental conditions, seedlings begin to emerge in 4 or 5 days.

In epigeal germination, which includes more than 90% of the dicot species, the hypocotyl is active and pulls the cotyledons above ground during its growth (Nelson and Larson, 1984). The environmental factors which affect germination are soil moisture, temperature, and oxygen supply. A moisture content of about 50% is required for germination of soybean seed, and soybeans fail to germinate if the soil moisture tension exceeds 6.6 atm (669 kPa). The optimum germination temperature is around 30–35°C. The germination and early seedling development stages are shown in Figure 12.2.

B. Vegetative Development

The important organs which develop during vegetative development of the soybean plant are roots, leaves, and stem. Vegetative development begins with the emergence of the young seedling from the soil surface and ends with the start of flowering. The vegetative growth supports the photosynthetic capacity of the plant, which supports the yield.

During vegetative growth, roots, leaves, and stems may compete for photosynthetic products. The partitioning of products may be controlled by plant hormones or by environmental factors. If there is water stress, plant shoots are affected more than roots. If soil temperature is more favorable for root growth than air temperature is for top growth, a greater percentage of dry matter may be diverted to root growth (Brown, 1984b).

Roots

Roots are important growth components in all plants because they anchor the plant and supply water and mineral nutrients. The soybean root system is

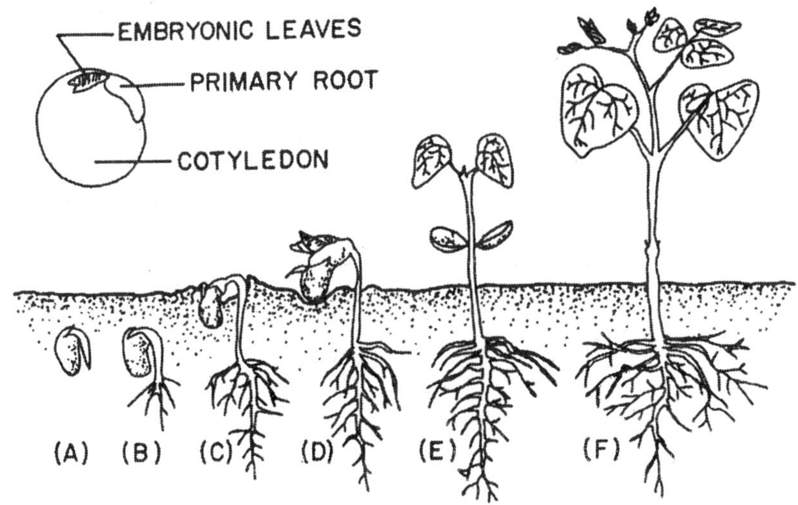

Figure 12.2 Germination and seedling development of soybean. Germination proceeds with emergence of the radicle to form the primary root (A), development of secondary (branch) roots (B), and elongation of the active hypocotyl with the hypocotyl arch penetrating through the soil surface (C). Seedling becomes erect due to action of light on auxins (D), with cotyledons attached to the first node providing photosynthate in addition to stored energy for a short period of time (E), prior to drying and falling from the autotrophic seedling (F). Inset of enlarged seed with one cotyledon removed shows the primary root and embryonic leaves that develop into the first true leaves attached to the second node. (From Nelson and Larson, 1984.)

characterized by a taproot that consists of lateral roots arising from the upper portion of the primary root. The soybean root system continues to grow throughout the life cycle of the plant except at physiological maturity, and the rate of root penetration is most rapid during early flowering (Kaspar et al., 1978). The major portion of the roots is concentrated in the top 15 cm of the soil profile (Mitchell and Russell, 1971), but soybean roots have been observed as deep as 1.5–2 m below the soil surface under normal field conditions (Kaspar et al., 1984).

The rate at which plant root systems grow downward partly determines the water available for uptake, especially in a drying soil profile. Soybean cultivars differ in their rate of downward growth during specific shoot development stages and in their maximum rooting depth on specific dates (Figure 12.3). Cultivars selected for rapid taproot elongation rates in a greenhouse trial were found to have greater rooting depths in rhizotron and field trials than cultivars selected for slow taproot elongation rates (Kaspar et al., 1984).

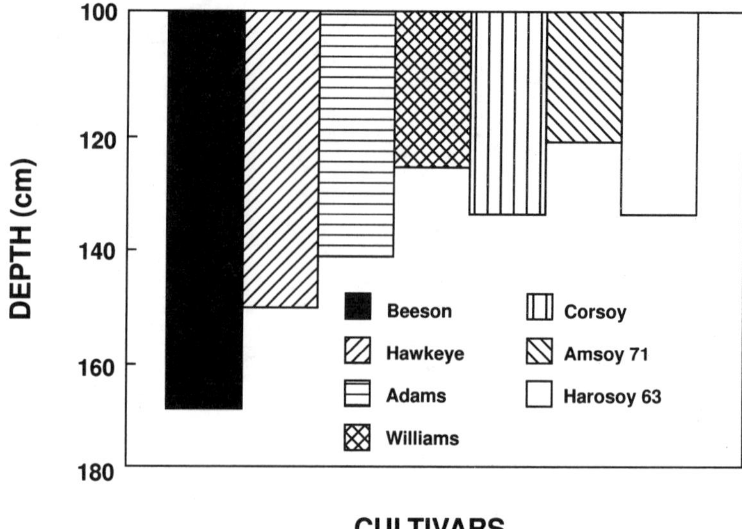

Figure 12.3 Differences in rooting depth of seven soybean cultivars. (From Klepper and Kaspar, 1994.)

Four stages of soybean root development under field conditions have been reported: 1) rapid root growth beneath plant rows during the vegetative stage, 2) branching of roots during early reproductive growth, 3) decreased root growth beneath rows and increased root growth between rows during pod set, and 4) cessation of root growth and root loss due to decomposition during physiological maturity (Brown, 1984a; Hoogenboom et al., 1987a,b).

Nodulation

As a legume, soybean has the capacity to form a symbiotic association with *Rhizobium japonicum* and fix atmospheric nitrogen. Many strains of this bacterium have been identified and some are more efficient than others (Whigham, 1983). When legume roots are infected by a particular *Rhizobium*, nodules are generally formed. Nodulation in legumes involves a series of biochemical interactions between the bacterium and the plant (Ciafardini and Barbieri, 1987). Bhuvaneswari et al. (1980) demonstrated that nodulation of soybean is limited to sites between the root tip and the smallest emergent root hair and does not occur on the walls of mature roots. Bergersen (1958) observed that nodules first appeared on Lincon soybean 9 days after planting and N_2 fixation began about 2 weeks later. Nodules produced with the first infections on the primary root of soybean have an average duration of 65 days (Bergersen, 1958). Since reinfection of younger roots of soybean may occur during the growing season, a mature soybean plant may have nodules of several age classes. A chronology of nodulation in soybean is given by Lersten and Carlson (1987). The quantity of N_2 fixed by soybean varies with environmental conditions and cultivar. According to Larue and Patterson (1981), there is not a single legume crop for which we have valid estimates of N_2 fixed in agriculture. For soybean, these authors have reported N_2 fixation values in the range of 15–162 kg N ha^{-1} in various countries.

Top Growth

The soybean is an annual plant and usually grows to 75–175 cm in height (Shibles et al., 1975). Soybean has two main growth habits, known as determinate and indeterminate. In the determinate growth habit, vegetative growth is nearly complete when the plant starts flowering. In the case of indeterminate growth, both vegetative and reproductive growth go on simultaneously. In the plant, flowers, fruit, and leaves are determinate structures, whereas roots and stems are indeterminate structures. Determinate soybeans usually have fewer nodes per plant and are shorter at maturity than indeterminate types. On the other hand, indeterminate plants are taller and have more nodes per plant, and the flowering structure is smaller, the pods fewer, and the leaflets smaller than those of the determinate types (Whigham, 1983). Except at the cotyledonary and second nodes, the soybean has a single trifoliate leaf at each node. Soybean plants increase in dry weight slowly at first and then more rapidly (Rosolem, 1980). Vegetative growth ceases at about the time seed enlargement starts. The dry weight of vegetative parts decreases during the latter part of grain development (Rosolem, 1980).

C. Reproductive Growth

The reproductive growth period is usually represented by flowering and pod and seed development. Initiation of flowering varies with genotype and environmental factors. Flowering may become visible at 25 days or may be delayed until 50 days when certain genotypes and environments interact (Whigham, 1983). Pods are normally visible about 10 days to 2 weeks after the start of flowering (Howell, 1960). Soybean is a predominately self-pollinating crop. Flowering may occur over 4–6 weeks, depending on the environment and the cultivar. After fertilization of the flower, the pods develop slowly for the first few days; then the rate of development increases until the pod reaches maximum length after 15–20 days (Whigham, 1983). The number of pods varies from 2 to more than 20 in a single inflorescence and up to 400 on a single plant (Carlson and Lersten, 1987). At maturity,

Table 12.2 Chronology of Development of Flower and Ovule of Soybean[a]

Days before flowering	Morphological and anatomical features
25	Initiation of floral primordium in axil and bract.
25	Sepal differentiation.
20–14	Petal, stamen, and carpel initiation.
14–10	Ovule initiation; maturation of megasporocyte; meiosis; four megaspores present.
10–7	Anther initiation; male archesporial cells differentiate; meiosis; microsporogenesis.
7–6	Functional megaspore undergoes first mitotic division.
6–2	Second mitotic division results in four-nucleate embryo sac.
	Third mitotic division results in eight-nucleate embryo sac.
	Cell walls develop around antipodals and egg apparatus, forming a seven-celled and eight-nucleate embryo sac.
	Polar nuclei fuse. Antipodal cells begin to degenerate.
	Nucellus begins to disintegrate at micropylar end and on sides of embryo sac.
	Single vascular bundle in ovule extends from chalaza through funiculus and joints with the carpellary bundle.
1	Embryo sac continues growth; antipodals disorganized and difficult to identify. Synergids with filiform apparatus; one synergid degenerating.
	Tapetum in anthers almost gone. Pollen grains mature; some are germinating.
	Nectary surrounding ovary reaches maximum height.
0	Flower opens; usually day of fertilization; resting zygote; primary endosperm nucleus begins dividing; nectary starts collapsing.

[a] The times are a compilation of data for several soybean cultivars studied by Kato et al. (1954), Murneek and Gomez (1936), Pamplin (1963), and Prakash and Chan (1976). The sequence of development is essentially the same regardless of cultivar, but the absolute times vary with environmental conditions and with cultivars. (From Carlson and Lersten, 1987.)

pods usually contain 2–3 seeds but can contain as many as five. Seeds may vary in shape from nearly spherical to somewhat flattened disks and in color from pale green and yellow to dark brown (Chapman and Carter, 1976). A chronology of development of flower, pod, and seed is presented in Tables 12.2 and 12.3. The data presented in these tables were compiled by Carlson and Lersten (1987) from several sources. According to these authors, the sequence of events remains the same regardless of cultivar or environmental conditions, but the absolute times between events may vary by several days as a function of cultivar and environment.

Table 12.3 Chronology of Development of Seed and Pod of Soybean[a]

Days after flowering	Morphological and anatomical features
0	Resting zygote. Several divisions of primary endosperm nucleus.
1	Two-celled proembryo. Endosperm with about 20 free nuclei.
2	Four- to eight-celled proembryo.
3	Differentiation into proembryo proper and suspensor. Endosperm in peripheral layer with large central vacuole.
4–5	Spherical embryo with protoderm and large suspensor. Endosperm surrounding embryo is cellular but elsewhere it is mostly acellular and vacuolate.
6–7	Initiation of cotyledons. Endosperm mostly cellular.
8–10	Rotation of cotyledons begins. Procambium appears in cotyledons and embryo axis. All tissue systems of hypocotyl present. Root cap present over root initials. Endosperm all cellular.
10–14	Cotyledons have finished rotation and are in normal position with inner surfaces or cotyledons parallel with sides of ovules.
	Cotyledons elongate toward chalazal end of ovule. Primary leaf primordia present. Endosperm occupies about half of seed cavity. Extensive vascularization of seed coat.
14–20	Continued growth of embryo and seed. Reduction in endosperm tissue by assimilation into cotyledons.
20–30	Primary leaves reach full size. Primordium of first trifoliate leaf present. Cotyledons reach maximum size. Endosperm almost gone.
30–50	Continued accumulation of dry matter and loss in fresh weight of seeds and pod. Maturation of pod.
50–80	Various maturity times depending on variety and environmental factors.

[a] The times are a compilation of data for several soybean cultivars studied by Bils and Howell (1963), Fukui and Gotoh (1962), Mengyuan (1963), Kammata (1952), Kato et al. (1955), Ozaki et al. (1956), Pamplin (1963), and Suetsugu et al. (1962). The sequence of development is essentially the same regardless of cultivar, but the absolute times vary with environmental conditions and with cultivars. (From Carlson and Lersten, 1987.)

D. Vegetative and Reproductive Growth Stages

The description of vegetative and reproductive growth stages given here is that described by Fehr et al. (1971) for soybeans. This growth stage development description applies to all soybean genotypes grown in any environment. Vegetative (V) stages are designated by the number of nodes on the main stem, beginning with the unifoliate node, that have or have had a completely unrolled leaf. Reproductive stages R_1 and R_2 are based on flow-

Table 12.4 Description of Vegetative and Reproductive Growth Stages in Soybean

Growth stages	Description
Vegetative (V)	
V_1	Completely unrolled leaf at the unifoliate node
V_2	Completely unrolled leaf at the first node above the unifoliate node
V_3	Three nodes on main stem beginning with the unifoliate node
V(N)	N nodes on the main stem beginning with the unifoliate node
Reproductive (R)	
R_1	One flower at any node
R_2	Flower at node immediately below the uppermost node with a completely unrolled leaf
R_3	Pod 0.5 cm long at one of the four uppermost nodes with a completely unrolled leaf
R_4	Pod 2 cm long at one of the four uppermost nodes with a completely unrolled leaf
R_5	Seeds beginning to develop (can be felt when the pod is squeezed) at one of the four uppermost nodes with a completely unrolled leaf
R_6	Pod containing full-size green seeds at one of the four uppermost nodes with a completely unrolled leaf
R_7	Pods yellowing; 50% of leaves yellow. Physiological maturity
R8	95% of pods brown. Harvest maturity.

Source: Fehr et al., 1971.

ering, R_3 and R_4 on pod development, R_5 and R_6 on seed development, and R_7 and R_8 on maturation. A detailed description of these growth stages is presented in Table 12.4.

IV. YIELD COMPONENTS

Soybean yield is a function of plants per unit area, pods per plant, seeds per pod, and weight per seed. Pod number is determined by pods per reproductive node and reproductive node number. The first parameter is determined by the difference between total pods per reproductive node initiated (pods at least 0.5 cm long) (Fehr and Caviness, 1977) and those aborted. Reproductive node number is determined by total node number and the percentage of these nodes becoming reproductive (percentage reproductive nodes). Although main stem node number in determinate cultivars is determined during the vegetative period, all other yield components contributing to pod number are formed mainly during R1 to R5 and possibly into the late reproductive period (R5 to R7). Contribution of each of these yield components to final seed yield can be summarized by the following equations (Board and Tan, 1995; Board et al., 1990):

seed yield (g m^{-2}) = *f*[pod number (no. m^{-2})]

pod number (no. m^{-2}) = pods per reproductive node (no.) × reproductive node number (no. m^{-2})

pods per reproductive node (no.) = total pods initiated per reproductive node (no.) – total pods aborted per reproductive node (no.)

reproductive node number (no. m^{-2}) = total node number (no. m^{-2}) × nodes becoming reproductive (%)

node number (no. m^{-2}) = main stem node number (no. m^{-2}) + branch node number (no. m^{-2})

branch node number (no. m^{-2}) = *f*[branch number (no. m^{-2})] and *f*[branch dry matter (g m^{-2})]

All these yield components are influenced by environmental conditions, management practices, and cultivar planted. Maximum grain yields have been reported from plant population ranging from 200,000 plants ha^{-1} to more than 600,000 plants ha^{-1} (Whigham, 1983). Hiebsch et al. (1995) studied the relationship between plant density and soybean yield and dry matter production of two cultivars (Figure 12.4). Optimum density was the same for seed yield (SY) and aboveground total dry matter yield (TY) for each soybean cultivar. The density producing 95% of maximum yield was 15 plants m^{-2} for the Cobb cultivar and 20 plants m^{-2} for the Davis cultivar. The number of pods per plant varies according to growth habit, and determinate types produce more pods than indeterminate types. Flower and pod abortion may range from 40 to 80%, and ovule or seed abortion after pod development may occur at a rate of 9–22% (Whigham, 1983). This abortion is a result of long photoperiods, high or low temperatures, or light stress. Seed size, measured as mass per seed, is an important yield component in soybean and other grain crops (Egli et al., 1987). Seed size in soybean is under genetic control, and genotypes are available with sizes ranging from 40 to 550 mg $seed^{-1}$ (Hartwig, 1973).

Seed size is also affected by environmental conditions during seed development (Shibles et al., 1975). Seeds in fruit developing from flowers that open early in the flowering period are larger than seeds in fruit developing from flowers that open late in the flowering period (Egli et al., 1978; Gbikpi and Crooksten, 1981). Spaeth et al. (1984) reported that seeds from lower

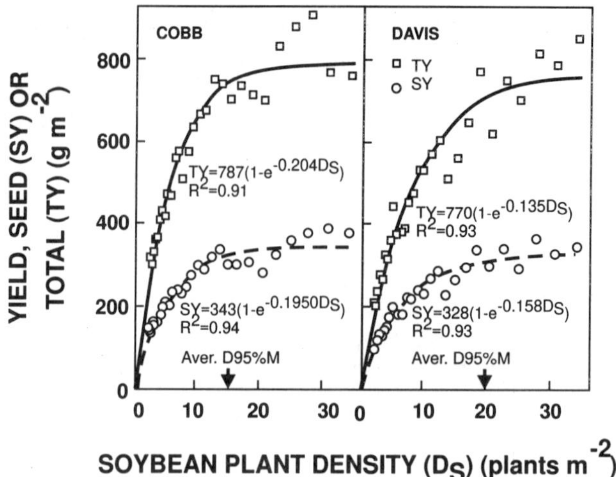

Figure 12.4 Seed yield and total dry matter yield of Cobb and Davis soybean cultivars as a function of soybean plant density. Arrows on the *x*-axes indicate the density for seed yield (SY) and aboveground total dry matter (TY) at 95% of maximum yield (D 95% M). (From Hiebsch et al., 1995.)

nodes were larger than seeds from the upper nodes on the main stem. These variations in seed size were related to variations in duration of seed fill resulting from differences in the time of beginning seed growth coupled with a relatively constant time of physiological maturity (Egli et al., 1987).

V. MAJOR YIELD-DETERMINING PHYSIOLOGICAL PARAMETERS

Important physiological parameters which are directly related to yield are dry matter production, crop growth rate, leaf area index (*LAI*), net assimilation rate, and harvest index.

A. Dry Matter

Accumulation of plant dry matter is a result of the net CO_2 exchange rate between a crop and the atmosphere. Measurements of crop net CO_2 exchange rates (CNCER) are therefore desirable in order to obtain information on physiological processes associated with plant growth and productivity and to validate crop growth models (Bugbee and Monje, 1992).

Crop net CO_2 exchange rate (CNCER, mg m^{-2} s^{-1}) can be defined as (Rochette et al., 1995):

$$\text{CNCER} = P_{gc} - R_r - R_{ag}$$

where P_{gc} is the gross crop photosynthesis and $R_r - R_{ag}$ are the respiration rates of the roots and aboveground parts of the crop, respectively. CNCER is considered positive when the crop is gaining C. CNCER cannot be measured directly under field conditions, but vertical CO_2 fluxes can be measured above the canopy ($F_{c,a}$) and at the soil surface ($F_{c,s}$):

$$C_{c,a} = -P_{gc} + R_r + R_{ag} + R_\mu$$

$$F_{cs} = R_\mu + R_r$$

where R_μ is the amount of $F_{c,s}$ that originates from respiratory activity of the soil fauna and flora. $F_{c,a}$ and $F_{c,s}$ are considered negative when fluxes are toward the field surface.

Soybean crop net CO_2 exchange rates were estimated based on measurements of CO_2 fluxes above the canopy and at the soil surface. Maximum daytime and minimum nighttime hourly values of CNCER were 1.48 and –0.23 mg m^{-2} s^{-1}, respectively (Rochette et al., 1995).

Soybean dry matter production is a function of environmental factors, cultivar planted, and management practices adopted. Figure 12.5 illustrates the cumulative dry weight of the different plant parts at progressive stages of plant development (Hanway and Weber, 1971a). Plant dry weight increased at an increasing rate until stages 3 to 5. After stage 5, the plant dry weight increased at a constant daily rate until stage 9 late in the season. Stem and pod dry weights decline during the late stages of seed filling as tissue loses dry matter because of respiration and mobilization to the seed. At maturity, dry matter consists of 28% fallen leaves, 15% fallen petioles, 17% stems, 11% pods, and 29% seeds. Figure 12.6 shows the influence of water on dry matter production of soybean. The deficit and dryland treatments produced significantly less dry matter (18 and 70%, respectively) than the well-irrigated treatment (Cox and Jolliff, 1986). A detailed discussion of the distribution of dry matter in different plant parts of soybean is given by Shibles et al. (1987).

B. Crop Growth Rate

Crop growth rate is related to light interception by the crop canopy, and the leaf area index for 95% light interception ranges from 3.1 to 4.5, depending on planting density and spatial arrangement (Shibles and Weber, 1966). The highest crop growth rate reported for irrigated soybeans is 30 g m^{-2} day^{-1}

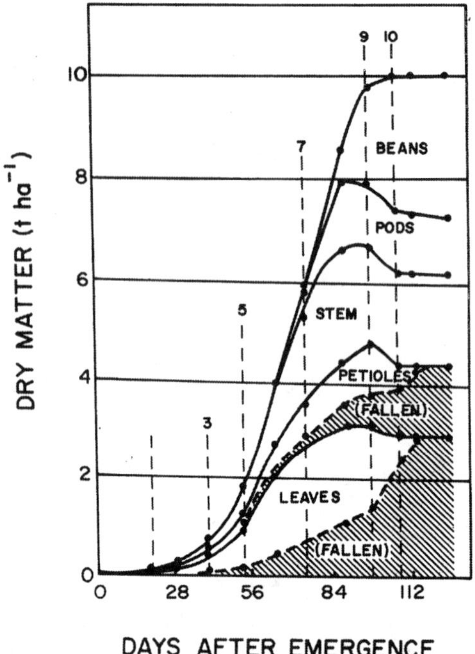

Figure 12.5 Dry matter accumulation of soybean plant parts at various stages of growth. (From Hanway and Weber, 1971a.)

(Cox and Jolliff, 1986). Figure 12.6 shows that irrigation has a significant effect on crop growth rate of soybean.

C. Leaf Area Index

Leaf area index is defined as the area of leaf surface per unit area of land surface. Maximum leaf area production for the determinate cultivars may occur near the beginning of flowering, but indeterminate cultivars reach a maximum leaf area near the end of flowering. Maximum *LAI* values may range from 5 to 8 in soybean (Whigham, 1983). Leaf area is critical for crop light interception and therefore has a substantial influence on crop yield. Shibles and Weber (1965) reported that an *LAI* of 3.2 was required for 95% light interception and 95% dry matter production. Figure 12.6 shows the *LAI* of soybean at different growth stages and the influence of irrigation treatments on *LAI*.

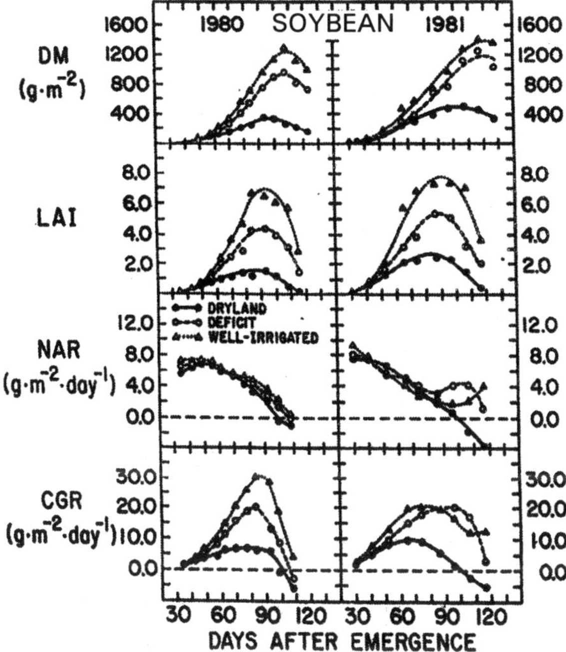

Figure 12.6 Seasonal patterns of total dry matter production (DM), leaf area index (*LAI*), net assimilation rate (*NAR*), and crop growth rate (CGR) in well-irrigated, deficit, and dryland treatments of soybean. (From Cox and Jolliff, 1986.)

D. Net Assimilation Rate

A useful measure of the photosynthesis efficiency of plants is net assimilation rate (*NAR*), defined as the rate of increase of dry matter per unit of leaf area. The *NAR* is generally constant in the beginning of crop growth and then decreases to a relatively constant rate (Figure 12.6). As *LAI* increases, the efficiency per unit leaf area is reduced as a result of competition for light, and *NAR* decreases. The other reason for *NAR* decrease is leaf abscission in the later part of the crop growth (Clawson et al., 1986).

E. Harvest Index

The harvest index, defined as the ratio of grain to total dry matter production in crops, is used as an index of dry matter distribution. It is now well established that partitioning of dry matter is related to increased yields in improved cultivars rather than total biomass production (Gifford and Evans, 1981). In soybeans, the competition for photosynthetic products varies during the growth cycle. Vegetative plant parts are the only sinks for assimilate prior to flowering, and during flowering and fruit set, vegetative and reproductive plant parts are competing sinks for assimilate. However, the fruit is the primary sink for assimilate during the seed-filling period (Egli, 1988).

Harvest index of a cultivar is a stable characteristic irrespective of substantial differences in environmental conditions during plant growth (Spaeth et al., 1984). Johnson and Major (1979) reported that harvest index for soybeans varied from 30 to 40%, depending on maturity of the cultivars and production practices adopted.

VI. NUTRIENT REQUIREMENTS

Nutrient requirements of a crop vary according to soil and climatic conditions, cultivar, yield level, cropping system, and management practices. Further, the quantity of fertilizer applied by farmers varies with the socioeconomic situation. Therefore, it is not possible to make fertilizer recommendations which are universally applicable. Basic principles of nutrient requirements for soybeans which have general applicability are discussed in this section. The principles can be used as a guideline for the mineral nutrition of soybean.

Soybeans can fix atmospheric N if the proper strain of *Rhizobium* bacteria is present in the soil or if the seed is properly inoculated. The plant starts to fix substantial amounts of atmospheric N approximately 4 weeks after germination. Most estimates show that soybean derives between 25 and 75% of its N from fixation (Deibert et al., 1979). The variable amount of N fixed by soybean is due to the several factors that affect fixation, including the length of time that a soybean variety actively supports N fixation (Hardy, 1977). Higher levels of mineral N retard early nodule development, but under some conditions, marked beneficial effects of mineral nitrogen on the total amount of N_2 fixed during the growing season have been recorded (Shibles et al., 1975). Early supplies of mineral N enable the plant to maintain a reasonable growth rate from the outset, enable nodule to develop more rapidly once the inhibition ceases, and, because of the increased plant size, may increase nodule mass (Shibles et al., 1975).

Imsande (1992) measured and calculated various N-dependent growth characteristics of soybean and their relation to seed yield and seed protein content. Twenty growth-yield characteristics, including yield, harvest index, N harvest index, and Kjeldahl analysis of N_2 fixation, were measured or calculated for each of the 384 plants examined. The highest seed yields, approximately 10 g plant^{-1}, and the highest seed N contents, approximately 560 mg plant^{-1}, were obtained when well-nodulated plants were provided some fertilizer-N during pod fill. Except for N content (%) of the dried plant, correlations between each pair of the 20 N-dependent growth-yield characteristics were generally positive. In the absence of fertilizer-N during pod fill, however, N content (%) of the seeds did not correlate with either harvest index or N harvest index, suggesting that insufficient N during pod fill interferes with the orderly mobilization of foliar-N to the developing seeds. New physiological parameters (seed yield merit, N yield merit, and merit of genotype) were proposed for the identification of genetic lines that produce both a high seed yield and a high seed protein content (Imsande, 1992).

Soybean response to N fertilizer is inconsistent, unpredictable, and often unexplainable (Small and Ohlrogge, 1973). Conflicting results of N application to soybean have been reported, from response to no response (Mengel et al., 1987). But soybean producers in the United States apply about 2–9 kg N ha^{-1} at planting, depending on the geographic region (Hargett and Berry, 1983). Similarly, in Brazil, which is the second largest producer of soybean, farmers apply about 10–20 kg N ha^{-1} at planting (Peter, 1980). Other essential nutrients such as P, K, Ca, Mg, and S and micronutrients should be applied in appropriate amounts to obtain optimum crop yield, if soils are deficient in these nutrients.

Phosphorus deficiency is widespread in acid soils for soybean production. An adequate level is generally applied on the basis of critical soil P level. Cox (1992) determined the critical P level of Mehlich 3 extracting solution with the linear-plateau; the function ranged from 33 to 39 mg P kg^{-1}, with a mean of 35 mg P kg^{-1}. With critical P levels calculated by exponential function at 95% of maximum yield, the range was 55–82 mg kg^{-1} with a mean of 64 (Figure 12.7). The soil used in this experiment was Typic Umbraquult.

Crop response to K fertilization is often correlated with exchangeable K. In soils of the cerrado region of Brazil, response to K fertilization is expected when the level of exchangeable K is less than 0.15 cmol kg^{-1}. Cox and Uribe (1992) determined soybean response to potassium in a tropical Ultisol. Critical exchangeable level was 75 kg ha^{-1} (37.5 mg K kg^{-1}) when K was extracted by $NaHCO_3$ extracting solution (Figure 12.8). Borkert et al. (1993) also determined soybean response to K fertilization in Brazilian Oxisol. Maximum grain yield was obtained at about 100 kg K ha^{-1} (Figure 12.9). These authors also calibrated a K soil test against soybean grain yield. Based upon the data of five years of experimentation, three classes of soil exchangeable potassium contents in the soil were defined: low, when K content is less than 23 mg kg^{-1}; medium, when it is between 23 and 40 mg kg^{-1}; and high, when it is higher than 40 mg kg^{-1} by Mehlich 1 extracting solution (Borkert et al., 1993).

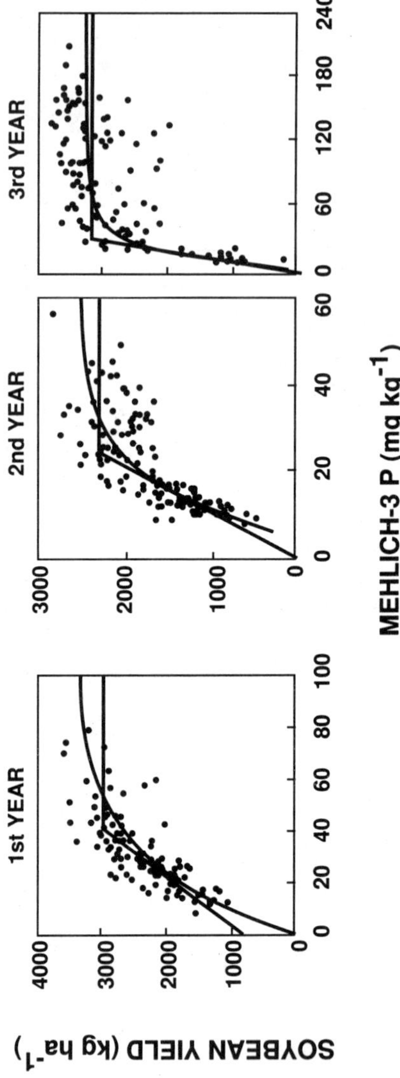

Figure 12.7 Response of soybean to Mehlich-3 P with linear-plateau and exponential prediction functions for three crops. (From Cox, 1992.)

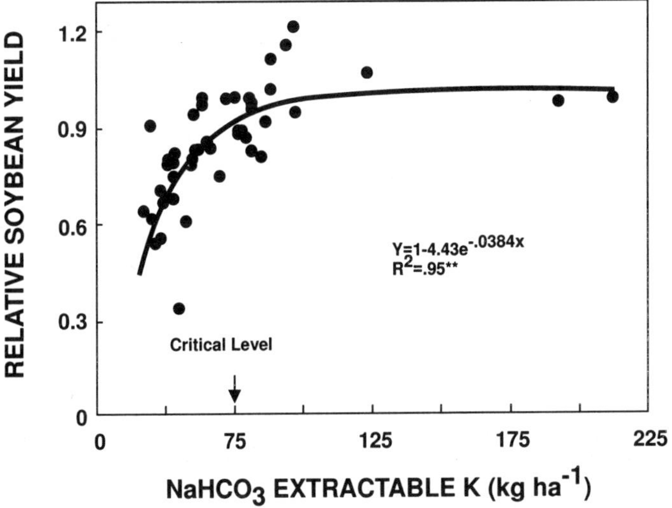

Figure 12.8 Relationship between $NaHCO_3$ extractable K and relative soybean yield. (From Cox and Uribe, 1992.)

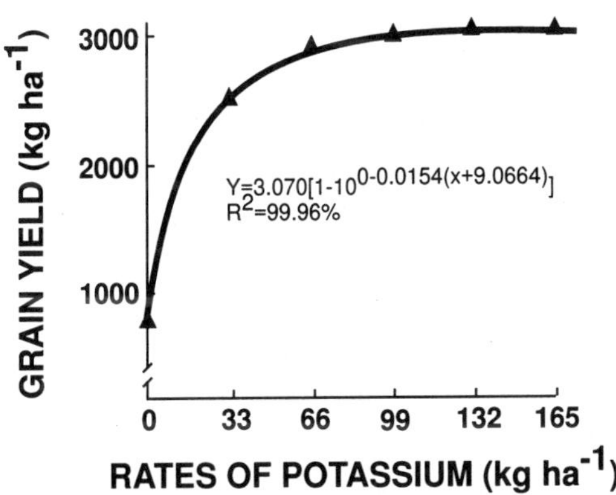

Figure 12.9 Relationship between soybean yield and rates of K applied as broadcast in an Oxisol. (From Borkert et al., 1993.)

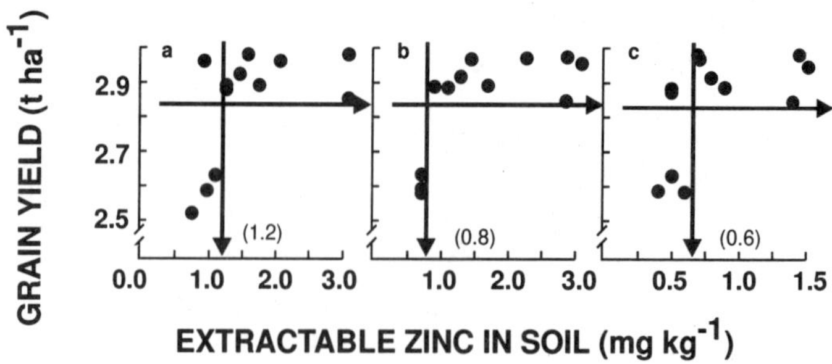

Figure 12.10 Relationship between soybean yield and extractable Zn in an Oxisol. Extractant of a = 0.1 N HCl, b = Mehlich 1, and c = DTPA. (From Galrão, 1993.)

Table 12.5 Deficient, Sufficient, and High Concentrations of Nutrients for Upper Fully Developed Trifoliate Prior to Pod Set

Nutrient	Deficient	Sufficient	High
		g kg^{-1}	
N	40	45–55	56–70
P	1.5	2.6–5	6–8
K	12.5	17–25	26–28
Ca	2.0	3.6–20	21–30
Mg	1.0	2.6–10	11–15
S	—	2.5–4	—
		mg kg^{-1}	
Fe	30	51–350	351–500
Mn	14	21–100	101–250
Zn	10	21–50	51–75
B	10	21–55	56–80
Cu	4	10–30	31–50
Mo	0.4	1–5	6–10
Al	—	200	201–400

Sources: Small and Ohlrogge, 1973; Rosolem, 1980.

Table 12.6 Approximate Nutrient Uptake by 1000 kg Soybean Grain Crop in Oxisol of Brazil

Nutrient	Quantity (kg)
N	81
P	6
K	27
Ca	17
Mg	11
S	2
	g
Fe	366
Mn	90
Zn	50
B	39
Cu	25
Mo	7
Cl	578

Source: Bataglia and Mascarenhas, 1977.

Galrão (1993) determined the critical zinc level in an Oxisol of central Brazil for soybean using three extracting solutions. These extracting solutions were 0.1 N HCl, Mehlich 1 (0.05 N HCl + 0.025 N H_2SO_4) and DTPA (diethyletriaminepentaacetic acid). The critical Zn levels were 1.2, 0.8, and 0.6 mg kg^{-1}, respectively, by these three extracting solutions (Figure 12.10).

A. Nutrient Concentration and Uptake

Information on nutrient concentration (content per unit dry matter) and uptake (concentration × dry matter) is important in understanding the mineral requirements of soybean. Nutrient concentration information is obtained through plant tissue analysis and can be used in the diagnosis of nutrient deficiency or sufficiency in the growth medium for crop production. Similarly, nutrient uptake data can be used to study nutrient removal from the soil in order to replenish to maintain or improve soil fertility for better yield. Nutrient concentration and uptake by soybean crop vary with environmental conditions, cultivar planted, yield level, and management practices. The data presented in Table 12.5 for concentration and in Table 12.6 for uptake provide some

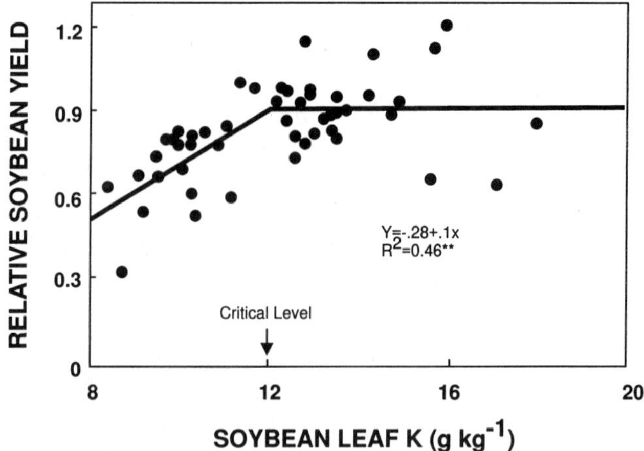

Figure 12.11 Relationship between K content of soybean leaf at flowering and relative grain yield. (From Cox and Uribe, 1992.)

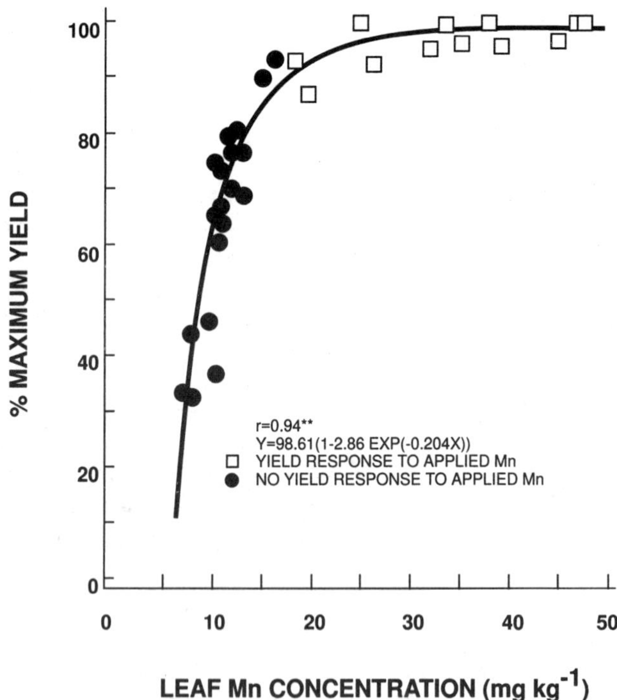

Figure 12.12 Soybean yield response curve for uppermost mature trifoliate leaf manganese concentrations sampled at early bloom. (From Gettier et al., 1985.)

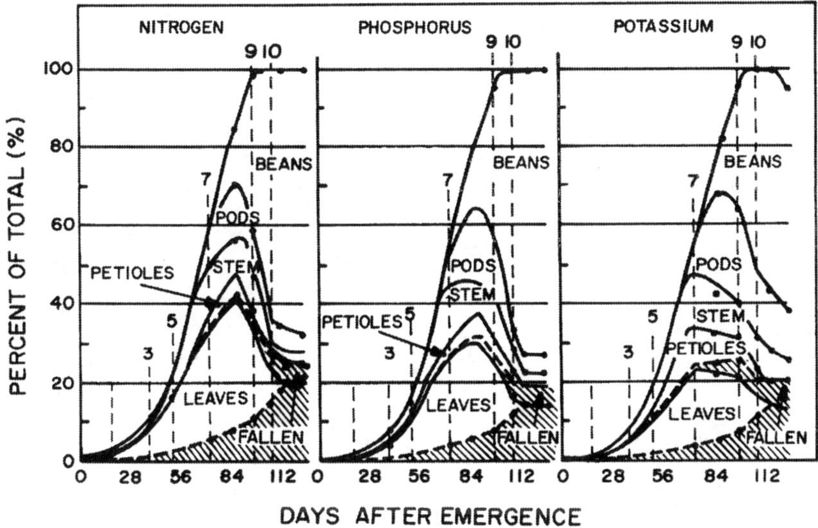

Figure 12.13 Relative amounts of N, P, and K in different plant parts of soybean at different growth stages. (From Hanway and Weber, 1971b.)

guidelines for understanding the mineral requirements of this crop. Nutrient concentrations below the minimum and above the maximum are easy to interpret, but interpretations in the middle range are more difficult. Cox and Uribe (1992) determined the K in soybean leaf at flowering, with 12 g kg^{-1} of dry weight as a critical level (Figure 12.11).

Manganese deficiency in soybean is frequently reported when soil pH is raised in acid soils by liming (Gettier et al., 1985). Further pressure is placed upon soil Mn reserves by high yields and multiple cropping systems. Gettier et al. (1985), based on 90% of maximum yield as the definition of the critical deficiency level, determined that the critical Mn levels for whole plant, leaf, and seed were 45, 17, and 20 mg kg^{-1}, respectively. A relationship between leaf Mn concentration and percent maximum yield is presented in Figure 12.12.

Hanway and Weber (1971b) determined the relative uptake of N, P, and K by indeterminate soybean under field conditions. The relative amounts accumulated in different plant parts are presented in Figure 12.13. Total accumulation of N, P, and K in the plants followed patterns similar to that of dry matter accumulation. Accumulation was slow early in the growth stage but then became rapid, and nutrients accumulated at constant daily rates between stages 5 and 9. Approximately 80% of the total accumulation of these nutrients occurred during the 46–day period between stages 5 and 9.

VII. SUMMARY

Soybean is the most important warm-season legume in temperate climates, and its use is growing in the tropics and subtropics. The species was probably domesticated in northeastern China over 1000 years ago. Its importance is due to the high protein and oil content of its grain, and its adaptability has increased as cultivars with different degrees of photoperiod sensitivity and better disease resistance have been developed. Maximum soybean yields exceed 6 Mg ha^{-1} and many countries report yields exceeding 4 Mg ha^{-1}, but world average yields are only about 1.9 Mg ha^{-1}. Soybeans are grown from 0 to 55° latitude and from below sea level to 2000 m elevation, but most commercial production is between 25 and 45° latitude and below 1000 m elevation. Soybeans can be grown on a wide range of well-drained soils, and the optimum soil pH is reported to be 6.0–6.5. Soybean is moderately tolerant of salinity. Like most grain crops, it is most sensitive to water stress during flowering and early grain development. Soybean is often grown in rotation with warm-season cereals such as corn in order to facilitate pest control and optimize available labor.

Soybean has efficient symbiotic nitrogen fixation that can provide over 80% of the nitrogen in the crop at maturity; however, in most cases fixation accounts for 25–75% of total plant nitrogen. Nitrogen fixation begins after nodulation by appropriate strains of *Rhizobium* (about 4 weeks after germination). Fixation is inhibited by high levels of mineral nitrogen in the soil, by drought stress, and by poor soil aeration. Though soybean's response to fertilizer N is inconsistent, small amounts are often applied at planting to stimulate plant growth until symbiotic nitrogen fixation begins.

One metric ton of soybean grain removes approximately 59 kg N, 60 kg P, and 19 kg K. To maintain soil fertility, nutrients (other than nitrogen) removed in the grain should be returned in fertilizers and/or manures. Both soil and plant analyses can be used to detect nutrient deficiencies and toxicities.

REFERENCES

Bataglia, O. C. and H. A. A. Mascarenhas. 1977. Nutrient absorption by soybean. Inst. Agron. Campinas Tech. Bull. 41.

Bergersen, F. J. 1958. The bacterial component of soybean root nodule: Changes in respiratory activity, cell dry weight, and nucleic acid content with increasing nodule age. J. Gen. Microbiol. 19: 312–323.

Bhuvaneswari, T. V., B. G. Turgeon, and W. D. Bauer. 1980. Early events in the infection of soybean (*Glycine max* L. Merr.) by *Rhizobium japonicum*. I. Localization of infectable root cells. Plant Physiol. 66: 1027–1031.

Bils, R. F. and R. W. Howell. 1963. Biochemical and cytological changes in developing soybean cotyledons. Crop Sci. 3: 304–308.

Board, J. E., B. G. Harville, and A. M. Saxton. 1990. Branch dry weight in relation to yield increases in narrow-row soybean. Agron. J. 82: 540–544.

Board, J. E. and Q. Tan. 1995. Assimilatory capacity effects on soybean yield components and pod number. Crop Sci. 35: 846–851.

Borkert, C. M., G. J. Sfredo, and D. N. Silva. 1993. Potassium calibration for soybean in a Latosssolo roxo distrofico (Oxisol). R. Bras. Ci. Solo 17: 223–226.

Brady, R. A., L. R. Stone, C. D. Nickell and W. L. Powers. 1974. Water conservation through proper timing of soybean irrigation. J. Soil Water Conserv. 29: 266–268.

Brown, D. M. 1960. Soybean Ecology. I. Development—temperature relationships from controlled environment studies. Agron. J. 52: 493–496.

Brown, D. A. 1984a. Characterizing root growth and distribution. Arkansas Farm Res. 33: 3.

Brown, R. H. 1984b. Growth of the green plant, pp. 153–174. *In*: M. B. Tesar (ed.). Physiological basis of crop growth and development. Am. Soc. Agron., Madison, Wisconsin.

Bugbee, B. and O. Monje. 1992. The limits of crop productivity. Bioscience 42: 494–502.

Carlson, J. B. and N. R. Lersten. 1987. Reproductive morphology, pp. 95–134. *In*: B. E. Caldwell (ed.). Soybeans: Improvement, production, and uses, 2nd Ed. Monogr. 16, Am. Soc. Agron., Madison, Wisconsin.

Carter, J. L. and E. E. Hartwig. 1962. The management of soybean. Adv. Agron. 14: 359–412.

Chapman, S. R. and L. P. Carter. 1976. Soybean, pp. 345–357. *In*: Crop Production: Principles and practices. Freeman, San Francisco.

Ciafardini, G. and C. Barbieri. 1987. Effects of cover inoculation of soybean on nodulation, nitrogen fixation, and yield. Agron. J. 79: 645–648.

Clawson, K. L., J. E. Specht, and B. L. Blad. 1986. Growth analysis of soybean isolines differing in pubescence density. Agron. J. 78: 164–172.

Cox, F. R. 1992. Range in soil phosphorus critical levels with time. Soil Sci. Soc. Am. J. 56: 1504–1509.

Cox, F. R. and E. Uribe. 1992. Potassium in two humid tropical Ultisols under a corn and soybean cropping system: I. Management. Agron. J. 84: 480–484.

Cox, W. J. and G. D. Jolliff. 1986. Growth and yield of sunflower and soybean under soil water deficits. Agron. J. 78: 226–230.

Deibert, E. J., M. D. Jeriego, and R. A. Olson. 1979. Utilization of ^{15}N fertilizer by nodulating and nonnodulating soybean isolines. Agron. J. 71: 717–723.

Doorenbos, J. and W. O. Pruit. 1977. Guidelines for predicting crop water requirements. FAO Irrigation and Drainage Paper 24, Food and Agricultural Organization of the United Nations, Rome.

Dornbos, D. L., Jr. and R. E. Mullen. 1991. Influence of stress during soybean seed fill on seed weight, germination, and seedling growth rate. Can. J. Plant Sci. 71: 373–383.

Egli, D. B. 1988. Alterations in plant growth and dry matter distribution in soybean. Agron. J. 80: 86–90.

Egli, D. B., J. E. Leggett, and J. M. Wood. 1978. Influence of soybean seed size and position on the rate and duration of filling. Agron. J. 70: 127–130.

Egli, D. B., R. A. Wiralaga, and E. L. Ramseur. 1987. Variation in seed size in soybean. Agron. J. 79: 463–467.

Fehr, W. R. 1987. Breeding methods for cultivar development. *In*: J. R. Wilcox (ed.) Soybeans: Improvement, production, and uses. 2nd Ed. Agron. Monogr. 16. ASA, CSSA, and SSSA, Madison, Wisconsin.

Fehr, W. R., and C. E. Caviness. 1977. Stages of soybean development. Iowa Agric. Exp. Stn. Spec. Res. 80, Ames.

Fehr, W. R., C. E. Caviness, D. T. Burmood, and J. S. Pennington. 1971. Stages of development descriptions of soybeans (*Glycine max* L. Merrill). Crop Sci. 11: 929–931.

Fukui, J. and T. Gotoh. 1962. Varietal difference of the effects of day length and temperature on the development of floral organs in the soybean. I. Developmental stages of floral organs of the soybean. Japan J. Breed. 12: 17–27.

Gabel, M. 1979. Ho-Ping. Food for everyone. Anchor Books, Garden City, New York.

Galrão, E. Z. 1993. Critical levels of zinc on soybeans grown in a cerrado clayey red Latosol. R. Bras. Ci. Solo 17: 83–87.

Gettier, S. W., D. C. Martens, and S. J. Donohue. 1985. Soybean yield response prediction from soil test and tissue manganese levels. Agron. J. 77: 63–67.

Gbikpi, P. J. and R. K. Crooksten. 1981. Effect of flowering date on accumulation of dry matter and protein in soybean seeds. Crop Sci. 21: 652–655.

Gibson, L. R. and R. E. Mullen. 1996. Influence of day and night temperature on soybean seed yield. Crop Sci. 36: 98–104.

Gifford, R. M. and L. T. Evans. 1981. Photosynthesis, carbon partitioning and yield. Annu. Rev. Plant Physiol. 32: 485–509.

Hanway, J. J. and C. R. Weber. 1971a. Dry matter accumulation in soybean (*Glycine max* L. Merrill) plants as influenced by N, P, and K fertilization. Agron. J. 63: 263–266.

Hanway, J. J. and C. R. Weber. 1971b. Accumulation of N, P, and K by soybean (*Glycine max* L. Merrill) plants. Agron. J. 63: 406–408.

Hardy, R. W. F. 1977. Rate-limiting steps in biological reproductivity, pp. 369–399. *In*: A. Hollagender (ed.). Genetic engineering for nitrogen fixation. Plenum, New York.

Hargett, N. L. and J. T. Berry. 1983. Fertilizer summary data. TVA Bull. Y-165.

Hartwig, E. E. 1973. Varietal development. *In*: B. E. Caldwell (ed.). Soybeans. Agronomy 16: 187–210.

Hiebsch, C. K., F. T. Kagho, A. M. Chirembo, and F. P. Gardner. 1995. Plant density and soybean maturity in a soybean-maize intercrop. Agron. J. 87: 965–969.

Hiler, E. A., T. A. Howell, R. B. Lewis, and R. T. Boos. 1974. Irrigation timing by the stress day index method. Trans. ASAE 17: 393–398.

Hoogenboom, G., C. M. Peterson, and M. G. Huck. 1987a. Shoot growth rate of soybean as affected by drought stress. Agron. J. 79: 598–607.

Hoogenboom, G., M. G. Huck, and C. M. Peterson. 1987b. Root growth rate of soybean as affected by drought stress. Agron. J. 79: 607–614.

Howell, R. W. 1960. Physiology of the soybean. Adv. Agron. 12: 265–310.

Imsande, J. 1992. Agronomic characteristics that identify high yield, high protein soybean genotypes. Agron. J. 84: 409–414.

Johnson, D. R. and D. J. Major. 1979. Harvest index of soybeans as affected by planting date and maturity rating. Agron. J. 71: 538–541.

Johnson, H. W. and R. L. Bernard. 1962. Soybean genetics and breeding. Adv. Agron. 14: 149–221.

Kamata, E. 1952. Studies on the development of fruit in soybean. Crop Sci. Soc. Japan Proc. 20: 296–298.

Kanemasu, E. T., L. R. Stone, and W. L. Powers. 1976. Evapotranspiration model tested for soybean and sorghum. Agron. J. 68: 569–572.

Kaspar, T. C., C. D. Stanley, and H. M. Taylor. 1978. Soybean root growth during the reproductive stages of development. Agron. J. 70: 1105–1107.

Kaspar, T. C., H. M. Taylor, and R. M. Shibles. 1984. Taproot-elongation rates of soybean cultivars in the glasshouse and their relation to field rooting depth. Crop Sci. 24: 916–920.

Kato, I., S. Sakaguchi, and Y. Naito. 1954. Development of flower parts and seed in soybean plant *Glycine max*. Tokai-Kinki Nat. Agric. Exp. Stn. Bull. 1: 96–114.

Kato, I., S. Sakaguchi, and Y. Naito. 1955. Anatomical observations on fallen buds, flowers, and pods of soybean *Glycine max*. Tokai-Kinki Nat. Agric. Exp. Stn. Bull. 2: 159–168.

Klepper, B. and T. C. Kaspar. 1994. Rhizotrons: Their development and use in agricultural research. Agron. J. 86: 745–753.

Larue, T. A. and T. G. Patterson. 1981. How much nitrogen do legume fix? Adv. Agron. 34: 15–38.

Lersten, N. R. and J. B. Carlson. 1987. Vegetative morphology, pp. 49–94. *In*: J. R. Wilcox (ed.). Soybeans: Improvement, production and uses, 2nd Ed. Monogr. 16, Am. Soc. Agron., Madison, Wisconsin.

Maas, E. V. 1986. Salt tolerance of plants. Appl. Agric. Res. 1: 12–25.

Mason, W. K., G. A. Constable, and R. G. G. Smith. 1981. Irrigation for crops in a subhumid environment. II. The water requirement of soybeans. Irrig. Sci. 2: 13–22.

McBlain, B. A., J. D. Hesketh, and R. L. Bernard. 1987. Genetic effects on reproductive phenology in soybean isolines differing in maturity genes. Can. J. Plant Sci. 67: 105–116.

McLean, E. O. and J. R. Brown. 1984. Crop response to lime in the midwestern United States, pp. 267–303. *In*: F. Adams (ed.). Soil acidity and liming, 2nd Ed. Monogr. 12, Am. Soc. Agron. Madison, Wisconsin.

Mengel, D. B., W. Segars, and G. W. Rehm. 1987. Soil fertility and liming, pp. 461–496. *In*: J. R. Wilcox (ed.). Soybeans: Improvement, production and uses, 2nd Ed. Monogr. 16, Am. Soc. Agron., Madison, Wisconsin.

Meng-Yuan, H. 1963. Studies on the embryology of soybeans. I. The development of embryo and endosperm. Acta. Biol. Sinica 11: 318–328.

Mitchell, R. L. and W. J. Russell. 1971. Root development and rooting patterns of soybean (*Glycine max* L. Merrill) evaluated under field conditions. Agron. J. 63: 313–316.

Momen, N. N., R. E. Carlson, R. H. Shaw, and O. Arjinand. 1979. Moisture stress effects on yield components of two soybean cultivars. Agron. J. 71: 86–90.

Murneek, A. E. and E. T. Gomez. 1936. Influence of length of day photoperiod on development of the soybean plant, *Glycine max* Var. Biloxi Mo. Agric. Exp. Stn. Res. Bull. 242.

Musick, J. T., L. L. New, and D. A. Dusek. 1976. Soil water depletion-yield relationships of irrigated sorghum, wheat, and soybeans. Trans. ASAE 19: 489–493.

Nathanson, K., R. J. Lawn, P. L. M. Jabrun, and D. E. Byth. 1984. Growth, nodulation and nitrogen accumulation by soybean in saturated soil culture. Field Crops Res. 8: 73–92.

Nelson, C. J. and K. L. Larson. 1984. Seedling growth, pp. 93–129. *In*: M. B. Tesar (ed.). Physiological basis of crop growth and development. Am. Soc. Agron., Madison, Wisconsin.

Nicholaides, J. J. and M. I. Piha. 1987. A new methodology to select cultivars tolerant to aluminum and with high yield potential, pp. 103–116. *In*: CIAT (ed.) Sorghum for acid soils: Proc. of a Workshop on Evaluating Sorghum for Tolerance to Al Toxic Tropical Soils in Latin America, held in Cali, Colombia, May 28 to June 2, 1984, Cali, Colombia.

Ozaki, K., M. Saito, and K. Nitta. 1956. Studies on the seed development and germination of soybean plants at various ripening stages. Res. Bull. Hokkaido Natl. Agric. Exp. Stn. 70: 6–14.

Pamplin, R. A. 1963. The anatomical development of the ovule and seed in the soybean. Ph.D. Dissertation, University of Illinois, Urbana (Diss. Abstr. 63–5128).

Peter, A. V. 1980. Fertilizer requirements in developing countries. Proc. Fertil. Soc. 188: 1–58.

Prakash, N. and Y. Y. Chan. 1976. Embryology of *Glycine max*. Phytomorphology 26: 302–309.

Purseglove, J. W. 1987. Tropical crops: Dicotyledons. Longman, New York.

Reddy, M. R. and J. J. Dunn. 1987. Differential response of soybean genotypes to soil pH and manganese application. Plant and Soil 101: 123–126.

Reicosky, D. C. and D. E. Deaton. 1979. Soybean water extraction, leaf water potential, and evapotranspiration during drought. Agron. J. 71: 45–70.

Rochette, P., R. L. Desjardins, E. Pattey, and R. Lessard. 1995. Crop net carbon dioxide exchange rate and radiation use efficiency in soybean. Agron. J. 87: 22–28.

Rosolem, C. A. 1980. Mineral nutrition and fertilization of soybean. Tech. Bull. 6, Potassium and Phosphate Institute, Piracicaba, Brazil.

Seddigh, M. and G. D. Jolliff. 1984a. Effects of night temperature on dry matter partitioning and seed growth of indeterminate field grown soybean. Crop Sci. 24: 704–710.

Seddigh, M. and G. D. Jolliff. 1984b. Night temperature effects on morphology, phenology, yield and yield components of indeterminate field grown soybean. Agron. J. 76: 824–828.

Shibles, R. M. and C. R. Weber. 1965. Leaf area solar radiation interception and dry matter production by soybeans. Crop Sci. 5: 575–578.

Shibles, R. M. and C. R. Weber. 1966. Interception of solar radiation and dry matter production by various soybean planting patterns. Crop Sci. 6: 55–59.

Shibles, R. M., I. C. Anderson, and A. H. Gibson. 1975. Soybean, pp. 151–189. *In*: L. T. Evans (ed.). Crop physiology. Cambridge University Press, Cambridge.

Shibles, R., J. Secor, and D. M. Ford. 1987. Carbon assimilation and metabolism, pp. 535–588. *In*: J. R. Wilcox (ed.). Soybeans: Improvement, production and uses, 2nd Ed. Monogr. 16, Am. Soc. Agron., Madison, Wisconsin.

Sionit, N., B. R. Strain, and E. P. Flint. 1987. Interaction of temperature and CO_2 enrichment on soybean: Photosynthesis and seed yield. Can. J. Plant Sci. 67: 629–636.

Small, H. G. and A. J. Ohlrogge. 1973. Plant analysis as an aid in fertilizing soybeans and peanuts, pp. 315–327. *In*: L. M. Walsh and J. D. Beaton (eds.). Soil testing and plant analysis. Soil Sci. Soc. Am., Madison, Wisconsin.

Smith, K. J. and W. Huyser. 1987. World distribution and significance of soybean, pp. 1–22. *In*: J. R. Wilcox (ed.). Soybeans: Improvement, production and uses, 2nd Ed. Monogr. 16, Am. Soc. Agron., Madison, Wisconsin.

Spaeth, S. C., H. C. Randall, T. R. Sinclair, and J. S. Vendeland. 1984. Stability of soybean harvest index. Agron. J. 76: 482–486.

Suetsugu, I., I. Anaguchi, K. Saito, and S. Kumano. 1962. Development processes of the root and top organs in the soybean varieties, pp. 89–96. *In*: Bull. Hokuriki Agric. Exp. Stn. Takada 3.

Troedson, R. J., R. J. Lawn, D. E. Byth, and G. L. Wilson. 1986. Saturated soil culture, an innovative water management option for soybean in the tropics and subtropics, pp. 171–180. *In*: S. Shammugasundaram and E. W. Sulzberger (eds.) Soybean in tropical and subtropical cropping systems. Proceedings of a Symposium, September 1993. Tsukuba, Japan.

Troedson, R. J., R. J. Lawn, D. E. Byth, and G. L. Wilson. 1989. Response of field-grown soybean to saturated soil culture. 1. Patterns of biomass and nitrogen accumulation. Field Crops Res. 21: 171–187.

USDA (United States Department of Agriculture). 1952. Manual for testing agricultural and vegetable seeds. Agric. Handbook. No. 30.

USDA (United States Department of Agriculture). 1980. World oil seeds situation and outlook. Foreign Agric. Circ. FOP. 14–80, Washington, DC.

Van Doren, D. M., Jr. and D. C. Reicosky. 1987. Tillage and irrigation, pp. 391–428. *In*: J. R. Wilcox (ed.). Soybeans: Improvement, production and uses, 2nd Ed. Monogr. 16, Am. Soc. Agron., Madison, Wisconsin.

Wang, G., M. B. Peoples, D. F. Herridge, and B. Rerkasem. 1993. Nitrogen fixation, growth and yield of soybean grown under saturated soil culture and conventional irrigation. Field Crops Res. 32: 257–268.

Whigham, D. K. 1983. Soybean, pp. 205–225. *In*: International Rice Research Institute (ed.). Potential productivity of field crops under different environments. IRRI, Los Banos, Philippines.

Wilcox, J. A., W. J. Wiebold, T. L. Niblack, and K. D. Kephart. 1995. Growth and development of soybean isolines that differ for maturity. Agron. J. 87: 932–935.

13

Common Bean and Cowpea

I. INTRODUCTION

Common bean and cowpea are important grain legume crops and supply a large part of the daily protein requirement of the people of South America, the Caribbean, Africa, and Asia. Common bean is a principal source of protein for more than 500 million people in Latin America and Africa; when consumed as snap beans, it is an important dietary source of vitamins and minerals in Asia (Yan et al., 1995). Land area devoted to bean production in developing countries has increased steadily in the last several decades (CIAT, 1992). However, production has not kept pace with population growth and must increase 42% in Latin America and 72% in Africa by the year 2000 to satisfy expected demand (Janssen, 1989). In China alone, the demand for snap beans was 3.5 million t in 1989, and a demand of 4.3 million t is projected for the year 2000 (Henry, 1989). Bean production in developing countries is often on marginal land, and few developing countries have significant reserves of arable land that can be opened to bean cultivation, so increased bean production will largely have to come about through increased yield per hectare rather than expansion of area under cultivation. Average bean yields in most developing countries are < 20% of yield potential, which indicates that substantial improvement in bean production could be realized by increasing yields per unit land are (Yan et al., 1995). Common beans and cowpea seeds are rich in protein (20–25%) but, as in most legumes, the proteins are deficient in sulphur amino acids; however, the yield of these two important crops is quite low in most cropping systems. Average yields of common bean are less than 1 t ha^{-1} in most developing countries and less than 1.4 t ha^{-1} in most developed countries (Laing et al., 1984). Similarly, the average cowpea yield on a world level is 380 kg ha^{-1} (Summerfield et al., 1983). The main reasons for low yields are water deficit, high incidence of diseases and insects, and limited use of inorganic fertilizers. Further, very little research has been devoted to improving productivity of these crops.

II. COMMON BEAN

Over 50 species of *Phaseolus* have been reported from the Americas (Debouck, 1991). Common beans (*Phaseolus vulgaris* L.) are the most widely grown of the five cultivated species of *Phaseolus*. The other four species are *P. coccineus*, *P. lumatus*, *P. polyanthus*, and *P. acutifolius* var. Latifolius. Common beans are also known as dry, field, French, snap, navy, or kidney beans. Beans are usually grown in tropical countries for dry seeds and in temperate countries for dry seeds as well as for fresh pod consumption and for processing as frozen vegetables. For dry seed it is grown annually on over 13 million hectares in the world, with production of about 9 million tons (Singh, 1992). *Phaseolus vulgaris* is believed to have originated in Central America or South America. It was taken to Europe, Africa, and Asia in the sixteenth century by the Spanish and Portuguese. In North America beans spread through California and into the other western states. Numerous introductions from Europe into the eastern part of the United States were made throughout the late nineteenth century (Evans, 1976). *Phaseolus vulgaris* is now grown widely in many parts of the tropics, subtropics, and temperate regions (Purseglove, 1987). Maximum production of *P. vulgaris* is concentrated in South America and the Caribbean, followed by Africa, China, North America, and Europe (Laing et al., 1983). In tropical and subtropical Latin America, bean production is 46.7%; in sub-Saharan Africa, 24.1%; in North America, 11.6%; in Europe 10.4%; and in Asia and North Africa, 6.5% (Pachico, 1989).

Asia, Iran, and Turkey are major bean producers, while Brazil (2.2 million tons) and Mexico (1.1 million tons) are the largest producers and consumers in the world. The highest yearly per capita consumption (> 40 kg) of dry bean is in Rwanda and Burundi (CIAT, 1981).

There are strong preferences for seed types of dry bean in different countries and regions within countries (Voysest, 1989; Voysest and Dessert, 1991;

Vieira, 1988). In Brazil, for example, small-seeded (< 25 g/100–seed weight) black, cream, and cream-striped beans are popular. The latter two predominate in the northeastern states, whereas black beans are more popular in the southern region (Vieira, 1988). Similarly, small black and/or red beans are consumed in Central America, Mexico, Cuba, and Venezuela, whereas, in the Andean countries of Colombia, Ecuador, and Peru, large-seeded red, pink, beige, and cream types, both solid and with various patterns of mottling, speckling, and spotting, are preferred. These latter types also predominate in most of sub-Saharan Africa. In Europe, North Africa, and western Asia, white, red, and cream-mottled beans of different sizes and shapes (but mostly medium and large) are consumed (Singh, 1992).

A. Climate and Soil Requirements

Although common beans are a warm-season crop, they are grown in a wide range of climates. Yields are usually higher in temperate than in tropical zones, but in temperate areas they are grown in the warm season. Studies conducted at Centro Internacional de Agricultura Tropical (CIAT; 1980) showed that approximately 80% of all bean production in Latin America is in regions where the mean temperature during the growing season is between 18 and 25°C. Major production occurs in areas where the temperature is around 21°C (Laing et al., 1984). Bean seeds germinate poorly in soils colder than 15°C (Kooistra, 1971). Seedlings emerge after about 17 days when soil temperatures are 10–11°C, after 6–8 days at 13–14°C, and after only 5 days at 15–16°C (Scarisbrick et al., 1976). The optimum temperature range for the germination of common bean is 20–30°C (Association of Official Seed Analysis, 1981; Scully and Waines, 1987). In general, high temperatures (> 30°C) during flowering cause dropping of buds and flowers, which reduces the yield. High temperature during the night is more detrimental than during the day (Gross and Kigel, 1994). The bean plant is intolerant of frost and a short exposure to 0°C or below will kill bean tissue (Wallace, 1980).

Common bean (*Phaseolus vulgaris* L.) production in many regions occurs under rainfed conditions where water deficit limits yield and causes instability of production (White et al., 1994). Although agronomic practices are important under water deficit, cultivar improvement is usually seen as the most promising approach to increase yields. Recent studies indicate that direct selection for seed yield in common bean can be effective both for well-watered (Nienhuis and Singh, 1988; Singh et al., 1990) and deficit conditions (White et al., 1994). In the latter study, values for realized gain in seed yield of bulk F_3 populations (after selection among F_2 populations) ranged from 0.4 to 15.7% in four rainfed environments. Nonetheless, yield testing is difficult

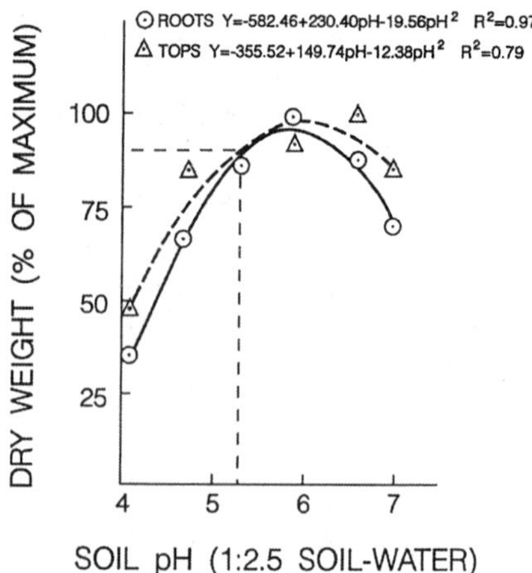

Figure 13.1 Relationship between dry weight of roots and tops and soil pH in an Oxisol. (From Fageria, 1996a.)

and costly, and gains from selection are sometimes low. This situation has led to extensive research on mechanisms of adaptation to water deficit not only in common bean but in many other crops (Ludlow and Muchow, 1990). Such studies hope to detect traits that either are more efficient as selection criteria than yield per se or aid in maximizing yield gains when selected simultaneously with yield (White et al., 1994).

Water requirements of common beans depend on soil and climatic factors, but the crop is considered to be poorly tolerant to water stress. Over 4 million ha of common bean are grown annually in the drought endemic areas of northeastern Brazil and the central highlands of Mexico alone. Drought in these regions is unpredictable in duration, intensity, frequency, and stages of crop growth affected. It can be predicted somewhat in areas where rains cease toward the end of the growing season or where it seldom rains (e.g., Central America and coastal Peru).

Drought tolerance in a broad sense, as defined by White and Singh (1991), encompasses all mechanisms that allow greater yields under soils moisture deficits; this includes a deep root system, earliness, and other traits. Genotypic differences, measured by seed yield per hectare, for response to moderate drought stress have been found in Brazil, Mexico (Singh, 1992), and Colombia (CIAT, 1985; White and Singh, 1991). These differences for seed yield have been observed in repeated evaluations. However, genetic studies and long-term selection experiments for drought tolerance have yet to be realized in the tropics and subtropics. A single gene responsible for heat-drought tolerance was reported in snap bean (Bouwkamp and Summers, 1982), but its value in dry bean improvement is not known, especially in the tropics and subtropics. The little attention dedicated to breeding for drought tolerance has been due to lack of information on its inheritance, difficulties in using seed yield as a selection criterion in early segregating generations, an overriding effect of local adaptation, and unavailability of any other dependable and easily usable selection criterion. Thus far, work has been restricted to systematic evaluations of germplasm accessions and advanced breeding lines under field conditions (Singh, 1992).

Water stress at flowering and at the early pod determination phase is especially harmful (Stoker, 1974). Soils at field capacity are optimum for bean crops; plant growth is reduced at a soil water potential of –0.03 MPa and ceases at about –0.5 MPa (Wallace, 1980). Beans seem to be more sensitive to water stress than deep-rooted crops like corn. Cowan (1977) cited a higher threshold leaf water potential for bean (about –0.8 MPa) than for corn (about –1.7 MPa). Bonnano and Mack (1983) found that an average soil water potential of –0.31 MPa decreased bean yield to about 43% of potential yield. The use of water-conserving systems for beans offers the possibility of increasing yields when water is limited.

The use of crop residue as mulch, either alone or with tillage, has been reported to affect soil water evaporation (Es), water storage, erosion, and yield of various crops (Rajat De et al., 1983). Barros and Hanks (1993) evaluated the effect of mulch on bean evapotranspiration (ET), yield, and water use efficiency (WUE). Dry matter and seed yields were significantly greater for mulched than for bare plots. Mulched plots had a higher WUE (yield/ET) than did bare plots for a given irrigation level but increased as irrigation level increased. Seasonal differences in ET between bare and mulched plots were small. The yield-ET relation for mulch was linear but was distinctly different from bare soil, indicating a different partitioning of ET into soil water evaporation (Es) and transpiration (Tr). For the conditions of this experiment, mulch reduced Es by about 45 mm, at the same ET, and Tr was increased by 45 mm. However, for the same irrigation level, ET was lower for mulched than for bare plots, indicating that not all of the water saved went to Tr.

Beans are also not suited to the very wet tropics, but they do well in areas of medium rainfall in tropical and temperate regions. Excessive rain causes flower drop and increases the incidence of diseases. Beans are also intolerant of poor soil aeration due to soil compaction and can tolerate a flooded soil for only about 12 hours. Beans can grow on a wide range of soils, from sandy to clay, provided water and drainage are adequate. The optimum soil pH for

Table 13.1 CIAT and Cambridge Systems of Growth Habit Classification in *Phaseolus vulgaris*

	Growth habit (CIAT classification)					
	I	II	III	IV		
Growth parameter	Determinate bush	Indeterminate bush	Indeterminate semiclimbing	Indeterminate climbing		
Plant height (cm)	44	92	103	160		
Nodes at flowering[a]	8	13	14	15		
Nodes at maturity	9	17	19	23		
Seed weight (mg/seed)	336	236	257	294		
Days to flowering	34	39	38	40		
Duration of flowering (days)	22	25	28	29		
	Cambridge classification					
	Indeterminate climber (1)	Indeterminate semiclimber (2)	Indeterminate bush (3)	Determinate multinoded (4.I)	Determinate intermediate noded (4.II)	Determinate bush (5)
Number of nodes[a]	13–35	11–30	7–23	11–15	7–10	3–6
Branching (scale 1–9)	2–6	7–9	2–6			
Main stem length (cm)	70–300	30–150	20–70			

[a] Nodes on main stem.

Sources: Compiled from Laing et al., 1983; Summerfield and Roberts, 1984a.

Table 13.2 Key for Identification of the Principal Growth Habits of Dry Beans (*Phaseolus vulgaris* L.)[a]

a	b	c	d	e	f	g	
Growth habit	Terminal bud	Growth	Stem and branch strength	Terminal guide	Climbing ability	Pod load distribution	Examples
Type I	Reproductive	Determinate	Strong, upright	Absent or small	Absent or weak climber	Along the length	Sanilac, Canario 101, Calima, Pompadour Checa, Albia Cerrillos INTA
Type II	Vegetative	Indeterminate	Strong, upright	Absent or small twining	Absent or weak climber	Along the length	Midnight, Porrillo, Sintetico, ICA Pijao Jamapa, Rio Tibago
Type III	Vegetative	Indeterminate	Weak, open, or prostrate	Small or medium twining	Weak or facultative climber	Concentrated in the basal portion	Pinto UI114, Flor de Mayo, Zamorano 2, Carioca, Cocacho
Type IV	Vegetative	Indeterminate	Very weak, twining	Very large, twining	Strong climber	Along the length or concentrated in the upper portion	Garbancillo Zarco, San Martin, Cargamanto Bola Roja, Caballeros

[a] Notes: 1) For growth habit identification the first evaluation needs to be made during flowering and the final evaluation 3 to 4 weeks later. 2) The type of terminal bud separates growth habit I (reproductive) from other three intermediate types (vegetative). The strength of stem and branches separates type II (strong and upright) from other indeterminate growth habits (weak, prostrate, or twining). It is the combination of characters *e*, *f*, and *g* which separates type III from growth habit IV. For evaluation of the latter two types, varieties need to be grown with and without support. 3) Types I, II, and III do not require support, and their grain yield is much higher in monoculture than with intercropping.

Source: Singh, 1982.

bean growth is in the range of 5.2–6.8 (Wallace, 1980), provided soil mineral nutrients are adequate. Figure 13.1 shows a relationship between relative dry weights of roots and tops and soil pH in an Oxisol. Maximum root weights of tops as well as roots were obtained at a pH of about 6. Beyond this value there was a decrease in dry weights. Bean is considered to be a salinity-susceptible crop, with a salinity threshold of about 1 dS m^{-1} (Maas and Hoffman, 1977).

B. Growth and Development

Growth and development from germination to maturity involve important changes in morphology and apply to components as well as the whole plant. Germination of common bean is epigeal, and if climatic conditions (soil moisture and temperature) are favorable and the seed is of good quality, germination is complete in 4–5 days. Mature seeds of common bean do not have a dormancy period. Under normal environmental conditions, most of the *P. vulgaris* cultivars cultivated for dry seeds complete their life cycle in 70–85 days.

Roots

Common beans, like many other legumes, have a taproot system with extensive lateral roots. The roots may grow to a depth of 1 m, but the lateral root system is mainly confined to the top 25 cm of the soil profile. The roots have nodules if inoculated with appropriate inoculants or if nitrogen-fixing *Rhizobium phaseoli* are present in the soil. Root growth is influenced by environmental factors. In addition, root characters are undoubtedly important in edaphic adaptation. White and colleagues have shown that drought tolerance in bean is related to depth of rooting (Sponchiado et al., 1989; White et al., 1990). Soil exploration by roots is associated with nutrient acquisition, especially in the case of immobile nutrients such as P. Genetic differences have been reported in common beans for root biomass and root to shoot ratio and root biomass distribution among distinct root types (Stofella et al., 1979; Lynch and Beem, 1993).

Tops

Common bean plants may be either bushy or trailing. These two types of common bean plants have been classified into different categories based on their growth habit. One classification has been proposed at the International Center for Tropical Agriculture (CIAT), Colombia, and another at Cambridge, England. These two classification schemes are summarized in Table 13.1. Singh (1982) simplified the CIAT classification and suggested a key for identification of four principal growth habits (Table 13.2). He gave a brief description of each of these growth habits and subtypes as follows:

Table 13.3 Vegetative and Reproductive Growth Stages in Common Bean

Growth stage	Description
Vegetative (V)	
V_1	Completely unfolded leaves at the primary (unifoliate) leaf node, average 10 days from sowing and having one node.
V_2	First node above primary leaf node, count when leaf edges no longer touch, average 19 days from sowing and having two nodes.
V_3	Three nodes on the main stem including the primary leaf node; secondary branching begins to show from branch of V_1 and average 27 days from sowing and having three nodes on main stem.
V_n	*n* nodes on the main stem, but with blossom clusters still not visibly opened. A new node each 3 days.
V_5	Bush (determinate) plants may begin to exhibit blossoms and become stage R_1. Average 50 days from sowing and five nodes on main stem.
V_8	Vine (indeterminate) plants may begin to exhibit blossoms and become stage R_1. Average 40 days from planting and eight nodes on main stem.
Reproductive (R) bush type	
R_1	One blossom open at any node. Average 50 days from sowing and having six nodes.
R_2	Pods about 1.25 cm long at first blossom position. Average 53 days from sowing and usually two to three nodes.
R_3	Pods 2.5 cm long at first blossom position. Secondary branching at all nodes. Average 56 days after sowing and plant one-half bloom.
R_4	Pods about 8 cm long, seeds not discernible and average 59 days after sowing.
R_5	Seeds discernible, and average 64 days after sowing.
R_6	Seeds at least 0.6 cm over long axis, average 66 days before sowing.
R_7	Oldest pods have developed seeds. Other parts of plant will have full-length pods with seeds almost as large as first pods. Pods will be developing over the whole plant. Average 72 days from sowing.
R_8	Leaves yellowing over half of the plant, very few small pods and these in axils of secondary branches, small pods may be drying. Point of maximum production has been reached. Average 90 days from sowing.
R_9	Mature, at least 80% of the pods showing yellow and mostly ripe. Only 40% of the leaves still green color. Average 105 days from sowing.

Table 13.3 (Continued)

Growth stage	Description
Reproductive (R) vine type	
R_1	One blossom open at any node. Tendril will begin to show. Average 40 days from sowing and eight nodes on main stem.
R_2	Pods 1.25 cm long at first blossom position. Blossom would have just suffed, nine nodes on main stem and average 43 days from sowing.
R_3	Pods about 2.5 cm long at first blossom position. Half bloom. 10 nodes on main stem and average 46 days from sowing.
R_4	Pods about 5 cm long at first blossom position, 11 nodes on main stem and average 50 days from sowing.
R_5	Pods more than 7.5 cm long, seeds discernible by feel, 12 nodes on main stem and average 56 days from sowing.
R_6	Pods 10 to 12.5 cm long with spurs. Seeds at least 0.75 cm in long axis. Average 80 days from sowing.
R_7	Oldest pods have fully developed green seeds. Other parts of plant will have full-length pods with seeds near same size. Pods to the top and blossom on tendril. Average 70 days from sowing.
R_8	Leaves yellowing over half of plant. Very few small new pods per blossom developing, small pods may be drying. Point of maximum production has been reached. Average 82 days from sowing.
	Mature, at least 80% of the pods showing yellow and mostly ripe. Only 30% of leaves are still green and average 94 days from sowing.

Source: Compiled from Lebaron, 1974.

Type I Determinate growth habit; reproductive terminal buds on the main stem and branches; limited or no further node and leaf production after flowering commences.

Branches and main stem generally strong and upright (Ia).

Branches and main stem weak, prostrate, and possess some ability to climb (Ib).

Type II Indeterminate growth habit; vegetative terminal bud on main stem and branches; node and leaf production occur after flowering commences. Both main stem and branches strong and upright.

The terminal guide or leader (excessively elongated and weak internodes) absent, thus lacking climbing ability (IIa).

Terminal guide of varying length present and hence possesses some climbing ability (IIb).

Type III Indeterminate growth habit. Branches relatively weak and open, semiprostrate, or twining. Pod load largely concentrated in the basal part of the plant. The maximum yield is realized in monoculture.

Branches relatively short, guide on the main stem and/or branches is small when present and possesses weak climbing ability (IIIa).

Branches long, often prostrate or twining, with relatively long main stem guide and moderate climbing ability (IIIb).

Type IV Indeterminate growth habit. Stem and branches very weak and excessively long, possessing strong climbing ability. Support essential for maximum production.

Pod load distributed all along the length of the plant (IVa).

Pod load mostly borne on the upper part of the plant (IVb).

Table 13.4 Average Dry Matter Yields (g per Two Plants) of Tops and Roots of Three Bean Cultivars at Different Growth Stages Under Greenhouse Conditions

Age (days)	Carioca		CNF10		CNF4856	
	Tops	Roots	Tops	Roots	Tops	Roots
17	0.29	0.12	0.31	0.14	0.24	0.09
31	1.45	0.34	1.73	0.43	1.49	0.39
45	5.58	0.88	4.67	1.03	5.17	0.82
60	14.45	1.17	11.92	1.05	12.65	1.29
80[a]	18.33	1.29	14.38	1.36	19.24	1.48

[a] Cultivar CNF10 was harvested at 68 days.

Source: Fageria, 1989.

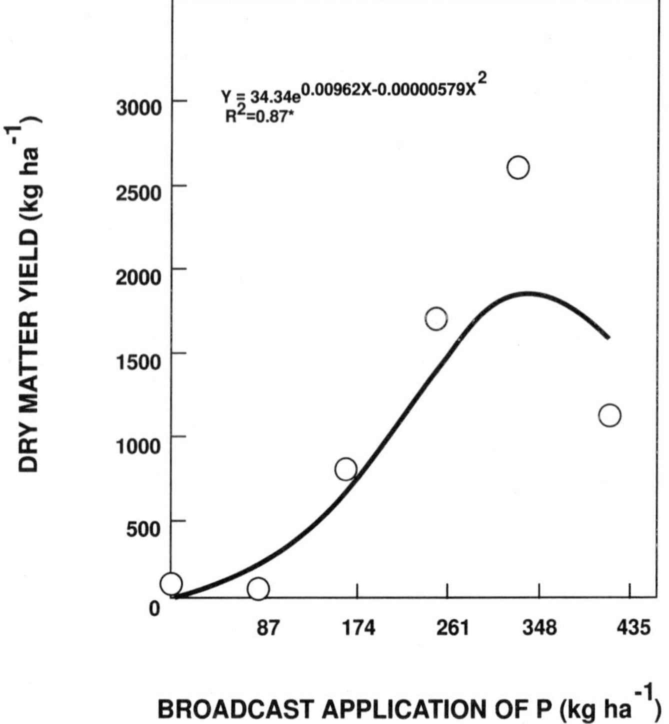

Figure 13.2 Dry matter yield of common bean as a function of P rates in an Inceptisol. (From Fageria, 1996b.)

Common bean stems and branches are slender, twisted, angled, and ribbed; in climbing forms they have more nodes, which are further apart than in determinate bush types. Leaves are alternate, trifoliate, and somewhat hairy, and each has a well-developed pulvinus at the base (Summerfield and Roberts, 1984a). Common beans are normally selfpollinated and less than 1% natural crossing occurs.

Growth Stages

An understanding of growth stages is needed for better management practices and improved crop yields. Growth and development of common beans are divided into vegetative and reproductive stages. The vegetative (V) stages are defined on the basis of number of nodes on the main stem, including the

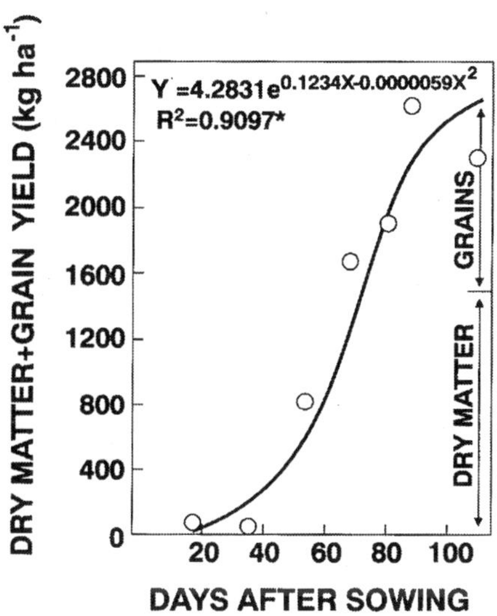

Figure 13.3 Dry matter and grain yield of common bean during crop growth in an Inceptisol. (From Fageria, 1996b.)

primary leaf node, whereas reproductive (R) stages are defined on the basis of pod and seed characteristics in addition to nodes. The different vegetative and reproductive growth stages are presented in Table 13.3 (based on the work of Lebaron, 1974).

Dry Matter

Photosynthesis provides 90–95% of plant dry weight (Kueneman et al., 1979). Thus, net photosynthesis of the entire plant canopy integrated over a growing season should mainly determine total plant dry weight and thereby indirectly determine the yield. *Phaseolus vulgaris* is a C_3 plant. Maximum values of net photosynthetic rate during the ontogeny of individual leaves range from 25 to 40 mg CO_2 dm^{-2} h^{-1} (Louwerse and Zweerde, 1977; Tanaka and Fujita, 1979). In southern Australia, Sale (1975) reported bean photosynthetic rates of 35–40 mg dm^{-2} h^{-1} at a saturation irradiance of 600–650 W m^{-2} and a leaf area index (*LAI*) of ~4.5. No change in net photosynthetic rate was recorded over a range of *LAI* values from 4.5 to 7.3. Silveira et al. (1996) determined the *LAI* of two Brazilian cultivars, Safira and Carioca, during the crop growth cycle, and maximum *LAI* was obtained at an age of about 55

Table 13.5 Correlation Between Grain Yield, Pod Dry Weight, and Dry Matter Production in Common Bean

Variable		Grain yield[a]	Pod weight[a]
Root dry weight	17 DAS[b]	–0.12NS[c]	–0.15NS
	31 DAS	0.22*	0.23*
	45 DAS	0.48**	0.47**
	60 DAS	0.72**	0.74**
	Maturity[d]	0.68**	0.71**
Top dry weight	17 DAS	0.12NS	0.10NS
	31 DAS	0.51**	0.51**
	45 DAS	0.84**	0.85**
	60 DAS	0.83**	0.85**
	Maturity[d]	0.93**	0.96**

[a] Values are across three common bean cultivars: Carioca, CNF4856, and CNF10.
[b] DAS, Days after sowing.
[c] NS, Not significant; * and **, significant at 5% and 1% levels of probability, respectively.
[d] Cultivars Carioca and CNF4856 matured in 80 days; cultivar CNF10 matured in 68 days.
Source: Fageria, 1989.

days after emergence. Dry matter production of roots and tops of three common bean cultivars grown in an Oxisol of central Brazil in a greenhouse experiment is presented in Table 13.4. In general, the dry weight of tops and roots increased with increasing age of the crop. This increase was greater in the tops than in roots. Maximum dry matter accumulation of tops as well as roots occurred in all the cultivars in the time period of 30–60 days after sowing. The cultivars Carioca and CNF4856 produced higher dry matter of tops than CNF10. This may be related to the longer growth cycle of the two cultivars.

Fageria (1996a) determined dry matter yield of common bean during crop growth under field conditions in Brazilian Inceptisol. Dry matter yield increased with increasing P levels up to 348 kg P ha^{-1} applied as broadcast, then it was decreased (Figure 13.2). Dry matter yield of the same crop was also determined during the crop growth cycle (Figure 13.3). Dry matter yield increased up to an age of about 68 days of growth, then decreased. The decrease in dry matter at the later growth stage may be related to more photosynthate translocation to grains and falling of old leaves. Table 13.5 shows the correlation between grain yield and dry weight of pods of common bean and dry matter production at different growth periods. Dry matter production of roots, as well as tops, was significantly related to grain yield and pod dry weight at all growth stages except 17 days after sowing.

C. Yield Components and Yield

Both physiological and morphological characteristics of the bean plant are thought to play a major and interdependent role in determining yields (Denis and Adams, 1978). Seed yield in common bean can be expressed as the product of three components: pods per plant, seeds per pod, and seed weight. Among these yield components, pods per plant has often been recommended as an indirect selection criterion for increasing yield, primarily because of its higher and more consistent correlation with yield (Bennet et al., 1977; Sarafi, 1978). Nienhuis and Singh (1986) measured the following morphological characteristics of common bean plant in relation to grain yield:

1. Branches per plant: a branch was defined as any appendage of the main stem with at least one node.
2. Number of nodes per branch: the total number of nodes on branches divided by the number of branches per plant.
3. Nodes on the main stem: the number of nodes on the main stem beginning with the cotyledonary node and ending on the node site of the last, fully expanded trifoliate leaf.
4. Nodes per plant: the number of nodes on the main stem plus the number of nodes on the branches.
5. Main stem length: the distance from the fully expanded, trifoliate leaf.
6. Main stem internode length: the main stem length divided by the number of nodes on the main stem.

The conclusions drawn by Nienhuis and Singh (1986) from this study were that yield was positively correlated with pods m^{-2}, seeds per pod, and all architectural traits except branches per plant. In contrast, seed weight was negatively correlated with yield, nodes per branch, nodes per plant, and nodes on the main stem, and was positively correlated with main stem internode length and main stem length. The genetic correlations among traits suggest that selection for increased main stem internode length and main stem length should result in simultaneous improvement in both yield and seed weight; however, the resulting plant type may prove too viny and prostrate for monoculture cropping systems.

Scully and Wallace (1990) evaluated a diverse set of 112 common bean accessions for variation in eight traits (days to flower, days to pod fill, days to maturity, biomass, harvest index, seed growth rate, biomass growth rate and economic growth rate) related to yield. The growth rates, biomass, and days of pod fill were linearly and positively related to yield. Biomass and the growth rates explained a large amount of the variation in yield, with r^2 values between 0.71 and 0.84; days of pod fill explained the least, with $r^2 = 0.09$. Yield followed a curvilinear relationship with days to flower and days to maturity; yield was maximized at 48.5 days to flower and 112.2 days to maturity. Yield was a quadratic function of harvest index and maximized at 57.2%. Among these three curvilinear traits, days to flower explained 80% of the variation in yield, while days to maturity and harvest index accounted for 25% and 12.5%, respectively. The "ideal" genotype for New York was defined at these maximum values for harvest index, days to maturity, days to flower, and at 63.7 days of pod fill. Additionally, a simple equation is proposed to aid breeders in the selection of common bean accessions with strong sink strength. It is defined as "relative sink strength" (RSS):

$$\mathrm{RSS} = \frac{\text{seed growth rate}}{\text{biomass growth rate}}$$

Values > 1.0 implied strong sink capacity in common beans (Scully and Wallace, 1990).

D. Biological Nitrogen Fixation

The genus *Rhizobium* contains bacteria which are able to form morphologically distinct nodules on the roots of members of the Leguminosae. In the case of *P. vulgaris*, the *Rhizobium* species which preferentially invades the roots for nodule formation and nitrogen fixation is *R. phaseoli*. An overall reaction involved in the reduction of nitrogen to ammonia in *Rhizobium* can be summarized as follows, although the nature of the reductant is not firmly established (Rawsthorne et al., 1980):

$$N_2 + n\mathrm{ATP} + 6\mathrm{NADPH} + 2\mathrm{H}^+ \rightarrow 2\mathrm{NH}_4^+ + n\mathrm{ADP} + n\mathrm{P_i} + 6\mathrm{NADP}^+ + 6\mathrm{e}^-$$

where n = 6.0–6.9 or 6.5 ATP/NH_4^+, depending on whether cell-free or cell mass balance figures, respectively, are used (Bergersen, 1971).

Common beans are generally considered to be weak in N_2 fixation and show a variable response to inoculation (Vincent, 1974). Acetylene reduction rates were estimated for 18 dry bean cultivars at Kimberly, Idaho (Westermann and Kolar, 1978). The calculated nitrogen fixation was about 10 kg N ha^{-1}, a small part of the total N uptake of 150–400 kg ha^{-1}.

Poor N_2 fixation by *P. vulgaris* has been attributed to the difficulty of establishing effective symbioses in the field and to genetic variability in the capacity to fix N (Graham, 1981). N_2 fixation by common beans is usually unreliable and N fertilization of field-grown plants is recommended (Graham, 1981; Westermann et al., 1981; Piha and Munns, 1987). Singh (1992) reviewed problems and prospects of nitrogen fixation in the tropics and subtropics. Bean cultivar and *Rhizobium* strain interaction, competition among efficient and inefficient strains, suppression of nodulation by residual soil nitrogen, high demand for phosphorus and photosynthates by *Rhizobium*, sensitivity to moisture stress, and interactions of these factors with environments have slowed the development of high nitrogen fixation cultivars in the tropics and subtropics. A recessive gene mutation responsible for supernodulation was reported. Although its usefulness in tropical and subtropical soils remains to be determined, in Canada the mutant yielded significantly less than the parental cultivar (Singh, 1992). However, work carried out at CIAT (1981) suggested that there are prospects for increasing biological nitrogen fixation in *P. vulgaris* through breeding.

E. Nutrient Requirements

Application of adequate amounts of nutrients is a prerequisite for higher bean production (Table 13.6). In Latin America, where almost 50% of the world bean crop is produced, phosphorus deficiency is the main yield-limiting factor. At least 50% of the beans in Africa are grown on severely P-deficient soils (Yan et al., 1995). Nitrogen deficiency may also seriously limit yields in soils with low organic matter or in soils in which biological N_2 fixation is not effective due to high temperatures or soil restrictions. Potassium deficiency seldom occurs in Latin America. Beans are extremely susceptible to Al/Mn toxicity, which frequently occurs in many of the acidic soils. Among the minor-element problems, boron and zinc deficiencies are most commonly observed in soils of high pH or soils with a very low content of weatherable minerals (Schwartz et al., 1978). Figure 13.4 shows the growth response of three common bean cultivars to P fertilization in Oxisols of central Brazil under greenhouse conditions. Similarly, common bean crop also responded to P fertilization under field conditions in an Inceptisol (Fig. 13.5).

Low P availability is a primary constraint to bean production in the tropics and subtropics. Fertilizers are not always available or affordable to farmers in the tropics and may be only marginally effective because of P fixation by Fe and Al oxides and allophane, making applied and native P unavailable to plants. Under these situations, the use of P-efficient bean cultivars may be a viable alternative or complement to fertilization. It is now well known that bean genotypes differ substantially in adaptation to low-P soils.

Table 13.6 Dry Matter and Grain Yield of Common Bean Under Different Fertility Treatments on an Oxisol

Fertility level	1st crop	2nd crop	3rd crop	Average
	Dry matter (kg ha^{-1})			
Low	958a	476b	817b	750c
Medium	1761a	2342a	1216a	1773a
High	1877a	2705a	1361a	1981a
Medium + green manure	1336a	1142b	1066ab	1181b
F-test (treatment)	NS	**	**	**
F-test (year)				**
F-test (T × Y)				**
	Grain (kg ha^{-1})			
Low	1935b	866c	890c	1230c
Medium	2382a	1831ab	1242ab	1818b
High	2568a	2432a	1486a	2162a
Medium + green manure	2344a	1202bc	1065bc	1537b
F-test (treatment)	**	**	**	**
F-test (year)				**
F-test (T × Y)				**

Low fertility level = without addition of fertilizers; medium fertility level = 50 kg N ha^{-1}, 26 kg P ha^{-1}, 33 kg K ha^{-1}, and 30 kg ha^{-1} fritted glass material as a source of micronutrients; high fertility level = all the nutrients applied were double the medium fertility level. *Cajnas cajan* L. was used as a green manure crop at the rate of 25.6 t ha^{-1} green dry matter.

*, ** Significant at the 5 and 1% probability levels, respectively. NS = not significant. Values followed by the same letter in the same column are not significantly different at 5% level by Tukey's test.

Source: Fageria and Souza, 1995.

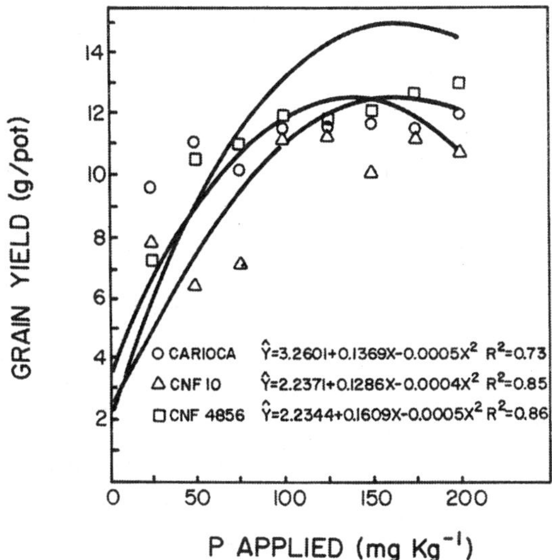

Figure 13.4 Grain yields of three common bean cultivars as influenced by P fertilization under greenhouse conditions in an Oxisol of central Brazil. (From Fageria, 1989.)

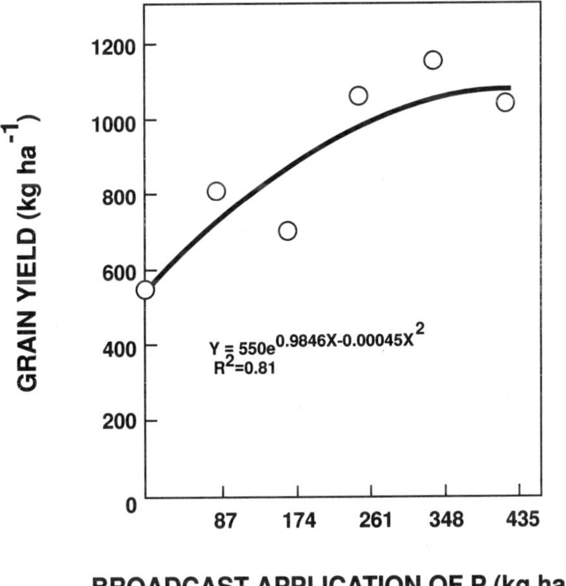

Figure 13.5 Relationship between common bean grain yield and broadcast P application under field conditions in an Inceptisol. (From Fageria, 1996b.)

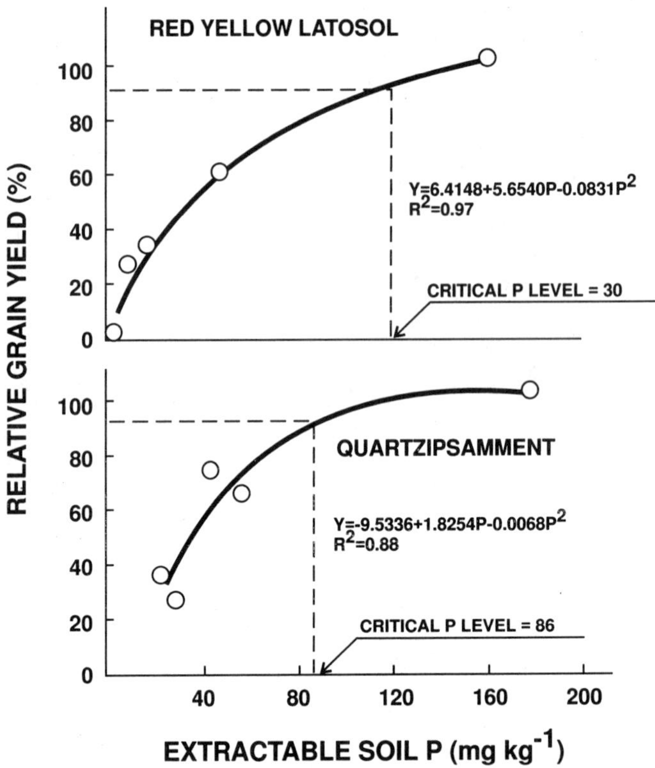

Figure 13.6 Relationship between relative grain yield of common bean and soil extractable P by Mehlich 1 extracting solution in two Brazilian (red-yellow Latosol and Quartzipsamment) soils under greenhouse conditions. (From Fageria and Carvalho, 1996.)

Fertilizer recommendations for beans should be based on soil and plant tissue analyses. Figures 13.6 and 13.7 show the relationship of common bean crop to grain yield to extractable P in an Oxisol under greenhouse conditions and (in Fig. 13.8) in an Inceptisol under field conditions. The critical P level under greenhouse varied from soil to soil. It was lowest in Dusky and Latosol and highest in quartzipsamment soil. Under field conditions, the extractable P and relative yield relationship was established (Figure 13.8). When relative yield was 70%, the corresponding extractable P level was about 0–5.3 mg kg^{-1} (very low); when relative grain yield was 70–90%, the extractable P level was 5.3–7.1 mg kg^{-1} (low); when relative yield was 90–100%, the extractable P level was 7.1–9.0 mg kg^{-1} (medium); and when relative yield was more than 100%, the extractable P level was more than 9.0 mg kg^{-1}

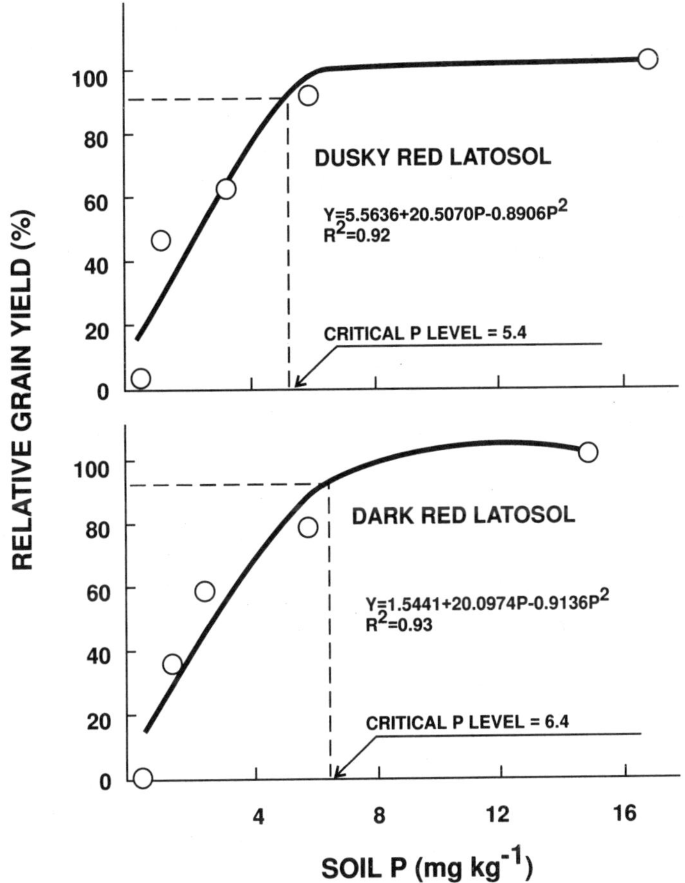

Figure 13.7 Relationship between relative grain yield of common bean and soil extractable P by Mehlich 1 extracting solution in two Brazilian (dusky red Latosol and dark red Latosol) soils under greenhouse conditions. (From Fageria and Carvalho, 1996.)

(high). These results are very useful in interpretation of soil analysis results for P fertilizer recommendations for bean crops in Inceptisols.

Nutrient Concentrations and Uptake

A knowledge of nutrient concentrations and distribution in plant parts is important for a basic understanding of plant nutrition. The distribution of nutrients in different plant parts of common bean at harvest is presented in Figure 13.9 for Brazilian cultivar CNF4856. Accumulations of N, P, K, Ca,

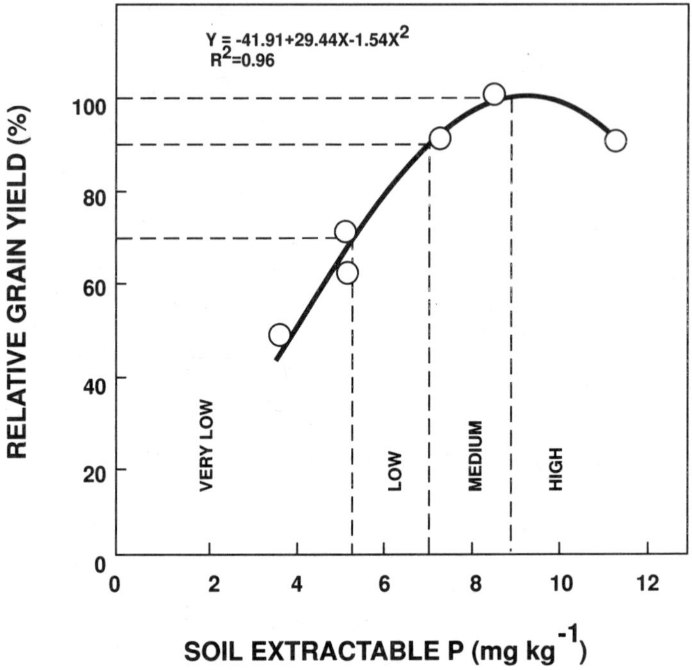

Figure 13.8 Relationship between relative grain yield of common bean and soil extractable P by Mehlich 1 extracting solution under field conditions in an Inceptisol. (From Fageria, 1996b.)

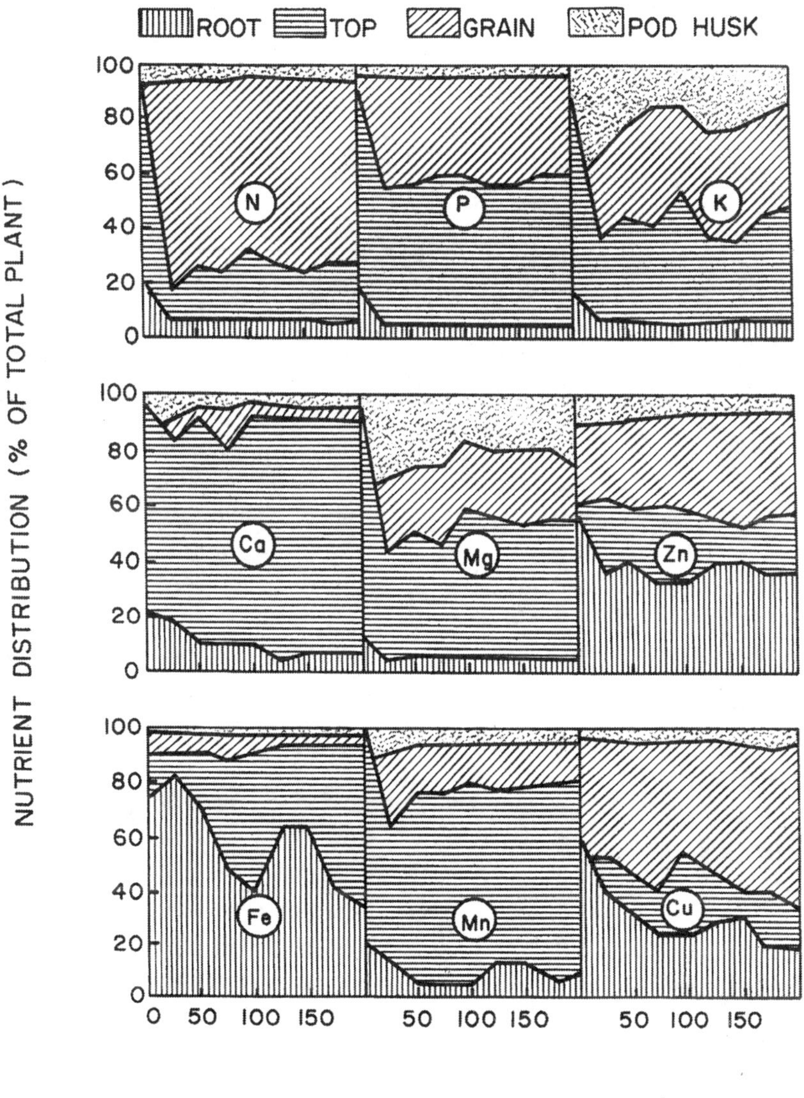

Figure 13.9 Distribution of nutrients in different plant parts of the bean cultivar CNF4856. (From Fageria, 1989.)

Table 13.7 Concentrations of Nutrients in Different Parts of the Bean Plant as Influenced by P Application[a]

Nutrient	Plant part	P applied ($mg\ kg^{-1}$)								
		0	25	50	75	100	125	150	175	200
N ($g\ kg^{-1}$)	Root	12.2	25.1	25.0	25.3	27.3	28.3	27.5	28.0	27.4
	Top	36.5	18.4	17.2	18.3	20.7	19.0	19.5	20.8	18.7
	Grain	5.1	40.8	41.5	39.5	37.4	39.5	40.2	40.5	40.3
	Pod husk	4.1	10.1	9.8	9.6	6.9	7.6	8.1	7.5	10.9
P ($g\ kg^{-1}$)	Root	1.5	2.3	2.2	2.7	3.0	3.9	2.7	3.2	2.7
	Top	1.7	1.2	1.6	2.0	2.2	2.2	1.9	2.0	2.1
	Grain	2.0	3.7	3.9	3.8	4.1	3.9	4.0	3.8	3.9
	Pod husk	0.8	0.8	0.8	1.0	0.7	0.8	0.8	0.7	1.0
K ($g\ kg^{-1}$)	Root	8.8	5.4	4.4	6.4	12.0	8.8	8.7	12.8	9.4
	Top	16.9	19.4	18.6	20.1	19.2	18.3	19.3	20.2	19.8
	Grain	7.0	13.7	13.4	13.1	13.6	14.5	13.2	12.9	12.7
	Pod husk	9.1	28.2	29.4	30.5	28.0	28.3	31.1	30.7	27.9
Ca ($g\ kg^{-1}$)	Root	9.8	13.1	12.0	11.2	10.8	10.5	9.4	10.9	10.1
	Top	22.2	36.4	32.8	32.1	37.0	36.0	36.1	33.6	33.2
	Grain	1.0	1.3	1.1	1.1	1.1	1.1	1.1	1.0	1.0
	Pod husk	2.1	6.3	6.5	4.7	4.9	4.8	4.6	4.3	4.7

Mg (g kg^{-1})	Root	1.6	1.6	1.9	2.5	2.7	2.8	2.7	3.9	3.1
	Top	4.7	5.6	5.3	5.0	4.8	5.1	5.3	4.5	4.8
	Grain	0.7	1.6	1.7	1.7	1.7	1.7	1.7	1.7	1.7
	Pod husk	1.6	5.5	5.4	5.2	5.1	5.3	5.2	5.1	5.1
Zn (mg kg^{-1})	Root	256	306	265	263	299	308	371	297	313
	Top	51	43	28	39	38	34	30	41	27
	Grain	17	36	39	39	37	47	39	39	40
	Pod husk	16	40	36	37	34	33	33	39	35
Fe (mg $kg^{-1} \times 10^3$)	Root	6.02	11.66	13.72	11.73	7.31	9.14	12.9	5.58	11.79
	Top	0.68	0.67	0.86	1.12	1.01	0.87	1.71	1.51	0.91
	Grain	0.03	0.06	0.07	0.07	0.06	0.07	0.06	0.06	0.06
	Pod husk	0.04	0.07	0.06	0.07	0.07	0.05	0.06	0.07	0.08
Mn (mg kg^{-1})	Root	84	102	125	112	91	136	152	117	135
	Top	36	93	88	102	136	122	110	150	103
	Grain	8	17	16	18	19	18	17	19	16
	Pod husk	11	28	29	24	28	28	24	28	25
Cu (mg kg^{-1})	Root	126	68	51	53	45	51	50	33	41
	Top	25	13	10	7	8	10	6	6	4
	Grain	6	12	10	10	10	13	9	8	8
	Pod husk	8	6	5	7	5	4	5	7	6

[a] Values are averages for three bean cultiars at harvest.

Source: Fageria, 1989.

and Mg in roots were similar, being in the range of 6–10% of the amounts in the total plant. The average percentages of plant totals of these nutrients in tops were: Ca 81%, P and Mg 52%, K 42%, and N 26%. This indicates that Ca is concentrated in the tops. In grain, the averages were: N 71%, P 40%, K 33%, Ca 4%, and Mg 22%. Very little N, P, or Ca was concentrated in pod husks (2–5%), but approximately 20% of both K and Mg was located in husks. The average distributions of micronutrients among the plant parts were as follows: roots—Fe 58%, Zn 40%, Cu 31%, and Mn 12%; tops—Fe 35%, Zn 23%, Cu 20%, and Mn 69%; grain—Fe 6%, Zn 30%, Cu 45%, and Mn 14%; pod husks—Fe 1%, Zn 7%, Cu 4%, and Mn 5%.

The concentrations of nutrients in different plant parts at various P levels are presented in Table 13.7. With a few exceptions, fertilization increased the total nutrient concentrations. The concentration of Zn in the tops and the concentration of Cu in roots and tops, however, decreased with P fertilization. This indicates that in Oxisols with high P fertilization, deficiencies of Zn and Cu may occur if these elements are not present in adequate concentrations. Adequate nutrient levels at harvest in the common bean plant are presented in Table 13.8.

Uptake of P in roots of three bean cultivars at different growth stages and at two P levels is presented in Figure 13.10. Phosphorus uptake in the roots was significantly increased with application of 25 mg P kg^{-1} of soil as compared with control plants. This increase is related to an increase in root weight with P fertilization. Uptake of P was highest in the cultivar CNF10 and lowest in the cultivar CNF4856. In case of the cultivars Carioca and CNF4856, P uptake at the higher P level increased with increasing age up to 60 days and was constant with further increases in plant age. In the case of cultivar CNF10, P uptake increased with increasing age in the P-supplemented treatment. The concentration and uptake of nutrients by common bean crop under field conditions are shown in Table 13.9. Concentration as well as uptake were significantly affected by soil fertility treatments. Nutrient uptake followed the order of K > N > Ca > Mg > P > Fe > Mn > Zn > Cu. To produce one ton of grains, it is necessary to accumulate 27 kg N, 2.1 kg P, 28 kg K, 16.3 kg Ca, 3.4 kg Mg, 47 g Zn, 7 g Cu, 90 g Mn, and 567 g Fe in the aereal parts at flowering.

The accumulation of P in the tops of three cultivars under greenhouse conditions is shown in Figure 13.11. Uptake of P was increased with age as well as with increasing levels of P. There was a great difference in P uptake between the control and the 25 mg P treatment. At the higher levels of P, the P uptake increased at a constant rate. At the highest P level, cultivar Carioca absorbed the most P and cultivar CNF10 the least. Cultivar CNF4856 was more or less equal to Carioca in P uptake at the 200 mg P level. The uptake of P in tops is related to dry matter production. There was practically no

Table 13.8 Adequate Levels of Nutrients in the Common Bean Plant

Nutrient	Growth stage	Plant part	Adequate concentration
			g kg^{-1}
N	Early flowering	UMB[a]	52–54
	Peak harvest	Pods	31
P	Early flowering	UMB	4–6
	Peak harvest	Pods	3
K	Early flowering	UMB	15–35
	Peak harvest	Pods	26
Ca	Early flowering	UMB	15–25
Mg	Early flowering	UMB	4–8
S	Early growth	WS[b]	1.6–6.4
	Peak harvest	Pods	1.7
			mg kg^{-1}
Fe	Early flowering	UMB	100–300
	Peak harvest	Pods	70
Mn	Early flowering	UMB	50–400
	Peak harvest	Pods	27
Zn	Early flowering	UMB	35–100
	Peak harvest	Pods	34
Cu	Early flowering	UMB	5–15
	Peak harvest	Pods	5
B	Early flowering	UMB	10–50
	Peak harvest	Pods	28
Mo	56 DAS[c]	WS	0.4

[a] UMB, Uppermost blade.
[b] WS, Whole shoot.
[c] DAS, Days after sowing.
Source: Compiled from Piggott, 1986.

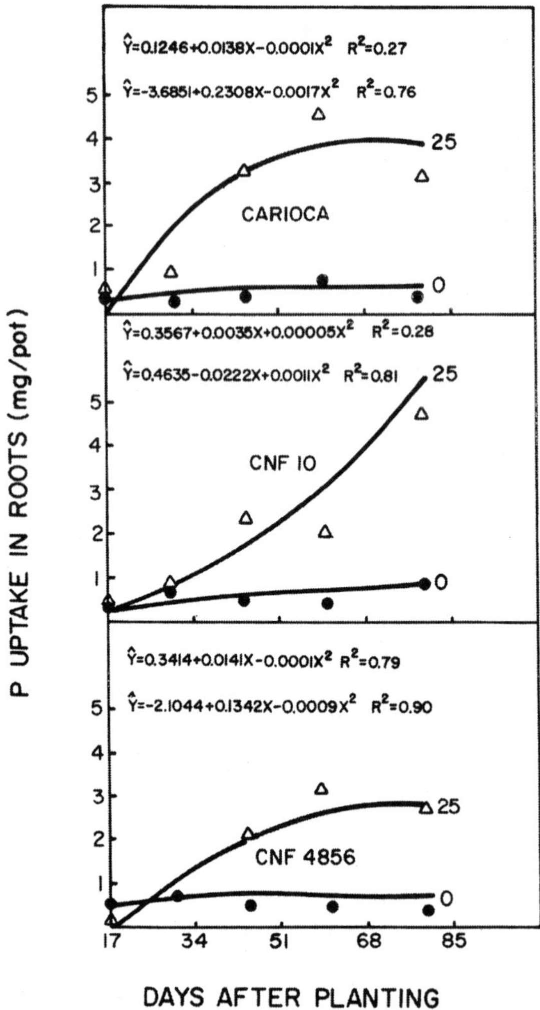

Figure 13.10 Uptake of P by roots of three bean cultivars at different growth periods at two P levels (0 and 25 mg kg^{-1}) in an Oxisol. (From Fageria, 1989.)

Table 13.9 Nutrient Concentrations and Uptake by Common Bean Crop at Flowering Growth Stage at Different Fertility Levels on an Oxisol Across Three Years

Fertility level	N	P	K	Ca	Mg	Zn	Cu	Mn	Fe
	Concentration (macronutrients are in g kg^{-1} and micronutrients in mg kg^{-1})								
Low	32.4	2.6a	34.7a	20.3a	4.1b	60a	10a	110a	828a
Medium	32.1a	2.5a	31.7b	19.7a	4.0b	53a	9a	105a	805a
High	34.4a	2.6a	33.3ab	19.5a	4.1b	56a	9a	100a	669a
Medium + green manure	31.5a	2.6a	32.2ab	19.5a	4.5a	61a	9a	121a	773a
F-test (treatments)	**	**	**	**	**	**	**	**	**
F-test (year)	NS	NS	*	NS	**	NS	NS	NS	NS
F-test (T x Y)	*	NS	NS	NS	NS	NS	**	NS	NS
	Uptake (macronutrients are in kg ha^{-1} and micronutrients in g ha^{-1})								
Low	24d	1.98b	24c	16b	3.21c	46b	7b	75c	620c
Medium	54b	4.48a	59c	35a	7.25a	95a	15a	197a	1211a
High	67a	4.81a	68a	36a	7.46a	109a	16a	203a	1094ab
Medium + green manure	37c	2.99b	39b	23b	5.02b	68b	11b	134b	903b
F-test (treatment)	**	**	**	NS	**	**	**	**	**
F-test (year)	**	**	**	**	**	**	**	**	**
F-test (T × Y)	*	NS	**	**	**	NS	NS	NS	NS

Low fertility level = without addition of fertilizers; medium fertility levels = 50 kg N ha^{-1}, 26 kg P ha^{-1}, 33 kg K ha^{-1}, and 30 kg ha^{-1} fritted glass material as a source of micronutrients; high fertility level = all the nutrients applied were double the medium level.

Cajnas cajan L. was used as a green manure crop at the rate of 25.6 t ha^{-1} green matter.

*, ** Significant at the 5 and 1% probability levels, respectively. NS = not significant. Values followed by the same letter in the same column are not significantly different at 5% level by Tukey's test.

Source: Fageria and Souza, 1995.

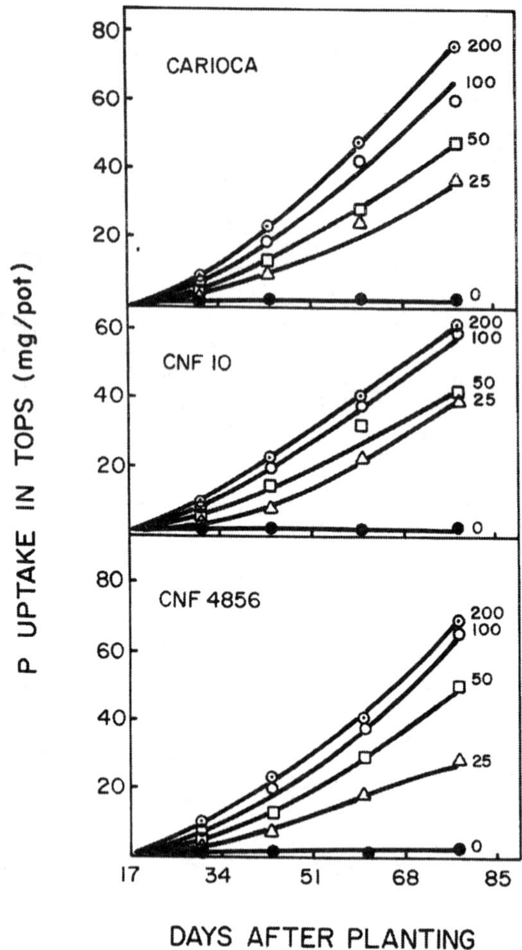

Figure 13.11 Uptake of P by tops of three bean cultivars as a function of different growth stages at five P levels (mg kg^{-1}) in an Oxisol. (From Fageria, 1989.)

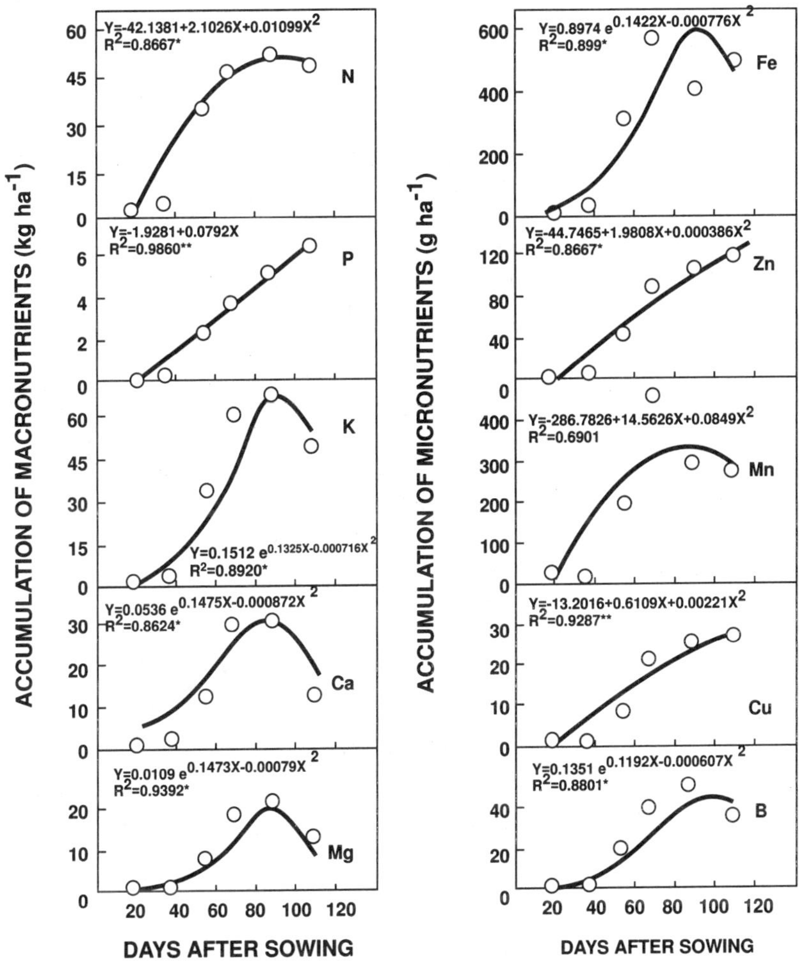

Figure 13.12 Accumulation of macro- and micronutrients in common bean during crop growth under field conditions in an Inceptisol. (From Fageria, 1996b.)

change in P uptake by the control plants at different growth stages. Accumulation of macro- and micronutrients under field conditions is presented in Figure 13.12. Accumulation of all the nutrients was quadratic except for P, which was linear during the growth cycle of the crop.

The different P uptake by bean cultivars may be due to genetic differences in nutrient requirements and uptake capabilities or to differences in redistribution of nutrients to plant parts.

III. COWPEA

Cowpeas (*Vigna unguiculata* L. Walp.) are also commonly called black eye pea, black eye bean, southern pea, China pea, kaffir pea, and marble pea. The cowpea differs from the common pea in that the style and keel are usually straight rather than spirally twisted. In *Phaseolus* the style does not have an apical appendage, whereas in *Vigna* the style has a distinct beak. *Phaseolus* stipules are not prolonged, but in *Vigna* the stipules are prolonged. Further, in *Phaseolus* the first pair of leaves are petiolate but in *Vigna* the first pair of leaves are sessile (Evans, 1980). Five subspecies of *V. unguiculata* are now recognized: *unguiculata*, *cylindrica*, *sesquipedatis*, *dekindtiana*, and *mensensis* (Verdcort, 1970).

West Africa and India are centers of diversity for cowpeas. Cowpeas are cultivated as a seed, vegetable, and fodder legume in diverse agricultural systems in many countries of Asia, Africa, and Latin America. The highest cowpea seed-producing nations are India, Brazil, Nigeria, and other West African countries (Summerfield et al., 1983). Cowpeas are also produced in the southern United States and Australia (Sellschop, 1962).

Cowpeas are an important source of protein for larger numbers of people in developing countries. Chemically mature cowpea seeds contain an average of 23% protein, 60% starch, and 2% oil (Aykroyd and Doughty, 1964). The high-quality protein of cowpeas is also a natural supplement to that of staple grain crops because of its high lysine content, but like other legumes, cowpeas are deficient in the sulfur amino acids methionine and cystine.

Besides human consumption of cowpea seeds, the whole plant is used as a livestock feed and for soil improvement. Cowpeas are usually used as a hay or grazing crop where paucity of rainfall or lack of irrigation water does not permit the production of fine-stemmed and easily cured fodder such as alfalfa and clover. Cowpea hay and alfalfa hay are equally digestible, except that the fiber of cowpea hay is more digestible than that of alfalfa hay (Sellschop, 1962). Cowpea hay is poorer in digestible proteins and richer in total digestible nutrients than alfalfa hay (Van Wyk, 1955). The best stage for cowpea harvest for fodder is when the first pods turn yellow. At this stage, the leaves and pods still contain more than 60% of the total quantity of crude protein of the plant and there is less danger of loss of leaves in drying and handling the hay.

A. Climate and Soil Requirements

Cowpea can be grown under a wide range of climatic conditions, but production is mostly concentrated in the tropics and subtropics. It is sensitive to cold and killed by frost. The cowpea can tolerate heat and relatively dry conditions and can grow with less rainfall and under more adverse conditions than *P. vulgaris* (Purseglove, 1987). Temperatures greater than 28°C lead to abnormal pollen development and anther indehiscence, and low temperatures (around 20°C) reduce the yield of cowpeas (Warring and Hall, 1984). A study conducted by Bagnall and King (1987) over a temperature range of 21/16 to 33/28°C (day/night) showed that maximum yields of cowpeas were obtained at 27/22°C (day/night) temperatures. These authors also concluded that grain yield was influenced by temperature after flowering in two ways: high temperatures decreased pod number, while temperatures below 21°C decreased seed number and size.

Adequate water is essential for higher production, and the effects of water deficit on crop growth and yield depend on the degree of stress and the developmental stage at which the stress occurs. The most drought-sensitive growth stages are flowering and pod filling (Shouse et al., 1981). According to Labanauskas et al. (1981), water stress during flowering and pod filling reduced seed yield by 44 and 29%, respectively, when compared to a control treatment. Water stress during the vegetative stage had no significant effects on seed productivity. Similar results were obtained by Turk et al. (1980) in their studies of cowpea yields. Cowpea is a relatively drought-tolerant crop as compared to common bean. It is generally considered that dehydration in cowpea is markedly delayed due to morphological or physiological modifications that reduce transpiration or increase adsorption. Plants in which dehydration is delayed can survive during a long period of time under drought conditions, and when water becomes available after stress, growth recovery is rapid, so that the net reduction in yield is minimized (Itani et al., 1993). Under drought conditions cowpea can maintain a high leaf and xylem water potential by complete stomatal closure (Itani et al., 1992b). Further, cowpea can withstand drought due to the deep root system of the plant (Itani et al., 1992a).

Cowpeas are adapted to a wide range of soils from sands to clays. The primary soil requirements are good drainage and the presence of or inoculation with the proper nitrogen-fixing bacteria (Martin and Leonard, 1967). Cowpeas can thrive well on acid soils, and good yields were obtained around pH 6 in an Oxisol of central Brazil (Fageria, 1991). In a field trial in Colombia on an Oxisol, at pH 4.2 and 78% Al saturation, yields were essentially doubled by an application of 0.5 t $CaCO_3$ ha^{-1}, which reduced Al saturation to 67% and gave a Ca saturation of 22% (Spain et al., 1975). Similarly, application of 0.5 t $CaCO_3$ ha^{-1} in Ultisols of Nigeria resulted in 85% of maximum growth for both Al-tolerant and Al-sensitive cowpea cultivars (Edwards et al., 1981). Cowpeas are considered to be moderately susceptible to soil salinity, with an initial yield decline (threshold) at around 1.3 dS m^{-1}. Yields decrease about 14% per unit of salinity increase beyond this threshold (Maas and Hoffman, 1977).

B. Growth and Development

The cowpea is an annual herbaceous legume. Germination is epigeal and no seed dormancy has been reported in cultivated cowpeas. When environmental conditions are favorable and seeds are sown at the appropriate depth (3–4 cm), germination is complete in 3–4 days. Although, germination is epigeal, cowpea cotyledons do not persist. They can lose as much as 91% of their dry weight by emergence (Ndunguru and Summerfield, 1975). Such effective mobilization of cotyledonary reserves probably contributes to rapid hypocotyl elongation and might improve emergence in adverse edaphic conditions (Wien, 1973). Under favorable conditions, cowpeas develop strong taproot systems and have many spreading laterals in the surface soil. Growth habits of cowpea are erect, semierect, trailing, prostrate, and climbing; determinate types with terminal inflorescences are rare (Summerfield and Roberts, 1984b).

The first pair of leaves above the cotyledonary node are simple and opposite, but they exhibit considerable variation in size and shape (Faris, 1965). The trifoliate leaves arise alternately, and the terminal leaflet is frequently long and of greater area than the asymmetrical lateral leaflets (Summerfield et al., 1974). The inflorescence of cowpea consists of a peduncle on which 4–6 units of flowers are formed alternately in acropetal succession. Each unit is a modified simple raceme consisting of 6–12 buds, and the entire inflorescence may be regarded as a panicle (Ojehomon, 1968). Cowpea may come into flower as early as 22–30 or not until after 90 days, and the crop may require a total duration of 210–240 days to mature (Summerfield and Roberts, 1984b). The cowpea is largely selfpollinated, and there is always a small proportion of outcrosses, especially in humid climates. As in other grain legumes, there is considerable shedding of unopened buds and premature fruits. Ojehomon (1968) reported that 70–88% of the buds are shed before anthesis and that flower abortion is one of the important factors responsible for low yields of cowpeas.

Numerous factors have been reported to be responsible for loss in productive potential of the crop. Among these are nucleic acids synthesis before and after anthesis, abnormal pollen formation, poor pollen germination, a large proportion of assimilates sequestered in older flowers and pods to the detriment of young ones, production of hormones (e.g., abscisic acid) by older pods that promote absorption of younger ones, warm and dry weather, and insect damage (Summerfield and Roberts, 1984b).

C. Dry Matter

Dry matter production of a crop is generally related to economic yield if all environmental factors are at optimum level during crop growth. Table 13.10 shows the correlation between grain yield and pod dry weight and dry matter

Table 13.10 Correlations Between Grain Yield Pod Dry Weight, and Dry Matter Production of Roots and Tops in Cowpea

Variable		Grain yield[a]	Pod dry weight[a]
Root dry weight	17 DAS[b]	0.43*	0.47*
	30 DAS	0.85**	0.78**
	43 DAS	0.88**	0.86**
	56 DAS	0.86**	0.86**
	76 DAS	0.87**	0.73**
Top dry weight	17 DAS	0.59*	0.62**
	30 DAS	0.86**	0.85**
	43 DAS	0.93**	0.93**
	56 DAS	0.95**	0.95**
	76 DAS	0.96**	0.95**

[a] * and **, Significant at 5% and 1% levels of probability, respectively.
[b] DAS, Days after sowing; 76 DAS correspond to maturity of cultivar CNC 0434.
Source: Fageria, 1991.

of roots and tops of cowpea for the data obtained in a greenhouse experiment in an Oxisol of central Brazil (Fageria, 1991). Root and shoot dry weights were significantly correlated with grain yield and pod dry weight throughout the growth cycle of the crop. Correlation coefficient values increased with age. This means that cowpea dry matter production at maturity can be a good indicator of grain yield. Correlation values for dry weight of tops were higher than for dry weight of roots. Table 13.11 shows the dry matter production of roots and tops of cowpea at different growth stages. Top dry weights increased with increasing age, but root dry weights increased up to 56 days after sowing

Table 13.11 Dry Matter of Roots and Tops of Cowpea at Different Growth Periods

Days after sowing	Roots (g per two plants)	Tops (g per two plants)
17	0.22	0.99
30	0.73	5.03
43	1.59	18.48
56	2.74	31.62
76	1.84	33.03

Source: Fageria, 1991.

and then decreased slightly. The variation in dry weight of tops was much greater than that of roots during the growing period of the crop.

The rate of dry matter production by cowpeas is closely dependent on the amount of incoming radiation intercepted by the crop canopy (Wien and Summerfield, 1984). In broad-leaf crops, canopies are generally more planophile and require a lower leaf area for complete light interception than in cereals (Monteith, 1969). In addition, crop growth rates, dry matter, and grain yields are generally lower for broad-leaf than cereal crops (Wien, 1982). If the available light could be distributed over a large leaf area by changing leaf angles, the productivity of broad-leaf crops with high leaf area indices might be increased (Duncan, 1971). Cowpea leaf area development was not limiting under field growing conditions when mean temperature varied between 23 and 28°C. These conditions allow the canopy to attain a leaf area index of 3 by 35–40 days after emergence in stand densities of 7–16 plants m^{-2} (IITA, 1973; Littleton et al., 1979; Chaturvedi et al., 1980). The cowpea is a C_3 plant, and Littleton et al. (1979) measured maximum growth rates of 16–25 g m^{-1} day^{-1} by a determinate, erect cultivar growing at equidistant spacing. Kassam and Kowal (1973) presented data which suggest that average crop growth rates might be expected to range between 18–25 and from 25 to 29 g m^{-2} day^{-1} in the forest and savanna zones, respectively, if the maximum rates of photosynthesis of individual cowpea leaves are taken to be 40–50 mg CO_2 dm^{-2} h^{-1}.

The partitioning of dry matter to fruits is fairly constant for a cultivar grown under different climatic conditions (Summerfield et al., 1978). Cowpea harvest index values ranging from 0.57 to 0.64 have been reported by Summerfield (1977).

D. Yield Components and Yield

The three components which determine yield of cowpea are number of pods per unit area, number of seeds per pod, and weight per seed. These three yield components are affected by environmental factors and genotypes. Figure 13.13 shows how these three components increased in Oxisols of central Brazil with the application of phosphorus in a P-deficient soil. Among the three yield-controlling characteristics, 100–seed weight was affected the most and pod number the least by P fertilization.

Grain yields of cowpeas are quite low in all farming systems. Average world cowpea yields are about 380 kg ha^{-1} (Summerfield et al., 1983), but under good management and as a monoculture, cowpeas can produce yields of 1000–4000 kg ha^{-1} (Wien and Summerfield, 1984). This means that the low productivity of cowpeas is due to poor management practices rather than low yield potential.

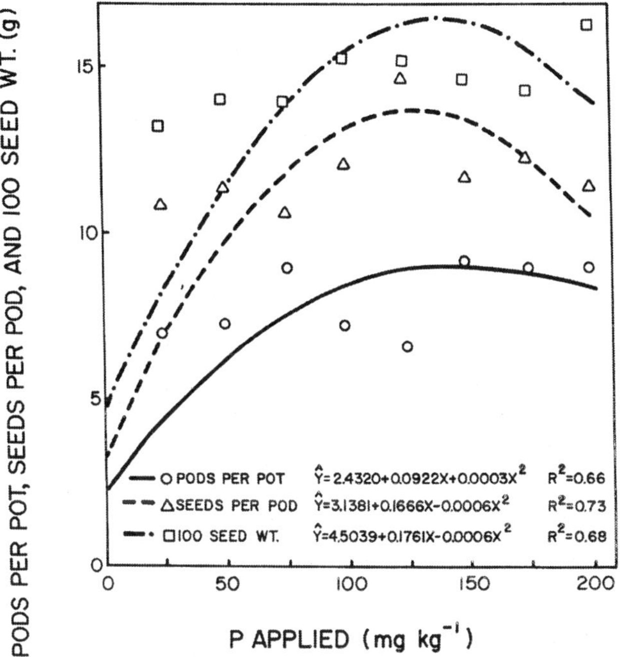

Figure 13.13 Influence of P on number of pods, seeds per pod, and 100–seed weight in cowpeas grown under greenhouse conditions in an Oxisol of Brazil. (From Fageria, 1991.)

E. Biological Nitrogen Fixation

Nitrogen fixation by bacteria is commonly known as biological N_2 fixation (Bezdicek, 1979). This biological process, which is distinguished from fixation by chemical or electrical means, involves conversion of atmospheric N_2 to ammonia (NH_3) in organisms and combination of the NH_3 with metabolic intermediates to form amino acids and proteins. A simplified version of the N_2–fixing reaction for all organisms is as follows (Bezdicek, 1979):

$$N_2 = \frac{6H + \text{energy}}{\text{nitrogenase}} \rightarrow 2NH_3$$

The actual N_2 fixation process occurs within the nodules. The initial phase of nodulation occurs when a specific *Rhizobium* bacterium increases in number near the root environment and causes curling of the root hair (Dart, 1974). Cowpeas are nodulated by the bacteria *R. japonicum*. This *Rhizobium* species also forms nodules in soybean, peanut, lespedeza, crotalaria, kudzu, and lima bean (Alexander, 1977).

The quantity of nitrogen fixation by cowpea varies with environmental conditions and genotypes. Most of the nitrogen fixation data available are from greenhouse experiments, and quantitative data obtained under field conditions are lacking. According to Eaglesham et al. (1977), maximum rates of N assimilation occurred during pod-fill, and symbiotic fixation supplied over 80% of total plant N throughout growth and contributed significantly to seed N during late pod-fill, when nutrient N assimilation was negligible. The average value of N fixation by cowpea in properly conducted field experiments is about 198 kg N ha^{-1} per year (Nutman, 1976). This value seems to be quite high, and further research is needed to quantify N fixation by cowpea under different agroecological conditions.

F. Nutrient Requirements

Generally, farmers in Latin America, Asia, and Africa do not use fertilizers for cowpeas. But with fertilization, yields of this legume crop can be substantially increased if sufficient water is available during the growing season. Figure 13.14 shows that application of P fertilizer in adequate combination with other nutrients can increase yields of this crop significantly. Phosphorus is the most yield-limiting nutrient in acid soils in Latin America, Africa, and Asia, where cowpea is largely produced.

Smyth and Cravo (1990) determined the critical P level in an Oxisol for a cowpea crop (Figures 13.15 and 13.16). Mehlich 1 (1: 10) critical P level was 8 mg kg^{-1}, and Bray 1 critical P level was 13 mg kg^{-1}. Similarly Cox and Uribe (1992) determined the critical K level in a humid tropical Ultisol (Figure 13.17). The extractable (modified Olsen) critical level for cowpea was 0.10 cmol K kg^{-1}. These results are useful in the interpretation of soil analyses and making P and K fertilizer recommendations for cowpea crops.

Data on uptake of nutrients by cowpea crops under different agroecological conditions are scarce. According to Rachie and Roberts (1974), 1000 kg ha^{-1} grain yield of cowpea removed approximately 40 kg N ha^{-1}, 7.4 kg P ha^{-1}, 40 kg K ha^{-1}, 11.4 kg Ca ha^{-1}, 9 kg Mg ha^{-1}, and 4 kg S ha^{-1}.

Concentrations of nutrients in roots, tops, seeds, and pod husks at maturity are presented in Table 13.12. Seeds had the highest P concentration, followed by roots, tops, and pod husks. The concentration of K was highest in tops and seeds and was similar in roots and pod husks. Calcium concentrations were also highest in tops but were lowest in the seeds. Magnesium concentration occurred in the order: roots > tops > pod husks > seeds. Concentrations of Zn and Fe were highest in the roots and lowest in the pod husks. Concentrations of Mn were highest in the tops and lowest in the seeds. Copper concentrations were highest in the roots and lowest in the tops. Adequate nutrient concentrations for the growth of cowpeas are given in Table 13.13.

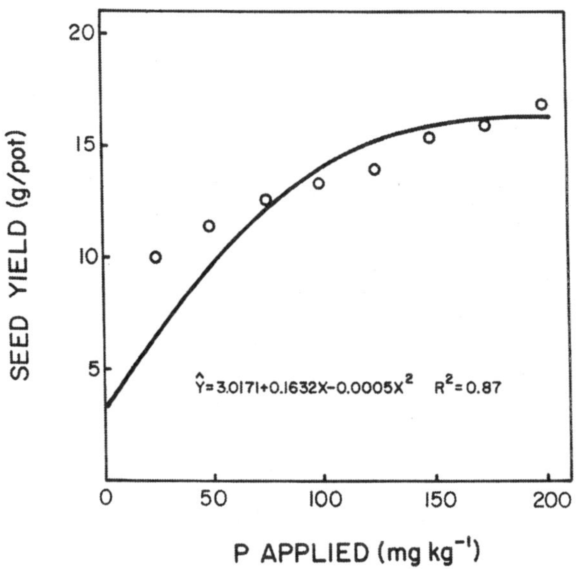

Figure 13.14 Influence of P fertilization on seed yields of cowpea. (From Fageria, 1991.)

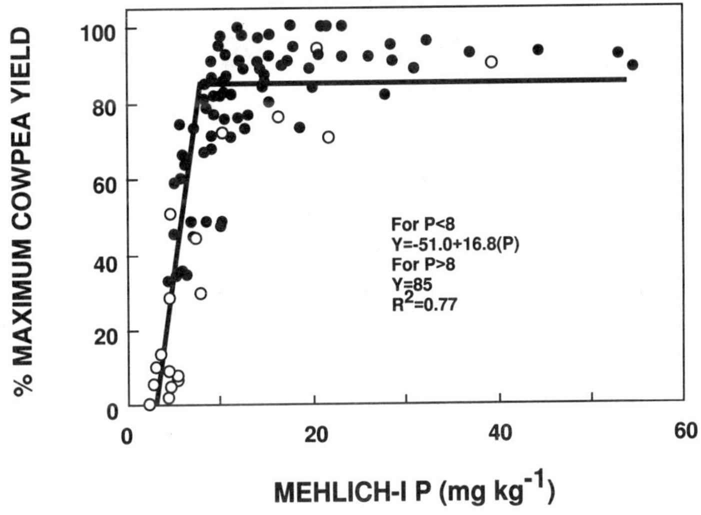

Figure 13.15 Relationship between cowpea yield and Mehlich 1 extracting soil P. (From Smyth and Cravo, 1990.)

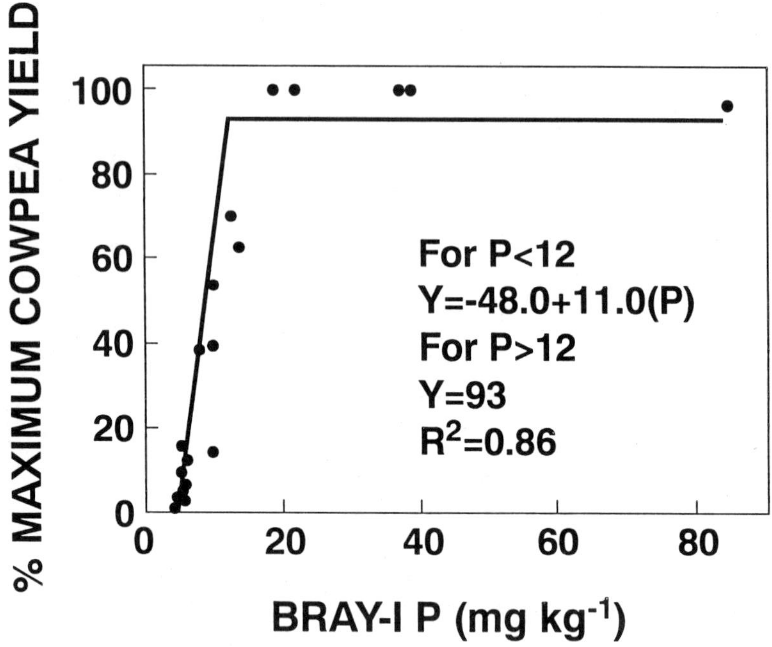

Figure 13.16 Relationship between cowpea yield and Bray 1 extracting soil P. (From Smyth and Cravo, 1990.)

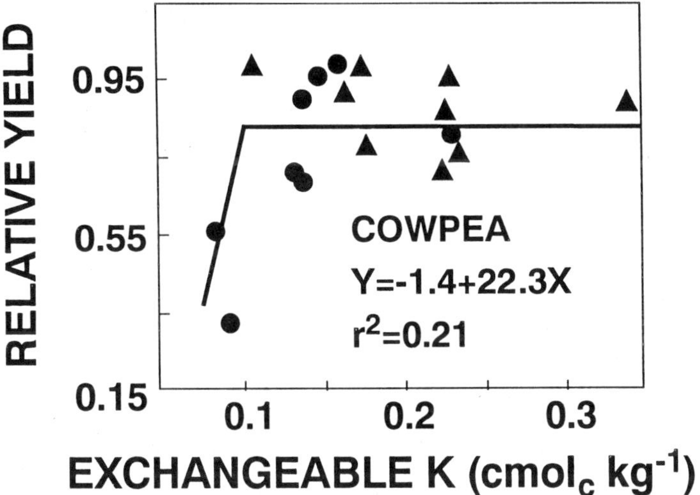

Figure 13.17 Relationship between cowpea yield and modified Olsen exchangeable K. (From Cox and Uribe, 1992.)

Table 13.12 Concentrations of Nutrients in Different Plant Parts of Cowpea at Harvest as Influenced by P Levels

Nutrient	Plant part	P levels (mg kg^{-1})								
		0	25	50	75	100	125	150	175	200
P (g kg^{-1})	Root	1.0	1.9	1.9	2.1	2.6	1.7	2.1	2.5	2.0
	Top	1.5	0.8	1.2	1.1	1.1	1.2	1.1	1.4	1.4
	Seed	—	3.3	4.0	4.2	4.4	4.7	4.5	4.4	4.5
	Pod husk	—	0.8	0.7	1.0	0.9	1.7	0.7	0.6	0.7
K (g kg^{-1})	Root	24.0	17.7	10.7	10.3	13.0	9.0	10.0	11.7	11.0
	Top	43.2	31.0	25.0	21.3	21.0	14.3	16.7	18.3	16.0
	Seed	—	15.0	14.7	15.0	15.3	15.0	15.0	14.0	14.5
	Pod husk	—	10.3	10.3	10.2	9.9	12.7	8.9	10.0	8.9
Ca (g kg^{-1})	Root	4.0	5.1	4.4	4.3	4.4	3.4	4.3	4.3	4.2
	Top	12.3	11.9	12.0	12.9	9.6	10.1	12.5	11.7	12.3
	Seed	—	0.6	0.5	0.6	0.7	0.7	0.7	0.7	0.6
	Pod husk	—	3.2	2.7	2.3	2.7	3.2	2.8	2.8	2.9
Mg (g kg^{-1})	Root	1.8	4.4	4.5	4.5	5.1	3.6	4.6	5.8	5.4
	Top	2.8	3.8	4.3	4.1	3.9	3.4	3.8	4.2	4.5
	Seed	—	1.8	1.8	1.7	1.8	1.8	1.8	1.8	1.7
	Pod husk	—	3.5	3.4	3.3	3.4	3.4	3.2	3.2	3.4

Zn (mg kg^{-1})	Root	108	136	125	133	143	83	111	152	119
	Top	58	34	26	29	24	22	22	27	23
	Seed	—	49	46	49	49	47	46	45	45
	Pod husk	—	22	19	21	16	27	15	15	13
Fe (mg kg^{-1} × 10^2)	Root	23	48	31	35	47	34	38	53	47
	Top	2	2	1	2	1	1	1	1	2
	Seed	—	0.75	0.75	0.76	0.78	0.88	0.82	0.80	0.80
	Pod husk	—	0.5	1	0.6	0.5	0.5	0.5	0.4	0.5
Mn (mg kg^{-1} × 10^2)	Root	30	39	36	42	49	36	38	55	49
	Top	48	35	41	41	42	55	54	55	69
	Seed	—	9	9	8	8	8	10	9	10
	Pod husk	—	15	16	18	19	22	20	20	21
Cu (mg kg^{-1})	Root	41	24	17	19	22	14	17	22	16
	Top	10	8	5	4	4	4	4	4	4
	Seed	—	7	9	9	7	6	6	5	6
	Pod husk	—	7	8	7	8	8	7	6	6

Source: Fageria, 1991.

Table 13.13 Adequate Nutrient Concentrations in the Cowpea Plant

Nutrient	Growth stage[a]	Plant part[a]	Adequate concentration g kg^{-1}
N	39 DAS	Whole tops	28
	Early flowering	PUMB	11–17
P	56 DAS	Whole tops	3
	Early flowering	PUMB	1.2–4.0
K	42 DAS	PUMB	3.5–6.0
	62–90 DAS	All LB	2.7–6.0
Ca	39 DAS	Whole tops	9
	Early flowering	PUMB	7.2–10.0
Mg	Early flowering	PUMB	1.7–3.1
			mg kg^{-1}
Zn	42 DAS	LB	27–32
Mn	35 DAS	Whole tops	< 1000
Fe	56 DAS	Whole tops	> 100

[a] DAS, Days after sowing.
[b] PUMB, Petiole of Uppermost mature leaf blade; LB, leaf blade (excluding sheath or petiole).
Source: Compiled from Reuter, 1986.

The distribution of these nutrients in different plant parts as a percentage of the total plant is presented in Figure 13.18. On an average, 9% P was in roots, 53% in tops, 35% in seeds, and 3% in pod husks. The potassium distribution was 5% in roots, 49% in tops, 36% in seeds, and 10% in pod husks. The Ca distribution was 5% in roots, 82% in tops, 5% in seeds, and 8% in pod husks, and the Mg distribution was 9% in roots, 50% in tops, 21% in seeds, and 20% in pod husks.

Zinc was distributed 19% in roots, 28% in tops, 46% in seeds, and 7% in pod husks. The iron content was highest in roots and lowest in pod husks; it was 72% in roots, 16% in tops, 9% in seeds, and 3% in pod husks. The manganese distribution was 10% in roots, 65% in tops, 14% in seeds, and 11% in pod husks. Similarly, 19% of the Cu was in roots, 28% in tops, 36% in seeds, and 17% in pod husks. These results suggest that cowpea roots retain maximum Fe and minimum K and Ca. Tops retain maximum Ca; almost 50% of the P, K, and Mg; and 65% Mn. Large amounts of P, K, Zn, and Cu are translocated to seeds. Pod husks contain a high amount of Mg and Cu.

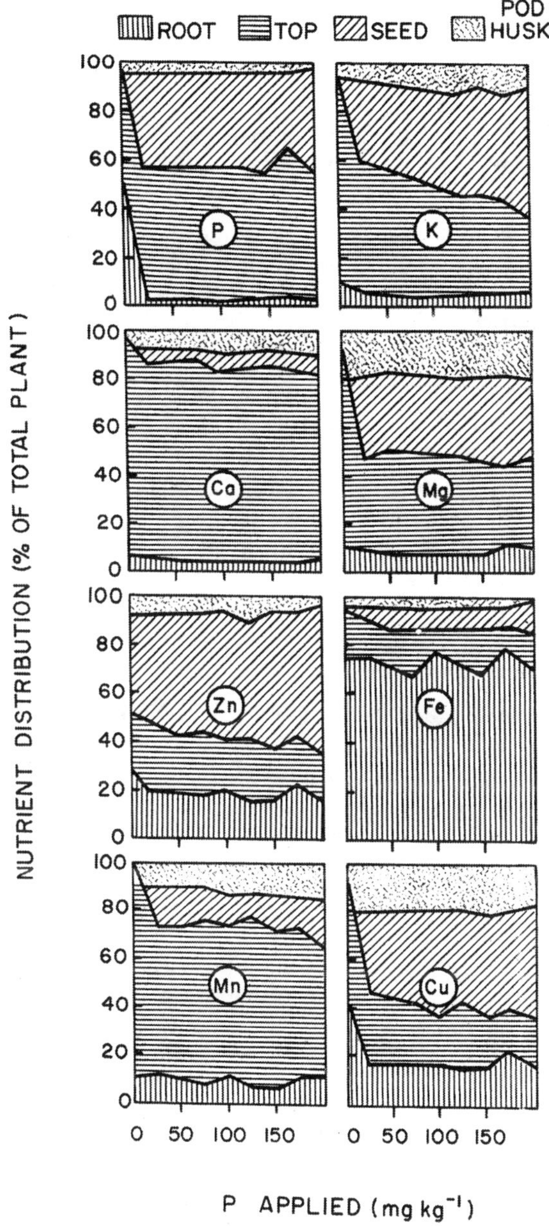

Figure 13.18 Distribution of nutrients in different parts of cowpeas grown in an Oxisol under greenhouse conditions. (From Fageria, 1991.)

IV. SUMMARY

Common bean and cowpea are important warm-season grain legumes in temperate, tropical, and subtropical regions. Common bean was domesticated in Central America and/or northern South America. Cowpea probably originated in Africa. For both crops, many cultivars with different plant architectures and seed characteristics are available. Both crops can be grown on a wide range of well-drained soils and produce grain rich in protein but deficient in sulfur amino acids. Cowpea is more tolerant of acid soils, drought stress, and salinity than common bean. It also fixes a larger fraction of total nitrogen accumulated by the crop than common bean, and its total fixation of nitrogen compares favorably with that of soybean.

Flowering of both species is sensitive to both temperature and photoperiod, and considerable genotypic variation exists in plant response to both factors. In the tropics, common beans and cowpeas are grown both as monocultures and as intercrops or relay crops. In Latin America, most common bean production is found in areas with mean growing season temperatures of 18–25°C, but the climatic and cultural conditions of particular cultivars are highly specialized.

Farmers' yields of both crops are commonly low due to poor control of diseases, insects, water stress, mineral nutrition, and other adverse soil conditions. Substantial increases in farmers' yields will require improving levels of inputs required to minimize plant stress and developing cultivar tolerance to stress. Breeding for production-limiting factors helps recover yield potential of commercial cultivars, minimize production losses, reduce production costs, and stabilize yield. It also permits subsistence farmers to take advantage of improved cultivars, minimize the risk of spreading pathogen populations, reduce dependence on chemical pesticides and fertilizers, and increase water use efficiency, thus maintaining a cleaner environment and conserving natural resources.

One metric ton of cowpea or common bean seed contains about 38 kg N, 5 kg P, and 15 kg K. To maintain soil fertility, nutrients (other than fixed nitrogen) should be returned in fertilizers or manure. Soil and tissue analyses can be used to diagnose and correct nutrient deficiencies and toxicities. Increased emphasis on breeding for nitrogen fixation may be needed in common bean.

REFERENCES

Alexander, M. 1977. Introduction to soil microbiology. Wiley, New York.

Association of Official Seed Analysis. 1981. Rules for testing seeds. J. Seed Technol. 6: 1–126.

Aykroyd, W. R. and J. Doughty. 1964. Legumes in human nutrition. FAO Nutritional Studies, No. 19, FAO, Rome.

Bagnall, D. J. and R. W. King. 1987. Temperature and irradiance effects on yield in cowpea (*Vigna unguiculata*). Field Crops Res. 16: 217–229.

Barros, L. C. G. and R. J. Hanks. 1993. Evapotranspiration and yield of bean as affected by mulch and irrigation. Agron. J. 85: 692–697.

Bennet, J. P., M. W. Adams, and C. Burga. 1977. Pod yield component variation and intercorrelation in *Phaseolus vulgaris* L. as affected by planting density. Crop Sci. 17: 73–75.

Bergersen, F. J. 1971. The central reaction of nitrogen fixation. Plant Soil Special Volume: 511–524.

Bezdicek, D. F. 1979. Nitrogen fixation, pp. 325–332. *In*: R. W. Fairbridge and C. W. Finkl, Jr. (eds.). Dowden, Hutchinson and Ross, Stroudsburg, Pennsylvania.

Bouwkamp, J. C. and W. L. Summers. 1982. Inheritance of resistance to temperature drought stress in the snap bean. J. Hered. 73: 385–386.

Chaturvedi, G. S., P. K. Aggarwal, and S. K. Sinha. 1980. Growth and yield of determinate and indeterminate cowpeas in dryland agriculture. J. Agric. Sci. 94: 137–144.

CIAT (Centro Internacional de Agricultura Tropical). 1980. Annual Report 1979. Bean Program CIAT, Cali, Colombia.

CIAT (Centro Internacional de Agricultura Tropical). 1981. Annual Report 1980. Bean Program CIAT, Cali, Colombia.

CIAT (Centro Internacional de Agricultura Tropical). 1981. Potential for field beans in eastern Africa. Proc. regional workshop held in Lilongwe, Melawi, March 9–14, 1980. CIAT, Cali, Colombia.

CIAT. 1985. Potentials for field beans in west Asia and North Africa. Proc. regional workshop held in Aleppo, Syria, May 21–23, 1983. CIAT, Cali, Colombia.

CIAT. 1992. Trends in CIAT commodities, 1992. Working Document 111. CIAT, Cali, Colombia.

Cowan, I. R. 1977. Stomatal behavior and environment. Adv. Bot. Res. 4: 117–228.

Cox, F. R. and E. Uribe. 1992. Management and dynamics of potassium in a humid tropical Ultisol under a rice-cowpea rotation. Agron. J. 84: 655–660.

Dart, P. J. 1974. The infection process, pp. 381–429. *In*: A. Quispel (ed.). The biology of nitrogen fixation. American Elsevier, New York.

Debouck, D. G. 1991. Systematics and morphology, pp. 55–118. *In*: A. Van Schoonhoven and O. Voysest (eds.) Common beans research for crop improvement. C.A.B. Int. Wallingford, UK and CIAT, Cali, Colombia.

Denis, J. C. and M. W. Adams. 1978. A factor analysis of plant variables related to yield in dry beans. I. Morphological traits. Crop Sci. 18: 74–78.

Duncan, W. G. 1971. Leaf angles, leaf area and canopy photosynthesis. Crop Sci. 11: 482–485.

Eaglesham, R. J., F. R. Minchin, R. J. Summerfield, P. J. Dart, P. A. Huxley, and J. M. Day. 1977. Nitrogen nutrition of cowpea (*Vigna unguiculata*). III. Distribution of nitrogen within effectively nodulated plants. Exp. Agric. 13: 369–380.

Edwards, D. G., B. T. Kang, and S. K. A. Danso. 1981. Differential response of six cowpea (*Vigna unguiculata* L. Walp.) cultivars to liming in an Ultisol. Plant Soil 59: 61–73.

Evans, A. M. 1976. Beans, pp. 168–172. *In*: N. W. Simmonds (ed.). Evolution of crop plants. Longman, New York.

Evans, A. M. 1980. Structure, variation, evolution and classification in *Phaseolus*, pp. 337–347. *In*: R. J. Summerfield and A. H. Bunting (eds.). Advances in legume science. Royal Botanic Gardens, Kew, England.

Fageria, N. K. 1989. Effects of phosphorus on growth, yield and nutrient accumulation in the common bean. Trop. Agric. 66: 249–255.

Fageria, N. K. 1991. Response of cowpea to phosphorus on an Oxisol. Trop. Agric. 68: 384–388.

Fageria, N. K. 1996a. Annual report of the project study of liming and fertilization of rice and common bean in cerrado region. EMBRAPA-CNPAF, Goiania, Brazil.

Fageria, N. K. 1996b. Evaluation of techniques for rice and common bean cultivation of lowland soils of central-west and north region. EMBRAPA-CNPAF, Goiania, Brazil.

Fageria, N. K. and A. M. Carvalho. 1996. Response of common bean to phosphorus on acid soils. Commun. Soil Sci. Plant Anal. 27: 1447–1458.

Fageria, N. K. and M. P. Souza. 1995. Response of rice and common bean crops in succession to fertilization in cerrado soil. Pesq. Agropec. Bras., Brasilia 30: 359–368.

Faris, D. G. 1965. The origin and evolution of the cultivated forms of *Vigna sinensis*. Can. J. Cytol. 7: 733–752.

Graham, P. H. 1981. Some problems of nodulation and symbiotic nitrogen fixation in *Phaseolus vulgaris* L.: A review. Field Crops Res. 4: 93–112.

Grass, Y. and J. Kigel. 1994. Differential sensitivity to high temperature of stages in the reproductive development of common bean (*Phaseolus vulgaris* L.). Field Crops Res. 36: 201–212.

Henry, G. 1989. Snap beans: Their constraints and potential for the developing world. Trop. Agric. Res. Soc. 23: 158–166.

IITA (International Institute of Tropical Agriculture). 1973. Grain legume improvement. Annual Report for 1972. IITA, Ibadan, Nigeria.

Itani, J., N. Utsunomiya, and S. Shigenaga. 1992a. Drought tolerance of cowpea. 1. Studies on water adsorption ability of cowpea. J. Trop. Agric. 36: 37–44.

Itani, J., N. Utsunomiya, and S. Shigenaga. 1992b. Drought tolerance of cowpea. 2. Comparative study on water relations and photosynthesis among cowpea, soybean, common bean and greengrain under water stress conditions. Japan J. Trop. Agric. 36: 265–274.

Itani, J., N. Utsunomiya, and S. Shigenaga. 1993. Effect of soil water deficit on leaf longevity in cowpea. 3. Effect of soil water deficit on leaf longevity in cowpea. Japan J. Trop. Agric. 37: 107–114.

Janssen, W. 1989. A socioeconomic perspective on earliness in beans, pp. 135–155. *In*: S. Beebe (ed.). Current topics in breeding of common bean. Working Document 47. Bean Program, CIAT, Cali, Colombia.

Kassam, A. H. and J. M. Kowal. 1973. Productivity of crops in the savanna and rain forest zones in Nigeria. Savanna 2: 39–49.

Kooistra, E. 1971. Germinability of bean (*Phaseolus vulgaris* L.) at low temperatures. Euphytica 20: 209–213.

Kueneman, E. A., D. H. Wallace, and P. M. Ludford. 1979. Photosynthetic measurements of field-grown dry beans and their relation to selection for yield. J. Am. Soc. Hortic. Sci. 104: 480–482.

Labanauskas, C. K., P. Shouse, and L. H. Stolzy. 1981. Effects of water stress at various growth stages on seed yield and nutrient concentrations of field grown cowpeas. Soil Sci. 131: 249–256.

Laing, D. R., P. J. Kretchmer, S. Zuluaga, and P. G. Jones. 1983. Field bean, pp. 227–248. *In*: IRRI (ed.). Potential productivity of field crops under different environments. IRRI, Los Banos, Philippines.

Laing, D. R., P. J. Jones, and J. H. C. Davis. 1984. Common bean (*Phaseolus vulgaris* L.), pp. 305–251. *In*: P. R. Goldsworthy and N. M. Fisher (eds.). The physiology of tropical field crops. Wiley, New York.

Lebaron, M. J. 1974. A description: Developmental stages of the common bean plant. University of Idaho, College of Agriculture, Current Information Series, No. 228.

Littleton, E. J., M. D. Dennett, J. Elston, and J. L. Monteith. 1979. The growth and development of cowpeas (*Vigna unguiculata*) under tropical field conditions. 1. Leaf Area. J. Agric. Sci. 93: 291–307.

Louwerse, W. and W. D. Zweerde. 1977. Photosynthesis, transpiration and leaf morphology of *Phaseolus vulgaris* and *Zea mays* growth at different irradiances in artificial and sunlight. Photosynthetica 11: 11–21.

Ludlow, M. M. and R. C. Muchow. 1990. A critical evaluation of traits for improving crop yields in water limited environments. Adv. Agron. 43: 107–153.

Lynch, J. and J. J. Beem. 1993. Growth and architecture of seedling roots of common bean genotypes. Crop Sci. 33: 1253–1257.

Maas, E. V. and G. J. Hoffman. 1977. Crop salt tolerance—current assessment. J. Irrig. Drainage Div. Am. Soc. Civil Eng. 103: 115–134.

Martin, J. H. and W. H. Leonard. 1967. Cowpeas, pp. 663–671. *In*: Principles of field crop production, 2nd Ed. Macmillan, New York.

Monteith, J. L. 1969. Light interception and radiative exchanges in crop stands, pp. 89–109. *In*: J. D. Eastin, F. A. Haskins, C. Y. Sullivan, and C. H. M. Vanbavel (eds.). Physiological aspects of crop yield. Am. Soc. Agron., Madison, Wisconsin.

Ndunguru, B. J. and R. J. Summerfield. 1975. Comparative laboratory studies of cowpea (*Vigna unguiculata*) and soybean (*Glycine max*) under tropical temperature conditions. II. Contribution of cotyledons to early seedling growth. East Afr. Agric. For. J. 41: 65–71.

Nienhuis, J. and S. P. Singh. 1986. Combining ability analyses and relationships among yield, yield components and architectural traits in dry bean. Crop Sci. 26: 21–27.

Nienhuis, J. and S. P. Singh. 1988. Genetics of seed yield and its components in common bean (*Phaseolus vulgaris* L.) of middle-America origin. II. Genetic variance, heritability and expected response from selection. Plant Breeding 101: 155–163.

Nutman, P. S. 1976. Alternative sources of nitrogen for crops. J. R. Agric. Soc. Eng. 137: 86–94.

Ojehomon, O. O. 1968. Flowering, fruit production and abscission in cowpea, *Vigna unguiculata* L. Walp. J. West Afr. Sci. Assoc. 13: 227–234.

Pachico, D. 1989. Trends in world common bean production, pp. 1–8. *In*: H. F. Schwartz and M. A. Pastor-Corrales (eds.) Bean production problem in the tropics. 2nd Ed. CIAT, Cali, Colombia.

Piggott, T. J. 1986. Vegetable crops, pp. 148–187. *In*: D. J. Reuter and J. B. Robinson (eds.). Plant analysis: An interpretation manual. Inkata Press, Melbourne.

Piha, M. I. and D. N. Munns. 1987. Nitrogen fixation capacity of field grown bean compared to other grain legumes. Agron. J. 79: 690–696.

Purseglove, J. W. 1987. Tropical crops. Dicotyledons. Longman, New York.

Rachie, K. O. and L. M. Roberts. 1974. Grain legumes of the lowland tropics. Adv. Agron. 26: 1–132.

Rajat De, D. V., S. Bhujanga Rao, Y. Yogeswara Rao, L. G. Giri Rao, and M. Ikramullah. 1983. Modification of irrigation requirement of wheat through mulching and foliar application of transpiration suppressants. Irrig. Sci. 4: 215–223.

Rawsthorne, S., F. R. Minchin, R. J. Summerfield, C. Cookson, and J. Combs. 1980. Carbon and nitrogen metabolism in legume root nodules. Phytochemistry 19: 341–355.

Reuter, D. J. 1986. Temperate and sub-tropical crops, pp. 38–99. *In*: D. J. Reuter and J. B. Robinson (eds.). Plant analysis: An interpretation manual. Inkata Press, Melbourne.

Sale, P. J. M. 1975. Productivity of vegetable crops in a region of high solar input. IV. Field chamber measurements on french beans (*Phaseolus vulgaris* L.) and cabbages (*Brassica oleracea* L.). Aust. J. Plant Physiol. 2: 461–470.

Sarafi, A. 1978. A yield component selection experiment involving American and Iranian cultivars of common bean. Crop Sci. 18: 5–7.

Scarisbrick, D. H., M. K. V. Carr, and J. M. Wilks. 1976. The effect of sowing date and season on the development and yield of navy beans (*Phaseolus vulgaris*) in southeast England. J. Agric. Sci. 86: 65–76.

Schwartz, H. F., G. E. Galvez, A. V. Schoonhoven, R. H. Howeler, P. H. Graham, and C. Flor. 1978. Field problems of beans in Latin America. CIAT, Cali, Colombia.

Scully, B. and J. G. Waines. 1987. Germination and emergence response of common and tepary beans to controlled temperature. Agron. J. 79: 287–291.

Scully, B. and D. H. Wallace. 1990. Variation in and relationship of biomass, growth rate, harvest index, and phenology to yield of common bean. J. Amer. Soc Hort. Sci. 115: 218–225.

Sellschop, J. P. F. 1962. Cowpea, *Vigna unguiculata* (L.) Walp. Field Crop Abstr. 15: 259–266.

Shouse, P., S. Dasberg, W. A. Jury, and L. H. Stolzy. 1981. Water deficit effects on water potential, yield and water use of cowpeas. Agron. J. 73: 333–336.

Silveira, P. M., L. F. Stone, G. P. Rios, t. Cobucci, and A. M. Amaral. 1996. Irrigation and common bean crop. EMBRAPA-CNPAF, Document 63, Goiania, Brazil.

Singh, S. P. 1982. A key for identification of different growth habits of *Phaseolus vulgaris* L. Bean improvement coop. New York. 25: 92–94.

Singh, S. P. 1992. Common bean improvement in the tropics. Plant Breeding Reviews 10: 199–269.

Singh, S. P., R. Lepiz, J. A. Gutierrez, C. Urrea, A. Molina, and H. Teran. 1990. Yield testing of early generation populations of common bean. Crop Sci. 30: 874–878.

Smyth, T. J. and M. S. Cravo. 1990. Critical phosphorus levels for corn and cowpea in a Brazilian Amazon Oxisol. Agron. J. 82: 309–312.

Spain, J. M., C. A. Francis, F. H. Howeler, and F. Calvo. 1975. Differential species and varietal tolerance to soil acidity in tropical crops and pastures, pp. 308–329. *In*: E. Bornemisza and A. Alvarado (eds.). Soil management in tropical America. North Carolina State University, Raleigh.

Sponchiado, B. N., J. W. White, J. A. Castillo, and P. G. Jones. 1989. Root growth of four common bean cultivars in relation to drought tolerance in environments with contrasting soil types. Exp. Agric. 25: 249–257.

Stofella, P. J., R. F. Sandsted, R. W. Zobel, and W. L. Hymes. 1979b. Root characteristics of black beans. II. Morphological differences among genotypes. Crop Sci. 19: 826–830.

Stoker, R. 1974. Effect on dwarf beans of water stress at different phases of growth. N. Z. J. Exp. Agric. 2: 13–15.

Summerfield, R. J., P. A. Huxley, and W. Steele. 1974. Cowpea (*Vigna unguiculata* L. Walp.). Field Crop Abstr. 27: 301–312.

Summerfield, R. J., F. R. Minchin, K. A. Stewart, and B. J. Ndunguru. 1978. Growth, reproductive development and yield of effectively nodulated cowpea plants in contrasting aerial environments. Annu. Appl. Biol. 90: 277–291.

Summerfield, R. J., F. R. Minchin, E. H. Roberts, and P. Hadley. 1983. Cowpea, pp. 249–280. *In*: IRRI (ed.). Potential productivity of field crops under different environments. IRRI, Los Banos, Philippines.

Summerfield, R. J. 1977. Vegetative growth, reproductive ontogeny and seed yield of selected tropical grain legumes, pp. 251–271. *In*: N. A. McFarlane (ed.). Crop protection agents: Their biological evaluation. Academic Press, London.

Summerfield, R. J. and E. H. Roberts. 1984a. *Phaseolus vulgaris*, pp. 139– 148. *In*: A. H. Haveley (ed.). Handbook of flowering, Vol. 1. CRC Press, Boca Raton, Florida.

Summerfield, R. J. and E. H. Roberts. 1984b. *Vigna unguiculata*, pp. 171–184. *In*: A. H. Haveley (ed.). Handbook of flowering. Vol. 1. CRS Press, Boca Raton, Florida.

Tanaka, A. and Fujita, K. 1979. Growth, photosynthesis and yield components in relation to grain yield of the field bean. J. Fac. Agric., Hokkaido Univ. 59: 145–238.

Turk, K. J., A. E. Hall, and C. W. Asbell. 1980. Drought adaptation of cowpea. I. Influence of drought on seed yield. Agron. J. 72: 413–420.

Van Wyk, H. P. D. 1955. The nutritive value of South African feeds. Dept. Agric. South Africa Sci. Bull. No. 354.

Verdcort, B. 1970. Studies in the Leguminosae-Papilionoideae for the flora of tropical East Africa. IV. Kew Bull. 24: 507–569.

Vieira, C. 19888. Phaseolus genetic resources and breeding in Brazil, pp. 467–483. *In*: P. Gepts (ed.). Genetic resources of phaseolus beans. Kluwer, Dordrecht, The Netherlands.

Vincent, J. M. 1974. Root nodule symbioses with *Rhizobium*, pp. 266–341. *In*: A. Guispel (ed.). The biology of nitrogen fixation. North Holland, Amsterdam.

Voysest, O. 1989. Market classes of dry edible beans consumed in Latin America. Annu. Report Bean Improv. Coop. 32: 66–67.

Voysest, O. and M. Dessert. 1991. Bean cultivars: Classes and commercial seed types, pp. 119–162. *In*: A. Van Schoonhoven and O. Voyset (eds.). Common beans: research for crop improvement. C.A.B. Int. Wallington, UK, and CIAT, Cali, Colombia.

Wallace, D. H. 1980. Adaptation of *Phaseolus* to different environments, pp. 349–357. *In*: R. J. Summerfield and A. H. Bunting (eds.). Advances in legume science. Royal Botanic Gardens, Kew, England.

Warring, M. O. A. and A. E. Hall. 1984. Reproductive responses of cowpea (*Vigna unguiculata* L. Walp.) to heat stress: I. Response to soil and day temperatures. Field Crops Res. 8: 3–16.

Westermann, D. Y. and J. J. Kolar. 1978. Symbiotic $N_2(C_2H_2)$ fixation by bean. Crop Sci. 18: 986–990.

Westermann, D. T., G. E. Kleinkopf, L. K. Porter, and G. E. Leggett. 1981. Nitrogen sources for bean seed production. Agron. J. 73: 660–664.

White, J. W., J. A. Castilllo, and J. Ehleringer. 1990. Association between productivity, root growth and carbon isotope discrimination in *Phaseolus vulgaris* under water deficit. Aust. J. Plant Physiol. 17: 189–198.

White, J. W., J. A. Castilllo, J. R. Ehleringer, J. A. Garcia, and S. P. Singh. 1994. Relations of carbon isotope discrimination and other physiological traits to yield in common bean (*Phaseolus vulgaris*) under rainfed conditions. J. Agric. Sci. Cambridge 122: 275–284.

White, J. W. and S. P. Singh. 1991. Breeding for adaptation to drought, pp. 501–560. *In*: A. V. Schoonhoven and O. Voysest (eds.). Common beans: research for crop improvement. C.A.B. Int. Wallingford, UK, and CIAT, Cali, Colombia.

Wien, H. C. 1973. Yield physiology of cowpea (*Vigna unguiculata* L. Walp.), a review of work at IITA, pp. 74–76. *In*: Proc. physiological program formulation workshop. IITA, Ibadan, Nigeria.

Wien, H. C. 1982. Dry matter production, leaf area development, and light interception of cowpea lines with broad and narrow leaflet shape. Crop Sci. 22: 733–737.

Wien, H. C. and R. J. Summerfield. 1984. Cowpea (*Vigna unguiculata* L. Walp.), pp. 353–383. *In*: P. R. Goldsworthy and N. M. Fisher (eds.). The physiology of tropical field crops. Wiley, New York.

Yan, X., J. P. Lynch, and S. E. Beebe. 1995. Genetic variation for phosphorus efficiency of common bean in contrasting soil types: I. Vegetative response. Crop Sci. 35: 1086–1093.

14

Peanut

I. INTRODUCTION

The peanut (*Arachis hypogaea* L.), commonly known as groundnut, earthnut, monkeynut, pinder, or goober, is both an oilseed crop and a food grain legume (Krapovickas, 1969). The genus *Arachis* contains a rich diversity of plant types. Both annuals and perennials are known, and although most species reproduce by seed, some are rhizomatous and reproduce largely through vegetative means. The species occur in regions as different as poorly drained, swampy areas near sea level, to drought conditions, to mountainous regions at elevations up to 1600 m. The genus *Arachis* is naturally restricted to the countries of Brazil, Argentina, Bolivia, Paraguay, and Uruguay in South America (Singh and Simpson, 1994). It is generally believed that the peanut cultivated for food and oil around the world, *Arachis hypogaea* L., originated in southern Bolivia or northern Argentina (Gregory et al., 1980.) This crop was probably brought to Africa from Brazil by the Portuguese early in the sixteenth century and somewhat later was transported from the west coast of South America to Asia. They may have reached the United States by way of slave ships from West Africa, although precisely when and where they were introduced is not known (Gibbons, 1980). The cultivated peanut is found throughout the tropical and temperate regions of the world; however, wild species of *Arachis* are found only in South America. Peanuts are grown on an estimated 19 million ha in 82 countries for use as food, oil, and a high-protein meal (Wynne and Gregory, 1981). Peanut seeds contain 25–30% protein, about 50% oil, 20% carbohydrate, and 5% fiber and ash. Properties of peanut oil are determined by the fatty acid composition. Approximately 90% of peanut oil is composed of palmitic acid (16 carbons, no double bonds; 16:0), oleic acid (18:1), and linoleic acid (18:2). Although many studies have identified genetic differences in fatty acid composition in peanuts, most have examined a limited number of genotypes (Knauft and Wynne, 1995).

India, China, and Indonesia have the largest peanut-growing areas in Asia, while in Africa the major producers are Nigeria, Senegal, and Sudan. In the western hemisphere, the United States, Brazil, and Argentina are the leading peanut producers.

Seventy percent of the world peanut production occurs in the semiarid tropics. The average yield of peanuts in the semiarid tropics (around 0.8 t ha^{-1} of dried pods) is lower than the world average (estimated at 1.1 t ha^{-1} of dried pods in 1985) and much lower than yields of over 3 t ha^{-1} obtained in the developed countries (ICRISAT, 1987). Drought, diseases, and insects are the main yield-limiting factors in these regions. Smartt (1994) gave a global view of production practices and noted that they vary considerably. In the United States, Australia, and portions of South America, the crop is grown with intense management, generally with high levels of mechanical and chemical input. The crop is grown in mixtures with other species, mainly to provide food and cooking oil for the farmer, in parts of Africa and Southeast Asia. In many countries the crop is grown in monoculture as a cash crop, primarily for export. The intensity of management varies considerably around the world, depending on the economic return for the crop or the role of peanuts in farm subsistence.

II. CLIMATE AND SOIL REQUIREMENTS

Although peanuts are predominantly a crop of the tropics, the approximate limits of present commercial production are between latitudes 40°N and 40°S, where rainfall during the growing season exceeds 500 mm (Gibbons, 1980). Peanuts perform well in the dry temperature range between 24 and 33°C but can survive up to 45°C if adequate moisture is maintained (Saxena et al., 1983).

Although the greatest recorded yield for the crop is 9.6 t ha^{-1} (Hildebrand, 1980), current commercial yields are 3.0–4.0 t ha^{-1} in many countries and as low as 1.0 t ha^{-1} in others. While potential yield of peanut at any given location is unlikely to equal record yield, it is likely that in most production regions current commercial crops fail to approach their potential. Climatic variability is a major cause of inability to achieve potential yield in irrigated and dryland production. Climatic variability can generate substantial production variability. The production risks associated with climatic variability can have severe consequences on individual farmers and on regionally based industries. A better quantification of potential yield and production risks for possible new production environments would facilitate planning for the location of production areas and determine production possibilities (Hammer et al., 1995).

In Asia and Africa, peanuts are mostly grown as a rainfed crop. Kassam et al. (1975) showed that, from sowing to harvest, a rainfed crop in Nigeria used 438 mm of water to produce a seed yield of 1.6 t ha^{-1} in 4 months. The crop water use efficiency was 489 g water g^{-1} dry matter produced. Reddy (1977) gives a similar estimate of 444 mm water used by a high-yielding crop. Results reported in various studies showed that 500–600 mm of water, if reasonably well distributed, can produce satisfactory peanut yields (Gorbett and Rhoads, 1975; Stansell et al., 1976; Pallas et al., 1977).

Daily water use by peanuts is low during the early growth stages and increases with increasing leaf canopy. The maximum daily water use rate ranged from 0.5 to 0.6 cm day^{-1} (Stansell et al., 1976). Maximum water use rates occurred at 70, 80, and 95 days after planting for Tifspan, Florunner, and Florigiant cultivars, respectively (Stansell et al., 1976). The period of maximum evapotranspiration (ET) for Spanish peanuts was reported 67–77 days after planting by Kassam et al. (1975) and at 55–80 days by Vivekanandan and Gunasena (1976). Kassam et al. (1975) observed that peak ET occurred shortly before peak leaf area index (*LAI*) was achieved.

In the United States, irrigation to obtain high peanut yields is becoming popular. In Georgia 45% of the allotted peanut acreage was under irrigation, with new installations increasing steadily (Henning et al., 1979b). According to Boote et al. (1982), optimum water management appears to be scheduling irrigation to maintain less than 50% soil water deficit in the top 30 cm during early growth and possibly irrigating at 25% soil water deficit during pod formation and seed growth. Some authors suggested that if soil water potential is measured in the top 15–30 cm of soil, irrigation should be scheduled to maintain soil water potential above –0.6 bar or –60 kPa (on sand or sandy loam), although irrigating to maintain soil water potential above –0.25 to –0.50 bar or –25 to –50 kPa may be desirable during long, dry, hot periods occurring during the sensitive growth stages of pegging, pod formation, and early pod fill.

Identification of physiological traits contributing to superior performance of crop plants under drought conditions has been a long-term goal of plant scientists. Water-use efficiency (WUE) is one such trait which can contribute to productivity when water resources are scarce. Reviews of the literature have suggested that intraspecific variations in WUE are small and are likely to be increased only by crop management or modifying the environment (Tanner and Sinclair, 1983). However, variation in WUE was shown to exist between and within species (Wright et al., 1994).

The most suitable soils for peanut production are well-drained, light sandy loams with an ample supply of calcium and moderate organic matter (Gibbons, 1980; Saxena et al., 1983). Adams and Hartzog (1980) summarized results from 78 field experiments in Alabama and found no correlation between soil pH and yield response of peanuts to liming, but yield was highly correlated with exchangeable Ca. Adams (1981) reported that peanuts are one of the most acid-tolerant crops, with a critical pH range of 5–5.5. Peanuts are considered moderately susceptible to soil salinity, and a salinity threshold (salinity at initial yield decline) of 3.2 dS m^{-1} has been reported (Maas and Hoffman, 1977). In practice, peanuts are grown on a range of soils from alkaline to acid and from clays to fine sands (Gibbons, 1980). But peanuts are highly susceptible to waterlogging, and harvesting can be a problem in heavily textured soils.

III. GROWTH AND DEVELOPMENT

Peanuts are a selfpollinated, annual, herbaceous legume and belong to the family Papilionaceae. The plant is erect or prostrate, sparsely hairy, and 15–60 cm high or higher. *Arachis hypogaea* describes the most peculiar trait of the species, underground fruit formation (*hypo* means under, and *agaea* means ground). The special feature of the peanut plant is that the fruit begins as a fertilized flower aboveground, but pod and seed mature in the ground. *Arachis hypogaea* consists of two subspecies, *hypogaea* and *fastigiata*, which differ in branching habit and seed dormancy. Each subspecies is further subdivided into two botanical varieties that correspond to market types. The two botanical varieties are hypogaea and hirsuta. The hypogaea variety has alternate branching, a spreading or bunching habit, and a long maturation period. Seeds undergo a period of dormancy after maturity and correspond to the Virginia market types. The botanical variety hirsuta is the Peruvian runner type. The *fastigiata* subspecies comprises two botanical varieties, fastigiata and vulgaris. The botanical variety fastigiata corresponds to the Valencia market type and botanical variety vulgaris to the Spanish market type (Gregory and Gregory, 1976).

A. Germination and Seedling Growth

Germination of peanuts is neither epigeal nor hypogeal but intermediate. The hypocotyl carries the cotyledons to the soil surface and remains there. The length of the hypocotyl is dependent on the depth of planting, but 10–15 cm length is normal. Work carried out at ICRISAT (1987) showed that 5 cm was the best sowing depth for peanuts, since deep sowing (10 cm) reduced pod yields by 30% due to lower crop growth rates after emergence. The first visible evidence of germination is the emergence of the radicle. Radicle emergence occurs by 24 hours or earlier for vigorous Spanish-type seed, but requires 36–48 hours in the Virginia types. During the first few days, the developing seedling depends on food reserves in the cotyledons for energy. After 5–10 days, depending on type of peanut and environmental conditions, the seedling becomes autotrophic (Ketring et al., 1982). The optimum temperature for peanut germination is about 30–35°C.

B. Roots

The peanut plant has a relatively deep taproot system with a well-developed lateral root system. Nodulation is governed by proper inoculation, as well as type of soil, genotype, and climatic factors. Rooting depth controls water extraction from the soil profile. The maximum depth of groundnut roots recorded varies with the soil and cultivar but ranges from more than 2.5 m in Florida sands (Boote et al., 1982) to only 1.2 m in an Alfisol in India (Gregory and Reddy, 1982). Root density in the upper horizons (30 cm) is substantial (up to 800 g m^{-3}), but below this depth root density is considerably less (±25% of the upper horizon). This distribution pattern seems to be an important contributor to the drought tolerance of this crop. The water in the top horizon is freely available and is utilized rapidly at a rate determined largely by the leaf area index and evapotranspiration. However, the water below this horizon is utilized much more slowly (at 50% or less of the potential evapotranspiration). Peanut roots can grow to a depth of about 200 cm in a well-drained sandy soil (Hammond et al., 1978). Root densities in this soil were 1.5 cm cm^{-3} in the 0–30 cm zone and 0.1–0.4 cm cm^{-3} at greater depths.

Robertson et al. (1979) reported that peanut roots showed a uniform distribution pattern under the row and laterally 46 cm from the row. The maximum root density was within the top 30–cm depth, which was included in a region above a tillage pan. Roots, which did penetrate the pan, extended to depths greater than 150 cm. McCloud (1974) reported peanut root weight in the top 15 cm of soil was 37% of the total crop dry matter at 21 days after planting (DAP) but only 1.5% at harvest. This was due to the large amount of dry weight at harvest. Root dry weight in the top 15 cm of soil reached a maximum by 78 DAP. Ketring et al. (1982) minimized the importance of extracting the complete peanut root system for calculations of total plant dry matter comparison because the root dry weight is usually a small percentage of total plant biomass. Even though the fibrous peanut roots below the usual harvesting depth of 30 cm constitute a small fraction of root weight, they may be important in water uptake (Boote et al., 1982; Meisner and Karnok, 1992).

C. Tops

Most peanut varieties have a bunch habit; the lateral branches rarely exceed the length of the main stem in Valencia, though they usually do so in Virginia bunch varieties (Gibbons et al., 1972). Peanut leaves are alternately arranged in a spiral (2/5 phyllotaxy on the n axis). They are pinnate, with two opposite pairs of leaflets 2 cm long. The petioles are 3–7 cm long. Individual flowers of peanuts emerge in sequence on the inflorescences. After fertilization, an intercalary meristem at the base of the ovary generates a stalk-like structure, the peg, which soon becomes positively geotropic and may extend to as much as 20 cm (Bunting and Elston, 1980). The ovary matures underground into a pod of the common unshelled peanut. The plant has an indeterminate growth habit. Flowers first appear about 30 or 40 days after sowing, and the plant may continue to produce flowers throughout much of its remaining growth (Ashley, 1984).

D. Dry Matter

The dry matter accumulation pattern in peanuts is similar to that of most other annual field crops. It is slow in the beginning, increases sharply in the late vegetative and early pod-filling stages, and reaches a plateau during late pod filling. Early top growth is composed mostly of main stem elongation and leaf production, but lateral branches account for the bulk of later growth (Ketring et al., 1982). Most of the dry matter produced occurs in the leaf blades and stems in the early stages and in the pods and stems at a later stage (Enyi, 1977).

Maximum crop growth rate values ranging from 13 to 24 g m^{-2} day^{-1} have been reported under different environmental conditions and for different cultivars (Williams et al., 1975; Enyi, 1977; Duncan et al., 1978). Duncan et al. (1978) reported that selection for higher yield in peanuts has not resulted in a corresponding increase in crop growth rate. A large part of the yield difference among peanut cultivars with nearly identical growth rates is associated with differences in partitioning of daily photosynthate to fruits.

The crop growth rate of legumes is generally less than that of cereals because both the capacity (leaf area) and intensity (rate of increase in dry weight per unit of leaf area) components are smaller (Bunting and Elston, 1980). This may be associated with an indeterminate fruiting habit, which compels the crop from an early stage of development to devote an increasing proportion of its assimilate to filling fruit rather than to making more leaves

(Bunting and Elston, 1980). Another reason is that legumes have a shorter vegetative period than cereals, and dry matter in legumes may also be lost by respiration associated with dinitrogen fixation.

Bell et al. (1993) found that variation among genotypes for dry matter production was accounted for by the effects of differing leaf area duration on cumulative, intercepted, photosynthetically active radiation. Their work also found genetic differences in sensitivity to night temperatures. They suggested that cultivars adapted to cool environments may provide opportunity for higher yields.

E. Leaf Area Index (*LAI*)

Leaf area is one of the principal growth parameters used to measure the functioning of the photosynthetic system of a crop species or cultivar. The leaf area index varies with environmental conditions, cultural practices, and stage of crop growth. The *LAI* should be related to growth stage when comparing different crop species or cultivars within species. For determinate crops, *LAI* measured at the beginning of the reproductive growth stage is more meaningful, whereas for indeterminate crops the upper limit of *LAI* may be used as a criterion for comparison. Maximum *LAI* values for peanuts range from 3.3 to 7.0 (Williams et al., 1975; Saxena et al., 1983). The *LAI* reaches a maximum value 65–75 days after sowing in cultivars with a 115– to 125–day growth cycle (Williams et al., 1975; Kassam et al., 1975; Enyi, 1977; Saxena et al., 1983). Misa et al. (1994) reported that the optimum *LAI* among 11 peanut genotypes ranged from 3.2 to 4.0. Duncan et al. (1978) reported that *LAI* continued to increase to more than 7.0 in some cultivars, but light interception reaches a maximum at an *LAI* of about 3.0. Further increases in *LAI* were assumed to have no measurable effect on crop growth rate.

The photosynthetic unit of the peanut plant is the tetrafoliate, pinnately compound leaf with two opposite pairs of leaflets. Photosynthetic response to different levels of light irradiance is strongly influenced by the light irradiance under which the plants were previously grown or pretreated (Bourgeios and Boote, 1992). Young leaves have higher net photosynthesis rates than older leaves. When exposed to higher light irradiances, lower leaves in the canopy have lower photosynthetic potential than higher leaves in the canopy (Henning et al., 1979a). Maximum apparent photosynthesis rates for peanut leaves range from 0.6 to 1.8 mg CO_2 m^{-2} s^{-1} (Henning et al., 1979b) with a mean value of 1.06 mg CO_2 m^{-2} s^{-1} (Pallas, 1973; Ketring et al., 1982). Pallas and Samish (1974) evaluated photosynthesis of peanut leaves at light irradiances from 180–1546 μmol m^{-2} s^{-1}. Leaf photosynthesis was not light-saturated in that range. This wide variability is mainly due to genotypic differences and variation in environmental conditions.

F. Harvest Index

The harvest index (ratio of the weight of kernels to total dry matter aboveground at harvest) varies between 0.35 and 0.5 in peanuts (Bunting and Elston, 1980). However, if allowance is made for the high energy value of the oil component of the seeds compared with the rest of the plant, the harvest index approaches 0.6 (Ashley, 1984). Sequentially branched forms in which plant structure is smaller and formed over a shorter time, and in which ratios of leaf growth to total growth decline earlier, tend to have a larger harvest index than the longer-lived and more freely branched alternate forms, in which vegetative growth is not arrested so soon by the internal regulation of partitioning (Bunting and Elston, 1980). The shelling percentage is about 80% for the early-maturing branch types compared with 60–75% for the spreading cultivars (Saxena et al., 1983).

IV. YIELD COMPONENTS

In the peanut crop, kernel yield per unit is the product of pod number, number of grains per pod, and weight of individual kernels. Kernels per pod varies from 2 to 6, pods per plant varies from 50 to 104, and 100–kernel weight varies from 28 to 62 g (Enyi, 1977; ICRISAT, 1987). This variation is related to cultivar, spacing, fertilizer, and climatic conditions. All three yield components are most sensitive to environmental stress during the flowering and kernel-filling growth stages.

V. NUTRIENT REQUIREMENTS

An adequate supply of essential nutrients is necessary to obtain high yields of peanuts. A balanced fertility program with particular emphasis on adequate levels of P, K, Ca, and Mg is essential to high yields (Henning et al., 1982). Peanuts, like other legumes, can fix nitrogen, but small amounts of N should be applied as starter fertilizer. Peanuts can fix atmospheric N if the correct strains of nitrogen-fixing bacteria are present in the soil. Inoculation of peanuts grown in rhizobia-free soil resulted in significant yield increases (Reddy and Tanner, 1980; Kremer and Peterson, 1983). However, Nambiar et al. (1984) reported that inoculation with *Rhizobium* increased peanut yields in fields where the crop had been previously grown at ICRISAT, India. Peanuts are nodulated by a large group of *Rhizobium* strains classified as the cowpea miscellany (Buchanan and Gibbons, 1974).

Soils often contain rhizobia that are highly competitive against those applied in inoculants. Selected strains of peanut *Rhizobium* failed to increase yields in the presence of high populations of indigenous rhizobia (Diatloff and Langford, 1975). Adverse environmental conditions cause inoculant quality to deteriorate. Hot and dry conditions at planting can cause a rapid decrease in rhizobia applied to seeds (Hardaker and Hardwick, 1978). Successful inoculation requires carriers capable of delivering high numbers of effective rhizobia under adverse conditions to ensure nodulation of the host legume (Kremer and Peterson, 1983).

Rhizobia differ in their ability to fix N_2, however, and the presence of nodules on roots of the groundnut plant does not necessarily mean that sufficient N_2 is being fixed for maximum growth of the host plant (Nambiar, 1985). Hence, it may be necessary to introduce superior strains of *Rhizobium* and apply some nitrogen to ensure an adequate nitrogen supply for maximum growth and yield of the host plant. Peanut studies indicate that at very high yield levels, the N requirements of nodulated peanut cannot be met from symbiotic N_2 fixation alone (Nambiar et al., 1986).

Peanut cultivars may also be developed with tolerance to soil deficiencies and toxicities. Mineral deficiencies can cause yield and quality problems in peanut production. Research on the mineral nutrition of peanuts was summarized by Gascho and Davis (1994). One of the most common production problems occurs in soils with low calcium. When peanuts are grown in these soils without additional calcium, pod rot and poorly filled pods are common (Gascho and Davis, 1994). Genetic differences exist in response to effects of varying soil calcium concentrations, with large-seeded types generally requiring higher levels of calcium (Walker and Keisling, 1978). Gascho and Davis (1994) summarized many studies on genotypic response to soil calcium concentrations and found that large-seeded types required approximately twice the calcium concentration of small-seeded types. Seed calcium content depends on the diffusion of calcium from the soil solution. Several studies examining genetic differences in seed and pod calcium levels have identified the ratio of pod surface area to seed weight as the cause of the greater calcium requirements in large-seeded types, rather than a genetic difference in the amount of calcium required per unit weight of seed (Gascho, 1992). Adams et al. (1993) identified genetic differences in requirements for ambient soil solution calcium concentrations. At present, the low cost of calcium supplements to the soil have made genetic manipulation for this trait a low priority.

VI. NUTRIENT CONCENTRATION AND UPTAKE

The importance of plant tissue analysis has become widely recognized with respect to soil fertility and plant nutrition. For annual crops, the primary function of plant analysis is to diagnose problems or to determine or monitor the nutrient status during the growing season (Dow and Roberts, 1982). Sahrawat et al. (1987) studied concentration changes in peanut leaves with time.

Table 14.1 Adequate Concentrations of Nutrients in Upper Stems and Leaves of Peanut Plant at Early Pegging

Nutrient	Adequate concentration
	g kg^{-1}
N	35.0–45.0
P	2.0–3.5
K	17.0–30.0
Ca	12.5–17.5
Mg	3.0–8.0
S	2.0–3.0
	mg kg^{-1}
Fe	100–250
Mn	100–350
Zn	20–50
B	20–50
Cu	10–50
Mo	1–5

Source: Compiled from Small and Ohlrogge, 1973.

Concentrations of N, P, K, Cu, Mn, and Zn in the leaves of the cultivar RMV2 generally decreased with increasing age. The concentrations of Ca increased markedly with leaf age, and Mg concentrations tended to increase as well. Adequate concentrations of macro- and micronutrients in peanut plants are presented in Table 14.1.

Nutrient uptakes (concentration × dry matter) by two peanut genotypes are presented in Table 14.2. Amounts of nutrient elements in the plant parts were greatest for N, followed by K, Ca, Mg, P, Fe, Mn, and Zn in descending order (Sahrawat et al., 1988).

VII. SUMMARY

The peanut is an important warm-season oilseed crop and food grain legume. It originated in the lowlands of South America and is now grown in the tropics, subtropics, and warm temperate regions worldwide. Peanut seeds contain about 50% oil and 25–30% protein and make a substantial contribution to human nutrition.

Table 14.2 Yield and Total Nutrients Harvested in Two Peanut Genotypes at Two N Levels

		0 kg N ha^{-1}		200 kg N ha^{-1}		SE±	
		Robut 33-1	Non-nod	Robut 33-1	Non-nod	a	b
Haulm (kg ha^{-1})		3249	1099	2660	1681	108.7	127.8
Pod (kg ha^{-1})		2099	553	2438	1053	267.9	198.4
Total nutrients harvested	N	176.6	23.9	143.9	51.3	11.98	8.47
(kg ha^{-1})	P	15.1	10.3	11.8	11.6	1.05	0.75
	K	55.0	21.0	43.0	35.0	3.88	2.74
	Ca	34.2	14.4	28.1	21.9	2.53	1.79
	Mg	27.0	7.3	24.0	15.0	2.25	1.59
	Fe	6.32	1.42	4.95	1.99	0.695	0.492
	Zn	0.25	0.12	0.19	0.15	0.019	0.013
	Mn	0.29	0.08	0.25	0.11	0.020	0.014

[a] Standard error of mean for comparing between the genotypes and nitrogen levels.
[b] Standard error of mean for comparing between the nitrogen levels.
Source: Sahrawat et al., 1988.

Average growing season temperatures of 24–33°C are adequate for peanut production. The crop can grow on a wide range of soils, but well-drained sandy loams with adequate supplies of calcium are ideal. Peanut is very tolerant of acid soils as long as adequate calcium is present in the surface soil to support peg and fruit development. It is intolerant of poor soil aeration and is moderately susceptible to salinity.

Under favorable conditions, symbiotic nitrogen fixation can account for much of the crop's nitrogen requirements. One metric ton of unshelled seeds removes approximately 40 kg N, 3 kg P, and 5 kg K. Since nitrogen fixation is normally significant, fertilizer requirements are modest compared with those of cereals. Both soil and plant analyses can be used to diagnose and correct nutrient deficiencies.

REFERENCES

Adams, F. 1981. Alleviating chemical toxicities: Liming acid soils, pp. 269–301. *In*: G. F. Arkin and H. M. Taylor (eds.). Modifying root environment to reduce crop stress. Monogr. 4, Am. Soc. Agr. Eng., St. Joseph, Michigan.

Adams, F. and D. Hartzog. 1980. The nature of yield response of Florunner plants to lime. Peanut Sci. 7: 120–123.

Adams, J. F., D. R. Hartzog, and D. P. Nelson. 1993. Supplemental calcium application on yield, grade and seed quality of runner peanut. Agron. J. 85: 86–93.

Ashley, J. M. 1984. Groundnut, pp. 453–494. *In*: P. R. Goldsworthy and N. M. Fisher (eds.). The physiology of tropical field crops. Wiley, New York.

Bell, M. J., G. C. Wright, and G. R. Harch. 1993. Environmental and agronomic effects on the growth of four peanut cultivars in a sub-tropical environment. I. Dry matter accumulation and radiation use efficiency. Exper. Agric. 29: 473–490.

Boote, K. J., J. R. Stansell, A. M. Schubert, and J. F. Stone. 1982. Irrigation, water use, and water relations, pp. 164–205. *In*: H. E. Pattee and C. T. Young (eds.). Peanut science and technology. Peanut science and technology. Am. Peanut Res. Educ. Soc., Yoakum, Texas.

Bourgeois, G. and K. J. Boote. 1992. Leaflet and canopy photosynthesis of peanut affected by late leaf spot. Agron. J. 84: 359–366.

Buchanan, R. E. and N. E. Gibbons, N. E. 1974. Bergey's manual of determinative bacteriology, 8th ed. Williams and Wilkins, Baltimore.

Bunting, A. H. and J. Elston. 1980. Ecophysiology of growth and adaption in the groundnut: An essay on structure, partition and adaption, pp. 495–500. *In*: R. J. Summerfield and A. H. Bunting (eds.). Advances in legume science. Royal Botanic Gardens, Kew, England.

Diatloff, A. and S. Langford. 1975. Effective natural nodulation of peanuts in Queensland. Queensland J. Agric. Anim. Sci. 32: 95–100.

Dow, A. I. and S. Roberts. 1982. Proposal: Critical nutrient ranges for crop diagnosis. Agron. J. 74: 401–403.

Duncan, W. G., D. E. McCloud, R. L. McGraw, and K. J. Boote. 1978. Physiological aspects of peanut yield improvement. Crop Sci. 18: 1015–1020.

Enyi, B. A. C. 1977. Physiology of grain yield in groundnuts (*Arachis hypogaea*). Exp. Agric. 13: 101–110.

Gascho, G. J. 1992. Groundnut, pp. 209–242. *In*: D. J. Halliday, M. E. Trenkel, and W. Wichman (eds.). IFA *World Fertilizer Use Manual*. Int. Fert. Ind. Assoc., Paris.

Gascho, G. J. and J. G. Davis. 1994. Mineral nutrition, pp. 214–254. *In*: J. Smartt (ed.) The groundnut crop: a scientific basis for improvement. Chapman and Hall, London.

Gibbons, R. W. 1980. Adaption and utilization of groundnuts in different environments and farming systems, pp. 483–493. *In*: R. J. Summerfield and A. H. Bunting (eds.). Advances in legume science. Royal Botanical Gardens, Kew, England.

Gibbons, R. W., A. H. Bunting, and J. Smartt. 1972. The classification of varieties of groundnut *Arachis hypogaea*). Euphytica. 21: 78–85.

Gorbett, D. W. and F. M. Rhoads. 1975. Response of two peanut cultivars to irrigation and Kylar (succinic acid 2,2 dimethylhydrazide, growth regulator). Agron. J. 67: 373–376.

Gregory, W. C. and M. P. Gregory. 1976. Groundnut, pp. 151–154. *In*: N. W. Simmons (ed.). Evolution of crop plants. Longman, New York.

Gregory, W. C., A. Krapovickas, and M. P. Gregory. 1980. Structures, variation, evolution and classification in *Arachis*, pp. 469–481. *In*: R. J. Summerfield and A. H. Bunting (eds.) Advances in legume science. Royal Botanic Gardens, Kew, United Kingdom.

Gregory, P. J. and M. S. Reddy. 1982. Root growth in an intercrop of pearl millet/groundnut. Field Crops Res. 5: 241–252.

Hammer, G. L., T. R. Sinclair, K. J. Boote, G. C. Wright, H. Meinke, and M. J. Bell. 1995. A peanut simulation model: I. Model development and testing. Agron. J. 87: 1085–1093.

Hammond, L. C., K. J. Boote, R. J. Varnell, and W. K. Robertson. 1978. Water use and yield of peanuts on a well-drained sandy soil. Proc. Am. Peanut Res. Educ. Soc. 10: 73.

Hardaker, J. M. and R. C. Hardwick. 1978. A note on *Rhizobium* inoculation of beans (*Phaseolus vulgaris*) using the fluid drill technique. Exp. Agric. 14: 17–21.

Henning, R. J., R. H. Brown, and D. A. Ashley. 1979a. Effects of leaf position and plant age on photosynthesis and translocation in peanut. I. Apparent photosynthesis and ^{14}C translocation. Peanut Sci. 6: 46–50.

Henning, R. J., J. F. McGrill, L. E. Samples, C. Swann, S. S. Thompson, and H. Womack. 1979b. Growing peanuts in Georgia: A package approach. Bull. 640, University of Georgia Cooperative Extension Service.

Henning, R. J., A. H. Allison, and L. D. Tripp. 1982. Cultural practices, pp. 123–138. *In*: H. E. Pattee and C. T. Young (eds.). Peanut science and technology. Am. Peanut Res. Educ. Soc., Texas.

Hildebrand, G. L. 1980. Groundnut production, utilization, research problems and further research needs in Zimbabwe, pp. 290–296. *In*: Proc. Int. Workshop on Groundnut, ICRISAT, India, Oct. 13–17, 1980. ICRISAT, Patancheru, India.

ICRISAT (International Crop Research Institute for the Semiarid Tropics). 1987. Annual Report. 1986. Andra Pradesh, India.

Kassam, A. H., J. M. Kowal, and C. Harkness. 1975. Water use and growth of groundnut at Samaru, north Nigeria. Trop. Agric. (Trinidad) 52: 105–112.

Ketring, D. L., R. H. Brown, G. A. Sullivan, and B. B. Johnson. 1982. Growth physiology, pp. 411–457. *In*: H. E. Pattee and C. T. Young (eds.). Peanut science and technology. Am. Peanut Res. Educ. Soc., Yoakum, Texas.

Knauft, D. A. and J. C. Wynne. 1995. Peanut breeding and genetics. Adv. Agron. 55: 393–445.

Krapovickas, A. 1969. The origin, variability and spread of the groundnut (*Arachis hypogaea*), pp. 427–444. *In*: P. J. Ucko and G. W. Dimbley (eds.). The domestication and exploitation of plants and animals. Duckworth, London.

Kremer, R. J. and H. L. Peterson. 1983. Field evaluation of selected *Rhizobium* in an improved legume inoculant. Agron. J. 75: 139–143.

Maas, E. V. and G. J. Hoffman. 1977. Crop salt tolerance—current assessment. J. Irrig. Drainage Div. Am. Soc. Civil Eng. 103: 115–134.

McCloud, D. E. 1974. Growth analysis of high yielding peanuts. Proc. Soil Crop Sci. Soc. Florida 33: 24–26.

Meisner, C. A. and K. J. Karnok. 1992. Peanut root response to drought stress. Agron. J. 84: 159–165.

Misa, A. L., A. Isoda, H. Nojima, Y. Takasaki, and T. Yoshimura. 1994. Plant type and dry matter production in peanut (*Arachis hypogaea* L.) cultivars. Japan J. Crop Sci. 63: 289–297.

Nambiar, P. T. C. 1985. Response of groundnut (*Arachis hypogaea* L.) to *Rhizobium* inoculation in the field: Problems and prospects. MIRCEN J. 1: 293–309.

Nambiar, P. T. C., P. J. Dart, B. S. Rao, and H. N. Ravishankar. 1984. Response of groundnut (*Arachis hypogaea* L.) to *Rhizobium* inoculation Oleagineux 39: 149–154.

Nambiar, P. T. C., T. J. Rego, and B. S. Rao. 1986. Comparison of the requirements and utilization of nitrogen by genotypes of sorghum (*Sorghum bicolor* L. Moench) and nodulating and nonnodulating groundnut (*Arachis hypogaea* L.). Field Crop Res. 15: 165–179.

Pallas, J. E., Jr. 1973. Diurnal changes in transpiration and daily photosynthetic rate of several crop plants. Crop Sci. 13: 82–84.

Pallas, J. E., Jr. and Y. B. Samish. 1974. Photosynthetic response of peanut. Crop Sci. 14: 478–482.

Pallas, J. E., Jr., J. R. Stansell, and R. R. Bruce. 1977. Peanut seed germination as related to soil water regime during pod development. Agron. J. 69: 381–383.

Reddy, M. N. 1977. Influence of plant type, soil moisture level and geometry of planting on growth and yield of groundnut (*Arachis hypogaea* L.). Mysore J. Agric. Sci. 11: 116.

Reddy, V. M. and J. W. Tanner. 1980. The effects of irrigation, inoculants and fertilizer nitrogen in peanuts (*Arachis hypogaea* L.). I. Nitrogen fixation. Peanut Sci. 7: 114–119.

Robertson, W. K., L. C. Hammond, J. T. Johnson, and G. M. Prine. 1979. Root distribution of corn, soybeans, peanuts, sorghum, and tobacco in fine sands. Proc. Soil Crop Sci. Soc. Florida 38: 54–59.

Sahrawat, K. L., J. K. Rao, and J. R. Burford. 1987. Elemental composition of groundnut leaves as affected by age and iron chlorosis. Plant Soil 10: 1041–1051.

Sahrawat, K. L., B. S. Rao, and P. T. C. Nambiar. 1988. Macro- and micronutrient uptake by nodulating and nonnodulating peanut lines. Plant Soil 104: 291–293.

Saxena, N. P., M. Natrajan, and M. S. Reddy. 1983. Chickpea, pigeonpea and groundnut, pp. 281–305. *In*: IRRI (ed.). Potential productivity of field crops under different environments. IRRI, Los Banos, Philippines.

Singh, A. K. and C. E. Simpson. 1994. Biosystematics and genetic resources, pp. 96–137. *In*: The groundnut crop: a scientific basis for improvement. Chapman and Hall, London.

Small, H. G. and A. J. Ohlrogge. 1973. Plant analysis as an aid in fertilizing soybeans and peanuts, pp. 315–327. *In*: L. M. Walsh and J. D. Beaton (eds.). Soil testing and plant analysis. Soil Sci. Soc. Am., Madison, Wisconsin.

Smartt, J. 1994. The groundnut in farming systems and the rural economy—a global view, pp. 664–699. *In*: The groundnut crop: a scientific basis for improvement. Chapman and Hall, London.

Stansell, J. R., J. L. Shepherd, J. E. Pallas, Jr., R. R. Bruce, N. A. Minton, D. K. Bell, and L. W. Morgan. 1976. Peanut response to soil water variables in the southeast. Peanut Sci. 3: 44–48.

Tanner, C. B. and T. R. Sinclair. 1983. Efficient water use in crop production: Research or research, pp. 1–28. *In*: H. Taylor (ed.) Limitations to efficient water use in crop production. ASA, CSSA, and SSSA, Madison, Wisconsin.

Vivekanandan, A. S. and H. P. M. Gunasena. 1976. Lysimeter studies on the effects of soil moisture tension on the growth and yield of maize (*Zea mays* L.) and groundnut (*Arachis hypogaea* L.). Beitr. Trop. Landwirtsch. Veterinaermed. 14: 369–378.

Walker, M. E. and T. C. Keisling. 1978. Response of five cultivars to gypsum fertilization on soils varying in calcium content. Peanut Sci. 5: 57–60.

Williams, J. H., J. H. Wilson, and G. C. Bate. 1975. The growth and development of four groundnut (*Arachis hypogaea* L.) cultivars in Rhodesia. Rhod. J. Agric. Res. 13: 131–144.

Wright, G. C., R.C. Nageswara Rao, and G. D. Fraquhar. 1994. Water-use efficiency and carbon isotope discrimination in peanut under water deficit conditions. Crop Sci. 34: 92–97.

Wynne, J. C. and W. C. Gregory. 1981. Peanut breeding. Adv. Agron. 34: 39–72.

15

Sugarcane

I. INTRODUCTION

Sugarcane (*Saccharum* spp. hybrid) is the world's most important sugar crop. It is an erect, very robust, tillering, perennial C_4 grass. It is grown primarily for sugar (sucrose), but molasses, ethyl alcohol, and fiber (bagasse) are important by-products. In 1984–1985, slightly over 60% of world raw sugar production was from sugarcane. Average fresh cane and raw sugar yields were 58 and 5.1 Mg ha^{-1}, respectively (USDA, 1985).

Commercial sugarcane cultivars are complex hybrid of *S. robustum*, *S. officinarum*—the "noble" canes, and *S. spontaneum*—a freely tillering wild species used as a source of vigor and disease resistance. The "noble" canes may have been selected from *S. robustum* by stone-age cultures in New Guinea. They were spread throughout the Pacific and Southeast Asia prior to the arrival of European man. Cane was taken by the Spanish and Portuguese to the New World to form the basis of sugarcane culture in the sixteenth century. In the late eighteenth century, more desirable cultivars of *S. officinarum* were introduced. Modern sugarcane breeding began at the end of the nineteenth century, when viable true seeds were discovered (Jones, 1985).

India, Cuba, and Brazil are major sugarcane producers, and there are significant productions also in the southern United States, the East Indies, several central and South American countries, Egypt, Indonesia, China, Philippines, and Australia.

II. CLIMATE AND SOIL REQUIREMENTS

Sugarcane is adapted to a range of tropical and subtropical climates. It is grown from 37°N in southern Spain to 31°S in the Republic of South Africa. It cannot tolerate freezing temperatures, and growth essentially ceases at mean minimum temperatures below about 12°C (Ryker and Edgerton, 1931). Maximum photosynthetic rates occur at air temperatures of about 34°C (Alexander, 1973), and intact plants can survive temperatures in excess of 52°C (Irvine, 1983). The ideal climate for a 1–year crop would include at least 4–5 months with mean daytime temperatures of 30–35°C to stimulate growth and 1.5–2 months of cooler temperatures prior to harvest to enhance sucrose accumulation (Gascho and Shih, 1982).

Sugarcane is successfully grown under a wide range of temperature, solar radiation, rainfall, and soil conditions. If soil, water, and plant nutrition are adequate, then temperature and/or solar radiation can be used to predict cane growth rates (Allen et al., 1978). When water is limiting, rainfall and/or irrigation may be correlated with yields (Jones, 1980; Early, 1974; Thompson, 1976). Sugarcane can be grown on a wide variety of soils and will also tolerate soil pH values ranging from about 4 to 9, though nutritional problems may occur at the extremes. Although some cultivars tolerate moderate salinity and seasonal flooding, good drainage and salinity management are required for high yields. Sugarcane is considered moderately susceptible to soil salinity; the salinity threshold (initial yield decline) has been reported to be 1.7 dS m^{-1}, and the yield decrease per unit of increase in salinity beyond the threshold is 5.9% (Maas and Hoffman, 1977). Serious yield reductions occur at a conductivity of 4–8 dS m^{-1}, and very little cane growth or death occurs above 10 dS m^{-1} (Bresler et al., 1982; Valdivia, 1981).

III. GROWTH AND DEVELOPMENT

Sugarcane is propagated vegetatively by planting stem cuttings (setts) from which auxiliary buds grow to produce erect primary stalks (main stems). Secondary and tertiary stalks (tillers) are produced at the base of the primary stalk. Sugarcane leaf laminae are 700–1200 mm long and up to 100 mm wide. Internodes are up to 250 mm long and 20–60 mm in diameter. About 10 fully expanded leaves are usually present on a stalk, and both lamina and sheath are shed when they senesce. The inflorescence, a panicle (also known as the arrow or tassel), is produced under certain environmental conditions. Genotype, photoperiod, temperature, nutrition, and water stress all affect panicle initiation and growth. Since the stalk ceases to grow and eventually deteriorates after flowering, genotypes that flower readily under field conditions are avoided (Clements, 1980).

Maximum dry matter accumulation occurs only with near-optimum temperatures, high solar radiation, complete ground cover, and minimal nutrient and water stresses. Under these conditions, short-term (1–2 months) production of aboveground biomass can reach 40–44 g m^{-2} day^{-1} (Irvine, 1983; Shih and Gascho, 1981; Thompson, 1978). Incomplete interception of light during canopy development, low solar radiation, suboptimal temperatures, reduced growth during ripening, and a variety of soil and biotic stresses typically reduce full-season growth rates of commercial fields and experimental plots to 6–25 and 20–32 g m^{-2} day^{-1} (Irvine, 1983).

A. Shoots

Sugarcane has the capacity to tiller rapidly. Stem numbers increase exponentially with time until a maximum of 20–30 stalks m^{-2} is reached at 4–6 months (Bull and Glaziou, 1975; Gosnell, 1968). Coale et al. (1993) studied the dry weight accumulation of sugarcane, and characterized it by a logistic growth model (Figure 15.1). The model included data from all locations and crops because the 95% confidence intervals for the models for single locations, crops, or combinations of location and crop overlapped. Therefore, dry matter accumulation models for any one location (L1 or L2) or crop (C1 or C2), or combination of location and crop (L1 C1, L1 C2, etc.) were not significantly different ($P < 0.05$). In Florida, the grand growth period (GGP) for sugarcane is usually defined as beginning June 1 (152nd day of year) and ending October 15 (288th day of year). Data collected over the four crop-years of this study corroborated this definition. Dry weight accumulation averaged 0.15 t ha^{-1} d^{-1} during the GGP. During this period of rapid growth, 64% of total crop dry matter was produced (Coale et al., 1993). Leaf area index (*LAI*) increases in a similar exponential manner. When *LAI* approaches 2.0–3.0 at 4–6 months, many younger tillers begin to die, possibly as a result of shading by older tillers, and tiller number normally stabilizes at 10–20 stems m^{-2}.

Several studies suggest that the maximum *LAI* for sugarcane is 7–8 m^{-2} leaf m^{-2} (Bull and Tovey, 1974; Irvine, 1983; Irvine and Benda, 1980). This value is attained only in vigorously growing canopies of erect cane. The *LAI* typically declines as the crop approaches harvest maturity (Bull and Tovey, 1974; Glover, 1972; Gosnell, 1968), especially when nitrogen and water stresses or chemical ripeners are used to slow expansion growth and increase sugar storage in the stem.

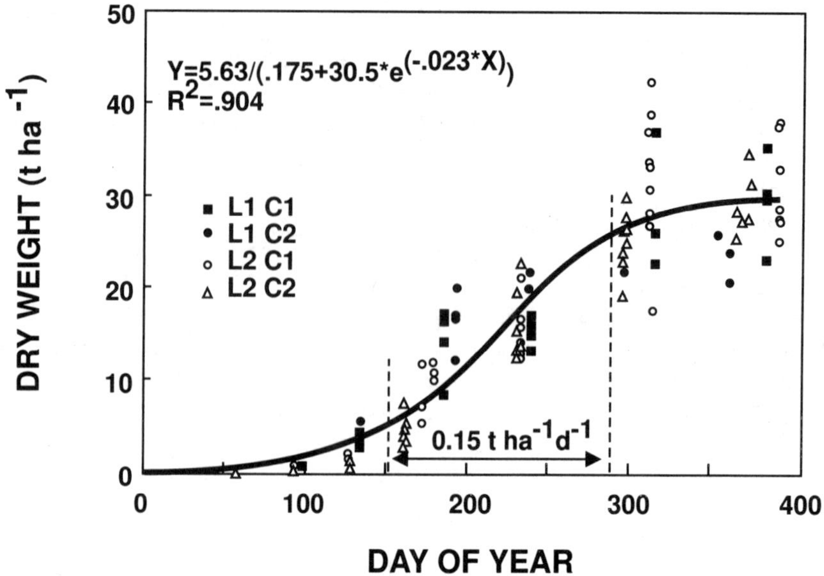

Figure 15.1 Dry weight accumulation by sugarcane grown on organic soils of the Everglades agricultural area (L1 = location 1; L2 = location 2; C1 = crop 1; C2 = crop 2). (From Coale et al., 1993.)

Since flowering is normally suppressed, the growth stages of commercial sugarcane crops are unrelated to flowering. However, more or less distinct periods of vegetative growth can be described. Gascho and Shih (1982) divided sugarcane vegetative growth into four stages: germination and emergence, tillering and canopy establishment, grand growth, and ripening. Duration of the stages is highly dependent on climate. Low temperatures or drought stress can delay germination, emergence, tillering, and canopy development. In subtropical climates, the first two stages may last up to 5 months, and warm temperatures and plentiful water and nutrients can promote vegetative growth and delay ripening. In Hawaii, where cane is grown for 2 years, the grand growth period can be extended by providing adequate water and nutrients until the crop is 20–22 months old.

B. Roots

The early growth of sugarcane roots has been described by Clements (1980), Dillewijn (1952), and Glover (1967). Under favorable environmental conditions, the axillary buds on setts become active within 3 days of planting, and sett roots begin to grow from the root band at the base of the internode. Sett roots grow at a maximum rate of 24 mm day^{-1} and stop elongating when they are 150–250 mm long. They turn dark, decompose rapidly, and disappear within 2 months after planting. Shoot roots begin to grow from the short basal internodes of the shoot at about the time it emerges from the soil. The first shoot roots are much thicker than the sett roots, their rate of growth is more rapid, they produce few branches, and they penetrate the soil at a steep angle.

Shoot roots produced later are finer and branch more freely than earlier shoot roots. Their maximum growth rate is 75 mm day^{-1} for periods of 1–2 days or 40 mm day^{-1} when their growth is averaged over a week (Glover, 1967). Wood and Wood (1968) used radioactive phosphorus uptake from different depths of a deep sandy soil and concluded that the rooting front reached 0.9 m in 112 days, 1.5 m in 161 days, and 2.1 m in 189 days.

The distribution of roots in the soil is strongly dependent on soil characteristics, cultivars, and soil water content. For example, Paz-Vergara et al. (1980) reported that for 11 furrow-irrigated fields in Peru, the percentage of roots in the 0.30–m horizon is 48–68%; from 0.3 to 0.6 m, 16–36%; from 0.6 to 0.9 m, 3–12%; from 0.9 to 1.20 m, 4–7%; from 1.2 to 1.5 m, 1–7%; and from 1.5 to 1.8 m, 0–4%.

Short irrigation intervals encourage roots to develop near the soil surface (Baran et al., 1974; Kingston, 1977). However, poor soil aeration restricts root growth. For example, Gosnell (1971) reported that sugarcane roots stop growth 50–100 mm above the water table. Root systems growing in deep sands tend to be finer, more highly branched, and deeper than those growing in heavy clay soils (Glover, 1968; Lee, 1926c; Thompson, 1976). Glover (1968) found an extensive, fine, well-branched root system to extend more than 140 cm in a sand. In a disturbed clay soil, the thick primary root system was well developed, but secondary branches were poorly developed. In an undisturbed clay soil, even the primary roots were poorly developed below the plow layer.

Genotypic variation in sugarcane root systems is well documented (Dastane, 1957; Lee, 1926a,b; Raheja, 1959). Some cultivars produce roots with a higher degree of branching than others (Stevenson and McIntosh, 1935). Root gravitropism also varies among cultivars, and cultivars with weakly gravitropic (more horizontal) root orientation are more resistant to lodging than other cultivars (Mukerji and Alan, 1959; Stevenson and McIntosh, 1935).

Environmental conditions can affect the expression of genotypic differences in root growth. For example, Rostron (1974) reported that the root distributions of two cultivars were similar under good conditions but differed under dry conditions.

IV. CROP CULTURE

Cultural practices associated with sugarcane production vary widely as a result of economic, social, climatic, and soil conditions. In areas with high labor costs, essentially all cultural practices are mechanized. In other areas a great deal of hand labor continues to be used.

Land preparation varies with soil type. Deep plowing, deep ripping, or subsoiling is often used to disrupt compacted layers. Soil may be formed into beds approximately 1.5 m apart to facilitate furrow irrigation, to improve surface drainage, or both. However, such beds increase land preparation costs and, where possible, flat culture is used.

A. Planting

Setts with two or more buds are cut by hand or mechanically, often from special areas maintained to minimize disease infestation. Setts may receive hot water and/or fungicide treatments to further reduce disease problems. They are planted in furrows and are covered with soil. Fertilizer may be applied broadcast, in the furrow, or through the irrigation system. In areas where poor drainage is common, the setts are planted on top of beds. Where furrow irrigation is used, they are often planted in the furrow between beds. Flat culture is practiced whenever possible to reduce land preparation. Ratoon tillage operations are designed to remove post-harvest compaction between rows and to control weeds.

In much of the world, sugarcane row spacing is about 1.5 m to facilitate mechanization. However, work on row spacing in Louisiana (Irvine and Benda, 1980) and Florida (Gascho and Shih, 1981; Shih and Gascho, 1980) suggests that narrow row spacing could result in more rapid canopy development, increased light interception, and higher yields in areas with short growing seasons.

B. Water Requirements, Irrigation, and Drainage

The sugar industries in Hawaii, Florida, Australia, South Africa, and Taiwan historically have relied on pan evaporation as an indicator of potential evapotranspiration.

Several studies suggest that potential evapotranspiration from a well-developed sugarcane canopy during the grand growth stage is approximately equal to evaporation from a standard U.S. Weather Bureau class A pan (Ekern, 1971; Fogliata, 1974; Hardy, 1966; Thompson, 1965; Thompson et al., 1963). When calculated on a monthly basis, the ratio of evapotranspiration to pan evaporation usually varies from 0.8 to 1.2, even though ratios as low as 0.63 and as high as 1.59 have been reported (Kingston and Ham, 1975).

Low ratios of potential evapotranspiration to pan evaporation are often, though not always, found as the crop nears maturity in winter months (Kingston and Ham, 1975; Moberly, 1974; Thompson, 1976) and in ratoon crops (Hardy, 1966; Moberly, 1974; Shih and Gascho, 1980; Thompson, 1976), presumably due to greater stomatal resistance of slowly growing crops (Thompson, 1986). Lodging can reduce evapotranspiration up to 30% until a uniform canopy is reestablished (Ekern, 1971).

Transpiration is strongly affected by the amount of solar radiation intercepted by the crop canopy. When the soil surface is dry, actual evapotranspiration is limited by canopy cover. For example, Chang et al. (1965) and Ekern (1971) reported that actual evapotranspiration of adequately watered sugarcane increases from 0.3 to 0.6 of pan evaporation during the first month after planting to 0.8–1.0 of pan evaporation at 4–5 months. Kingston (1973) recommends using pan factors of 0.4, 0.6, 0.8, 0.9, and 1.0 for ground cover fractions of 0–0.25, 0.25–0.50, 0.50–0.75, 0.75–1.0, and 1.0, respectively.

If the crop canopy is complete, evapotranspiration proceeds at the potential rate until 60–70% of the total plant-extractable water is removed from the soil profile (Koehler et al., 1982; Moberly, 1974). Thereafter, the ratio of evapotranspiration to potential evapotranspiration declines until evapotranspiration ceases in the completely senescent crop. Soil type strongly affects the amount and distribution of plant-extractable water in the soil profile. Gosnell and Thompson (1965) found that sugarcane extracted water to at least 2.2 m in one soil. However, more water was extracted deep in a sandy soil profile than deep in a sandy loam or a shallow clay loam overlying decomposing shale (Hill, 1966; Thompson et al., 1967). Several other factors, including subsoil aluminum toxicity, presence of layers with high soil strength, or fragmental or cemented horizons, can limit root growth in the subsoil, thereby reducing water available for transpiration and increasing the susceptibility of the crop to drought stress.

Sugarcane is most susceptible to drought stress during the first 3–4 months after planting, when severe stress can reduce stands and make replanting necessary. During tillering and canopy development, young tillers are more susceptible to drought than the main stem or older tillers. Leaves on young tillers begin to roll earlier than those on the primary stem, probably due to the less developed root systems of the tillers (Clements, 1980).

After a tiller begins to produce elongated internodes, the rate of stem elongation and final internode lengths are convenient means of assessing the effects of drought stress. Internodes which elongate during stress are permanently shortened relative to those of well-watered plants, and they serve as a record of the timing, length, and severity of drought stress.

Drought stress has long been known to increase nonstructural carbohydrates in sugarcane leaves and stems (Clements and Kubota, 1943). For example, early-morning total sugar concentration of young leaf sheaths falls as low as 5% in rapidly growing, well-watered plants; however, it frequently increases to more than 10% during drought stress.

High correlations are found among leaf sheath water content and leaf nitrogen and potassium concentrations (Clements, 1980; Samuels, 1971; Samuels et al., 1953). All decrease during drought stress, and in some areas plant analyses used to detect nitrogen and potassium deficiencies are adjusted to take tissue water content into account (Clements, 1980).

Irrigation by flooding, furrow, sprinklers, drip (trickle), and subirrigation (by water table adjustment) are used for sugarcane. Several reviews of sugarcane irrigation practices are available (Finkel, 1983; Gibson, 1974; Gosnell and Pearse, 1971; Leverington and Ridge, 1975; Thompson, 1977).

Four methods have been widely used to schedule irrigation: resistance blocks, tensiometers, the water balance, and tissue moisture content. All are based on their ability to predict when incipient drought stress will occur, usually as indicated by a decrease in stem or leaf elongation. Experiments which compare cane and sugar yields from plots irrigated with different frequencies and/or amounts are used to validate these methods. Regardless of which method is used to select ideal irrigation schedules, actual irrigation practices are modified in response to availability of water, personnel, and equipment.

Jones (1980) analyzed several irrigation experiments conducted in Hawaii. After eliminating treatments in which excessive water had reduced yields, presumably due to poor soil aeration and/or leaching of nutrients, a linear relationship was found between relative water use (*RWU*, the ratio of effective water applied to class A pan evaporation) and relative cane yield (*Y*, the ratio of actual yield to maximum yield in the experiment) in three experiments.

$$Y = 1.01RWU + 0.03 \qquad (r^2 = 0.90, n = 14)$$

Thompson (1976) selected data from studies in South Africa (Boyce, 1969; Thompson and Boyce, 1967, 1968, 1971; Thompson and De Robillard, 1968), Australia (Kingston and Ham, 1975), Hawaii (Campbell et al., 1959), and Mauritius (Hardy, 1966) for which evapotranspiration had been estimated. For total crop evapotranspiration (*ET*) ranging from 660 to 3840 mm and cane yields (*Y*) ranging from 57 to 342 Mg ha^{-1}, he found the following relationship:

$$Y = 0.969ET - 2.4 \qquad (r^2 = 0.90, n = 91)$$

The slope of the relationship is similar to that reported by Jones (1980) for furrow- and sprinkler-irrigated experiments in Hawaii (0.103 Mg cane mm^{-1} effective water).

In many areas sugarcane is grown on nearly level soils in which the water table reaches the root zone during at least part of the year. In Louisiana, sugarcane is grown on low, nearly level land near large bodies of water. High water tables are common, especially during the winter months, when sugarcane leaf area and evapotranspiration are low. Response to supplemental irrigation is rare (Carter and Floyd, 1973), probably because the crop can extract much of its water requirement from the water table. A similar situation occurs in western Taiwan, where water tables frequently rise near the soil surface in the rainy season. In the dry season the crop obtains 25–50% of its water requirement from a water table at approximately 1.6 m. This dramatically reduces irrigation demands (Chang and Wang, 1983; Hunsigi and Srivastava, 1977; Yang and Chang, 1976).

The most important use of water tables for irrigation occurs on organic soils of the south Florida everglades. During the dry season, growers pump supplemental water from a primary canal system into a system of farm canals, lateral ditches, and field ditches. Lateral movement of water from the field ditches into the field is often facilitated by mole drains 15 cm in diameter, 70–90 cm deep, and 2–3 m apart. The organic soils in this region are quite permeable, lateral movement is rapid, and the water table can easily be raised to provide supplemental water.

If the water table remains near the soil surface, poor aeration may cause setts to decay, reduce ratoon stalk populations (Carter et al., 1985), reduce root growth (Gosnell, 1971; Banath and Monteith, 1966), and cause the production of aerotropic (Srinivasan and Batcha, 1962) or floating (Sartoris and Belcher, 1949) roots. Rudd and Chardon (1977) reported a yield decline of 0.46 Mg cane ha^{-1} for each day the water table rose above 0.5 m.

Louisiana sugarcane farmers plant cane on high (0.30–0.45 m) ridges 1.8 m apart to enhance surface drainage and provide a small volume of aerated soil at almost all times. When winter rainfall is relatively low, these practices provide adequate aeration, but in many years excess rainfall raises water tables and reduces yields (Carter, 1977a; Carter and Camp, 1983; Carter and Floyd, 1975).

A 7–year lysimeter study on fine-textured Louisiana soils indicated that drainage systems should reduce the water table to 1.2 m within 4 days in order to maximize cane and sugar yields (Carter, 1977b). This maintains high redox potentials in the upper 0.5 m of soil (Carter, 1980), allows good ratoon growth early in the spring, and increases stalk populations.

Sugarcane cultivars differ in sensitivity to high water tables (Andreis, 1976; Srinivasan and Batcha, 1962). Cultivars tolerant to waterlogging respond with production of aerotropic or floating roots (Shah, 1951; Srinivasan and Batcha, 1962). Some Florida cultivars exhibit better growth with water tables at 0.32 m than at 0.61 or 0.84 m; however, the growth of other cultivars is greatly inhibited by the 0.32–m water table (Gascho and Shih, 1979).

C. Nutrient Requirements

Because of its long growing season and high maximum growth rate, sugarcane has large nutrient requirements, but since it is a high-value crop, most sugarcane industries apply large amounts of fertilizers in an attempt to avoid nutrient deficiencies. Both soil and plant analyses have long been used to assess crop nutrition and soil fertility.

Coale et al. (1993) studied the N, P, K, Ca, and Mg seasonal accumulation patterns by sugarcane grown on organic soils of the Everglades (Figures 15.2, 15.3, 15.4, and 15.5). The period of most rapid nutrient uptake corresponded with the grand growth period, and the rates of nutrient uptake during different phases of crop development was defined. At harvest 71% of total dry matter and 55, 63, 64, 25, and 38% of total accumulated N, P, K, Ca, and Mg, respectively, were removed from the field as millable sugarcane. Phosphorus and K removed from the field by crop harvest was equivalent to 179 and 201%, respectively, of added P and K.

Nitrogen accumulation by sugarcane closely paralleled biomass accumulation (Figure 15.2). A single logistic model best characterized N accumulation

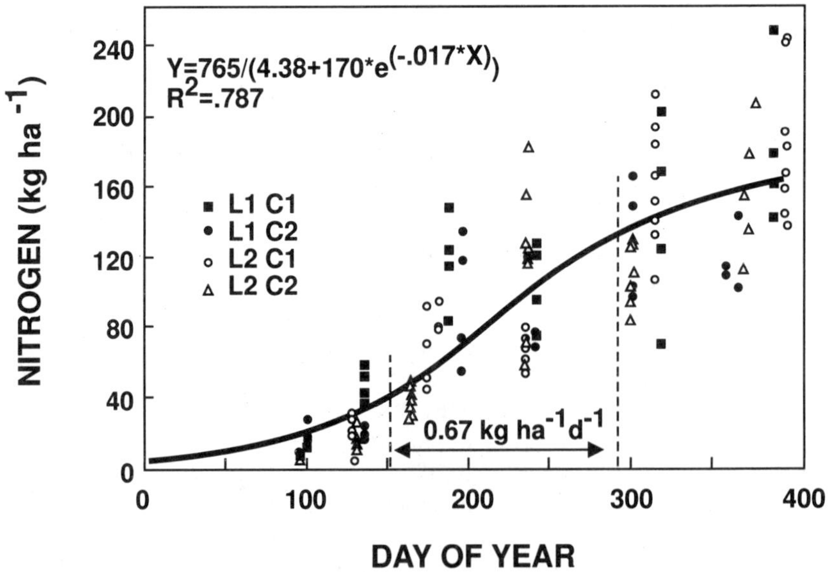

Figure 15.2 Nitrogen accumulation by sugarcane grown on organic soils of the Everglades Agricultural Area (L1 = location 1; L2 = location 2; C1 = crop 1; C2 = crop 2). (From Coale et al., 1993.)

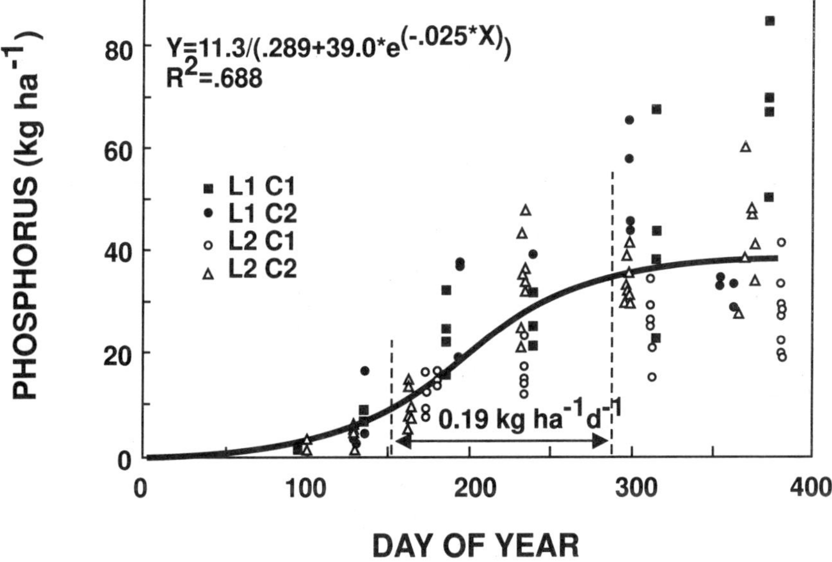

Figure 15.3 Phosphorus accumulation by sugarcane grown on organic soils of the Everglades Agricultural Area (L1 = location 1; L2 = location 2; C1 = crop 1; C2 = crop 2). (From Coale et al., 1993.)

over the four crop-years. During the GGP, N uptake averaged 0.67 kg ha^{-1} d^{-1} and accounted for 54% of total N accumulation. Average N uptake during the GGP was equivalent to only 15–20% of the estimated rate of soil N mineralization (Terry, 1980).

Positive yield responses to P and K fertilization of sugarcane grown on organic soils have been documented (Gascho and Kidder, 1979). Seasonal P and K accumulation were also described by logistic models (Figures 15.3 and 15.4, respectively). Phosphorus and K uptake during GGP contributed 67 and 68% of total P and K accumulation, respectively. The K uptake rate was very high (4.14 kg ha^{-1} d^{-1}) during the first 60 days of GGP. Again, the 95% confidence intervals for the P and K accumulation models for any one location, crop, or combination of location and crop overlapped, and therefore the overall model included all locations and crops.

Phosphorus and K fertilizers are applied at the time of planting for plant-cane crops and after initiation of ratoon regrowth for ratoon crops. For optimum sugarcane production on organic soils, fertilizer applications of 0–37 kg P ha^{-1} and 0–233 kg K ha^{-1} are recommended, depending on soil-test extractable P and K determined prior to planting (Sanchez, 1990).

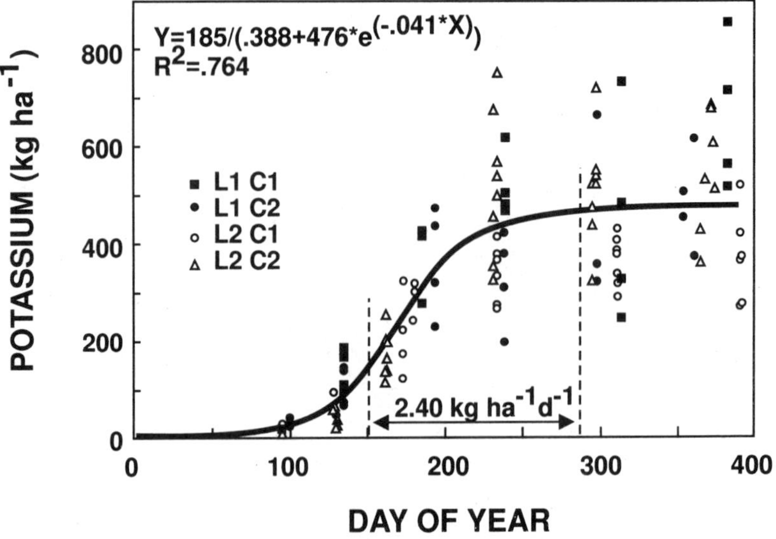

Figure 15.4 Potassium accumulation by sugarcane grown on organic soils of the Everglades Agricultural Area (L1 = location 1; L2 = location 2; C1 = crop 1; C2 = crop 2). (From Coale et al., 1993.)

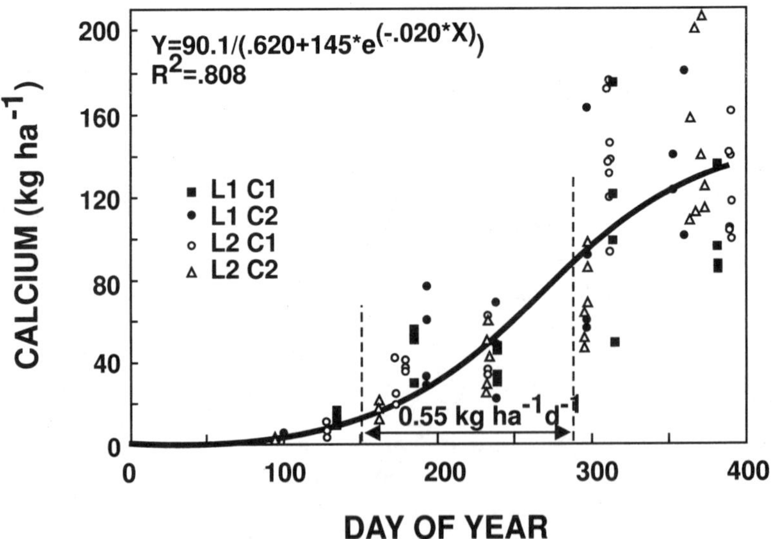

Figure 15.5 Calcium accumulation by sugarcane grown on organic soils of the Everglades Agricultural Area (L1 = location 1; L2 = location 2; C1 = crop 1; C2 = crop 2). (From Coale et al., 1993.)

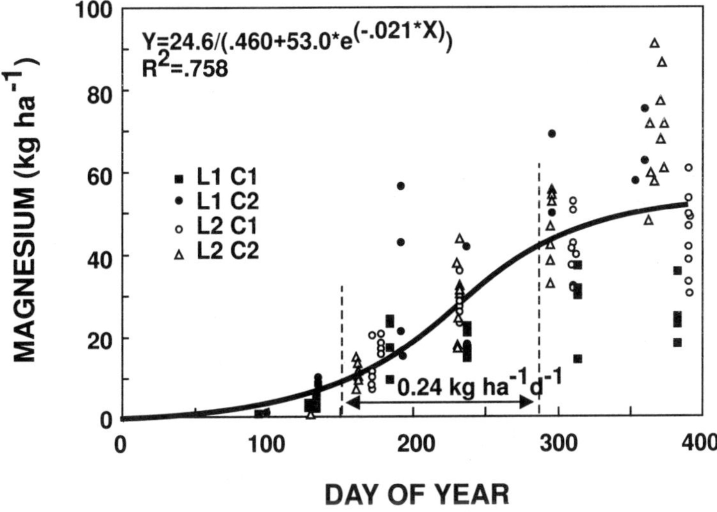

Figure 15.6 Magnesium accumulation by sugarcane grown on organic soils of the Everglades Agricultural Area (L1 = location 1; L2 = location 2; C1 = crop 1; C2 = crop 2). (From Coale et al., 1993.)

Calcium accumulation by sugarcane was described by a logistic model similar to the models derived for N, P, and K uptake (Figure 15.5). One notable difference was the absence of a well-defined stationary phase for crop Ca accumulation at crop maturity. As with the other macronutrients, Mg uptake was most rapid during the GGP, when 62% of total crop Mg accumulation occurred (Figure 15.6).

Barnes (1974) reported removal of 0.7–0.9 kg N, 0.22–0.26 kg P, and 1.28 kg K per ton of cane. In contrast, nutrient accumulation values in Hawaii (Clements, 1980) per ton of cane were 0.48 kg N, 0.09–0.33 kg P, and 0.75 kg K. These differences may be accounted for by the amount of trash (leaves and roots) transported to the mill, age of sugarcane at harvest, and nutrient uptake differences among cultivars (Gascho et al., 1993). Pereira et al. (1995) studied the relationship between P rates and sugarcane yield in a Brazilian Vertisol (Figure 15.7). Maximum yield was obtained at about 92 kg P ha^{-1} (211 kg P_2O_5 ha^{-1}).

Table 15.1 gives adequate levels of sugarcane nutrient concentrations. As with other crops, sugarcane nutrient concentrations vary with age and among plant parts. The highest shoot N concentrations occur in young plants and tillers. The concentration declines with age due to the production of internodes with low N concentrations (Table 15.2). As leaves age, their N concentrations

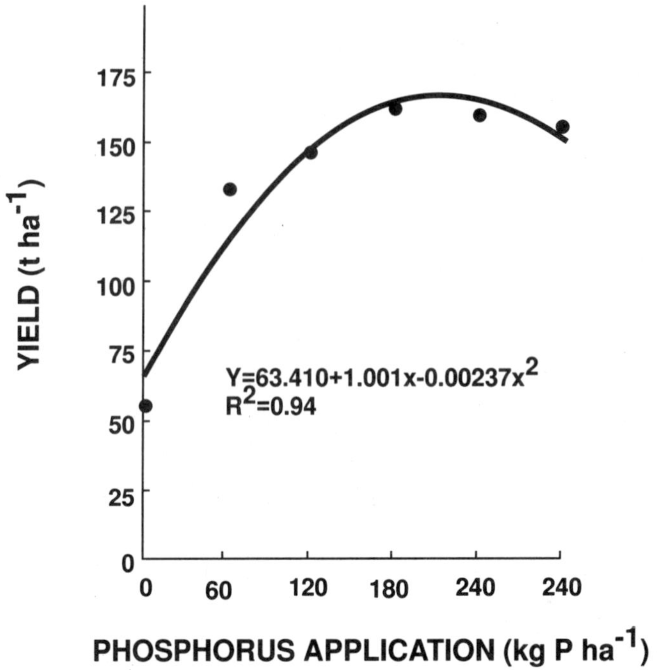

Figure 15.7 Sugarcane yield as a functon of phosphorus rates in a Brazilian Vertisol. X-axis and regression equation values of P are expressed as P_2O_5. (From Pereira et al., 1995.)

decline due to production of structural tissues and retranslocation of nutrients to active meristems. For example, Ayres (1936) reported that N concentration declines from 1.2% in the youngest leaves to about 0.4% in the oldest green leaves and 0.2% in senescent leaves still attached to the plant.

Environmental conditions that reduce crop growth can cause tissue N concentrations to increase. Thus, cool wet weather often results in higher N concentrations than normal (Clements, 1980; Dillewijn, 1952).

Nitrogen deficiency has several effects on sugarcane growth. Leaf expansion decreases, the interval of leaf appearance increases, and leaf senescence increases. Thus, N-deficient sugarcane tillers may have only 4–6 green leaves instead of the 12–14 that are found on normal tillers (Clements, 1980).

Photosynthesis is less sensitive to N stress than expansion growth. As a result, sucrose accumulates in the leaves and internodes of N-deficient sugarcane, and fertilizer N is normally withheld from sugarcane prior to harvest in order to increase stalk sucrose concentration (Clements, 1980).

Table 15.1 Adequate Concentration of Nutrients in Sugarcane Plants

Nutrient	Growth stage[a]	Plant part[a]	Adequate concentration
			g kg^{-1}
N	3 month (plant)	TVD	24–25
	6 month (plant)	TVD	19
	3 month (plant)	TVD	21
	4–5 month (ratoon)	TVD	19
	Early rapid growth	Leaves 3–6	15–27
P	3–6 month (plant)	TVD	2.1–3.5
	10.3 month (plant)	3rd LB below A	2.1–3.0
	2–4.5 month (ratoon)	TVD	2.1–3.5
	7 month (ratoon)	3rd LB below A	2.1–3.0
	Early rapid growth	Sheath 3–6	0.5–2.0
K	3–6 month (plant)	TVD	12.5–20.0
	10.3 month (plant)	3rd LB below A	13.0–20.0
	2–4.5 month (ratoon)	TVD	12.5–20.0
	6–7 month (ratoon)	TVD	11.0–18.0
	Early rapid growth	Sheath 3–6	22.5–60.0
	7–14 month (plant)	Internodes 8–10	10.0
Ca	3 month (plant)	TVD	1.4–1.8
	4.5–6 month (plant)	TVD	1.5–2.0
	2–3 month (ratoon)	TVD	1.6–2.0
	5 month (ratoon)	TVD	2.0–2.4
	Early rapid growth	Sheath 3–6	1.0–2.0
Mg	3 month (plant)	TVD	0.9–1.2
	4.5–6 month (plant)	TVD	1.2–1.8
	2–3 month (ratoon)	TVD	1.0–1.8
	5 month (ratoon)	TVD	1.2–1.8
	Early rapid growth	Sheath 3–6	1.5–10.0
S	35 DAS	Sheath 3–6	6.1
	70 DAS	Sheath 3–6	0.8
	7 month	TVD	1.3

Table 15.1 (continued)

Nutrient	Growth stage[a]	Plant part[a]	Adequate concentration
			mg kg^{-1}
Cu	Rapid growth	TVD	4.0–15.0
	6–7 month (ratoon)	TVD	4.2–12.2
	Rapid growth	Blades 3–6	5.0–100.0
Zn	Rapid growth	TVD	15.0–50.0
	6–7 month (ratoon)	TVD	12–50
	Rapid growth	Blades 3–6	20–100
Mn	Rapid growth	TVD	20–200
	6–7 month (ratoon)	TVD	15–200
	Rapid growth	Blades 3–6	20–400
Fe	Rapid growth	TVD	5–100
	7 month	TVD	49–915
	Rapid growth	Blades 3–6	20–600
B	Rapid growth	TVD	2–10
	7 month	TVD	1.6–10
	Rapid growth	Blades 3–6	2.0–30
Mo	Rapid growth	TVD	0.08–1
	Rapid growth	Blades 3–6	0.05–4.0

[a] DAS, days after sowing; TVD, top visible dewlap, which is approximately the third leaf from the shoot apex; LB, leaf blade (excluding sheath); A, apex.
Source: Compiled from Reuter, 1986.

Growing plant meristems need a continuous supply of P for incorporation into new tissue. Soil P supply is not constant due to variation in soil water content and root development; therefore, plants have developed mechanisms for taking up and storing excess P and subsequently using it for new growth. Sugarcane accumulates inorganic P in the stem when the soil P supply permits. During periods of inadequate uptake, this inorganic P can be translocated to meristematic tissues (Hartt, 1972). Since much of the P in sugarcane internodes is in the highly mobile inorganic form, internode P concentration is more sensitive to P nutrition than is leaf P, which is predominantly organic (Hartt, 1972).

As in the case of N, shoot P concentration decreases with age as a result of a decrease in the ratio of meristematic tissues with high P concentrations to structural tissues with low P concentrations (Clements, 1980).

The concentration of K in sugarcane varies among plant parts and with time (Tables 15.1 and 15.3). Since K remains largely in solution rather than

Table 15.2 Tissue Nitrogen Concentration of Sugarcane in Hawaii at 12 Months of Age

Plant part		N concentration ($g\ kg^{-1}$)
Meristem		17.7
Young blades		11.8
Old blades		8.9
Young sheaths		4.5
Old sheaths		3.1
Internodes	10–12	1.7
	7–9	1.1
	4–6	1.4
	1–3	1.5

Source: Modified from Clements, 1980.

immobilized in protein or structural components of the cell, Clements (1980) proposed that sugarcane K concentration be expressed as a percentage of tissue water content rather than dry weight.

Unlike that of some grain crops, the K content of sugarcane shoots (including dead leaves) increases steadily with time, and the stalk tissue usually contains more total K than the green leaves (Jones, 1985). Since whole sugarcane stalks (sometimes including leaves) are removed from the field,

Table 15.3 Typical Potassium Concentrations of Sugarcane Tissues Prior to Harvest

Plant part		K concentration ($g\ kg^{-1}$)
Meristem and expanding cane		35.5
Young blades		16.5
Old blades		13.1
Young sheaths		24.8
Old sheaths		20.5
Internodes	45–47	10.7
	24–26	7.5
	1–3	7.2

Source: Modified from Clements, 1980.

sugarcane has a high fertilizer K requirement. Most of the K removed from the field in the cane stalks is concentrated in the molasses; therefore, little is returned to the field in filter mud, bagasse, or furnace ash.

The effect of plant age on sugarcane Ca concentration has long been recognized. Gosnell and Long (1971) reported that in Uganda the Ca content of the youngest fully expanded leaf decreases from 0.37% at 1 month to 0.25% at 5 months. Similarly, in Hawaii the Ca content of immature sheaths declines from above 0.3% in the first 4 months to below 0.2% after 18 months (Bowen, 1975). Drought stress tends to increase the Ca concentration of leaf tissues (Gosnell and Long, 1971).

In Hawaii, Florida, and several other parts of the world, sugarcane is grown on highly weathered or organic soils that are low in soluble silicates. Plant silicon occurs in the form of opal (SiO_2nH_2O) and plays a role in strengthening cell walls. As the plant ages, total Si concentrations of blades and sheaths increase as uptake, translocation, and deposition occur. The total Si concentration of leaf sheaths is normally higher than that of blades, and the highest concentrations are found in the leaf tip and margins, where trichomes and marginal sclerenchyma are heavily silicified.

Silicon-deficient plants develop small, elongated, chlorotic spots on the leaves. These spots eventually coalesce, and affected leaves die prematurely. Calcium silicate can be applied to provide adequate soluble Si for plant growth, and nearly two-fold increases in cane yields have been attributed to addition of silicates (Elawad et al., 1982a,b).

D. Ripening

Ripening of sugarcane refers to the gradual increase in stalk sucrose content (on a dry weight basis) as harvest approaches. Numerous studies have shown that cool temperatures, high solar radiation, moderate nitrogen and/or drought stress, and use of chemical ripeners can stimulate ripening (Clowes and Inman-Bamber, 1980; Lonsdale and Gosnell, 1974; Mason, 1976; Rostron, 1977). The combined stresses reduce expansion growth before they reduce photosynthesis (Inman-Bamber and de Jager, 1986a,b). Therefore, more sucrose is available for translocation to storage tissues in the stem, and the total amount of sucrose in the crop increases.

E. Harvest and Ratoon Growth

Cane is usually burned before harvest to reduce the amount of trash and green leaves hauled to the mill. Harvesting may be labor-intensive or highly mechanized. Good harvesting systems minimize cane breakage, transport little trash and other extraneous material to the mill, and minimize the time between harvest and cane processing. For many years, manual harvest of unburned cane was practiced in South Africa. Cane tops and trash are left on the soil surface to conserve moisture and control erosion. Mechanized harvest of unburned cane is being evaluated in Australia, where cane trash is needed for systems designed to reduce costs and conserve soil moisture.

At harvest, the dry weight of a sugarcane tiller usually consists of about 50–60% millable cane, 30–40% tops (leaves and immature stem), and 10% roots and stubble. Crops grown for more than 1 year usually have higher percentages of millable cane. Of the millable cane's fresh weight, about 70% is water; of its dry weight, about 50% is sucrose. Efficient factories can recover about 85% of the sucrose in the cane. Depending on the sucrose content of the cane and its recovery, 8–15% of the millable cane harvested is recovered as raw (unrefined) sucrose (Clements, 1980; Jones, 1985).

Most of the world's sugarcane is harvested once a year, and several ratoon (stubble regrowth) crops usually follow the plant crop. Planting operations and seed (stalks for vegetative propagation) costs constitute the largest input of sugarcane production (Salassi and Giesler, 1995). Inadequate ratoon crop yields limit the economic production of sugarcane in semitropical regions (e.g., Louisiana), where ratoon crop yields typically decrease with age (Johnson et al., 1993; Shrivastava et al., 1992). The reasons for this decline are complex, but primarily relate to diseases, insects, weed competition, management practices, and winter kill (Shrivastava et al., 1992). Also, genotypes can vary substantially in their ratoon crop yields (Chapman, 1988; Chapman et al., 1992).

Ratooning ability can be enhanced by indirect selection for disease or insect resistance, or by direct selection of genotypes with high ratoon crop yields. Traits such as high stalk number, bud viability, vigorous root formation, high biomass accumulation, and high light-use efficiency have been suggested as being indicative of better ratooning cultivars (Sundara, 1989; Ferraris et al., 1993). The importance of maintaining stalk weight in older crops has also been noted (Chapman, 1988; Chapman et al., 1992).

Ratooning ability can be defined in either absolute or relative terms. In absolute terms, a good ratooning cultivar is one that produces high ratoon crop yields or several profitable ratoon crops. Relative to other cultivars, a good ratooning cultivar is one whose ratoon crop yields are a higher percentage of its plant cane or other younger crops' yields. Cultivars with high plant cane yields commonly produce high and numerous ratoon crop yields, but exceptions exist (Chapman, 1988; Sundara, 1989; Milligan et al., 1996). One-year sugarcane is bred and managed to avoid lodging. However, stems of 2–year crops can reach a length of 10 m or more by growing upward, lodging, and then turning upward again. At harvest the 2–year crop consists of a mat of tangled stems which have lodged several times (Clements, 1980).

V. SUMMARY

Sugarcane is the world's most important sugar crop. It probably was domesticated in New Guinea in prehistoric times, and it is widely grown in tropical and subtropical environments from 36°N to 31°S latitude. Sugarcane's greatest asset is its ability to produce sugar efficiently and in great quantity per unit land area. As a result of its C_4 photosynthetic pathway, maximum crop growth rates are over 40 g m^{-2} day^{-1}, and total dry matter production in many areas exceeds 40 metric tons ha^{-1} yr^{-1}. Maximum fresh cane yields are over 200 tons ha^{-1} yr^{-1}, with sucrose concentrations of 7–13%. Commercial yields are usually less than half the maximum yields.

Most sugarcane is grown as a one-year crop, and several ratoon crops are normally produced. The ideal environment for a one-year crop would include at least 4–5 months with high solar radiation and mean daytime temperatures of 30–35°C to stimulate growth. Prior to harvest, 1.5–2 months of cooler temperatures help increase sucrose concentrations. Little growth occurs when mean air temperatures are below 20°C.

Sugarcane can be grown in a wide variety of soils and tolerates soil pH from 4 to 9, but the optimum pH range is 5.8–7.2. Cane and sugar yields are sensitive to drought, poor soil aeration, salinity, and nutrient deficiencies and toxicities. Because of its high value, sugarcane producers usually attempt to minimize stresses by application of irrigation, fertilizers, soil amendments, and pesticides. However, moderate drought and nitrogen stresses are often imposed for 1–2 months prior to harvest to increase cane sucrose concentrations.

Because of its long duration, high growth rates, and harvest of all millable cane (and sometimes tops and leaves), large amounts of nutrients are removed from the field. One metric ton of fresh millable cane contains about 1.6 kg N, 1 kg P, and 3.4 kg K. Since yields of 50–100 tons millable cane are common, fertilization is required to achieve adequate yields and maintain soil fertility. Both soil and plant analyses are used to detect and correct nutrient deficiencies and toxicities.

REFERENCES

Alexander, A. G. 1973. Sugarcane physiology. Elsevier, Amsterdam.

Allen, R. J., Jr., G. Kidder, and G. J. Gascho. 1978. Predicting tons of sugarcane per acre using solar radiation, temperature and percent plant cane, 1971 through 1976. Proc. Am. Soc. Sugar Cane Technol. 7: 18–22.

Andreis, H. J. 1976. A water table study on everglades peat soil. Sugar J. 39: 8–12.

Ayers, A. 1936. Factors influencing the mineral composition of sugarcane. Rep. Assoc. Hawaii. Sugar Technol. 15: 29–41.

Banath, C. L. and N. H. Monteith. 1966. Soil oxygen deficiency and sugar cane root growth. Plant Soil 25: 143–149.

Baran, R., D. Bassereau, and N. Gillet. 1974. Measurement of available water and root development on an irrigated sugar cane crop in the Ivory Coast. Proc. Int. Soc. Sugar Cane Technol. 15: 726–735.

Barnes, A. C. 1974. The sugarcane. Wiley, New York.

Bowen, C. E. 1975. Micronutrient composition of sugarcane sheaths as affected by age. Trop. Agric. 52: 131–137.

Boyce, J. P. 1969. First ratoon results of two irrigation experiments at Pongola. Proc. S. Afr. Sugar Technol. Assoc. 43: 35–46.

Bresler, E., B. L. McNeal, and D. L. Carter. 1982. Saline and sodic soils. Springer-Verlag, Berlin.

Bull, T. A. and K. T. Glaziou. 1975. Sugar cane, pp. 51–72. *In*: L. T. Evans (ed.) Crop physiology: some case histories. Cambridge University Press, London.

Bull, T. A. and D. A. Tovey. 1974. Aspects modelling sugarcane growth by computer simulation. Proc. Int. Soc. Sugar Cane Technol. 15: 1021–1032.

Campbell, R. B., J.-H. Chang, and D. C. Cox. 1959. Evapotranspiration of sugarcane in Hawaii as measured by in-field lysimeters in relation to climate. Proc. Int. Soc. Sugar Cane Technol. 10: 637–645.

Carter, C. E. 1977a. Excess water decreases cane and sugar yields. Proc. Am. Soc. Sugar Cane Technol. 6: 44–51.

Carter, C. E. 1977b. Drainage parameters for sugarcane in Louisiana. Proc. Third National Drainage Symp., Am. Soc. Agric. Eng., pp. 135–138.

Carter, C. E. 1980. Redox potential and sugarcane yield relationship. Trans. ASAE 23: 924–927.

Carter, C. E. and C. R. Camp. 1983. Subsurface drainage of an alluvial soil increased sugarcane yields. Trans. ASAE 26: 426–429.

Carter, C. E. and J. M. Floyd. 1973. Subsurface drainage and irrigation for sugarcane. Trans. ASAE 16: 279–281, 284.

Carter, C. E. and J. M. Floyd. 1975. Inhibition of sugarcane yields by high water table during dormant season. Proc. Am. Soc. Sugar Cane Technol. 4: 14–18.

Carter, C. E., J. E. Irvine, V. McDaniel, and J. Dunckelman. 1985. Yield response of sugarcane to stalk density and subsurface drainage treatments. Trans. ASAE 28: 172–178.

Chang, J. H., R. B. Campbell, H. W. Brodie, and L. D. Baver. 1965. Evapotranspiration research at the HSPA Experiment Station. Proc. Int. Soc. Sugar Cane Technol. 12: 10–24.

Chang, Y. T. and P. L. Wang. 1983. Water consumption and irrigation requirements of sugarcane. Taiwan Sugar Res. Inst. Annu. Rep. 1982–1983.

Chapman, L. S. 1988. Constraints to production in ratoon crops. Proc. Australian Soc. Sugarcane Technol. 10: 189–192.

Chapman, L. S., R. Ferrias, and M. M. Ludlow. 1992. Ratooning ability of cane varieties: Variation in yield and yield components. Proc. Australian Soc. Sugarcane Technol. 14: 130–138.

Clements, H. F. 1980. Sugarcane Crop Logging and Crop Control: Principles and Practice. University of Hawaii Press, Honolulu.

Clements, H. F. and T. Kubota. 1943. The primary index—Its meaning and application to crop management with special reference to sugarcane. Hawaiian Planters' Record 47: 257–297.

Clowes, M. St. J. and N. G. Inman-Bamber. 1980. Effects of moisture regime, amount of nitrogen applied and variety of the ripening response of sugarcane to glyphosates. Proc. S. Afr. Sugar Technol. Assoc. 54: 127–133.

Coale, F. J., C. A. Sanchez, F. T. Jzuno, and A. B. Bottcher. 1993. Nutrient accumulation and removal by sugarcane grown on Everglades Histosols. Agron. J. 85: 310–315.

Dastane, N. G. 1957. Evaluation of root efficiency in moisture extraction. J. Aust. Inst. Agric. Sci. 23: 223–226.

Dillewijn, C. van. 1952. Botany of sugarcane. Chronica Botanica, Waltham, Massachusetts.

Early, A. C. 1974. The yield response of sugarcane to irrigation in the Philippines. Proc. Int. Soc. Sugar Cane Technol. 15: 679–693.

Ekern, P. C. 1971. Use of water by sugarcane in Hawaii measured by hydraulic lysimeters. Proc. Int. Soc. Sugar Cane Technol. 14: 805–812.

Elawad, S. H., G. J. Gascho, and J. J. Street. 1982a. Response of sugarcane to silicate source and rate. I. Growth and yield. Agron. J. 74: 481– 484.

Elawad, S. H., J. J. Street, and G. J. Gascho. 1982b. Response of sugarcane to silicate source and rate. II. Leaf frickling and nutrient content. Agron. J. 74: 484–487.

Finkel, H. J. 1983. Irrigation of sugar crops, pp. 119–135. *In*: Handbook of irrigation technology. CRC Press, Boca Raton, Florida.

Fogliata, F. A. 1974. Sugarcane irrigation in Tucuman. Proc. Int. Soc. Sugar Cane Technol. 15: 655–667.

Gascho, G. J., D. L. Anderson, and J. E. Bowen. 1993. Sugarcane, pp. 37–42. *In*: W. F. Bennett (ed.) Nutrient deficiencies and toxicities in crop plants. Amer. Phytopath. Soc., St. Paul, Minnesota.

Gascho, G. J. and G. Kidder. 1979. Responses to phosphorus and potassium and fertilizer recommendations for sugarcane in south Florida. Florida Agric. Exp. Stn. Bulletin 809 (technical).

Gascho, G. J. and S. F. Shih. 1979. Varietal response of sugarcane to water table. I. Lysimeter performance and plant response. Soil Crop Sci. Soc. Florida Proc. 38: 23–27.

Gascho, G. J. and S. F. Shih. 1981. Row spacing effects on biomass and composition of sugarcane in Florida. Proc. Am. Soc. Sugar Cane Technol. 8: 72–76.

Gascho, G. J. and S. F. Shih. 1982. Sugarcane, pp. 445–479. *In*: I. D. Teare and M. M. Peet (eds.). Crop-water relations. Wiley, New York.

Gibson, W. 1974. Hydraulics, mechanics, and economics of subsurface and drip irrigation of Hawaiian sugarcane. Proc. Int. Soc. Sugar Cane Technol. 15: 639–648.

Glover, J. 1967. The simultaneous growth of sugarcane roots and tops in relation to soil and climate. Proc. S. Afr. Sugar Technol. Assoc. 41: 143–159.

Glover, J. 1968. Further results from the Mount Edgecombe Root Laboratory. Proc. S. Afr. Sugar Technol. Assoc. 42: 123–132.

Glover, J. 1972. Practical and theoretical assessments of sugarcane yield potential in Natal. Proc. S. Afr. Sugar Technol. Assoc. 46: 138–141.

Gosnell, J. M. 1968. Some effects of increasing age on sugarcane growth. Proc. Int. Soc. Sugar Cane Technol. 13: 499–513.

Gosnell, J. M. 1971. Some effects of water-table on the growth of sugarcane. Proc. Int. Soc. Sugar Cane Technol. 14: 841–849.

Gosnell, J. M. and A. C. Long. 1971. Some factors affecting foliar analysis of sugarcane. Proc. S. Afr. Sugar Technol. Assoc. 45: 1–16.

Gosnell, J. M. and T. L. Pearse. 1971. Methods of surface irrigation in sugarcane. Proc. Int. Soc. Sugar Cane Technol. 14: 875–885.

Gosnell, J. M. and G. D. Thompson. 1965. Preliminary studies on depth of soil moisture extraction by sugarcane using neutron probe. Proc. S. Afr. Sugar Technol. Assoc. 39: 158–165.

Hardy, M. 1966. Water consumption of the cane plant. Annu. Rep. Mauritius Sugar Ind. Res. Inst., pp. 95–101.

Hartt, C. E. 1972. Translocation of carbon-14 in sugarcane plants supplied with or deprived of phosphorus. Plant Physiol. 49: 69–71.

Hill, J. N. S. 1966. Availability of soil water to sugarcane in Natal. Proc. S. Afr. Sugar Technol. Assoc. 40: 276–282.

Hunsigi, G. and S. C. Srivastava. 1977. Modulation of ET values of sugarcane because of high water table. Proc. Int. Soc. Sugar Cane Technol. 16: 1557–1564.

Inman-Bamber, N. G. and J. M. de Jager. 1986a. Effect of water stress on growth, leaf resistance and canopy temperature in field-grown sugarcane. Proc. S. Afr. Sugar Technol. Assoc. 60: 156–161.

Inman-Bamber, N. G. and J. M. de Jager. 1986b. The reaction of two varieties of sugarcane to water stress. Field Crops Res. 14: 15–28.

Irvine, J. E. 1983. Sugarcane, pp. 361–381. *In*: International Rice Research Institute (ed.). Biological basis, physical environment, and crop productivity. IRRI, Los Banos, Philippines.

Irvine, J. E. and G. T. A. Benda. 1980. Genetic potential and restraints in *Saccharum* as an energy source, pp. 1–9. *In*: A. G. Alexander (ed.) Alternate uses of sugarcane for development in Puerto Rico. CEER Pub. B-52, University of Puerto Rico, San Juan.

Johnson, J. L., A. M. Heagler, H. O. Zapata, and R. Ricaud. 1993. The impact of succession planting and a third ratoon crop on economic efficiency in sugarcane production in Louisiana. Amer. Soc. Sugar Cane Technol. 13: 28–32.

Jones, C. A. 1980. A review of evapotranspiration studies in irrigated sugarcane in Hawaii. Hawaiian Planters' Record 59: 195–214.

Jones, C. A. 1985. C_4 grasses and cereals. Growth, development, and stress response. Wiley, New York.

Kingston, G. 1973. The potential of "Class A" pan evaporation data, for scheduling irrigation of sugar cane at Bandaberg. Proc. Queensland Soc. Sugar Cane Technol. 40: 151–157.

Kingston, G. 1977. The influence of accessibility on moisture extraction by sugar cane. Proc. Int. Soc. Sugar Cane Technol. 16: 1239–1250.

Kingston, G. and G. J. Ham. 1975. Water requirements and irrigation scheduling of sugar cane in Queensland. Proc. Queensland Soc. Sugar Cane Technol. 42: 57–65.

Koehler, P. H., P. H. Moore, C. A. Jones, A. Dela Cruz, and A. Maretzki. 1982. Response of drip-irrigated sugarcane to drought stress. Agron. J. 74: 906–911.

Lee, H. A. 1926a. Progress report on the distribution of cane roots in the soil under plantation conditions. Hawaiian Planters' Record 30: 511–519.

Lee, H. A. 1926b. A comparison of the root weights and distribution of H109 and D1135 cane varieties. Hawaiian Planters' Record 30: 520–523.

Lee, H. A. 1926c. The distribution of the roots of sugar cane in the Hawaiian Islands. Plant Physiol. 1: 363–378.

Leverington, K. C. and D. R. Ridge. 1975. A review of sugar cane irrigation. Proc. Queensland Soc. Sugar Cane Technol. 42: 37–43.

Lonsdale, J. E. and J. M. Gosnell. 1974. Monitoring maturity of sugarcane during drying-off. Proc. Int. Soc. Sugar Cane Technol. 15: 713–725.

Maas, E. V. and G. J. Hoffman. 1977. Crop salt tolerance: Current assessment. J. Irrig. Drainage Div. Am. Soc. Civil Eng. 103: 115–134.

Mason, G. F. 1976. Chemical ripening in Trinidad of variety B41227 with glyphosine. Proc. West Indies Sugar Technol., pp. 130–138.

Milligan, S. B., K. A. Gravois, and F. A. Martin. 1996. Inheritance of sugarcane ratooning ability and the relationship of younger crop traits to older crop traits. Crop Sci. 36: 45–50.

Moberly, P. K. 1974. The decline in rate of evapotranspiration of fully canopied sugarcane during a winter stress period. Proc. Int. Soc. Sugar Cane Technol. 15: 694–700.

Mukerji, N. and M. Alan. 1959. Roots as indicator to varietal behavior under different conditions. Indian J. Sugarcane Res. Dev. 3: 131–134.

Paz-Vergara, J. E., A. Vasquez, W. Iglesias, and J. C. Sevilla. 1980. Root development of the sugarcane cultivars H32–8560 and H57–5174, under normal conditions of cultivation and irrigation in the Chicama Valley. Proc. Int. Soc. Sugar Cane Technol. 17: 534–540.

Pereira, J. R., C. M. B. Faria, and L. B. Margado. 1995. Effect of phosphorus application and of its residue on the yield of sugarcane in a Vertisol. Pesq. Agropec. Bras., Brasilia 30: 43–48.

Raheja, P. C. 1959. Performance of sugarcane varieties in relation to ecological habitat. Indian J. Sugarcane Res. Dev. 3: 127–130.

Reuter, D. J. 1986. Temperate and subtropical crops, pp. 38–99. *In*: D. J. Reuter and J. B. Robinson (eds.). Plant analysis: an interpretation manual. Inkata Press, Melbourne.

Rostron, H. 1974. Radiant energy interception, root growth, dry matter production and the apparent yield potential of two sugarcane varieties. Proc. Int. Soc. Sugar Cane Technol. 15: 1001–1010.

Rostron, H. 1977. A review of chemical ripening of sugarcane with ethrel in southern Africa. Proc. Int. Soc. Sugar Cane Technol. 16: 1605–1616.

Rudd, A. V. and C. W. A. Chardon. 1977. The effects of drainage on cane yields as measured by watertable heights in the Macknade mill area. Proc. Queensland Soc. Sugar Cane Technol. 44: 111–117.

Ryker, T. C. and C. W. Edgerton. 1931. Studies on sugar cane roots. Louisiana Agric. Exp. Stn. Bull. 223.

Salassi, M. E. and G. G. Giesler. 1995. Projected costs and returns—sugarcane, Louisiana, 1995. Dept. Agric. Economics and Agribusiness, AEA Info. Series No. 132, LAES, LSU Agric. Ctr., Baton Rouge.

Samuels, G. 1971. Influence of water deficiency and excess on growth and leaf nutrient element content of sugarcane. Proc. Int. Soc. Sugar Cane Technol. 14: 653–656.

Samuels, G., B. G. Cap, and I. S. Bangdiwala. 1953. The nitrogen content of sugarcane as influenced by moisture and age. J. Agric. Res. Univ. Puerto Rico 37: 1–12.

Sanchez, C. A. 1990. Soil-testing and fertilization recommendations for crop production on organic soils in Florida. Florida Agric. Exp. Stn. Bulletin 809 (technical).

Sartoris, G. B. and B. A. Belcher. 1949. The effect of flooding on flowering and survival of sugar cane. Sugar 44(1): 36–39.

Shah, R. 1951. Negatively geotropic roots in water-logged canes. Sugar 46(1): 39.

Shih, S. F. and G. J. Gascho. 1980. Relationships among stalk length, leaf area, and dry biomass of sugarcane. Agron. J. 72: 309–313.

Shih, S. F. and G. J. Gascho. 1981. Sugarcane biomass production and nutrient content as related to climate in Florida. Proc. Am. Soc. Sugar Cane Technol. 8: 77–83.

Shrivastava, A. K., A. K. Ghosh, and V. P. Agnihotri. 1992. Sugar cane ratoons. Oxford and IBH Publishing Co. PVT. Ltd., New Delhi.

Srinivasan, K. and M. B. G. R. Batcha. 1962. Performance of clones of *Saccharum* species and allied genera under conditions of waterlogging. Proc. Int. Soc. Sugar Cane Technol. 11: 571–577.

Stevenson, G. C. and A. E. S. McIntosh. 1935. Investigations into the root development of the sugar cane in Barbados. I. Root development in several varieties under one environment. Bull. 5, British West Indies Cent. Sugar Cane Breed. Stn., Barbados.

Sundara, B. 1989. Improving sugarcane productivity under moisture constraints and through cropping systems, pp. 221–256. *In*: K. M. Naidu et al. (eds.) Proc. Intl. Symp. Sugarcane varietal improvement—present and future strategies, Sept. 3–7, 1987, Sugarcane Breeding Institute, Coimbatore, India.

Terry, R. E. 1980. Nitrogen mineralization in Florida Histosols. Soil Sci. Soc. Am. J. 44: 747–750.

Thompson, G. D. 1965. The relation of potential evapotranspiration of sugarcane to environmental factors. Proc. Int. Soc. Sugar Cane Technol. 12: 3–9.

Thompson, G. D. 1976. Water use by sugarcane. S. Afr. Sugar J. 60: 593–600.

Thompson, G. D. 1977. Irrigation of sugarcane. S. Afr. Sugar J. 61: 126–131, 161–174.

Thompson, G. D. 1978. The production of biomass by sugarcane. Proc. S. Afr. Sugar Technol. Assoc. 52: 180–187.

Thompson, G. D. 1986. Agrometeorological and crop measurements in a field of irrigated sugarcane. Mount Edgecombe Research Report No. 5, S. Afr. Sugar Assoc. Exp. Stn., Mount Edgecombe, South Africa.

Thompson, G. D. and J. P. Boyce. 1967. Daily measurements of potential evapotranspiration from fully canopied sugarcane. Agric. Meteorol. 4: 267–279.

Thompson, G. D. and J. P. Boyce. 1968. The plant crop results of two irrigation experiments at Pongola. Proc. S. Afr. Sugar Technol. Assoc. 42: 143–157.

Thompson, G. D. and J. P. Boyce. 1971. Comparisons of measured evapotranspiration of sugarcane from large and small lysimeters. Proc. S. Afr. Sugar Technol. Assoc. 45: 169–176.

Thompson, G. D. and P. J. M. De Robillard. 1968. Water duty experiments with sugarcane on two soils in Natal. Exp. Agric. 4: 295–310.

Thompson, G. D., J. M. Gosnell, and P. J. M. De Robillard. 1967. Responses of sugarcane to supplementary irrigation on two soils in Natal. Exp. Agric. 3: 223–238.

Thompson, G. D., C. H. O. Pearson, and T. G. Cleasby. 1963. The estimation of the water requirements of sugarcane in Natal. Proc. S. Afr. Sugar Technol. Assoc. 37: 137–141.

USDA (United States Department of Agriculture). 1985. Agricultural statistics 1985. U.S. Government Printing Office, Washington, DC.

Valdivia, V. S. 1981. Advances in Peruvian research on the effects of salinity of sugarcane production. Turrialba 31: 237–244.

Wood, G. H. and R. A. Wood. 1968. The estimation of cane root development and distribution using radiophosphorus. Proc. S. Afr. Sugar Technol. Assoc. 41: 160–168.

Yang, S. J. and Y. T. Chang. 1976. Studies on irrigation scheduling by using crop-soil-climatic data. Taiwan Sugar Res. Inst. Annu. Rep. 1975–1976, pp. 13–14.

16

Cassava and Potato

I. INTRODUCTION

Cassava and potato are the most important root tuber crops used as a human food source. These two crops are also used as livestock feed and as a raw material source for the manufacture of alcohol, flour and starch. Cassava is mostly grown and consumed in the tropics, while potatoes are grown commercially in most countries of the world. A synthesis of the information and selected references in this chapter will provide the readers authentic information to understand the production physiology and mineral requirements of these two crops.

II. CASSAVA

Contrary to views that cassava (*Manihot esculenta*) is only known in cultivation, an argument is made that wild accessions of the species grow over much of the American neotropics, in Brazil, Bolivia, Peru, Venezuela, Guyana, and Surinam (Allem, 1994). Three subspecies are recognized. *M. esculenta* is the domesticated subspecies and includes all cultivars known in cultivation.

The wild *M. esculenta peruviana* occurs in eastern Peru and western Brazil. The wild *M. esculenta flabellifolia* shows a wider distribution that ranges from the central Brazilian state of Goiás northward to Venezuelan Amazonia. The large area of distribution of the two wild subspecies makes it difficult to assign a place of initial domestication (Allem, 1994).

Cassava (*Manihot esculenta* Crantz) is a perennial shrub belonging to the family Euphorbiaceae, subfamily Crotonoideae and class/subclass Angiospermae (Dicotyledones). Cassava is commonly known as tapioca, manioc, mandioca, yuca and sagu, and in Africa more than a half-dozen vernacular names prevail. The exact origin of cassava is not known, but apparently it was first domesticated somewhere in South America. It was taken to Africa by Portuguese as early as 1558 and spread to Asia in the 17th century (Cock, 1984). The maximum production of cassava is in Africa, followed by Asia and South America. Although it is one of the world's most important food staples, cassava is used in North America and Europe almost solely in tapioca, an occasional dessert (Janick et al., 1974). The major cassava-producing countries are Brazil, Indonesia, Thailand, Zaire, Nigeria, India, and China. The average world yield is about 8.7 tons of fresh root ha^{-1}, which is far below the potential yield of 80 tons ha^{-1} produced under experimental conditions (Howeler, 1985). This large gap between potential and actual productivity is due to the fact that the crop is largely produced by subsistence farmers with low technology. Cassava's high yield potential has been attributed to a high total crop dry weight in relation to foliage development, leaf area duration, canopy architecture, and a high ratio of dry weight of storage roots to total dry weight, the so-called harvest index (Jose and Mayobre, 1982).

A. Climate and Soil Requirements

Although cassava is, by origin, a tropical crop, it is successfully grown in latitudes up to 25°S in South America, southern Africa, and in some trial commercial plantings in high latitudes of Australia (Harris, 1978a). Cook and Rosas (1975) and Cock (1983) suggested that cassava's ecological zone lies between latitudes 30°N and 30°S at elevations up to 2,300 m above sea level. The cassava plant is very sensitive to frost and cannot be grown successfully where the temperature is below 15°C (Cock, 1984). In areas with marked seasonal temperature changes, cassava is grown only when the annual mean temperature is greater than 20°C (Cock, 1983). Cassava is grown in areas with as little as 750 mm rainfall per year since the crop has an extremely conservative pattern of water use. Reduced leaf area and stomatal closure markedly reduce crop growth rates during periods of stress (Connor et al., 1981).

Cassava can grow on infertile soils varying in texture from light to heavy, and in pH from 3.5 to 7.8, but it will not tolerate excess water or high salinity. The highest yields occur on well-drained, medium to heavy texture, fertile soils with a pH of about 5.5–7 (Howeler, 1981). Abruna et al. (1982) found that Cassava is highly tolerant to soil acidity. Yields of cassava grown on a corozal clay (Aquic Tropudults) were affected only when the soil pH dropped below 4.3 and the exchangeable Al increased above 60% saturation. Cassava is more susceptible than other food crops to soil salinity and alkalinity, but large varietal differences in tolerance exist. Yield of this root crop are markedly reduced when the Na saturation is above 2–5% and the electrical conductivity is above 0.5–0.7 dS m^{-1} (Cock and Howeler, 1978).

B. Growth and Development

The basic constituents of the cassava plant are: 1) nodal units which consist of a leaf blade, petiole, and internode, and 2) thickened roots which form mainly at the base of the stem cutting that is used as planting material. The weight of the dry stem internode, which varies with variety, averages from 0.5 to 3.0 g/internode for a mature plant, and leaf blades together with petioles have an area-to-weight ratio of about 135 cm^2 g^{-1}. The plant generally shows strong apical dominance and does not generally produce leaves from the axillary buds (Cock et al., 1979).

Propagation of cassava is possible either through true seed or through cuttings. Planting through cuttings is the normal practice for commercial production, while planting through seed is practiced in breeding programs. Cutting length is normally 20 cm and preferably taken from the middle of the stems. The growth duration of the crop depends on environmental conditions. The period from planting to harvest is about 9–12 months in hot regions and 2 years in cooler or drier regions (Cock, 1984).

C. Roots

Roots are the main storage organs in cassava. Secondary thickening results in tuber development as swellings on adventitious roots a short distance from the stem. There are usually 5–10 tubers per plant. The tubers are cylindrical or tapering, 15–100 cm long, 3–15 cm across, and occasionally branched (Purseglove, 1987). Keating et al. (1982) reported that the final number of storage roots was generally reached within 90–135 days after planting and ranged from 10 to 14 storage roots per plant, independent of planting date and season. Campos and Sena (1974) found most roots concentrated in the top 30 cm, with some roots as far down as 140 cm; however, Connor et al. (1981) found roots at a depth of 250 cm. They also observed root character-

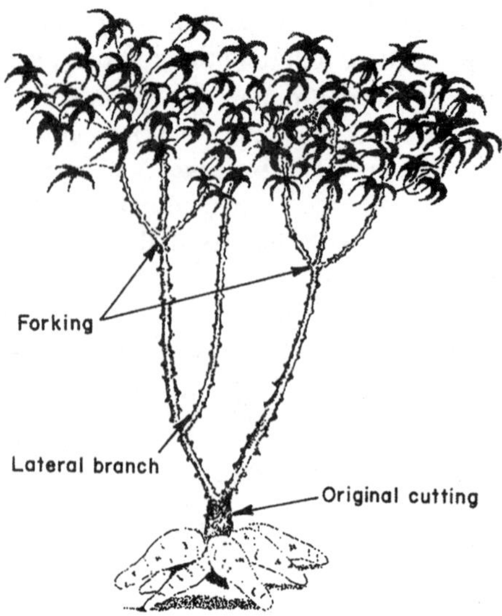

Figure 16.1 Cassava plant with vegetative and reproductive organs. (From Cock, 1984. © John Wiley & Sons, reproduced with permission.)

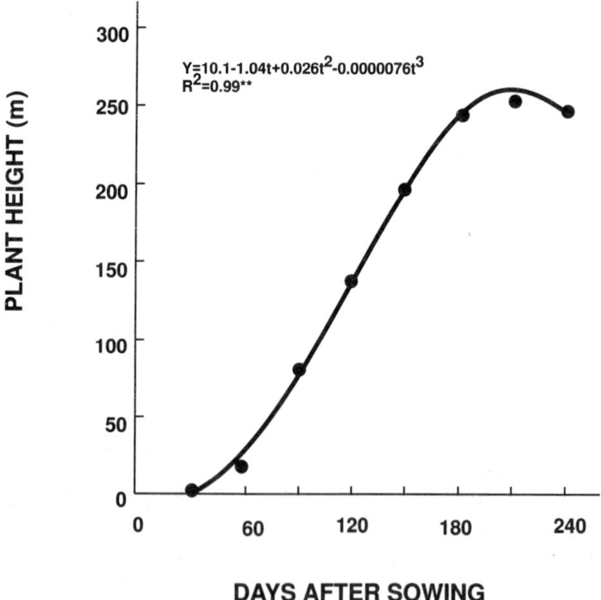

Figure 16.2 Cassava plant height during crop growth in a Cambisol of Brazil. (From Sangoi and Kruse, 1993.)

istics and found that cassava has rather thicker roots than most species (0.37–0.67 mm diameter), but very short roots (less than 1 km m^{-2} land surface).

Tops

Cassava plants vary in height and branching habit. Each nodal unit consists of a node, which subtends a leaf on an internode. The total number of nodes per plant depends on the number of nodes per shoot and the number of shoots, or apices, per plant (Cock, 1984). Cassava leaves are spirally arranged (phyllotaxis 2/5) and variable in size, color of stipules, petioles, midribs and laminae, number of lobes, depth of lobing, and in the shape and width of lobes (Purseglove, 1987). Figure 16.1 shows the cassava plant with roots, stem, branching, and leaves. Sangoi and Kruse (1993) measured plant height during the crop growth cycle on the highlands of Santa Catarina, Brazil. During the first 60 days of plant growth, plant height increase was slow (Figure 16.2). Maximum plant height increased during 60–180 days after planting. Maximum plant height was reached at about 210 days after sowing, after which it was more or less constant.

Leaf Area Index

Leaf area index (*LAI*) is an important growth parameter which determines the photosynthetic capacity of a crop. It is affected by climatic factors, soil fertility levels, and cultural practices. Leaf area index normally increases as the crop grows, reaches a maximum value, and declines in the latter part of the crop growth due to leaf abscission. The *LAI* of cassava increases in the first 4–6 months after planting and then declines (Williams, 1972; CIAT, 1979). The maximum *LAI* values reported in the literature are between 6 and 8 (Enyi, 1972). According to Keating et al. (1982) substantial leaf abscission begins at *LAI* values on the order of 5–6. Crop growth is related to *LAI*. Cassava is a C_3 plant that reaches a maximum growth rate of 120–150 g m^{-2} per week at a *LAI* of about 4 under solar radiation of about 450 cal cm^{-2} day^{-1} (Cock, 1983). Figure 16.3 illustrates the relationship between *LAI* and crop growth rate. There is a linear relationship between *LAI* and growth rate of leaves and stems. The relationship between *LAI* and the growth rate of roots was quadratic, with a maximum growth rate achieved at about 3 *LAI*, followed by a decline with further increase in *LAI*. Total plant growth rate increased up to 4 *LAI* and then remained constant. Maximum crop growth rates of cassava within the tropics are of the order of 120 g m^{-2} wk^{-1} (Cock, 1984).

Dry Matter

A study conducted by Irizarry and Rivera (1983) in Ultisol of Puerto Rico showed that dry matter in the stems, roots, and whole plant increased steadily

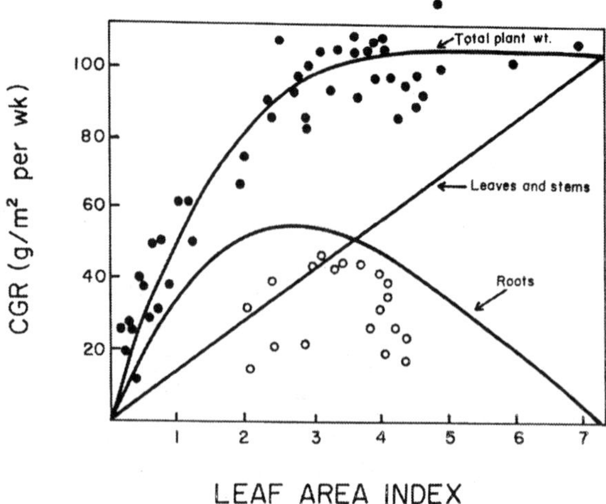

Figure 16.3 Leaf area index and crop growth rate. (From Cock, 1983.)

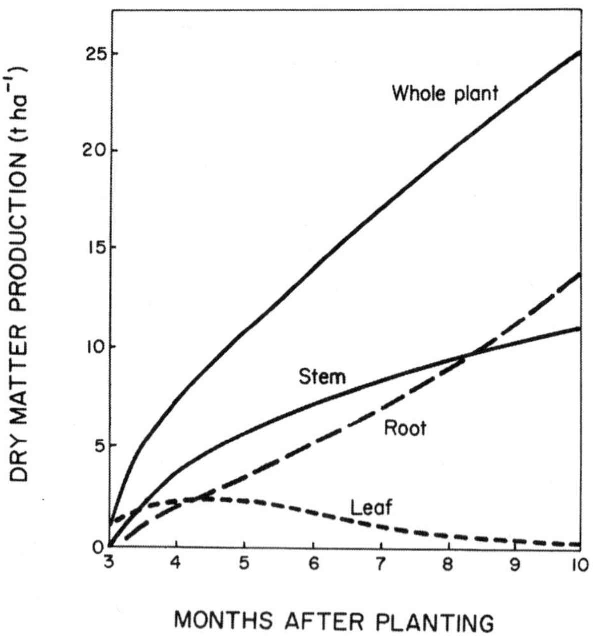

Figure 16.4 Dry matter accumulation among different plant parts of Cassava. (From Irizarry and Rivera, 1983.)

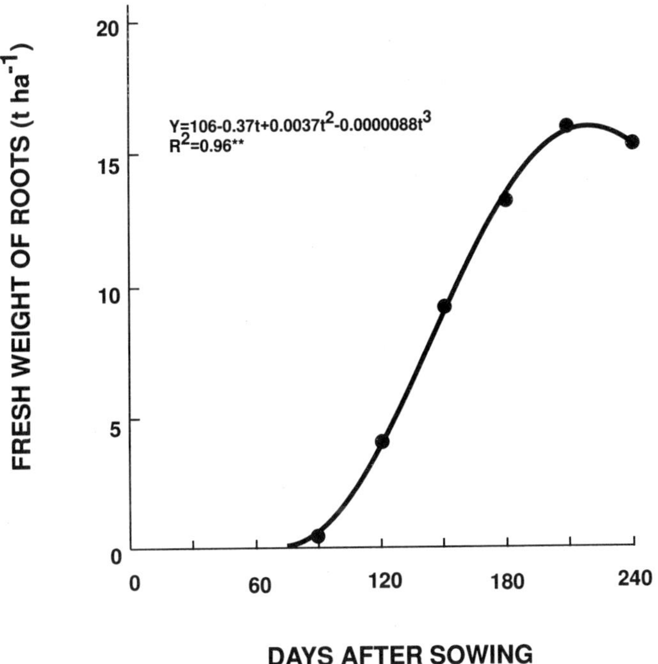

Figure 16.5 Fresh weight of cassava roots during crop growth cycle. (From Sangoi and Kruse, 1993.)

with plant age (Figure 16.4). Dry matter in the leaves declined sharply from 5 to 8 months after planting and thereafter remained stable. Total plant dry matter 10 months after planting was 23 t ha^{-1}, and yield of edible fresh roots was about 11 t ha^{-1}. Howeler and Cadavid (1983) reported in an experiment carried out in Colombia that cassava dry matter accumulation was slow during the first 2 months, increased rapidly during the next 4 months, and slowed down during the final 6 months as dry matter production was partly offset by leaf fall. At harvest (12 months) dry matter was present mainly in roots followed by stems, leaves, and petioles. Sangoi and Kruse (1993) determined fresh weight of roots in a Brazilian Cambisol during the crop growth cycle. At the age of about 210 days, maximum root weight of about 15 t ha^{-1} was achieved (Figure 16.5).

Harvest Index

Harvest index (HI) is the measure of distribution of dry matter to the economically useful plant parts. In cereals, this is measured by taking into consideration the grain yield and total aboveground plant parts. In cassava, the roots are the major economic part, and root to total dry matter production is used as an index to quantify economically useful plant parts. Cock (1976) obtained HI values after 1 year ranging from less than 0.30–0.57 for a number of clones when a population of 20,000 plants ha^{-1} was used. In one trial with M col.22, HI values of more than 0.7 were obtained, but HI decreased as population increased above 12,000 plants ha^{-1} (Cock et al., 1977). Harvest index can be used as one of the selection criteria for higher yield potential in cassava cultivars (Kawano, 1978). According to Iglesias et al. (1994), a harvest index value of 0.5–0.6 is the optimum level because at higher values of HI, root production potential is affected as a result of reduced photosynthetic area.

D. Nutrient Requirements

Cassava is considered to be tolerant to low-fertility conditions and grows well on acid soils where other crops cannot be grown satisfactorily (Howeler, 1981). Although this crop has been traditionally grown without the use of fertilizers, it is now well established that the cassava plant responds well to fertilization, and in order to obtain high yield, adequate nutrients should be supplied (Howeler, 1981; Howeler and Cadavid, 1983). Howeler (1981) summarized the response to fertilization and liming in the three most extensive tropic soils (Oxisols, Ultisols, and Inceptisols). Phosphorus is generally the element most limiting to yield. Cassava extracts large amounts of K from soil and may cause depletion of this element if grown continuously without adequate K fertilization. Compared to other crops, cassava has a low requirement for N, and high N application may lead to excessive top growth, a reduction in starch synthesis, and poor root thickening (Howeler, 1981). Cassava is acid tolerant, and the optimum pH is between 5.5–7.5. The crop often responds to a low rate of liming, but overliming may induce micronutrient deficiencies. The critical levels of various nutrients in the soil reported by Howeler (1981), based upon summarization of results from various studies, are: P, 7–9 mg kg^{-1}; K, 0.06–0.15 cmol kg^{-1}; Ca, 0.25 cmol kg^{-1}; Zn 1 mg kg^{-1}; Mn 5–9 mg kg^{-1}; and S, 8 mg kg^{-1} of soil.

Nutrient Concentration

Nutrient concentrations in plants vary with the plant part analyzed, stage of plant growth, soil fertility, climatic conditions, and management practices. Table 16.1 shows the average nutrient concentrations in various plant parts of 2– to 4–month-old cassava plants. These cassava cultivars had their greatest growth rate, and thus their greatest nutrient demand (Howeler and Cadavid,

Table 16.1 Concentration of Nutrients in Various Plant Parts of Fertilized and Nonfertilized Cassava[a]

			g kg^{-1}						mg kg^{-1}				
			N	P	K	Ca	Mg	S	B	Cu	Fe	Mn	Zn
With fertilizers[b]	Leaves	Upper	57.4	4.2	19.8	7.2	3.4	3.0	15.0	11.9	202	456	128
		Middle	51.8	2.7	18.0	10.1	3.8	2.8	16.6	12.1	251	610	149
		Lower	44.0	2.0	15.8	13.4	4.9	2.2	16.5	12.3	288	775	195
	Petioles	Upper	22.5	2.2	29.3	9.0	3.9	0.6	13.0	7.9	59	604	106
		Middle	14.1	1.4	23.5	11.3	3.9	0.2	12.7	6.9	57	824	146
		Lower	13.5	1.2	22.3	15.4	4.8	0.1	14.5	8.2	93	1456	237
	Stem	Upper	27.3	3.0	31.5	8.2	3.7	1.8	12.7	15.6	104	368	102
		Middle	22.1	2.7	22.1	10.2	3.8	1.6	10.5	20.6	108	408	146
		Lower	12.8	2.2	11.4	6.5	3.1	0.9	6.9	19.3	207	186	141
	Roots		15.2	1.8	15.6	2.4	1.4	0.5	6.1	9.2	423	147	64
Without fertilizers	Leaves	Upper	53.1	3.3	18.6	7.1	3.3	3.0	7.1	12.4	150	344	86
		Middle	44.4	1.9	16.0	9.5	3.7	2.7	7.8	12.0	224	435	90
		Lower	35.6	1.5	14.5	12.8	5.0	2.3	6.9	9.8	383	619	80
	Petioles	Upper	19.1	1.7	30.7	9.5	4.0	0.7	8.1	9.9	74	462	73
		Middle	15.2	0.8	23.2	12.2	4.1	0.2	6.9	7.3	54	847	108
		Lower	11.4	0.6	19.9	16.2	4.9	0.2	7.1	6.8	153	1484	143
	Stem	Upper	29.2	2.6	34.9	8.8	4.0	2.1	8.2	20.5	162	311	69
		Middle	25.6	1.8	20.0	9.6	3.9	2.0	6.6	24.8	105	350	94
		Lower	11.2	1.1	9.5	6.9	3.2	1.0	5.0	27.9	244	153	52
	Roots		13.3	1.2	16.1	2.7	1.5	0.4	5.9	12.1	592	209	67

[a] Data are averages of samples taken at 2, 3, and 4 months from M Col 22 and M Mex 59.
[b] Fertilization (kg ha^{-1}): 100 N, 131 P, 83 K, 20 S, 10 Zn, 1 B.
Source: Howeler and Cadavid, 1983.

Table 16.2 Nutrient Sufficiency Levels in Cassava Plant

Nutrient	Growth stage	Plant part	Sufficiency level
			$g\ kg^{-1}$
N	Vegetative	UMB[a]	50–60
P	Vegetative	UMB	3–5
K	100 DAS[a]	UMB	12–20
	200 DAS	UMB	6–10
	580 DAS	UMB	5–10
Ca	Vegetative	UMB	6–15
Mg	Vegetative	UMB	2.5–5
S	Vegetative	UMB	3–4
			$mg\ kg^{-1}$
Cu	Vegetative	UMB	7–15
Zn	Vegetative	UMB	40–100
Mn	Vegetative	UMB	50–250
Fe	Vegetative	UMB	60–200
B	Vegetative	UMB	15–20

[a] DAS, Days after sowing; UMB, uppermost mature leaf blade.
Source: Compiled from Reuter, 1986.

1983), in this time interval. Howeler (1981) recommended that tissue be sampled for diagnosis purposes during this period. The concentrations reported in Table 16.1 for fertilized plants can be considered nearly optimum. The N, P, and S concentrations were high in the leaves, while those of K, Ca, and Mg were high in petioles and stem. The Mn concentration was high in roots. Boron and Cu were evenly distributed within the plant.

Nutrient sufficiency levels in cassava tissue are summarized in Table 16.2. In general, a fertilizer response is not likely when the uppermost mature leaf blades contain 50–60 g kg^{-1} N, 3–5 g kg^{-1} P, 12–20 g kg^{-1} K, 6–15 g kg^{-1} Ca, 2.5–5 g kg^{-1} Mg, and 3–4 g kg^{-1} S. Similarly, the fertilizer response is not expected when the Cu concentration in plant tissue is 7–15 mg kg^{-1}, Zn 40–100 mg kg^{-1}, Mn 50–250 mg kg^{-1}, Fe 60–200 mg kg^{-1}, and B 15–20 mg kg^{-1}.

Irizarry and Rivera (1983), working in a Puerto Rican Ultisol, reported that six months after planting, optimum leaf concentrations were approximately 43, 1.2, 18, 14, and 4 g kg^{-1} for N, P, K, Ca, and Mg, respectively.

Nutrient Accumulation

The nutrient accumulation (concentration × dry matter) pattern can provide useful information about the nutrient requirements of a crop. Howeler and Cadavid (1983) studied nutrient accumulation in different plant parts of cassava in Colombia during different growth stages (Table 16.3). Most nutrients accumulated initially in leaves and stems, but were translocated to roots in the later part of the growth cycle. Only Ca, Mg and Mn accumulated more in stems than roots. Nutrient removal at the 12–month growth period was in the order of N > K > Ca > P > Mg > S > Fe > Mn > Zn > Cu > B. Irizarry and Rivera (1983) reported that in an Ultisol of Puerto Rico, 10 months after planting, cassava accumulated 204 kg N ha^{-1}, 13 kg P ha^{-1}, 222 kg K ha^{-1}, 86 kg Ca ha^{-1}, and 33 kg Mg ha^{-1}, with a total dry matter production of about 23 t ha^{-1} of leaves, stems, and roots. Howeler (1981) calculated nutrient removal by cassava roots based on various studies in the literature and concluded that, on the average, each ton of cassava roots removed 2.3 kg N, 0.5 kg P, 4.1 kg K, 0.6 kg Ca, and 0.3 kg Mg.

III. POTATO

The potato (*Solanum tuberosum* L.) ranks fourth after wheat, rice, and corn as an important source of food worldwide. In the United States, the potato has essentially moved to second place, replacing third-place rice and even second-place corn, as a direct food source because of its culinary adaptability and the popularity of french fries at fast-food establishments. The potato in its natural state is exceedingly low in sodium and relatively rich in potassium and vitamin C (Ulrich, 1993).

The potato (*Solanum tuberosum* L.) belongs to the family Solanaceae and is a herbaceous annual plant grown for edible tubers. Actually, tubers are underground stems and are high in carbohydrates. The origin of potato is believed to be the Andean region of Peru and Bolivia. Although many crops were brought to Europe by Columbus and others soon after the discovery of the New World in 1492, the potato arrived much later. This is because it is a cool-temperate crop of the high Andes of South America and were not discovered by the Spaniards until 1532. Potatoes were not recorded in the literature until 1537 in what is now Colombia and did not appear in published work until 1552 (Hawkes and Ortega, 1993). The potato was taken to Europe by the Spanish around 1570 and introduced into England in 1586. From Europe it spread to Africa and Asia. In the United States, potatoes are said to have been introduced in 1621, presumably via Bermuda, but it was not extensively planted in the United States until 1700. Nowadays it is planted in almost all countries. At the world level, the USSR, West and East Germany,

Table 16.3 Dry Matter (g/Plant) and Nutrient Content (mg/Plant) in Various Parts of Fertilized M Col 22, During a 12-Month Growth Cycle[a]

		Months after planting								
		1	2	3	4	5	6	8	10	12
DM	Leaves	1.8	22.7	76.0	100.6	56.2	100.2	50.5	58.7	67.0
	Petioles	0.2	4.9	21.5	38.2	19.0	27.4	8.6	12.1	11.5
	Stems	14.1	29.1	58.9	125.2	182.1	269.1	302.7	428.6	459.9
	Roots	0.1	7.1	80.5	229.6	360.0	571.9	782.6	942.4	1387.0
	Total	16.2	63.8	236.9	493.7	617.3	968.6	1144.4	1441.8	1925.4
N	Leaves	89	1231	4230	5300	2703	4877	2206	2702	3350
	Petioles	6	134	368	485	202	378	144	182	207
	Stems	117	422	1146	1919	3022	4191	4707	5984	6930
	Roots	—	125	1078	2250	4428	5605	7043	9424	9709
	Total	212	1912	6824	9954	10355	15051	14100	18292	20196
P	Leaves	5	71	267	227	137	288	136	147	174
	Petioles	—	10	35	34	16	31	10	20	18
	Stems	37	71	157	205	358	422	482	378	766
	Roots	—	11	153	344	576	629	861	1036	1387
	Total	42	163	612	810	1087	1370	1489	1581	2345
K	Leaves	24	337	1408	1716	507	1564	712	817	945
	Petioles	9	161	598	744	347	561	159	201	207
	Stems	58	213	872	1681	2581	2588	2817	3233	3676
	Roots	5	123	1248	2870	4176	4463	5635	6879	10402
	Total	96	834	4126	7011	7611	9176	9323	11130	15230

Ca	Leaves	15	157	583	924	525	857	424	452	435
	Petioles	4	68	212	393	248	420	125	165	186
	Stems	216	244	485	864	1061	1704	1986	2412	3083
	Roots	1	20	113	321	432	915	939	1508	1248
	Total	236	489	1393	2502	2266	3895	3474	4537	4952
Mg	Leaves	9	67	248	411	166	276	146	146	174
	Petioles	2	23	77	142	68	130	32	41	56
	Stems	93	125	216	401	424	586	707	746	1147
	Roots	—	9	72	230	288	400	626	660	693
	Total	104	224	613	1184	946	1392	1511	1593	2070
S	Leaves	2	61	203	335	185	256	101	88	241
	Petioles	—	4	5	—	14	30	7	7	14
	Stems	15	19	63	101	227	383	360	337	578
	Roots	—	5	8	—	216	171	391	283	555
	Total	17	89	279	436	642	840	859	715	1388
B	Leaves	0.02	0.29	0.98	1.44	1.02	1.00	0.40	0.73	0.62
	Petioles	—	0.07	0.29	0.52	0.32	0.32	0.08	0.17	0.13
	Stems	0.09	0.22	0.48	0.99	1.31	1.89	2.07	2.42	3.52
	Roots	—	0.06	0.50	1.19	1.58	1.37	2.50	2.83	3.88
	Total	0.11	0.64	2.25	4.14	4.23	4.67	5.11	6.15	8.15

Table 16.3 (Continued)

		Months after planting								
		1	2	3	4	5	6	8	10	12
Cu	Leaves	0.02	0.20	0.89	1.22	0.44	0.89	0.37	0.58	0.71
	Petioles	—	0.03	0.18	0.22	0.77	0.13	0.03	0.08	0.07
	Stems	0.40	0.59	1.09	2.14	2.04	2.74	2.36	3.87	3.80
	Roots	—	—	0.69	2.02	2.12	3.43	2.50	6.03	6.66
	Total	0.42	0.82	2.85	5.60	4.67	7.19	5.26	10.56	11.24
Fe	Leaves	0.6	4.3	32.4	22.3	14.6	20.6	10.2	12.9	11.7
	Petioles	—	0.3	2.1	1.7	1.1	1.4	0.5	1.0	0.8
	Stems	2.4	3.2	15.0	17.8	15.7	38.7	52.7	43.8	141.7
	Roots	—	4.3	14.6	28.7	32.4	60.0	165.1	125.3	142.9
	Total	3.0	12.1	64.1	70.5	63.8	120.7	228.5	183.0	297.1
Mn	Leaves	0.4	10.7	38.3	66.6	27.5	67.5	25.2	31.5	26.4
	Petioles	0.1	5.2	19.5	52.4	37.8	64.6	15.5	21.2	21.0
	Stems	0.8	3.4	16.9	40.7	46.3	67.8	74.1	106.7	108.4
	Roots	0.2	1.5	4.5	11.7	17.3	23.4	20.3	35.8	26.3
	Total	1.5	20.8	79.2	171.4	128.9	223.3	135.1	195.2	182.1
Zn	Leaves	0.19	2.75	7.37	13.56	5.39	8.23	3.54	4.29	4.76
	Petioles	0.04	0.86	3.09	5.15	1.64	2.61	0.86	1.33	1.50
	Stems	4.34	4.17	8.19	15.06	12.93	17.67	17.74	23.94	26.27
	Roots	0.02	0.52	3.78	10.79	9.00	14.30	14.09	16.02	16.64
	Total	4.59	8.30	22.43	44.56	28.96	42.81	36.23	45.58	49.17

[a] Fertilization (kg ha^{-1}): 100 N, 131 P, 83 K, 20 S, 10 Zn, 1 B.
Source: Howeler and Cadavid, 1983.

and Poland are the leading potato producers. In Asia, China is the main producer, while in South America, Argentina, Brazil, Colombia, and Peru are dominant producers. The potato is the most important tuber crop used as a source of human food. It ranks as the fifth major food crop of the world, after wheat, rice, corn, and barley. In North America, dry matter production of potatoes per unit of land area exceeds that of wheat, barley, and corn by factors of 3.04, 2.68, and 1.12, respectively. Potato protein yield per unit of land area exceeds that of wheat, rice, and corn by factors of 2.02, 1.33 and 1.20, respectively (Hooker, 1986). Besides being used for human consumption, potatoes are used as a feed for animals and for starch, spirits, and industrial alcohol. Potato tubers contain 70–80% water, 8–28% starch, and 1–4% protein, with traces of minerals and other food elements. The nutritive value of potatoes has been discussed in detail by Gray and Hughes (1978).

A. Climate and Soil Requirements

The potato is classified as a cool-season crop, but it is grown in the tropics at an elevation of approximately 2,000 m or higher. The highest average yields are obtained in countries where day length is 13–17 hours during the growing season, average temperatures around 15–18°C prevail, and rainfall or irrigation provides ample water (Haeder and Beringer, 1983). Various studies reported in the literature have shown that potatoes have a thermal optimum for growth that varies from 16 to 28°C depending on the variety, age, and plant part considered (Burton, 1972; Marinus and Bodlaender, 1975; Benoit et al., 1986). Benoit et al. (1983) reported that leaf expansion had an optimum temperature of about 25°C, stem elongation had an optimum temperature of 31°C at rhizome initiation when plants were 20 cm tall, and 27°C was optimum for early tuber bulking. Vegetative growth is favored by temperatures at the lower end of the range (Haun, 1975; Marinus and Bodlaender, 1975). Decreased plant growth occurs above and below the optimum temperatures, probably as a result of shifts that occur in the dynamic balance between photosynthesis and respiration (Gent and Enoch, 1983; Lahav and Trochoulias, 1982). Table 16.4 summarizes the overall effects of soil temperature on plant growth.

Potatoes are quite sensitive to water stress, and an adequate water supply is required from tuber initiation until near maturity for high yield and good quality (Van Loom, 1981; Hang and Miller, 1986). Water shortages during tuber bulking decreased yield to a larger extent than during other growth stages (Hang and Miller, 1986). According to Harris (1978b), the potato crop is very sensitive to small water deficits, and for near-maximum yields, this deficit should not be allowed to exceed approximately 50% of the available

Table 16.4 Relationship Between Mean Daily Soil Temperature at 20 cm Depth and Potato Growth

Restriction	Temperature (°C)	Characteristics
Extremely severe	≤ 0	Continuous frost-killing temperature. Lack of plant growth.
Severe	0–5	High risk of frost-killing temperatures. Very slow plant growth.
Moderate	5–10	Moderate risk of frost-killing temperatures. Slow plant growth.
Slight	10–15	Slight risk of frost-killing temperatures.
None	15–20	Optimum temperatures for plant growth.
None	20–25 NT[a] < 20	Optimum temperatures for plant growth.
Slight to moderate	20–25 NT 20–25	Delayed growing season and late tuber initiation.
Severe to extremely	> 25	Delayed growing season and lack of tuber initiation.

[a] NT = Night temperature.
Source: Manrique, 1992.

water within the root range of the crop. Water requirements for the potato may be assessed by measuring the soil water tension range at which water is readily available for plant growth. Optimal soil water tension for the potato has been reported to range between 20–60 kPa (Van Loom, 1981). Irrigation to attain soil water tension values lower than 20 kPa may reduce tuber yields through impaired aeration. Allowing the soil water tension to rise above 60 kPa may reduce production and translocation of assimilates (Manrique, 1992).

Potatoes can be grown on a wide range of soils. They are well suited to acidic soils. The optimum pH range reported for potato production is from 5.2 to 6.5 (McLean and Brown, 1984). The limited response of potato to liming suggests that soil acidity and Al toxicity are not serious constraints to potato growth in all but the most acid soils. The percentage Al saturation is a useful index of soil acidity constraints and lime requirements for different crops. Potato suffers no yield decline at an Al saturation below 20% (Manrique, 1992). Sandy, well-drained loams, and soils high in organic content, are generally best for potato growing. The potato plant is considered moderately tolerant to soil salinity and the salinity threshold for yield decline is reported to be about 1.7 dS m^{-1} (Maas and Hoffman, 1977).

B. Growth and Development

The potato is commercially propagated vegetatively and only planted through true seeds in breeding programs. The germination of potatoes is epigeal, and the radicle emerges from the micropylar end of the seed and develops as a tap root, which soon forms lateral roots (Cutter, 1978). The commercial potato plant contains one or more lateral branches, each arising from a bud on the seed tuber, and the roots are adventitious (Hooker, 1986). The potato crop is normally harvested from 3 to 4 months after planting. Figure 16.6 summarizes the development of the potato plant through the different growth stages.

Roots

The plants from true seeds have a primary tap root system, whereas commercial potato plants planted from seed tubers have adventitious root systems. But in both cases, a much-branched fibrous root system is formed (Cutter, 1978). Harris (1978c) summarized the work of various investigators and came to the conclusion that potato roots were essentially restricted to soil depths of 0–30 cm and decreased rapidly below this depth. Roots and stolons develop from the underground stem between seed tubers and the soil surface. Thus, the vegetative propagative unit should be planted sufficiently deep to permit adequate root and stolon formation (Hooker, 1986).

Tops

Potato is a herbaceous, branched annual, 0.3–1 m in height, with swollen stem tubers (Purseglove, 1987). Vegetative stems (or branches) of the potato plant are usually quite thick and erect and are greatly branched. They range from 30–60 cm in length. Leaves are compound; they have terminal leaflets and from two to four pairs of fairly oblong, pointed leaflets and two or more small leaflets at the base. Flowers are borne in compound terminal cymes; thus, the plant is determinate in growth habit (Chapman and Carter, 1976).

Tubers

The tubers, which are the harvested portion of the potato plant, are modified, underground stems with greatly extended internodes. The tuber is formed at the tip of the stolon (rhizome) as a lateral proliferation of storage tissue resulting from rapid cell division and enlargement which approximate a 64–fold cell volume increase (Hooker, 1986). Tuber composition varies with the cultivar and growing conditions. But on an average, fresh tuber constituents are: water, 63–87%; carbohydrates,, 13–30% (including a fiber content of 0.17–3.48%); protein 0.7–4.6%; fat, 0.02–0.96%; and ash, 0.44–1.9%. Additional constituents include sugars, nonstarchy polysaccharides, enzymes, ascorbic acid and other vitamins, phenolic substances, and nucleic acids (Hooker,

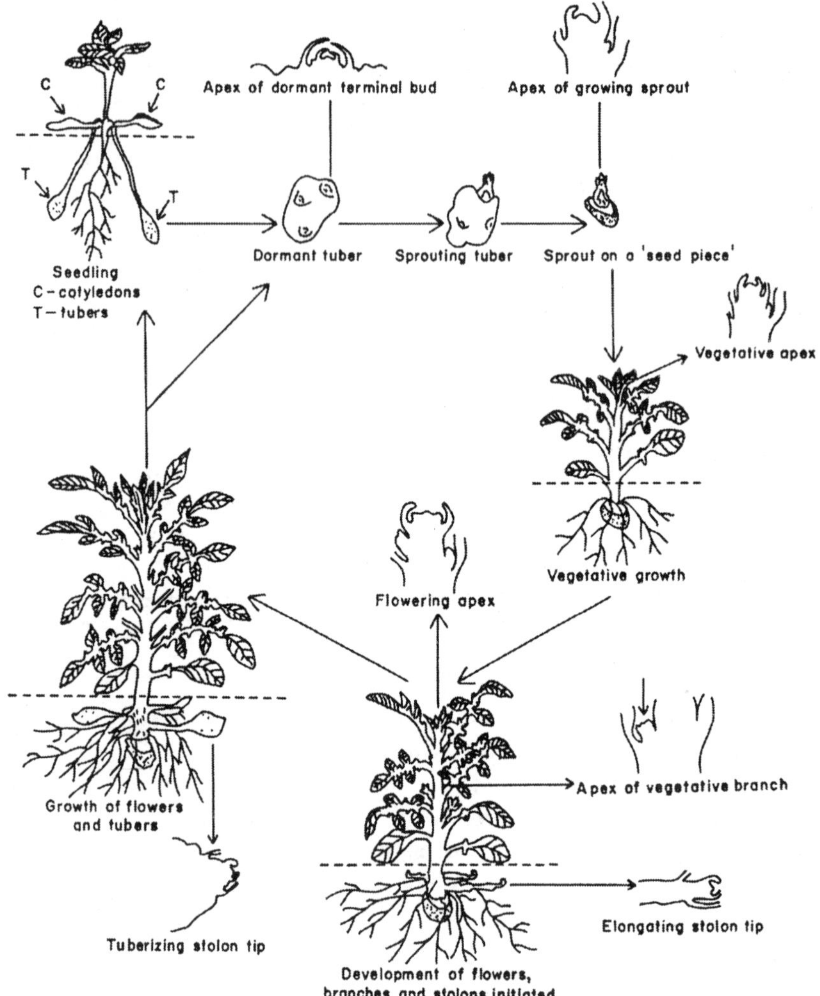

Figure 16.6 Diagrammatic representation of the development of a potato plant. Plantlets from seeds may form tuberizing stolons in the axils of cotyledons. Dormant tubers, as stored in the cold (6–7°C) after harvest, include several internodes with buds in the axils of scale leaves. Severed "seed pieces" produce sprouts from buds, and these give rise clonally to plants that develop first vegetatively then reproductively, as they form both sexual (i.e., flowers bearing seeds) and asexual (i.e., stolons and tubers bearing buds) reproductive organs. Throughout this cycle it is the shoot apices that constitute the main seat of the developmental events. (From Steward et al., 1981. $ Academic Press, reproduced with permission.)

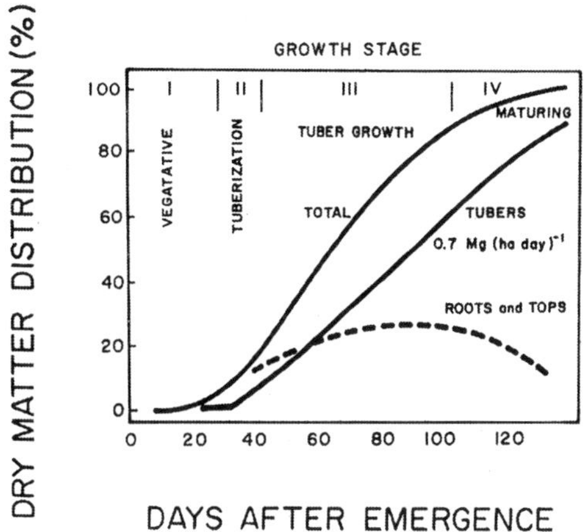

Figure 16.7 Dry matter distribution during four growth stages of russet Burbank potatoes. (From Kleinkopf et al., 1979.)

1986). The rate of growth of the tubers as a whole (the rate of bulking) is exponential for the first two to three weeks, but then becomes almost linear (Moorby and Milthorpe, 1973). Moorby (1968) reported maximum tuber growth rate of 25 g m^{-2} day^{-1} and 60 tubers m^{-2}.

Dry Matter

Dry matter production of potatoes is divided into three components: roots, tops, and tubers. Actually tubers are the principal organ which accumulates a major part of the total dry matter production. Figure 16.7 shows the distribution of dry matter in roots, tops, and tubers of potatoes at different growth stages. Dry matter production in growth Stages I and II is very small. Stage I, vegetative growth, is the period from emergence up to 30 days after emergence when tuberization is initiated (Kleinkopf et al., 1979). Stage II, in which early tuber bulking starts, covers 10–14 days following the vegetative stage when tubers form rhizome tips and there is some floral initiation. Most of the dry matter accumulation occurs in growth stage II, starting about 50 days after emergence. The tuber growth stage extends for 100 days or more (Lorenz et al., 1954). During growth Stage III, after floral initiation, photosynthate and minerals are rapidly translocated into the tubers (Haeder et al., 1973). In growth Stage IV, which is designated as maturing, the interval is marked by vine yellowing, leaf senescence, and slowed growth rate. This stage begins about 100 days after emergence and continues to the end of the growing

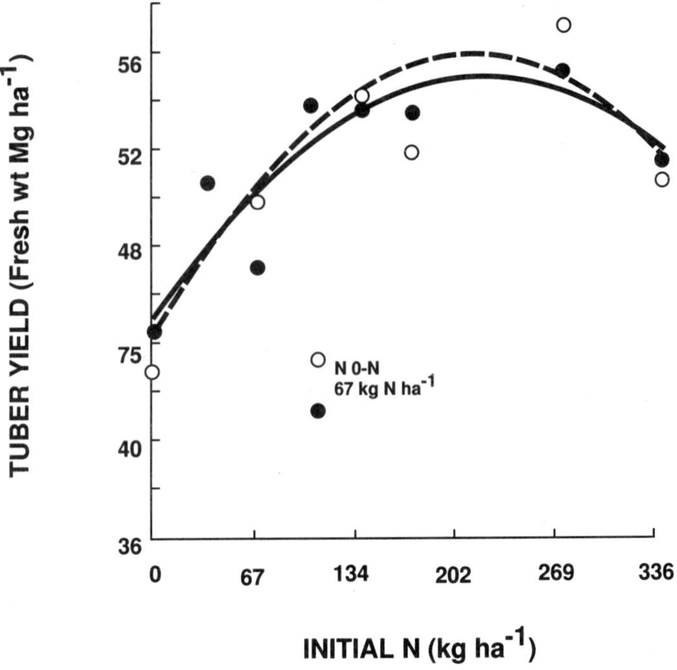

Figure 16.8 Potato yield response to initial application rates of NH_4NO_3–N for 2 years. Open circles and dashed lines represent treatments receiving additional application of 67 kg N ha^{-1} 21 days after tuber initiation. (From Westcott et al., 1991.)

season (Roberts and Dole, 1985). The most important factor which determines dry matter production is photosynthetic efficiency of the crop, which varies with cultivar, environmental conditions, and leaf area index (*LAI*). Generally a *LAI* of 4–5 might be sufficient for higher tuber yields (Haeder and Beringer, 1983).

The dry matter accumulation curve against time (Figure 16.7) shows very clearly the dominant role of tubers as a sink for assimilate over the greater part of the growing season.

C. Nutrient Requirements

A fair assessment of the nutrient requirement of a crop is the crop nutrient response curve. If a given soil is deficient in a nutrient, crop response to addition of the limiting nutrient is expected, provided other growth factors are not limiting. To determine the response curve, fertilizer trials are needed for each type of soil under each agro-ecological region. Figure 16.8 shows potato yield response to N application rates. The yield response was quadratic in nature and the highest yield was obtained with about 200 kg N ha^{-1}. Detailed information about mineral requirements of potatoes can be obtained from articles by Smith (1949), Harris (1978c), and Roberts and Dole (1985).

Nutrient Concentration

Considerable progress has been made in the use of plant analysis in determining nutrient requirements of crop plants, including potatoes. The fourth petiole of the most recently matured leaf from the growing tip is usually the plant part used for nutrient analysis in potatoes (Roberts and Dow, 1982). The evaluation of a plant's nutritional status is based upon a significant relationship between the nutrient in question and plant yields. This relationship is called a nutrient response curve and can identify nutrient concentrations that are deficient, adequate, and excessive (Westermann and Kleinkopf, 1985). Based on this concept, adequate levels of various nutrients in potato plant tissue are presented in Table 16.5.

Nutrient Accumulation

Information on nutrient accumulation by the potato plant is helpful in developing fertilizer programs (Roberts and Dole, 1985). The quantity of nutrients removed in the harvested portion of the crop will depend on the yield and the concentration of the nutrients, and both can vary from place to place and year to year. Table 16.6 shows the amount of nutrients removed by tops and tubers of potatoes. Nitrogen and potassium were removed in large quantities, both by tops and tubers. This means these two elements are most likely to be deficient in normal agricultural soils and must be supplied in adequate quantities to obtain higher yields.

IV. SUMMARY

Cassava and potato are the two most economically important tuber crops. These two crops are an important human food source in developed as well as developing countries. Both these crops have a high yield potential, but current average yields are much below this potential productivity. Factors which are responsible for low yield, especially in the developing countries, are water stress, low soil fertility, and higher incidence of diseases and pests. To overcome these constraints, a multidisciplinary research approach is necessary to develop a suitable technology.

Table 16.5 Adequate Nutrient Concentrations in Potato Plant

Nutrient	Growth stage	Plant part	Sufficiency level
			$g\ kg^{-1}$
N	42 DAE[a]	UMB + P[a]	40–50
	Early flowering	UMB + P	55–65
	Tubers half grown	UMB + P	30–50
P	42 DAE	UMB + P	2–4
	Early flowering	UMB + P	3.5–5.5
	Tubers half grown	UMB + P	2–4
K	42 DAE	UMB + P	35–50
	Early flowering	UMB + P	45–65
	Tubers half grown	UMB + P	40–80
Ca	42 DAE	UMB + P	6–9
	Early flowering	UMB + P	10–20
	Tubers half grown	UMB + P	15–25
Mg	42 DAE	UMB + P	8–11
	Early flowering	UMB + P	3–5
	Tubers half grown	UMB + P	5–8
S	Early flowering	MSL	3–5
	Tubers half grown	PUMB	1.9–3.6
			$mg\ kg^{-1}$
Fe	Tubers half grown	UMB + P	70–150
Mn	Tubers half grown	UMB + P	30–450
Zn	Tubers half grown	UMB + P	20–40
B	Tubers half grown	UMB + P	30–40
Mo	Early flowering	MSL	0.1–1.5

[a] DAE, Days after emergence; UMB + P, uppermost mature leaf blade + petiole; PUMB, petiole of uppermost mature leaf blade; MSL, midstem leaves.
Source: Compiled from Piggott, 1986.

Table 16.6 Nutrient Accumulation in Tops and Tubers for Russet Burbank Potatoes Yielding 79 tons ha^{-1} in an Irrigated Field of Central Washington

	Nutrient Accumulation ($kg\ ha^{-1}$)[a]		
Nutrient	Tops	Tubers	Total
N	140	282	422
P	7	40	47
K	200	320	520
Ca	60	12	72
Mg	34	16	50
Zn	3	3	6

[a] Nutrient accumulation corresponds to 95 days after emergence.
Source: Dow and Cline, 1980.

REFERENCES

Abruna, F., J. Vicente-Chandler, and J. Badillo. 1982. Effect of soil acidity components on yield and foliar composition of tropical root crops. Soil Sci. Soc. Am. J. 46: 1004–1007.

Allem, A. C. 1994. The origin of *Manihot esculenta* Crantz (*Euphorbiaceae*). Genetic Resources and Crop Evaluation 41: 133–150.

Benoit, G. R., C. D. Stanley, W. J. Grant, and D. B. Torrey. 1983. Potato top growth as influenced by temperatures. Am. Potato J. 60: 489–501.

Benoit, G. R., W. J. Grant, and O. J. Devine. 1986. Potato top growth as influenced by day-night temperature differences. Agron. J. 78: 264–269.

Burton, W. G. 1972. The response of the potato plant and tuber to temperature, pp. 217–233. *In*: A. R. Rees et al. (eds.). The crop processes in controlled environments. Academic Press, London.

Campos, H. R. and Z. F. Sena. 1974. Root depth of cassava (*Manihot esculenta* Crantz) as a function of plant age. Universidade Federal da Bahia. Escola de Agronomia, Cruz das Almos, Bahia, Brazil.

Chapman, S. R. and L. P. Carter. 1976. Potatoes, pp. 431–441. *In*: Crop production: principles and practices. W. H. Freeman and Company, San Francisco.

CIAT (Centro Internacional de Agricultura Tropical). 1979. Annual report of the CIAT, Cali, Colombia.

Cock, J. H. 1976. Characteristics of high yielding cassava varieties. Exp. Agric. 12: 135–143.

Cock, J. H. 1983. Cassava, pp. 341–359. In: IRRI (ed.) Potential productivity of field crops under different environments. IRRI, Los Banos, Philippines.

Cock, J. H. 1984. Cassava, pp. 529–549. *In*: P. R. Goldsworthy and N. M. Fisher (eds.). The physiology of tropical field crops. Wiley, New York.

Cock, J. H., D. W. Wholey, and O. G. Casas. 1977. Effects of spacing on cassava (*Manihot ecsulenta* Crantz). Exp. Agric. 13: 289–299.

Cock, J. H. and R. H. Howeler. 1978. The ability of cassava to grow on poor soils, pp. 145–154. *In*: G. A. Jung (ed.) Crop tolerance to suboptimal land conditions. Am. Soc. Agron. Special Publication No. 32, Madison, Wisconsin.

Cock, J. H., D. Franklin, G. Sandoval, and P. Juri. 1979. The ideal cassava plant for maximum yield. Crop Sci. 19: 271–279.

Connor, D. J., J. H. Cock and G. H. Parra. 1981. The response of cassava to water shortage. I. Growth and yield. Field Crops Res. 4: 181–200.

Cook, J. H. and C. R. Rosas. 1975. Ecophysiology of cassava, pp. 1–14. *In*: Symposium on ecophysiology of tropical crops. Communication Division of CEPLAC, Ilheus-Itabuna, Bahia, Brazil.

Cutter, E. G. 1978. Structure and development of the potato plant, pp. 70–152. *In*: P. M. Harris (ed.) The potato crop: the scientific basis for improvement. Chapman and Hall, London.

Dow, A. I. and T. A. Cline. 1980. Growth curve, yield, nutrient removal and petiole nutrient levels of potatoes grown on sandy soils under three center pivot irrigation systems, pp. 61–74. *In*: Proc. of the 31st Annual Northwest Fertilizer Conf., Salt Lake City. July 14–16. Northwest Plant Food Association, Portland, Oregon.

Enyi, B. A. C. 1972. The effects of spacing on growth, development and yield of single and multi shoot plants of cassava (*Manihot esculenta* Crantz). I. Root tuber yield and attributes. East Afr. Agric. For. J. 38: 23–26.

Gent, M. P. N. and H. Z. Enoch. 1983. Temperature dependence of vegetative growth and dark respiration: A mathematical model. Plant Physiol. 71: 562–567.

Gray, D. and F. C. Hughes. 1978. Tuber quality, pp. 504–544. *In*: P. M. Harris (ed.) The potato crop: the scientific basis for improvement. Chapman and Hall, London.

Hang, A. N. and D. E. Miller. 1986. Yield and physiological responses of potato to deficit, high frequency sprinkler irrigation. Agron. J. 78: 436–440.

Haeder, H. E., K. Mengel, and H. Forster. 1973. The effect of potassium on translocation of photosynthates and yield pattern of potato plant. J. Sci. Food Agric. 24: 1479–1487.

Haeder, H. E. and H. Beringer. 1983. Potato, pp. 307–317. *In*: IRRI (ed.) Potential productivity of field crops under different environments. IRRI, Los Banos, Philippines.

Harris, N. V. 1978a. The potential of cassava in coastal Queensland: Some observations at the Yandaran plantation, pp. 76–79. *In*: Proc. Conf. Alcohol Fuels. Int. of Chemical Engineers, N.S.W., Sydney.

Harris, P. M. 1978b. Water, pp. 244–277. *In*: The potato crop: the scientific basis for improvement. Chapman and Hall, London.

Harris, P. M. 1978c. Mineral nutrition, pp. 195–243. *In*: The potato crop: the scientific basis for improvement. Chapman and Hall, London.

Haun, J. R. 1975. Potato growth environment relationship. Agric. Meteorol. 115: 325–332.

Hawkes, J. G. and J. F. Ortega. 1993. The early history of the potato in Europe. Euphytica 70: 1–7.

Hooker, W. J. 1986. The potato, pp. 1–6. *In*: Compendium of potato diseases. Am. Phytopathological Society. St. Paul, Minnesota.

Howeler, R. H. 1981. Mineral nutrition and fertilization of cassava. CIAT, Cali, Columbia. 52 p.

Howeler, R. H. 1985. Potassium nutrition of cassava, pp. 819–841. *In*: R. D. Munson (ed.). Potassium in agriculture. Am. Soc. Agron., Madison, WI.

Howeler, R. H. and L. F. Cadavid. 1983. Accumulation and distribution of dry matter and nutrients during a 12–month growth cycle of cassava. Field Crops Res. 7: 123–139.

Iglesias, C. A., F. Calle, C. Hershey, G. Jaramillo, and E. Mesa. 1994. Sensitivity of cassava (*Manihot esculenta* Crantz) clones to environmental changes. Field Crops. Res. 36: 213–220.

Irizarry, H. and E. Rivera. 1983. Nutrient and dry matter contents of intensively managed cassava grown on an Ultisol. J. Agric. Univ. Puerto Rico 67: 213–220.

Janick, J., R. W. Schery, F. W. Woods, and V. W. Ruttan. 1974, pp. 455–500. *In*: Plant science: An introduction to world crops. 2nd Ed. W. H. Freeman and Company, San Francisco.

Jose, J. J. S. and F. Mayobre. 1982. Quantitative growth relationships of cassava (*Manihot esculenta* Crantz): Crop development in a savanna wet season. Ann. Bot. 50: 309–316.

Kawano, K. 1978. Genetic improvement of cassava (*Manihot esculenta* Crantz) for productivity. Trop. Agric. Res. Ser. 11: 9–21.

Keating, R. A., J. P. Evenson and S. Fukai. 1982. Environmental effects on growth and development of cassava (*Manihot esculenta* Crantz). I. Crop development. Field Crops Res. 5: 271–281.

Kleinkopf, G. E., D. T. Westermann, and C. G. Painter. 1979. Nitrogen effects and russet Burbank potato growth, pp. 143–150. *In*: Proc. 30th Annual Northwest Fertilizer Conf., Spokane, Washington. July 17–19. Northwest Plant Food Association, Portland, Oregon.

Lahav, E. and T. Trochoulias. 1982. The effect of temperature on growth and dry matter production of avocado plants. Aust. J. Agric. Res. 33: 549– 558.

Loon, C. D. Van. 1981. The effect of water status on potato growth, development and yield. Amer. Potato J. 58: 51–69.

Lorenz, O. A., J. C. Bishop, B. J. Hoyle, M. P. Zobel, P. A. Minges, L. D. Doneen, and A. Ulrich. 1954. Potato fertilizer experiment in California. Calif. Agric. Exp. Stn. Bull. 744.

Maas, E. V. and G. J. Hoffman. 1977. Crop salt tolerance-current assessment. J. Irrig. and Drainage. Div. Am. Soc. Civil Eng. 103: 115–134.

Manrique, L. A. 1992. Potato production in the tropics: Crop requirements. J. Plant Nutr. 15: 2679–2728.

Marinus, J. and K. B. A. Bodlaender. 1975. Response of some potato varieties to temperature. Potato Res. 103: 115–134.

McLean, E. O. and J. R. Brown. 1984. Crop response to lime in the midwestern United States, pp. 267–303. *In*: F. Adams (ed.) Soil acidity and liming, 2nd Ed. Am. Soc. Agron. Monograph 12, Madison, Wisconsin.

Moorby, J. 1968. The influence of carbohydrates and mineral nutrient supply on the growth of potato tubers. Ann. Bot. 32: 57–68.

Moorby, J. and F. L. Milthorpe. 1973. Potato, pp. 225–257. *In*: L. T. Evans (ed.) Crop physiology: some case histories. Cambridge University Press, London.

Piggott, T. J. 1986. Vegetable crops, pp. 148–187. *In*: D. J. Reuter and J. B. Robinson (eds.) Plant analysis: an interpretation manual. Inkata Press, Melbourne.

Purseglove, J. W. 1987. Tropical crops: Dicotyledons. Longman Scientific and Technical, New York.

Reuter, D. W. 1986. Temperate and subtropical crops, pp. 38–99. *In*: D. J. Reuter and J. B. Robinson (eds.) Plant analysis: an interpretation manual. Inkata Press, Melbourne.

Roberts, S. and A. J. Dow. 1982. Critical nutrient ranges for petiole phosphorus levels of sprinkler irrigated Russet Burbank potatoes. Agron. J. 74: 583–585.

Roberts, S. and R. E. Dole. 1985. Potassium nutrition of potatoes, pp. 799–818. *In*: R. D. Munson (ed.) Potassium in agriculture. Am. Soc. Agron., Madison, Wisconsin.

Sangoi, L. and N. D. Kruse. 1993. Accumulation and distribution of dry matter of cassava on the highlands of Santa Catarina, Brazil. Pesq. Agropec. Bras. Brasilia 28: 1151–1164.

Smith, O. 1949. Potato production. Adv. Agron. 1: 353–390.

Steward, F. C., V. Moreno, and W. M. Roch. 1981. Growth, form and composition of potato plants as affected by environment. Ann. Bot. 48: 1–45.

Ulrich, A. 1993. Potato, pp. 149–156. *In*: W. F. Bennett (ed.) Nutrient deficiencies and toxicities in crop plants. APS Press, The Amer. Phytopath. Soc. St. Paul, Minnesota.

Van Loom, C. D. 1981. Effect of water stress on potato growth, development and yield. Am. Potato J. 58: 51–69.

Westcott, M. P., V. R. Stewart, and R. E. Lund. 1991. Critical petiole nitrate levels in potato. Agron. J. 83: 844–850.

Westermann, D. T. and G. E. Kleinkopf. 1985. Phosphorus relationships in potato plants. Agron. J. 77: 450–494.

Williams, C. N. 1972. Growth and productivity of tapioca (*Manihot utilissima*). III. Crop ratio, spacing and yield. Exp. Agric. 8: 15–23.

17

Cotton

I. INTRODUCTION

There are four domesticated species of cotton: *Gossypium hirsutum* L., *G. barbadense* L., *G. herbaceum* L., and *G. arboreum* L. *Gossypium hirsutum,* known most widely as upland cotton, contributes about 90% of the current world production of 65 million bales of fiber weighing about 218 kg/bale (Lee, 1984). Lint length of *G. hirsutum* ranges from slightly less than 2.5 cm to 3 cm and is classified as medium to long staple. *Gossypium barbadense* lint length is 3 cm or over and classified as extra-long staple. *Gossypium herbaceum* and *G. arboreum* are the short-staple cotton, and lint length is less than about 2.2 cm (Phillips, 1976). *Gossypium hirsutum* is the most widely grown species because of its relatively high productivity and wide adaptability. Though cotton species are basically tropical perennials, cotton is grown as an annual crop in both temperate and tropical regions. The major *G. hirsutum*-producing countries are the United States and the USSR. Extra-long staple production is dominated by Egypt, Sudan, and the USSR, while India, Pakistan, and China produce virtually all the short-staple cotton (Phillips, 1976). Although cotton is mostly grown for fiber, the seeds are also important.

Cottonseed oil is used for culinary purposes, and the oil cake residue is a protein-rich feed for ruminant livestock.

II. CLIMATE AND SOIL REQUIREMENTS

Cotton is a warm-weather plant, and the cultivated species are not tolerant of freezing temperatures. Even so, production is not limited to the tropics. Adapted cultivars and effective production techniques permit successful culture in regions where the frost-free period is less than 180 days (Niles and Feaster, 1984). Cotton is grown virtually around the world in tropical latitudes and as far north as 43°N in the USSR and 45°N in the People's Republic of China.

Ancestors of commercial cotton varieties are of tropical origin and are naturally adapted to growth in hot environments. Derived cultivars retain the high optimal temperatures for growth characteristic of their progenitors. For example, the optimal temperature for photosynthesis in the cultivar Deltapine Smooth Leaf is 25°C, with a rapid decline at lower temperatures (Downton and Slatyer, 1972). Temperature is a primary environmental factor controlling plant growth and development rates, and it often determines the rate of seed germination, seedling establishment, time of flowering and fruit maturation, and growth throughout the life of the plant (Reddy et al., 1992).

Growth of cotton is very sensitive to suboptimal temperatures at all stages of development (Warner et al., 1993). Temperatures below 20°C result in reduced root and shoot growth, increased days to first bloom, decreased rate of fiber elongation, and decreased fiber quality (Gipson, 1986). Reduced temperatures over the range of 25°C down to 5°C delay boll development in Paymaster and Acala cultivars (Gipson and Joham, 1968). In many temperate regions of the United States, cotton grown for fiber production experiences suboptimal day and night temperatures throughout much of the growing season. It is important to note that even moderately cool temperatures at night alone limit growth and production. On the High Plains of Texas, irradiance and temperatures are optimal for cotton growth during the day, but at night temperatures are typically 15–20°, and biomass production is reduced (Burke et al., 1988).

The optimum temperature range for biochemical and metabolic activities of plants has been defined as the thermal kinetic window (TKW) (Burke et al., 1988). Plant temperatures either above or below the TKW result in stress that limits growth and yield. The TKW for cotton growth is 23.5–32°C, with an optimum temperature of 28°C, and biomass production is directly related to the amount of time that foliage temperatures are within the TKW (Burke et al., 1988; Warner and Burke, 1993).

Previous research has addressed some of the environmental components that can influence fiber quality (Pettigrew, 1995). For example, temperature influences both fiber elongation and secondary wall deposition. Gipson and Joham (1969) reported that the optimum range of night temperature for maximum fiber length was 15–21°C, with length reductions occurring for night temperatures outside this range. They also found that micronaire, an estimate of secondary wall deposition, was reduced at night temperatures lower than 25°C.

Taproot length of 10–day-old cotton seedlings grown in soil at 20°C was 50% of the length of cotton grown at 25–30°C, and there was no lateral root development at 20° (McMichael and Quisenberry, 1993). Cotton taproot length at 17°C was less than 40% of the length at 27°C, and root dry weight at 17°C was 68% of that at 27°C (Loffroy et al., 1983). This differential response of root system components at suboptimal temperatures changes the partitioning of growth substrates between plant shoot and root systems. In a review of soil temperature effects on root growth, Kaspar and Bland (1992) concluded that soil temperature often limits the rate of rooting depth increase and the maximum attainable depth by crops grown in temperate regions. Cotton root length at 80 h after radicle emergence increased nonlinearly to a maximum at 32°C (Pearson et al., 1970).

Available moisture is another key factor affecting growth and yield of cotton. A minimum of approximately 58 cm of moisture is needed to produce a crop of 0.75 bales ha^{-1} (Waddle, 1984). Proper distribution of rainfall during crop growing season is also important. The water use efficiency of harvested product averages 0.2 kg lint m^{-3}, but it can reach 0.4 kg lint m^{-3} (Hearn, 1979).

Vegetative growth is highly sensitive to water deficits. Cell photosynthesis and cell division continue at water potentials low enough to inhibit cell enlargement. In general, there is linear decline in leaf expansion with decreasing leaf water potential from –1.2 to –2.4 MPa and a linear decline in photosynthesis below –2.0 MPa leaf water potential (Ackerson et al., 1977; Sung and Krieg, 1979; Karami et al., 1980). As a result, leaf expansion is more sensitive to drought stress than photosynthesis. Boll growth is even more tolerant, and it is not affected until leaf water potential reaches –2.7 to –2.8 MPa (Grimes and Yamada, 1982). Hearn and Constable (1984) proposed the following classification of water stress based on leaf water potential (LWP): LWP > –1.5 MPa, minimal stress; LWP –1.5 to –2.0 MPa, mild stress; LWP –2.0 to –2.5 MPa, moderate stress; and LWP < –2.5 MPa, severe stress.

Water deficit may also lead to changes in the concentrations of several plant hormones known to induce leaf and boll abscission. With a boll load on the plant, stress at peak or late flowering increases boll abscission and square loss. Grimes et al. (1970) reported that the combined loss of bolls and squares due to severe stress at peak blooming caused a 32% yield loss. Severe stress either early or late in the blooming period gave a 20% yield loss. Therefore, severe stress should be avoided during this most sensitive period.

According to Grimes et al. (1969), maximum seasonal evapotranspiration (ET) of cotton in the San Joaquin Valley of California reached a maximum of 72 cm when 110 cm of irrigation was applied. Yield of lint was maximum at an ET value of 68 cm. Peak daily rates of soil moisture depletion were 1.07 cm, occurring near the time when maximum leaf area was obtained. In the southern desert of California and in Arizona, actual ET for maximum production is considerably higher. For example, at Mesa, Arizona, seasonal ET is 105 cm (Halderman, 1973), reflecting both a longer growing season and higher daily use rates. A similar seasonal ET is observed in the Imperial Valley (Grimes and El-Zik, 1982).

Cotton is very sensitive to waterlogging. Hodgson (1982) found that inundation during surface irrigation reduced numbers of bolls and yield by up to 20%. The reduction was correlated with duration of anaerobiosis. Three hours of anaerobiosis kills cotton roots, compared with 5 hours for soybeans (Huck, 1970).

High light intensities throughout the growing period are essential for satisfactory vegetative development, for minimal shedding of buds and bolls, and hence for high yields (Arnon, 1972). Reducing light intensity to one-third of normal reduced carbohydrate content of the leaves by 24%, of the stems by 38%, and of the bolls by 8%, the cotton yield being reduced by 47% (Eaton and Ergle, 1954).

Cotton can be grown on a variety of soils from light sandy soils to heavy alluvium and clays. Aeration, moisture, temperature, and a supply of nutrients are important soil factors affecting yield. Cotton is considered as moderately acid tolerant, and the critical pH range is 5.5–6.0 (Adams, 1981). The upper pH limit for cotton growth is about 8.0. The main criterion for suitability of soils within this pH range appears to be a depth of at least 0.6 m and freedom from prolonged waterlogging (Hearn, 1981).

Cotton is considered a salinity-tolerant crop, and salinity threshold (initial yield decline) is about 7.7 dS m^{-1} (Maas and Hoffman, 1977), and a 50% reduction in yield occurred at an EC_e of 17 ds m^{-1} (Ayers, 1977).

III. GROWTH AND DEVELOPMENT

Knowledge of plant structure and growth and development of cotton is essential to improve management practices and yield. The primary factors affecting growth are cultivar, climate, soil, pests, and cultural practices. Under favorable conditions, development of *G. hirsutum* follows a rather well-defined phenological pattern (Grimes and El-Zik, 1982). Table 17.1 presents the

Table 17.1 Phenology of Cotton (*G. hirsutum*) Cultivars (Acala SJ-2 and SJ-5)

Growth stage	Calendar days		Degree days (base 15.6°C)
	Range	Average	
Sowing to emergence	5–20	10	50
Emergence to square	40–60	50	450
Square to bloom	20–27	23	330
Bloom to open boil	45–80	58	950
Normal crop production	190–210	200	> 2800

Sources: El-Zik and Sevacherian, 1979; El-Zik et al., 1980.

phenology of Acala SJ-2 and Acala SJ-5 varieties and shows the range and average calendar days and degree days (base 15.6°C) needed for each growth stage. The concept of degree days utilizes temperature rather than calendar days in describing growth and development of an organism was discussed by Toscano et al. (1979).

A. Germination and Emergence

Poor germination and seedling establishment often limit cotton plant populations and yields. Good-quality seed is the first prerequisite for ideal plant population. The value of cotton seed used for planting is determined by germination and vigor tests. Most, if not all, states in the United States require that cotton seed sold for planting purposes exhibit 80% or higher germination at 30°C, as defined by the Association of Official Seed Analysts (AOSA, 1970). Seedling vigor is defined as the potential for rapid, uniform emergence and development of normal seedlings under a wide range of field conditions (McDonald, 1980).

Cotton seed germination is complicated by water uptake and chilling problems. The cotton seed is covered by a dense mat of fuzz fibers (linters), by a heavy seed coat with a chalazal cap-opening mechanism, and by a thin layer of endosperm (McArthur et al., 1975). An impermeable seed coat and impenetrability of seed coat fibers by water are the two major factors inhibiting cotton seed germination. It is generally recommended that fuzzy seeds should be delinted by machine, flame, or sulfuric acid (Cherry and Leffler, 1984). Soil crusting also adversely affects germination of cotton seeds.

Under favorable conditions, seedlings emerge 5–15 days after sowing. The rates of elongation of the radicle and hypocotyl are controlled by temperature and soil water potential. Wanjura et al. (1970) found that the temperature limits were 14–42°C with an optimum 34°C. At optimum temperatures, seedlings emerged in 5 days at –0.03 MPa, in 7 days at –0.3 MPa, and did not emerge in 13 days at –1.0 MPa (Wanjura and Buxton, 1972).

B. The Root System

Because the radicle is more massive in the seed than the epicotyl, the root develops more rapidly during germination. During the period before true leaves begin expansion, the primary root is penetrating deeply into the soil and branch roots are being formed (Mauney, 1984).

The cotton plant has a primary taproot with many lateral roots. The main absorption and anchoring structure of the cotton plant is the mass of roots that branch from the taproot. The depth of primary root penetration depends on soil, climate, and genotype. Borg and Grimes (1986) calculated the rooting depth of cotton based on various root growth data in the literature and concluded that maximum root depth of cotton was 150–300 cm under favorable conditions. Root length density and dry weight decrease exponentially with depth. Depletion of water in the upper profile can lead to proliferation of roots at depth, resulting in greater extraction of water at depth (Klepper et al., 1973). The general development of the cotton root system was described by Taylor and Klepper (1978), Grimes and El-Zik (1982), and McMichael (1986).

Root growth is a dynamic process responding to signals from both the soil environment and the shoot. Abiotic and biotic stresses reduce root growth, ultimately reducing shoot development and realization of yield potentials (Burke and Upchurch, 1995). Abiotic factors known to alter root growth include water, oxygen and nutrient availability, temperature, soil structure, and light penetration (Bowen, 1991). Genetic diversity in root growth between and within species has been identified, and the effect of environmental stimuli on the realization of these differences has been studied (Bland, 1993; Engels, 1994). Temperature is a universal factor for regulating cellular growth and development. Root elongation rates show distinct temperature sensitivities between species and have been used to identify optimal temperature ranges for root development. A better understanding of thermal limitations to root development can be obtained by comparing the temperature optimum for root development with spatial and temporal temperature fluctuations observed in the field during the growing season. Temperature influences not only root proliferation, but also root function. Changes in hydraulic conductivity associated with changes in root temperature have been found (Bolger et al., 1992). Altered nutrient uptake occurs in response to changing temperatures, and changes in the synthesis and transport of regulatory hormones have been shown (Clark and Reinhard, 1991; Engels, 1994).

C. Stem and Leaves

The main stem of the cotton plant is monopodial, with leaves and branches but no flowers. There are usually two axillary buds at each main stem node, the second branching off the first. Normally, only one bud develops. At lower nodes the first bud remains vegetative and may develop into a vegetative branch of monopodium which is a replica of the main stem (Hearn and Constable, 1984).

Usually the vegetative branches occur in a definite zone near the base of the plant, and the fruiting branches occur farther up the stem. The number of nodes from the base of the main stalk to the first fruiting branch varies considerably among cotton species and is affected by cultural practices.

Leaves of most American cotton varieties have five more or less clearly defined lobes. The blade is usually large, thin, and relatively hairy, although smooth leaf types exist. Its surface contains many stomata through which gases are exchanged between the plant and the surrounding atmosphere. Most of the stomata are on the underside of the leaf. The petiole is usually about as long as the leaf blade. At the point where it joins the stem or branch, it is flanked by two small stipules (Tharp, 1965).

D. Fruit Development

The development of cotton fruit involves a complex series of events and interactions that begin with the appearance of the first flower bud (square) and continue until the boll opens. Pollination of a cotton flower generally takes place on the first day the flower opens. The majority of the flowers are selfpollinated, but cross-pollination occurs if bees and other insects are active. Normally, cross-pollination ranges from 6 to 25%, but over 50% has been reported (Purseglove, 1982). After anthesis, the fruit, called a boll, develops from a superior ovary consisting of 3–5 united carpels into a tough-walled capsule. Internally it is divided into 3–5 loculi, each of which initially contains 10–12 ovules, 5–11 of which develop into coat seeds covered with fiber (Hearn and Constable, 1984). The contents of a mature loculus form a lock of seed cotton. The number of ovules per boll is influenced by genetic and physiological factors, especially nitrogen supply (Hughes, 1966). It is natural for some squares, blooms, and small bolls to be shed. Shedding is increased by such factors as moisture deficiency, an inadequate number of fertilized ovules, an insufficient nutrient supply, excessive heat or cold, and damage from insects and diseases (Grimes and El-Zik, 1982).

Flower and fruit (boll) retention is high early in the reproductive phase but usually decreases with increasing fruit load. Pronounced decreases in vegetative growth, flowering, and boll retention are commonly referred to as

Table 17.2 Developmental Events in Relation to Days Before and After Anthesis

Age	Events
–40	Floral stimulus.
–32	Carpel and anther number established.
–23	Ovule number established.
–22	Pollen mother cell meiosis; “pin-head square.”
–14	Megaspore mother cell meiosis.
–7	Begin exponential expansion of corolla.
–3	Begin fiber differentiation.
0	Flower open; pollen shed, germinates; fiber initiation; K accumulation in fiber.
+1	Fertilization of egg and polar nuclei; division of primary endosperm nucleus; zygote shrinks.
+2	Liquid endosperm developing; fibers begin elongating; most dry mass goes to fibers.
+3 to 4	Zygote divides.
+5 to 6	Ovule integument division stops; fuzz fibers uninitiated; globular embryo dividing but not increasing in size; ovule enlargement stage proceeding rapidly; dry mass to internal parts increases.
+12 to 13	Endosperm becomes cellular around embryo; palisade cells elongate; embryo differentiation begins.
+14 to 16	Secondary deposition in fibers, outer integument and palisade begins; embryo elongating, accumulates Ca and Mg; fibers begin slow accumulation of Ca; outer integument begins rapid weight increase.
+20	Endosperm completely cellular and at maximum weight; fiber elongation slows rapidly; P translocated from fiber; embryo begins accumulating protein; weight distribution about equal between fiber and embryo.
+25	Fiber elongation complete; bur weight maximum; cotyledons complete; embryo maximum length; endosperm declining; oil accumulation starts.
+30 to 32	Embryo enters period of grand weight gain; endosperm nearly depleted; maximum rate of cellulose deposition in fiber, oil, and protein in embryo; rapid P and K accumulation in embryo; fibers begin losing K.
+42	Dry weight of boll nearly maximum; some oil accumulation; fibers lose Mg; cellulose deposition stops.
+45 to 50	Internal changes in seed hormones and enzymes; seed coat hardens; boll sutures dehisce in response to ethylene.

Source: Stewart, 1986.

cutout (Patterson et al., 1978). If cutout occurs before the end of the growing season, yield may be below what it would have been if the crop had utilized the entire growing season. Conversely, cutout at the end of the season is desirable because it facilitates harvest and deprives insect pests of a food source before they enter diapause (Kittock et al., 1973). Fruit load appears to be a major factor affecting cutout. As plants become loaded with bolls, growth and flowering rates slow and boll retention decreases (Patterson et al., 1978). These effects could result from competition for photosynthate, a change in hormonal status, or both. Guinn (1982) suggested that a nutritional stress increases boll shedding (an important aspect of cutout) through an increase in ethylene production. Guinn (1985) also suggested that growth, flowering, and boll retention decrease when the demand for photosynthate increases and exceeds the supply. This means that an increase in photosynthesis should permit more bolls to be set before cutout.

Two types of fibers cover the seed surface of most cultivars of upland cotton (*G. hirsutum*): The long lint fibers that are used in the manufacture of fabrics and the short fuzzy fibers that form a dense mat near the surface of the seed (Quisenberry and Kohel, 1975). Each cotton fiber is formed from a single cell in the outer cell layer of the seed coat.

The growth and development of the cotton fiber consists of two phases. The first phase is a period of fiber elongation, or lengthening, and the second is a period of fiber thickening, or secondary wall development. After fiber initiation, the primary cell membrane of the long lint fibers, and to a lesser extent that of the short fuzzy fibers, elongates until final fiber length is reached (Quisenberry and Kohel, 1975). Fibers usually attain their full length within 18 days (15–25 days) after bloom. Cotton fiber is usually referred to as lint and is packed in bales weighing approximately 218 kg (Perkins et al., 1984). Lint generally forms 20–40% of the material transported to the gin, depending on whether the cotton is stripped or picked and on the type of harvester used.

An abbreviated summary of the sequence of events from flower induction to boll opening in relation to days before and after anthesis is presented in Table 17.2. The timing of events described in Table 17.2 is approximate and can change with environmental conditions.

E. Dry Matter Production

Dry matter production of cotton is influenced by climate, soil, and plant factors. Olson and Bledsoe (1942) determined the dry matter production of various growth stages of cotton grown on three different soil types. On all three soils, the largest percentage of dry matter was produced during the period from early boll formation to maturity. Dry matter production ranged from 5.6 to 10.9 t ha^{-1}. Oosterhuis et al. (1983) reported that about 60% of the total dry matter was produced between 10 and 16 weeks after sowing at two locations in Zimbabwe. Total dry matter production was 8.5 t ha^{-1} at maturity (150 days after sowing).

Most of the research devoted to cotton growth and development has centered on upland cotton rather than Pima cotton (*G. barbadense* L.) (Unruh and Silvertooth, 1996). Wells and Meredith (1984a,b,c) compared the growth of obsolete and modern upland cotton cultivars; they found that an increase in the number of harvestable bolls was the major component contributing to greater yield of modern upland cultivars (Wells and Meredith, 1984c). Wells and Meredith (1984b) found that modern upland cultivars produce larger lint yields primarily by: 1) a greater partitioning of dry matter to reproductive organs, and 2) an increased amount of reproductive development occurring when maximal leaf mass and area are present. Results of a later study (Meredith and Wells, 1989) suggested that further yield increases in upland cotton by conventional breeding methods would likely be achieved through continued partitioning of dry matter from vegetative to reproductive plant structures. Other studies have described dry matter accumulation by upland cotton under both dryland (Mullins and Burmester, 1990) and irrigated conditions (Bassett et al., 1970; Halevy, 1976).

When ground cover is complete, cotton growth rates range from 15 to 20 g m^{-2} day^{-1}. The growth rate of a square is exponential, its weight increasing from about 10 to 130 mg in the 4 weeks prior to flowering (Hearn and Constable, 1984). According to Mutsaers (1976), boll weight follows a sigmoid pattern, with a maximum growth rate of 0.28 g day^{-1} about 20 days after flowering,

At boll maturity, the total dry weight of a boll (4–10 g) is typically 18% boll wall and bracts, 48% seed, and 34% lint. Cotton seed is about 22% protein and about 22% oil (Hearn and Constable, 1984).

F. Leaf Area Index (*LAI*)

Experimental work carried out under field conditions in the high plains of Texas (USDA/ARS, 1988) showed that leaf area index of cotton increased in a sigmoid manner. Leaf area index values in excess of 6 developed on the sandy soil under 100% ETa replacement. The maximum *LAI* on the clay loam soil was slightly in excess of 3. Plants at both locations achieved maximum *LAI* in 70–80 days after sowing, a time which coincided with peak flower production. According to Hearn and Constable (1984), peak *LAI* occurs 3–5 weeks after the start of flowering and varies from 0.5 for a severely water-stressed crop to more than 6 for a well-fertilized and irrigated crop grown in a warm area.

G. Photosynthesis

The photosynthetic rate of a crop stand may be defined as the product of the amount of light intercepted and the efficiency of the intercepting tissue (Hesketh and Baker, 1967). Percent light interception depends on solar angle and stand geometry. Efficiency is often constant over considerable periods of time. However, most studies of cotton photosynthesis have been done under greenhouse or growth chamber conditions, and their rates may differ from rates measured under natural conditions (Bjorkman and Holmgrin, 1963).

The potential rate of net photosynthesis of individual cotton leaves is in the order of 130 mg CO_2 cm^{-2} s^{-1} (30 μmole CO_2 m^{-2} s^{-1}) for a recently fully expanded leaf on a plant well supplied with water and nutrients, at ambient CO_2levels, about 30°C leaf temperature, and at light saturation (quantum flux 2000 μE m^{-2} s^{-1} (Hearn and Constable, 1984). Hesketh (1967) showed that the photosynthetic rate of Acala B-54 cotton leaves was enhanced 38% by O_2-free air. This enhancement is due to the presence of photorespiration typical of C_3 plants. The CO_2 compensation point of cotton leaves is 70 μl l^{-1}, which indicates the presence of photorespiration (Benedict et al., 1972).

Growing cotton plants in atmospheres enriched with CO_2 indicates that yield can be regulated by photosynthesis (Mauney et al., 1978). Cotton plants grown at 620 mg kg^{-1} CO_2 showed an initial increase in CO_2 exchange rate of 65% (which later declined to 31%) as compared to plants grown at 330 mg kg^{-1} CO_2 (Benedict, 1984).

IV. YIELD AND YIELD COMPONENTS

The world cotton yield has increased rather consistently from 234 kg ha^{-1} in 1950 to an average of 424 kg ha^{-1} for the period 1976 through 1980 (Bowling, 1984). Production in the United States was about 294 kg ha^{-1} or slightly less, from around 1870 to 1935 (Lewis and Richmond, 1968). By 1968, yields of lint in the United States had reached 561 kg ha^{-1}, a twofold increase in 30 years (Lee, 1984). Since that time, the yield has leveled off. The increased yield per unit area may be attributed to increased irrigation as well as improved technology and production management practices. Breeding has played a significant role in modifying cotton plant structures and increasing pest resistance. The first step in yield improvement was the development of the annual habit in a perennial shrub. This was accompanied by loss of photoperiod sensitivity in most cases, lowering the nodal position of the first fruiting branch, and reduction in the amount of facultative bud shedding in wet weather (Hearn and Constable, 1984).

The yield components of cotton are bolls per plant, seed per boll, boll weight, and lint percentage. The number of bolls per plant is the most important yield variable. Both genotypic and environmental differences in cotton lint yield are often associated with the number of mature bolls per unit area rather than lint per boll (Wells and Meredith, 1984c; Heitholt et al., 1992; Heitholt, 1995). The number of mature bolls per unit area is a product of the number of flowers produced per unit area and the fraction of flowers that produce mature bolls (boll set or boll retention). Heitholt (1993) reported that high-yielding cultivars produced 16% more flowers (seasonal total number of flowers produced per unit area) than a low-yielding experimental line. Likewise, the high-yielding cultivars had higher boll retention (48%) than a low-yielding experimental line (42%). Cook and El-Zik (1993) reported that yield increases due to irrigation were associated with both increased flower production and boll retention.

Morrow and Kreig (1990) studied the cotton yield relationship with yield components as influenced by water and N supply. Lint yield (LY) was defined as a function of components including plant density, bolls per plant, and average boll size. Regression analysis was used to determine LY response to treatments. Lint yield was most highly correlated with boll number per unit ground area with equal contribution from plant density and bolls per plant. Water supply was most responsible for boll number; however, increasing N supply within each H_2O regime resulted in a positive response in boll number per plant. Multiple regression analysis revealed that LY responded to H_2O and N supplies during the fruiting period to a greater extent than to preflower supplies. Within any heat unit regime, LY was maximized as water supply increased by maintaining a constant ratio of 0.2 kg N ha^{-1} mm^{-1} H_2O.

Water, nutrients, insects, and diseases are major constraints on cotton productivity in major cotton-growing regions.

V. NUTRIENT REQUIREMENTS

Sound nutrition is one ingredient of high yield of field crops, including cotton. Nutrition affects the yield of cotton to a far greater extent than it affects lint quality, which is largely determined by genotype and weather (Hearn, 1981). Since cotton production covers a wide range of environments and economic circumstances, yields and hence nutritional requirements vary greatly.

Nutrition of the cotton plant is influenced by several characteristics that distinguish the nutrient requirements of cotton from those of other field crops. First, cotton is a tropical perennial shrub that is grown as an annual crop. Indeterminate growth habit and vegetative branching provide an infinite number of potential fruiting sites unless growth is limited by low temperatures, lack of water, insufficient supply of nutrients or carbohydrates derived from photosynthesis, or other limiting factors. Second, the cotton plant has a deep taproot system with unusually low root density in the surface soil layer where available nutrient levels are greatest. This rooting pattern makes the cotton plant more dependent on nutrient acquisition from subsoil than most other crop plants (Cassman, 1993). Diagnosis of nutrient disorders of cotton often requires evaluation of subsoil conditions, particularly where subsoils are low in potassium or calcium or toxic in aluminum. And third, unlike most annual field crops, it is the yield of lint (a cellulose fiber) rather than seed yield that determines the economic value of cotton. Deficiencies of certain nutrients reduce fiber quality as well as causing a reduction in plant growth and lint yield (Cassman, 1993).

Hearn (1981) summarized the relative frequency of nutrient deficiencies for different agroclimatic regions: N deficiency is very common; P and K deficiencies are common; Mg, S, Zn, B, and Mn deficiencies are occasional; deficiencies of Ca, Fe, and Cu are rare; and Mo and Cl deficiencies are unknown. Inadequate fertilization is one of the main reasons for low cotton yields, especially in developing countries. A sample survey on fertilizer practices in selected districts of various states in India showed that only 7–20% of the irrigated area received fertilizers. The average rate of N application was about 25 kg N ha-1, though the recommended rate was 56 kg N ha^{-1} (Seshadri and Prasad, 1979). Thus, there is considerable scope for increasing the cotton yield by improving fertilization.

Deficiency of essential nutrients reduces plant growth and yield. A deficiency of one group of nutrients (P, K, Ca, Mg, B, and Zn) limits fruit production to a greater extent than vegetative growth, whereas deficiency of a second group of nutrients (N, S, Mo, and Mn) restricts vegetative and fruiting growth to an equal extent (Benedict, 1984). Most of the nutrients in the first group may affect fruiting efficiency because they function in the control of carbohydrate translocation.

Potassium deficiency symptoms have recently been reported in increased frequency throughout the United States cotton production regions (Pettigrew et al., 1996). Potassium deficiency in cotton is responsible for yield reduction averaging 15–20% in California (Cassman et al., 1989) and is also a major concern of producers in the U.S. Southwest (Reeves and Mullins, 1995). This observation has coincided with the increased use of early-maturing and high-yielding genotypes, leading some researchers to speculate that genotype development has affected the K response (Oosterhuis et al., 1991). Because the root system of cotton is less dense than that of other crops (Gerik et al., 1987), the hypothesis has been offered that the rooting systems of these fast-fruiting genotypes were unable to supply the K requirements during boll growth because the bolls are a major sink for K (Leffler and Tubertini, 1976), and cotton root growth slows during boll development (Pearson and Lund, 1968).

Generally, fertilizer requirements are based on soil and/or plant analyses. These tests are usually specific to particular soils, climates, and yield levels and cannot be extrapolated easily to other areas. Therefore, in this section no attempt has been made to give detailed fertilizer recommendations. All cotton-growing states in the United States and essentially all cotton-growing countries of the world provide soil testing services for cotton growers (Melsted and Peck, 1973). However, there are some basic principles of mineral nutrition, nutrient uptake, and concentration that help to understand cotton nutrient requirements.

A. Nutrient Uptake

The soil-plant system is a dynamic system, and uptake of nutrients is influenced by several factors. It is very difficult to know the exact amounts of nutrients required by a crop. However, the amounts of nutrients removed in the economic yield are an indication of the crop's nutritional requirement. Nutrient uptake or accumulation is related to yield level. Nitrogen, phosphorus, and potassium are the nutrients removed in the greatest amounts. Cotton yielding 2.5 bales ha^{-1} (1 bale = 218 kg approximately) removes approximately 40 kg N ha^{-1}, 7 kg P ha^{-1}, 14 kg K ha^{-1}, 4 kg Mg ha^{-1}, and 3 kg Ca ha^{-1} (Berger, 1969). With 3.75 bales ha^{-1} yields, 62, 11, 22, 7, and 4 kg ha^{-1} of N, P, K, Mg, and Ca, respectively, are removed. A yield of 7.5 bales ha^{-1} removes approximately 125, 22, 43, 13, and 9 kg ha^{-1} of the same nutrients. According to Malavolta et al. (1962), for every 100 kg of fibers produced, the cotton crop will require approximately 19 kg of N, 8 kg of P, 15 kg of K, 15 kg of Ca, and 4 kg of Mg under Brazilian conditions.

Under field conditions, Mullins and Burmester (1992) evaluated the uptake of Ca and Mg by cotton grown on two nonirrigated soils containing adequate levels of Ca and Mg, using cultural practices that are normal for the Southeastern United States. Four cultivars were compared. Plants were collected at two-week intervals throughout the growing season, beginning at 15 d after emergence, partitioned into leaves, stems, burs, seed, and lint, and analyzed (except for lint) for Ca and Mg. Total Ca and Mg uptake, when averaged for both soils and all four cultivars, was 64 and 18 kg ha^{-1}, respectively (or 9.3 kg of Ca and 2.6 kg Mg per 100 kg lint produced). Total Ca uptake was significantly lower on the Norfolk soil (44 kg ha^{-1}) as compared to the Decatur soil (75 kg ha^{-1}) which had a higher level of extractable Ca. There were no cultivar differences in total Ca and Mg uptake or uptake within a given plant part. There were no consistent differences among the cultivars for the concentration of Ca in various plant parts. The concentration of Mg in leaves and burs was affected by the cultivar. These cultivars accumulated similar amounts of Ca and Mg as compared to older cultivars, although previous research has shown that modern cultivars partition dry matter differently from older cultivars (Mullins and Burmester, 1992).

Apparent recovery of N by a cotton crop averages about 40% of the applied N and is seldom > 50% (Weier, 1994). Factors identified as affecting recovery include water stress, crop rotation, waterlogging, and soil compaction, with the latter two having considerable influence on the loss of fertilizer N through biological denitrification.

Distributions of N, P, K, Ca, Mg, and Mn among the bur (carpel walls), seed, and fiber fractions of cotton bolls were measured between 10 days after flowering and boll maturity by Leffler and Tubertini (1976). During the initial 3-week phase of boll enlargement, the bur accumulated reserves of N, P, and Mg; these were presumably drawn upon by the seed and fiber during later development. The bur continuously accumulated K, which reached 5.5% at maturity. Concentrations of most minerals in the seed declined initially and then increased markedly. The major mineral accumulated by the seed was N. At boll opening, over 90% of boll N was in the seed. The fiber accumulated minerals during the first 5 weeks of development but lost most of them during the final 3 weeks. The most abundant mineral in the fiber was K. The data suggested that the boll remained a physiological continuum throughout development.

B. Nutrient Concentration

Nutrient concentrations in plant parts have been used to evaluate fertilizer practices and to investigate problems of poor growth. Their most promising role appears to be in assessing the adequacy of plant nutrient supply for cotton during the growing season (Sabbe and Mackenzie, 1973). Adequate nutrient concentrations in cotton plant parts at different growth stages are presented in Table 17.3. Since cotton culture involves the growth of many varieties in different soil, climatic, and management environments, it is reasonable to expect sufficiency levels to vary somewhat among regions.

VI. SUMMARY

Cotton is a major fiber crop in developing as well as developed countries. The cotton plant is a warm-season perennial shrub that is grown as an annual field crop between 47°N and 32°S latitude. The United States, USSR, People's Republic of China, India, and Pakistan produce approximately 75% of the world supply. Cotton provides about 25% of the fibers for the world market, while synthetic and other natural fibers provide the remaining 75%. Still, cotton is an agricultural and industrial commodity of worldwide importance, and growing world consumption ensures that it will continue to be a significant commodity in the future. In addition, interest in the potential of cotton

Table 17.3 Adequate Nutrient Concentration in Cotton (*Gossypium hirsutum*)

Nutrient	Growth stage[a]	Plant part[a]	Adequate concentration
			g kg^{-1}
N	45 DAS	LB at 3rd and 4th nodes below A	> 50
	1st flowering	YMB	38–45
P	45 DAS	LB at 3rd and 4th nodes below A	> 4
	1st flowering	YMB	3–5
	Early fruiting	YMB	3.1
	Late fruiting	YMB	3.3
	Late maturity	YMB	2.4
K	45 DAS	LB at 3rd and 4th nodes below A	> 32
	76 DAS	PYMB	49–62
	101 DAS	PYMB	46–60
	120 DAS	PYMB	25–40
Ca	1st flowering	YMB	22–30
Mg	1st flowering	YMB	5–9
	34 to 105 DAS	All LB	6–8
	34 to 105 DAS	All P	4–7
S	Midseason	YMB	6–10
			mg kg^{-1}
Zn	1st flowering	YMB	20–60
	43 DAS		17–18
Mn	1st flowering	YMB	50–350
	36 DAS	YMB	11–247
Fe	1st flowering	YMB	50–250
B	1st flowering	YMB	20–60
Mo	5 months	LB	2.4

[a] DAS, Days after sowing; LB, leaf blade (excluding sheath and petiole); YMB, youngest (uppermost) mature leaf blade; and P, petiole.
Source: Compiled from Reuter,, 1986.

seed as a source of edible vegetable protein has been stimulated by increased understanding of its constituents and processing methods.

The optimum temperature for cotton growth is 24–34°C. It can be grown on a variety of soils, from sandy to clayey, with pH from 4.3 to 8.4, though the optimum pH range is 5.0–7.0. Maximum lint yields may be greater than 2 metric tons ha^{-1} in irrigated areas with very long growing seasons; however, world average yields are normally less than 0.5 tons ha^{-1}.

Cotton lint is almost pure cellulose and contains negligible amounts of N, P, and K. However, cotton seed contains about 3.7%, 0.55% P, and 0.95% K. In addition, cotton burs contain significant amounts of nutrients. Thus fertilizers or manures are usually required to replace nutrients removed in the seed and burs. Both soil and plant analyses can be used to detect and correct nutrient deficiencies and toxicities.

REFERENCES

Ackerson, R. C., D. R. Kreig, C. L. Haring, and N. Chang. 1977. Effects of plant water status on stomatal activity, photosynthesis, and nitrate reductase activity of field grown cotton. Crop Sci. 17: 81–84.

Adams, F. 1981. Alleviating chemical toxicities liming acid soils, pp. 269–301. *In*: G. F. Arkin and H. M. Taylor (eds.). Modifying the root environment to reduce crop stress. Monogr. 4, Am. Soc. Agric. Eng., St. Joseph, Michigan.

AOSA (Association of Official Seed Analysts). 1970. Rules for testing seeds. Proc. Assoc. Off. Seed Anal. 60: 1–116.

Arnon, I. 1972. Crop production in dry regions. Vol. II. Systematic treatment of the principal crops. Barnes & Noble, New York.

Ayers, R. S. 1977. Quality of water for irrigation. J. Irrig. Main Div., Am. Soc. Civ. Eng. 103 (IR2): 135–154.

Bassett, D. M., W. D. Anderson, and C. H. E. Werkhoven. 1970. Dry matter production and nutrient uptake in irrigated cotton (*Gossypium hirsutum*). Agron. J. 62: 299–303.

Benedict, C. R. 1984. Physiology, pp. 151–200. *In*: R. J. Kohel and C. F. Lewis (eds.). Cotton. Monogr. 24, Am. Soc. Agron., Madison, Wisconsin.

Benedict, C. R., K. J. McCree, and R. J. Kohel. 1972. High photosynthesis rate of chlorophyll mutant of cotton. Plant Physiol. 49: 968–971.

Berger, J. 1969. The world's major fibre crops, their cultivation and manuring. Centre d'Edude de l'Azote, Zurich, Switzerland.

Bjorkman, O. and P. Holmgrin. 1963. Adaptability of the photosynthetic apparatus to light intensity in ecotypes from exposed and shaded habitats. Physiol. Plant. 16: 889–914.

Bland, W. L. 1993. Cotton and soybean root system growth in three soil temperature regimes. Agron. J. 85: 906–911.

Bolger, T. P., D. R. Upchurch, and B. L. McMichael. 1992. Temperature effects on cotton root hydraulic conductance. Environ. Exp. Bot. 32: 49–54.

Borg, H. and D. W. Grimes. 1986. Depth development of roots with time: an empirical description. Trans. ASAE 29: 194–197.

Bowen, G. D. 1991. Soil temperature, root growth, and plant function, pp. 309–330. *In*: Y. Waisel et al. (ed.) Plant roots: the hidden half. Marcel Dekker, New York.

Bowling, A. L. 1984. Marketing and economics, pp. 571–587. *In*: R. J. Kohel and C. F. Lewis (eds.). Cotton. Monogr. 24, Am. Soc. Agron., Madison, Wisconsin.

Burke, J. J. and D. R. Upchurch. 1995. Cotton rooting patterns in relation to soil temperatures and the thermal kinetic window. Agron. J. 87: 1210–1216.

Burke, J. J., J. R. Mahan, and J. L. Hatfield. 1988. Crop-specific thermal kinetic windows in relation to wheat and cotton biomass production. Agron. J. 80: 553–556.

Cassman, K. G. 1993. Cotton, pp. 111–119. *In*: W. F. Bennett (ed.) Nutrient deficiencies and toxicities in crop plants. ASP Press, Amer. Phytopath. Soc., St. Paul, Minnesota.

Cassman, K. G., T. A. Kerby, B. A. Roberts, D. C. Bryant, and S. M. Brouder. 1989. Differential response of two cotton cultivars to fertilizer and soil potassium. Agron. J. 81: 870–878.

Cherry, J. P. and H. R. Leffler. 1984. Seed, pp. 511–569. *In*: R. J. Kohel and C. F. Lewis (eds.). Cotton. Monogr. 24, Am. Soc. Agron., Madison, Wisconsin.

Clark, R. B. and N. Reinhard. 1991. Effects of soil temperature on root and shoot growth traits and iron deficiency chlorosis in sorghum genotypes grown on a low iron calcareous soil. Plant Soil 130: 97–103.

Cook, C. G. and K. M. El-Zik. 1993. Fruiting and lint yield of cotton cultivars under irrigated and nonirrigated conditions. Field Crops Res. 33: 411–421.

Downton, J. and R. O. Slatyer. 1972. Temperature dependence of photosynthesis in cotton. Plant. Physiol. 50: 518–522.

Eaton, F. M. and D. R. Ergle. 1954. Effects of shade and partial defoliation on carbohydrate levels and the growth, fruiting and fiber properties of cotton plants. Plant Physiol. 29: 39–49.

El-Zik, K. M. and V. Sevacherian. 1979. Modeling cotton growth and development parameters with heat units. Proc. Beltwide Cotton Prod. Res. Conf. National Cotton Council, Memphis, Tennessee.

El-Zik, K. M., H. Yamada, and V. T. Walhood. 1980. The effect of management on blooming boll-retention, and productivity of upland cotton (*Gossypium hirsutum* L.). Proc. Beltwide Cotton Prod. Res. Conf. National Cotton Council, Memphis, Tennessee.

Engels, C. 1994. Effect of root and shoot meristem temperature on shoot to root dry matter partitioning and the internal concentrations of nitrogen and carbohydrates in maize and wheat. Ann. Bot. (London) 73: 211–219.

Gerik, T. J., J. E. Morrison, and F. W. Chichester. 1987. Effects of controlled-traffic on soil physical properties and crop rooting. Agron. J. 79: 434–438.

Gipson, J. R. 1986. Temperature effects on growth, development and fiber properties, pp. 47–56. *In*: J. R. Mauney and J. M. Stewart (eds.) Cotton physiology. The Cotton Foundation, Memphis, Tennessee.

Gipson, J. R. and H. E. Joham. 1968. Influence of night temperature on growth and development of cotton (*Gossypium hirsutum* L.): I. Fruiting and boll development. Agron. J. 60: 292–295.

Gipson, J. R. and H. E. Joham. 1969. Influence of night temperature on growth and development of cotton. II. Fiber elongation. Crop Sci. 9: 127–129.

Grimes, D. W. and K. M. El-Zik. 1982. Water management for cotton. Bull. 1904, Cooperative Extension Division of Agricultural Sciences, University of California, Berkeley.

Grimes, D. W. and H. Yamada. 1982. Relation of cotton growth and yield to minimum leaf and water potential. Crop Sci. 22: 134–139.

Grimes, D. W., H. Yamada, and W. L. Dickens. 1969. Functions for cotton (*Gossypium hirsutum* L.) production from irrigation and nitrogen fertilization variables. I. Yield and evapotranspiration. Agron. J. 61: 769–776.

Grimes, D. W., R. J. Miller, and L. Dickens. 1970. Water stress during flowering of cotton. Calif. Agric. 24: 4–6.

Guinn, G. 1982. Fruit age and changes in abscisic acid content, ethylene production, and abscission rate of cotton fruits. Plant Physiol. 69: 349–352.

Guinn, G. 1985. Fruiting of cotton. III. Nutritional stress and cutout. Crop Sci. 25: 981–985.

Halderman, A. D. 1973. Cotton water use. Arizona Agric., File A-198.

Halevy, J. 1976. Growth rate and nutrient uptake of two cotton cultivars under irrigation. Agron. J. 68: 701–705.

Hearn, A. B. 1979. Water relations in cotton. Outlook Agric. 10: 159–166.

Hearn, A. B. 1981. Cotton nutrition. Field Crop Abstr. 34: 11–34.

Hearn, A. B. and G. A. Constable. 1984. Cotton, pp. 495–527. *In*: P. R. Goldsworthy and N. M. Fisher (eds.). The physiology of tropical field crops. Wiley, New York.

Heitholt, J. J. 1993. Cotton boll retention and its relationship to lint yield. Crop Sci. 33: 486–490.

Heitholt, J. J. 1995. Cotton flowering and boll retention in different planting configurations and leaf shapes. Agron. J. 87: 994–998.

Heitholt, J. J., W. T. Pettigrew, and W. R. Meredith, Jr. 1992. Light interception and lint of narrow-row cotton. Crop Sci. 32: 728–733.

Hesketh, J. 1967. Enhancement of photosynthesis CO_2 assimilation in the absence of oxygen, as dependent upon species and temperature. Planta 76: 371–374.

Hesketh, J. and D. Baker. 1967. Light and carbon assimilation by plant communities. Crop Sci. 7: 285–293.

Hodgson, A. S. 1982. The effects of duration, timing and chemical amelioration of short-term waterlogging during furrow irrigation of cotton in cracking grey clay. Aust. J. Agric. Res. 33: 1019–1028.

Huck, M. G. 1970. Variation in taproot elongation rates as influenced by composition of soil air. Agron. J. 62: 815–818.

Hughes, L. C. 1966. Factors affecting numbers of ovules per loculus in cotton. Cotton Grow. Rev. 43: 273–285.

Karami, E., D. R. Kreig, and J. E. Quisenberry. 1980. Water relations and carbon-14 assimilation in cotton with different leaf morphology. Crop Sci. 30: 421–426.

Kaspar, T. C. and W. L. Bland. 1992. Soil temperature and root growth. Soil Sci. 154: 290–299.

Kittock, D. L., J. R. Mauney, H. F. Arle, and L. A. Bariola. 1973. Termination of late season cotton fruiting with growth regulators as an insect control technique. J. Environ. Qual. 2: 405–408.

Klepper, B., H. M. Taylor, M. G. Huck, and E. L. Fiscus. 1973. Water relations and growth of cotton in drying soil. Agron. J. 65: 307–310.

Lee, J. A. 1984. Cotton as a world crop, pp. 1–25. *In*: R. J. Kohel and C. F. Lewis (eds.). Cotton Monogr. 24, Am. Soc. Agron., Madison, Wisconsin.

Leffler, H. R. and B. S. Tubertini. 1976. Development of cotton fruit. II. Accumulation and distribution of mineral nutrients. Agron. J. 68: 858–861.

Lewis, C. F. and T. R. Richmond. 1968. Cotton as a crop, pp. 1–21. *In*: F. C. Elliot, M. Hoover, and W. K. Porter, Jr. (eds.). Advances in the production and utilization of quality cotton. Iowa State University Press, Ames.

Loffroy, O. C., C. Hubac, and J. B. V. Da Silva. 1983. Effect of temperature on drought resistance and growth of cotton plants. Physiol. Plant. 59: 297–301.

Maas, E. V. and G. J. Hoffman. 1977. Crop salt tolerance: Current assessment. J. Irrig. Drainage Div. Am. Soc. Civil Eng. 103: 115–134.

Malavolta, E., F. P. Haag, F. A. F. Mello, and M. O. C. Brasil. 1962. On the mineral nutrition of some tropical crops. International Potash Institute, Bern, Switzerland.

Mauney, J. R. 1984. Anatomy and morphology of cultivated cottons, pp. 59–80. *In*: R. J. Kohel and C. F. Lewis (eds.). Cotton. Monogr. 24, Am. Soc. Agron., Madison, Wisconsin.

Mauney, J. R., K. E. Fry, and G. Guinn. 1978. Relationship of photosynthetic rate to growth and fruiting of cotton, soybean, sorghum and sunflower. Crop Sci. 18: 259–263.

McArthur, J. A., J. D. Hesketh, and D. N. Baker. 1975. Cotton, pp. 297–325. *In*: L. T. Evans (ed.). Crop physiology. Cambridge University Press, London.

McDonald, M. B., Jr. 1980. Vigor test subcommittee report. Assoc. Off. Seed Anal. Newsletter 54: 37–40.

McMichael, B. L. 1986. Growth of roots, pp. 29–38. *In*: J. R. Mauney and J. M. Steward (eds.). Cotton physiology. The Cotton Foundation. Memphis, Tennessee.

McMichael, B. L. and J. E. Quisenberry. 1993. The imapct of the soil environment on the growth of root systems. Environ. Exp. Bot. 33: 53–61.

Melsted, S. W. and T. R. Peck. 1973. Principles of soil testing, pp. 13–21. *In*: L. M. Walsh and J. D. Beaton (eds.). Soil testing and plant analysis. Soil Sci. Soc. Am., Madison, Wisconsin.

Meredith, W. R., Jr., and R. Wells. 1989. Potential for increasing cotton yields through enhanced partitioning to reproductive structures. Crop Sci. 29: 636–639.

Morrow, M. R. and D. R. Krieg. 1990. Cotton management strategies for a short growing season environment: Water-nitrogen considerations. Agron. J. 82: 52–56.

Mullins, G. L. and C. H. Burmester. 1990. Dry matter, nitrogen, phosphorus, and potassium accumulation by four cotton varieties. Agron. J. 82: 729–736.

Mullins, G. L. and C. H. Burmester. 1992. Uptake of calcium and magnesium by cotton grown under dryland conditions. Agron. J. 84: 564–569.

Mutsaers, H. J. W. 1976. Growth and assimilate conversion of cotton bolls. 2. Influence of temperature on boll maturation period and assimilation conversion. Ann. Bot. 40: 317–324.

Niles, G. A. and C. V. Feaster. 1984. Breeding, pp. 201–231. *In*: R. J. Kohel and C. F. Lewis (eds.). Cotton. Monogr. 24, Am. Soc. Agron., Madison, Wisconsin.

Olson, L. C. and R. P. Bledsoe. 1942. The chemical composition of the cotton plant and the uptake of nutrients at different stages of growth. Georgia Exp. Stn. Bull. 222.

Oosterhuis, D. M., J. Chipamaunga, and G. C. Bate. 1983. Nitrogen uptake of field grown cotton. I. Distribution in plant components in relation to fertilization and yield. Exp. Agric. 19: 91–101.

Oosterhuis, D. M., R. G. Hurren, W. N. Wiley, and R. L. Maples. 1991. Foliar-fertilization of cotton with potassium nitrate. Proc. 1991 Cotton Res. Mtg., Univ. Arkansas, Arkansas Agric. Exp. Stn., Spec. Rep. 149: 21–25.

Patterson, L. L., D. R. Buxton, and R. E. Briggs. 1978. Fruiting in cotton as affected by controlled boll set. Agron. J. 70: 118–122.

Pearson, R. W. and Z. F. Lund. 1968. Direct observation of cotton root growth under field conditions. Agron. J. 60: 442–443.

Pearson, R. W., L. F. Ratliff, and H. M. Taylor. 1970. Effect of soil temperature, strength, and pH on cotton seedling root elongation. Agron. J. 63: 243–246.

Perkins, H. H., Jr., D. E. Ethride, and C. K. Bragg. 1984. Fiber, pp. 437– 509. *In*: R. J. Kohel and C. F. Lewis (ed.). Cotton. Monogr. 24, Am. Soc. Agron., Madison, Wisconsin.

Pettigrew, W. T. 1995. Source-to-sink manipulation effects on cotton fibre quality. Agron. J. 87: 947–952.

Pettigrew, W., J. J. Heitholt, and W. R. Meredith, Jr. 1996. Genotypic interactions with potassium and nitrogen in cotton of varied maturity. Agron. J. 88: 89–93.

Phillips, L. L. 1976. Cotton, pp. 196–200. *In*: N. W. Simmonds (ed.). Evolution of crop plants. Longman, London.

Purseglove, J. W. 1982. Tropical crops: dicotyledons. Longman, London.

Quisenberry, J. E. and R. J. Kohel. 1975. Growth and development of fiber and seed in upland cotton. Crop Sci. 15: 163–467.

Reddy, K. R., H. F. Hodges, J. M. McKinion, and G. W. Wall. 1992. Temperature effects on Pima cotton growth and development. Agron. J. 84: 237–243.

Reeves, D. W. and G. L. Mullins. 1995. Subsoiling and potassium placement effects on water relations and yield of cotton. Agron. J. 87: 847–852.

Reuter, D. J. 1986. Temperate and sub-tropical crops, pp. 38–99. *In*: D. J. Reuter and J. B. Robinson (eds.). Plant analysis: an interpretation manual. Inkata Press, Melbourne.

Sabbe, W. E. and A. J. Mackenzie. 1973. Plant analysis as an aid to cotton fertilization, pp. 299–313. *In*: L. M. Walsh and J. D. Beaton (eds.). Soil testing and plant analysis. Soil Sci. Soc. Am., Madison, Wisconsin.

Seshadri, V. and R. Prasad. 1979. Influence of rates and sources of nitrogen on growth, nitrogen uptake and yield of cotton (*G. hirsutum* L.). Z. Pflanzenernaehr. Bodenkd. 142: 731–739.

Stewart, J. M. 1986. Integrated events in the flower and fruit, pp. 261–300. *In*: R. Mauney and J. M. Stewart (eds.). Cotton physiology. The Cotton Foundation, Memphis, Tennessee.

Sung, F. J. M. and Krieg, D. R. 1979. Relative sensitivity of photosynthetic assimilation and translocations of 14C carbon to water stress. Plant Physiol. 64: 852–856.

Taylor, H. M. and B. Klepper. 1978. The role of rooting characteristics in the supply of water to plants. Adv. Agron. 30: 99–128.

Tharp, W. H. 1965. The cotton plant: how it grows and why its growth varies. USDA Handbook No. 178, Washington, DC.

Toscano, N. C., V. Sevacherian, and R. A. Van Steenwyk. 1979. Pest management guide for insects and nematodes of cotton in California. Publ. 4089, Division of Agricultural Sciences University of California, Berkeley.

Unruh, B. and J. E. Silvertooth. 1996. Comparisons between an upland and a Pima cotton cultivar: I. Growth and yield. Agron. J. 88: 583–589.

USDA/ARS. 1988. Systematic analysis of genetic and cultural factors contributing to water use efficiency in cotton. Annual report to USDA/ARS Plant Stress and Water Conservation Program. Lubbock, Texas.

Waddle, B. A. 1984. Crop growing practices, pp. 233–263. *In*: R. J. Kohel and C. F. Lewis (eds.). Cotton. Monogr. 24, Am. Soc. Agron., Madison, Wisconsin.

Wanjura, D. F. and D. R. Buxton. 1972. Hypocotyl and radicle elongation of cotton as affected by soil environment. Agron. J. 64: 431–434.

Wanjura, D. F., D. R. Buxton, and N. N. Stapleton. 1970. A temperature model for predicting initial cotton emergence. Agron. J. 62: 741–743.

Warner, D. A. and J. J. Burke. 1993. Cool night temperatures after leaf starch and photosystem. II. Chlorophyll fluorescence in cotton. Agron. J. 85: 836–840.

Warner, D. A., A. S. Holaday, and J. J. Burke. 1995. Response of carbon metabolism to night temperature in cotton. Agron. J. 87: 1193–1197.

Weier, K. L. 1994. Nitrogen use and losses in agriculture in subtropical Australia. Fert. Res. 39: 245–257.

Wells, R. and W. R. Meredith, Jr. 1984a. Comparative growth of obsolete and modern cotton cultivars: I. Vegetative dry matter partitioning. Crop Sci. 24: 858–862.

Wells, R. and W. R. Meredith, Jr. 1984b. Comparative growth of obsolete and modern cotton cultivars: II. Reproductive dry matter partitioning. Crop Sci. 24: 863–868.

Wells, R. and W. R. Meredith, Jr. 1984c. Comparative growth of obsolete and modern cotton cultivars: III. Relationship of yield to observed growth characteristics. Crop Sci. 24: 868–872.

Wells, R. and W. R. Meredith, Jr. 1986. Normal versus okra leaf yield interactions in cotton: II. Analysis of vegetative and reproductive growth. Crop Sci. 28: 223–228.

18

Forage

I. INTRODUCTION

Literally, forages are any plants consumed by livestock. Usually the term is restricted to pasture and browse plants, hay, silage, and immature cereals or straw (Janick et al., 1969). Forage agriculture is also known as grassland agriculture and can be defined as a farming system which emphasizes the importance of grasses and legumes in livestock and land management (Ahlgren, 1956). The main feature of grassland agriculture is its dependence on grasses, legumes, and other forage for proper land use and increased animal profitability (Heath, 1982). Besides providing feed for livestock, grassland agriculture helps in renewal of soil organic matter, prevents soil erosion, improves soil tilth, and restores soil fertility. It is a very common practice in developing countries where population pressure is high and farmers cannot afford to apply fertilizers. Some part of the land used for farming is normally left in pasture for 2–3 years to restore soil fertility. Grass roots that develop deep in the soil layer play an important role in increasing the infiltration rate. This also dries the soil uniformly throughout the soil layer by evapotranspiration, improving infiltration and increasing potential water storage capacity for subsequent rainfall events (Hino et al., 1987).

II. CLIMATE AND SOIL REQUIREMENTS

Climatic and edaphic factors are the major determinants of the distribution and production of plants, in both natural and human-modified ecosystems (Andrew, 1978). Forage grasses and legumes can grow under a wide range of climatic and soil conditions, and it is impossible to discuss climatic and soil requirements for individual grass and legume species. Because there are about 5000 species of grasses and most herbaceous legumes, there are about 11,000 species in the family used for forages (Janick et al., 1969). However, there are some basic principles related to soil and climate of forages which have universal applicability.

Natural grasslands of the world were developed under restricted precipitation effectiveness (Figure 18.1). Precipitation effectiveness is determined by the ratio of precipitation to evaporation (Bula, 1982). Characteristics of important grass and legume species are listed in Table 18.1. These grasses and legume species are the important forage species in the tropical, subtropical, and temperate regions around the world.

Seasonal drought is a dominant feature of the grasslands environment, so that adapted forage species must be able to tolerate considerable soil water

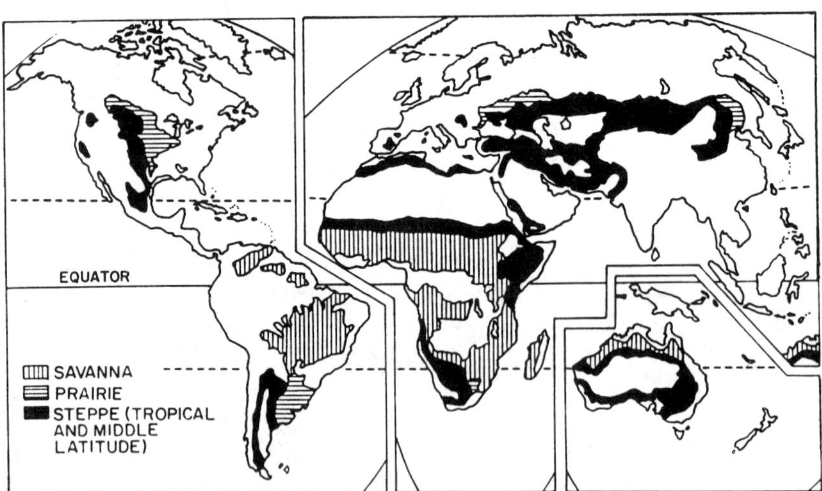

Figure 18.1 World distribution of native grasslands. (From Trewartha et al., 1967. © McGraw-Hill Book Company, reproduced with permission.)

Table 18.1 Some Important Forage Grasses and Legumes

	Common name	Scientific name	Annual (A) or Perennial (P)	Photores-piration	Cool (C) or warm (W) season
Grasses	Bahiagrass	*Paspalum notatum* Flugge	P	C_4	W
	Bermudagrass	*Cynodon dactylon* (L) Pers.	P	C_4	W
	Bromegrass	*Bromus inermis* Leyss.	P	C_3	C
	Buffelgrass	*Cenchrus ciliaris* (L) Link	P	C_4	W
	Dallisgrass	*Paspalum dilatatum* Poir.	P	C_4	W
	Gambagrass	*Andropogon gayanus* Kunth	P	C_4	W
	Guineagrass	*Panicum maximum* Jacq.	P	C_4	W
	Italian ryegrass	*Lolium multiflorum* Lam.	A	C_3	C
	Jaragua	*Hyparrhenia rufa* (Nees) Stapf.	P	C_4	W
	Kentucky bluegrass	*Poa pratensis* L.	P	C_3	C
	Kikuyugrass	*Pennisetum clandestinum* Hoschst.	P	C_4	W
	Kleingrass	*Panicum coloratum* Walt.	P	C_4	W
	Limpograss	*Hemarthria altissima* (Poir) Stapf. & Hubbard	P	C_4	W
	Napiergrass	*Pennisetum purpureum* Schumach	P	C_4	W
	Orchardgrass	*Dactylis glomerata* L.	P	C_3	C
	Pangolagrass	*Digitaria decumbens* Stent.	P	C_4	W
	Paragrass	*Brachiaria mutica* (Forsk) Stafp.	P	C_4	W
	Perennial ryegrass	*Lolium perenne* L.	P	C_3	C
	Reed canarygrass	*Phalaris arundinacea* L.	P	C_3	C
	Redtop	*Agrostis alba* L.	P	C_3	C
	Rhodesgrass	*Chloris gayana* Kunth.	P	C_4	W
	Speargrass	*Heteropogon contortus*	P	C_4	W
	Sudangrass	*Sorghum bicolor* drummondi	A	C_4	W
	Surinamgrass	*Brachiaria decumbens* Stapf.	P	C_4	W

Table 18.1 (Continued)

	Common name	Scientific name	Annual (A) or Perennial (P)	Photores-piration	Cool (C) or warm (W) season
Grasses (cont.)	Switchgrass	*Panicum virgatum* L.	P	C_4	W
	Tall fescue	*Festuca arundinacea* Schreb.	P	C_3	C
	Timothy	*Phleum pratense* L.	P	C_3	C
	Crested wheatgrass	*Agropyron cristatum* (L.) Gaertn.	P	C_3	C
Legumes	Alfalfa	*Medicago sativa* L.	P	C_3	C
	Birdsfoot trefoil	*Lotus corniculatus* L.	P	C_3	C
	Blue lupine	*Lupinus angustifolius* L.	A	C_3	C
	Centro	*Centrosema pubescens* Benth	P	C_3	W
	Common vetch	*Vicia sativa* L.	A	C_3	C
	Common lespedeza	*Lespedeza striata* Hook & Arn	A	C_3	W
	Flatpea	*Lathyrus sylvestris* L.	A	C_3	C
	Hairy vetch	*Vicia villosa* Roth	A	C_3	C
	Kudzu	*Pueraria thunbergiana* Benth	P	C_3	W
	Ladino clover	*Trifolium repens* L.	P	C_3	C
	Red clover	*Trifolium pratense* L.	P	C_3	C
	Silverleaf desmodium	*Desmodium uncinatum* Jacq	P	C_3	W
	Sub clover	*Trifolium subterraneum* L.	A	C_3	C
	Townsville stylo	*Stylosanthes humilis* H.B.K.	A	C_3	W
	White lupine	*Lupinus albus* L.	A	C_3	C
	White sweet clover	*Mclitus alba* Med.	P	C_3	C
	Yellow lupine	*Lupinus luteus* L.	A	C_3	C
	Yellow sweet clover	*Melilotus officinalis* Lam.	P	C_3	C

Table 18.2 Water Stress Tolerance of Forage Legumes

Legume species	Minimum water potential (bars)	Time to reach minimum water potential (weeks)
Stylosanthes capitata	–65.9	21.5
Desmodium ovalifolium	–65.7	14.8
Arachis pintoi	–61.9	13.3
Centrosema acutifolium	–60.8	13.2
Centrosema brasilianum	–28.9	16.8
Standard error of means	5.4*	1.0*

*P < 0.001.
Source: Rao et al., 1992.

deficits (Cochrane et al., 1985). Rao et al. (1992) reported that a number of most promising forage grasses, including *Brachiaria* spp., *Andropogon gayanus*, and *Panicum maximum*, are tolerant of seasonal droughts (CIAT, 1978; Rao et al., 1992). The ability of a range of forage legume species to tolerate water stress is shown in Table 18.2. The data presented are for the minimum water potential measured and the number of weeks to achieve it. Water potential is a measure of physiological water deficit in plant tissue (Hsiao, 1973). Well-watered plants have leaf water potentials around 0 to –5 bars. Most crop plants tolerate water deficits up to –20 bars in leaves. Tropical forage legumes, as compared to most field crops, seem to tolerate much lower water potentials. Both *Stylostanthes capitata* and *Desmodium ovalifolium* were outstandingly tolerant of water stress. In contrast, *Centrosoma brasilianum* maintained its water potential relatively high (less negative value). Tolerance of soil water deficits in *Arachis pintoi* was better than in *C. brasilianum*. These two species may have contrasting physiological mechanisms to tolerate (*A. pintoi*) or avoid (*C. brasilianum*) water deficits in soil. These data indicate a broad range of adaptation to water stress in the forage species.

Tropical grasses have a higher optimum temperature for dry matter production than temperate grasses. Optimum temperatures for tropical grasses are in the range of 30–40°C as compared to 20–30°C for many temperate grasses (Jones, 1985; Rotar and Kretschmer, 1985). Optimum temperatures for tropical pasture legumes vary from 25 to 30°C, while for many temperate legumes the optimum temperatures are in the range of 20–25°C (Rotar and Kretschmer, 1985). Growth of most tropical grasses and legumes is significantly reduced below 15°C, and these species are susceptible to frost. On the other hand, temperate forage species are tolerant to frost, although their growth rates are lower (Rotar and Kretschmer, 1985). Table 18.3 shows

Table 18.3 Tolerance of Selected Tropical Grass and Legume Species to Drought, Frost, and Waterlogging

Species		Tolerance rating			
		Frost	Drought	Waterlogging	Minimum annual rainfall (mm)
Grasses	*Andropogon gayanus*	Poor	Good	Good	400
	Brachiaria decumbens	Poor	Fair	Fair	1250
	Brachiaria mutica	Poor	Fair	Excellent	1000
	Cenchrus ciliaris	Fair	Excellent	Poor	300
	Cynodon dacrylon	Poor	Excellent	Poor	500
	Digitaria decumbens	Poor	Fair	Good	1000
	Eragrostis curvula	Fair	Good	Poor	500
	Hemarthria altissima	Poor	Fair	Good	1000
	Hyparrhenia rufa	Fair	Good	Poor	600
	Melinis minutiflora	Poor	Fair	Poor	900
	Panicum antidotale	Fair	Excellent	Fair	400
	Panicum maximum	Poor	Good	Fair	900
	Panicum purpureum	Fair	Good	Fair	1000
Legumes	*Centrosema pubescens*	Poor	Good	Fair	1250
	Desmodium uncinatum	Fair	Fair	Fair	900
	Desmodium heterophyllum	Poor	Fair	Good	1500
	Lablab purpureus	Fair	Good	Poor	500
	Leucaena leucocephala	Fair	Excellent	Poor	500
	Pueraria phaseoloides	Poor	Poor	Good	1000
	Stylosanthes guianensis	Fair	Good	Fair	700
	Stylosanthes humilis	Poor	Excellent	Fair	500

Source: Compiled from Rotar and Kretschmer, 1985.

Table 18.4 Grass and Legume Species Adapted to Various Soil pH Ranges

Grasses		
pH < 4.5	Weeping lovegrass	*Eragrostis curvula* (Schrad) Nees.
	Bermudgrass	*Cynodon dactylon* L. Pers.
	Switchgrass	*Panicum virgatum* L.
	Bentgrass	*Agrostis canina* L. spp.
	Deertongue	*Panicum clandestinum* L.
	Red top	*Agrostis alba* L.
	Chewing fescue	*Festuca rubra* var. commutata Gaud.
	Red fescue	*Festuca rubra* L.
	Broomsedge	*Andropogon virginicus* L.
pH 5–6	Orchardgrass	*Dactylis glomerata* L.
	Tall fescue	*Festuca arundinacea* Schreb.
	Bromegrass	*Bromus inermis* Leyss
	Ryegrass	*Lolium perenne* L.
	Timothy	*Phleum pratense* L.
	Little bluestem	*Andropogon scoparius* Michx.
	Kentucky bluegrass	*Poa pratensis* L.
	Reed canarygrass	*Phalaris arundinacea* L.
	Tall oatgrass	*Avena sativa* L.
Legumes		
pH 4–5	Birdsfoot trefoil	*Lotus corniculatus* L.
	Lespedeza species	*Lespedeza cuneata* (Dumont)
	Crownvetch	*Coronilla varia* L.
	Flatpea	*Lathyrus sylvestris* L.
	White clover	*Trifolium repens* L.
	Kura clover	*Trifolium ambiguum* Bieb
	Zigzag clover	*Trifolium medium* L.
pH > 5	White clover	*Trifolium repens* L.
	Crimson clover	*Trifolium incarnatum* L.
	Red clover	*Trifolium pratense* L.
	Alfalfa	*Medicago sativa* L.

Source: Bennett et al., 1986.

tolerance of some grass and legume forage species to drought, frost, and waterlogging, and gives their minimal annual rainfall requirement.

Most of the pasture grasses and legumes do well in deep and well-drained soils having high water-holding capacity. Forage grasses and legumes have shown a wide range of adaptability to soil pH. Table 18.4 shows various grass and legume species and their adaptability to various soil pH ranges. Optimum soil pH in general for forage legumes and grasses is in the range of 5–7.

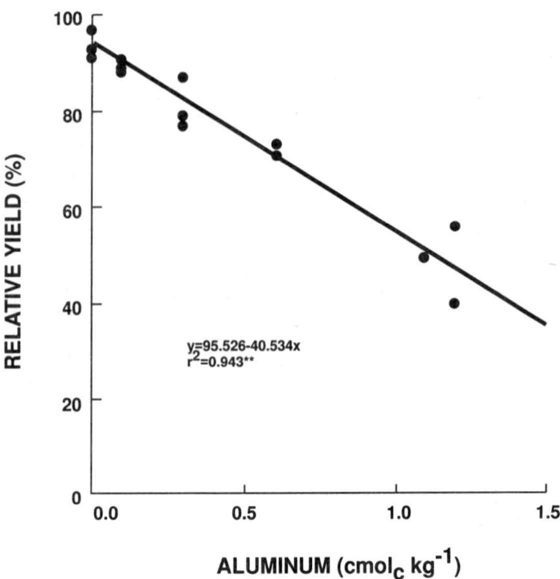

Figure 18.2 Effect of soil aluminum on relative dry matter yield of forage grass. (From Cruz, 1994.)

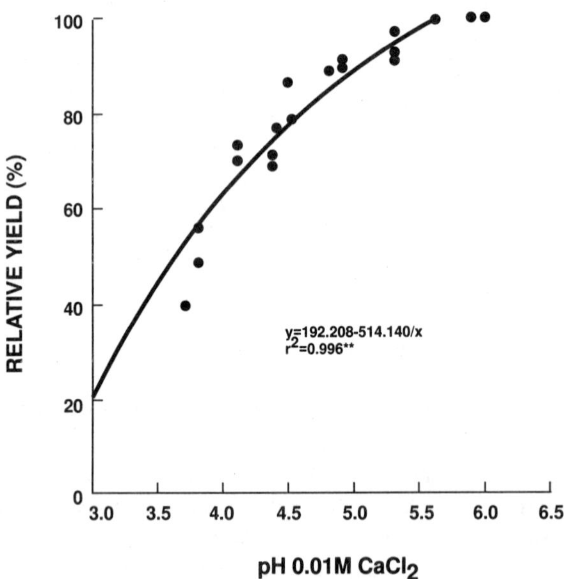

Figure 18.3 Relationship between soil pH and relative dry matter yield of forage grass. (From Cruz, 1994.)

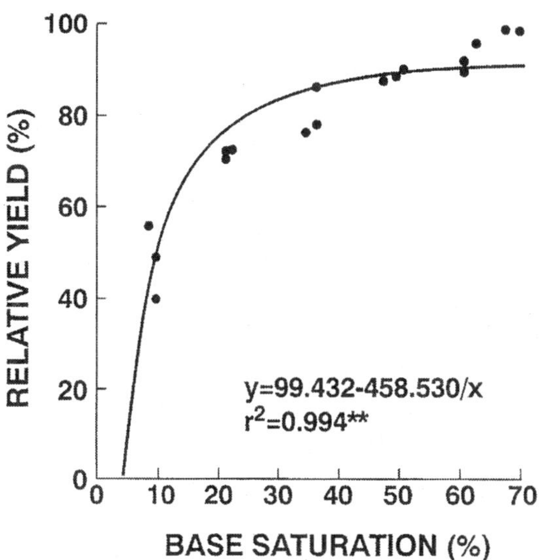

Figure 18.4 Relationship between base saturation and relative dry matter yield of forage grass. (From Cruz, 1994.)

Figure 18.2 shows that a higher level of Al under low pH can significantly reduce the growth of grasses. When Al concentration was more than 1 cmol kg^{-1}, of soil, relative yield was less than 50%. There was a quadratic response to forage grasses to increasing soil pH from 3 to 6 (Figure 18.3). Similarly, maximum productivity calculated on the basis of regression equation was obtained at a base saturation of about 50% (Figure 18.4). This means that, in acid soils, the use of liming is an important practice to improve dry matter yield of grasses as well as legume forages.

At the International Tropical Agriculture Center (CIAT), Colombian forage species were evaluated for soil acidity. Evaluation of germplasm introduction for acid soil tolerance has been conducted in tropical America at screening sites representing the major ecosystems where Ultisols or Oxisols predominate (i.e., the isohyperthermic savannas of Colombia and Venezuela, the isothermic savannas of the Brazilian cerrado; Miles and Lapointe, 1992). Some of the identified tropical grasses and legume species exhibit very high potential to produce good yield on acid soils (Table 18.5). A liming trial conducted at CIAT (1980) showed that there is a difference in dry matter yields of tropical legumes in relation to different liming levels (Table 18.6). Some species yield was increased with liming, while others showed no response to liming.

Table 18.5 Tropical Forage Species with Special Potential on Acid Soils

Grasses	Legumes
Andropogon gayanus	*Arachis pintoi*
Brachiaria brizantha	*Calopogonium mucunoides*
Brachiaria decumbens	*Centrosema acutifolium*
Brachiaria dictyoneura	*Centrosema brasilianum*
Brachiaria humidicola	*Centrosema macrocarpum*
Panicum maximum	*Centrosema pubescens*
	Cratylia arquentea
	Desmodium ovalifolium
	Flemingia macrophylla
	Pueraria phaseoloides
	Stylosanthes capitata
	Stylosanthes guianensis

Source: Zeigler et al., 1995.

Table 18.6 Effect of Lime on Dry Matter Yields (kg ha^{-1}) of Tropical Legumes, First Cutting

	Lime (t ha^{-1})			
Species	0	0.5	2	6
Centrosema plumieri CIAT 470	0	0	582	1698
Centrosema sp. CIAT 1787	445	912	2014	2769
Centrosema sp. 1733	356	1330	1568	1317
Centrosema pubescens	680	1729	1996	2035
Desmodium ovalifolium	1118	2302	2018	2480
Pueraria phaseoloides	1286	1688	1422	1434
Zornia sp. CIAT 728	3000	3108	2686	2628
Stylosanthes capitata CIAT 1019	2365	2361	3011	2458

Source: CIAT, 1980.

Table 18.7 Salinity Threshold and Salt Tolerance Rating of Selected Forage Grasses and Legumes

Species		Salinity threshold	Salinity rating[a]
Grasses	Bermudagrass	6.9	T
	Buffelgrass	—	MS
	Italian ryegrass	—	MT
	Reed canarygrass	—	MT
	Dallisgrass	—	MS
	Hardigrass	4.6	MT
	Kallagrass	—	T
	Lovegrass	2.0	MS
	Orchardgrass	1.5	MS
	Panicgrass	—	MT
	Perennial ryegrass	5.6	MT
	Rescuegrass	—	MT
	Rhodegrass	—	MS
	Sudangrass	2.8	MT
	Tall fescue	3.9	MT
	Timothy	—	MS
	Wheatgrass	3.5	MT
Legumes	Alfalfa	2.0	MS
	Common vetch	3.0	MS
	Ladino clover	1.5	MS
	Red clover	1.5	MS
	Sweet clover	—	MT
	White clover	—	MS

[a] T, Tolerant; MT, moderately tolerant; MS, moderately susceptible.
Source: Compiled from Maas, 1986.

Salt tolerance ratings of some selected forage grasses and legumes are given in Table 18.7. Salt tolerance of forage grasses and legumes varies considerably among species from susceptible to tolerant.

III. GROWTH AND DEVELOPMENT

A. The Grasses

Almost all grasses (Family Graminae) are herbaceous, annuals, or perennials. The grasses are monocotyledons and the legumes are dicotyledons (monocotyledons have only one cotyledon while dicotyledons have two). Grass plants are composed of leaves, stems, roots, and inflorescences and are widely divergent in size, shape, and growth habit.

Roots

Grasses have fibrous root systems. The grass root system may be divided into seminal roots, subcoleoptile internode roots, adventitious roots, and the primary root system (Figure 18.5). Seminal roots are produced from the embryo. The

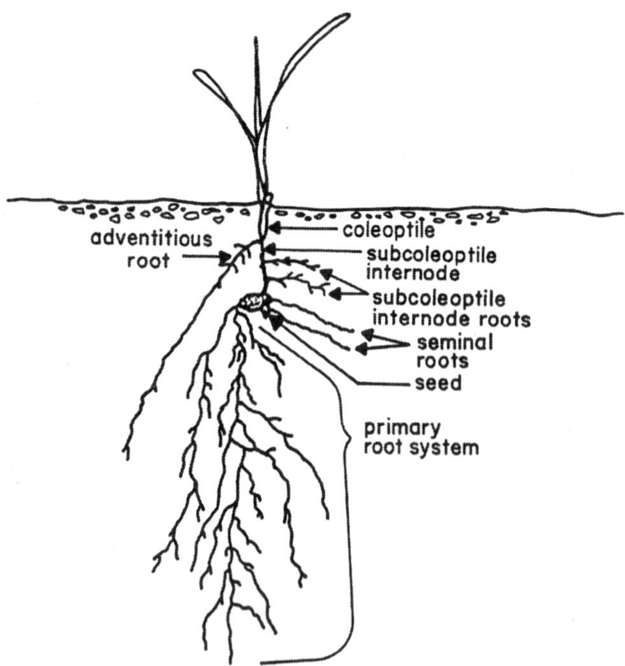

Figure 18.5 Grass seedling root system. (From Newman and Moser, 1988.)

adventitious root system of an established grass plant develops from the coleoptilar node and other nodes above the seed node. Coleoptile length and the amount of internode elongation between the seed node and coleoptilar node (subcoleoptile internode elongation) determine the position of the seedling crown, which is the source of adventitious roots (Newman and Moser, 1988).

The stolons and rhizomes produce roots at the nodes and bear scales (Purseglove, 1985). In general, grasses have longer, thinner, more finely branched roots than legumes, and these differences give the grasses a strong competitive advantage over legumes in nutrient and water uptake when they are in short supply.

Stems

The stems of a grass plant (also called culms) are distinctly divided into nodes and internodes. The node or joint is always solid, whereas the internodes may be hollow, pithy, or solid. In addition to the vertical flowering stems or culms, many grasses have horizontal underground stems called rhizomes and creeping stems above the ground called stolons (Metcalfe, 1982). Examples of grasses having rhizomes are Johnsongrass, Kentucky bluegrass, quackgrass, and many others. Two of the best known stoloniferous grasses are buffalograss and bermudagrass. Lateral buds arise in the axils of the leaves, and the leaves have their vascular connections with the stem at the node.

Leaves

The leaves consist of the sheath, the ligule, and the leaf blade. They are borne on the stem, alternately in two rows, one at each node. The blades are parallel-veined and typically flat, narrow, and sessile.

Inflorescence

The inflorescence is usually terminal on the culm and consists of groups or clusters of spikelets arranged in dense to loose panicles. The panicle is the most common type of grass inflorescence. The panicle may be either open, diffuse, or contracted (Metcalfe, 1982). The fruit of most grasses is a caryopsis or kernel.

B. The Legumes

Roots

Most of the legumes (Family Legumaceae), especially the herbaceous ones, have taproots. The taprooted plant has a rather narrow protruding crown, the taproot penetrating vertically into the ground with branch roots arising at intervals from it. Such plants are unable to spread sideways except to a limited degree by crown expansion as the plant ages (Heinrichs, 1963). The roots of many leguminous species become infected by *Rhizobium* sp. of bacteria and form nodules for atmospheric N fixation.

Tops

The tops of legumes consist of a main stem with axillary branches, usually with compound leaves. A characteristic feature of the legume family is the presence of a pulvinus at the base of the leaflets and of the petiole. Flowers usually are arranged in racemes as in peas, in heads as in clover, or in a spikelike raceme as in alfalfa (Metcalfe, 1982). The fruit is a pod containing one to several seeds. Each seed is enclosed in a testa or seed coat.

C. Growth Stages

Table 18.8 describes growth stages of grasses and legumes.

D. Growth Habit

The growth habit of pasture species is an extremely important characteristic because it governs the response of plants to defoliation. The proportion of tops of pasture species removed by grazing animals varies according to the plants' growth habit. The grazing animal removes a greater proportion of tops of upright plants than of plants of prostrate growth habit and hence, in the latter case, relatively more green leaf, stem, and sheath is left after grazing to facilitate regrowth (Haynes, 1980). Figure 18.6 shows the growth habit of upright and prostrate grasses and legumes and the relative amount of tops remaining after defoliation. Forage species with upright growth habits are more susceptible to damage by grazing than those with a prostrate or creeping growth habit. Growth habit is also important in competition of forage species in pastures. Red clover has an upright growth habit and has an immediate advantage over prostrate white clover whenever the grazing interval is long enough to allow the grasses to shade the clovers. However, because of its prostrate habit, white clover is able to recover very quickly from cutting or grazing in comparison with red clover (Haynes, 1980).

E. Dry Matter

Dry matter production of forage grasses and legumes depends on photosynthetic efficiency. Photosynthetic efficiency is determined by plant species, temperature, nutrient and water supply, solar radiation, and management practices. Perennial tropical grasses and legumes have the potential for year-round production of herbage.

The optimum leaf area index (*LAI*) for maximum growth varies greatly with species. Optimum *LAI* for *Panicum maximum* and *Cynodon dactylon* is about 4 (Alexander and McCloud, 1962; Humphreys, 1966). Similarly, the optimum *LAI* for white and red clovers is in the range of 3–5, and for ryegrasses from 5 to 6 (Brougham, 1960). Forage grasses and legumes have

Table 18.8 Growth Stages in Grasses and Legumes[a]

Growth stage			Description
Grasses	First growth	Vegetative	Leaves only, stems not elongated. (Specify extended leaf length and if seedling or older plants.)
		Stem elongation	Stems elongated. (Specify early or late jointing depending on less than or more than one-half the leaves exposed, respectively.)
		Boot	Inflorescence enclosed in flag leaf sheath and not showing.
		Heading	Inflorescence emerging or emerged from flag leaf sheath but not shedding pollen. (Specify proportion emerged.)
		Anthesis	Flowering stage, anthers shedding pollen. (Specify if early or late anthesis.)
		Milk stage	Seeds immature, endosperm milky. (Specify if early or late milk.)
		Dough stage	Well developed seeds, endosperm doughy. (Specify if early or late dough.)
		Ripe seed	Seeds ripe, leaves green to yellow brown.
		Postripe seed	Seeds postripe, some dead leaves and some heads shattered. (Specify amount of dead leaf tissue.)
		Stem cured	Leaves cured on stem, seeds mostly cast. (Specify if frosted.)
	Regrowth[b]	Vegetative	Leaves only, stems not elongated. (Specify extended leaf length.)
		Jointing	Green leaves and elongated stems. (Specify if before or after killing frost.)
		Late growth	Leaves and stems weathered. (State age of growth and time of year; specify if before or after killing frost.)

Table 18.8 (Continued)

Growth stage			Description
Legumes[c]	Spring and summer growth	Vegetative (or prebud)	No buds. (Specify plant height and if seedling or older plants.)
		Bud	No flowers. (Specify early or late bud based on condition of the floral buds.)
		First flower	First flowers appear on plants.
		Bloom (flower)	Plants flowering. (Specify percent of stems with one or more flowers; determine from 100 randomly selected stems.)
		Pod (or green seed) development	Green seedpods developing. (Specify percent of stems with one or more green seedpods formed; estimate amount of leaf loss, if any.)
		Ripe seed	Mostly mature brown seedpods with lower leaves dead and some leaf loss. (Estimate amount of leaf loss.)
	Fall recovery growth		Vegetative or with floral development. (Specify plant height and condition of floral development.)

[a]*Note*: Specify date of observation, cutting number, days of regrowth, and species and cultivar. Dry matter percentage and extended leaf growth measurements are reommended at all growth stages.
[b]If flowering occurs, use nomenclature outlined for first-growth forage.
[c]These growth stages best describe upright-growing legumes. For stoloniferous legumes like white clover, specify if leaves and floral stems only are included or if stolons are also.
Source: Metcalfe (1982). Reprinted with permission from Forages, 3rd ed., by M. Heath, D. Metcalfe, and R. Barnes. © 1982 by Iowa State University Press, Ames, Iowa.

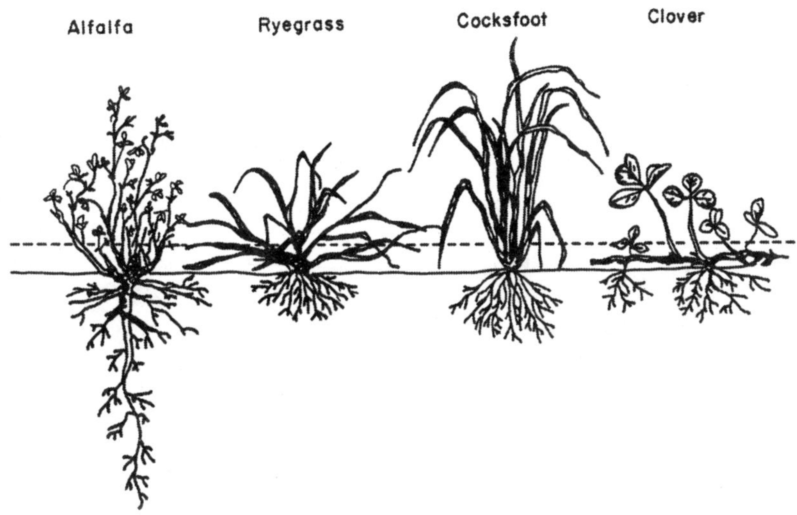

Figure 18.6 Growth habits of upright and prostrate legumes and grasses and their influence on the relative amount of shoot material left after defoliation. Dotted line represents height of defoliation. (From Haynes, 1980. © Academic Press, reproduced with permission.)

Table 18.9 Selected Forage Grasses and Legumes with High and Low Photosynthetic Capacity

Species	High photosynthetic	Low photosynthetic
Grasses	*Andropogon scoparium* Michx.	*Agropyron repens* (L.) Beauv.
	Andropogon virginicus L.	*Agrostis alba* L.)
	Cenchrus ciliaris L.	*Dactylis glomerata* L.
	Cynodon dactylon (L.) Pers.	*Festuca arundinacea* Schreb.
	Cyperus rotundus L.	*Lolium multiflorum* Lam
	Cyperus esculentus L.	*Panicum commutatum* Schult.
	Digitaria pentzel Stent.	*Phalaris arundinacea* L.
	Eragrostis chloromelas Strud.	*Phalaris canariensis* L.
	Panicum capillaare L.	*Poa pratensis* L.
	Panicum maximum Jacq.	
	Panicum virgatum L.	
	Paspalum dilatatum Poir.	
Legumes	*Atriplex rosea* L.	*Cassia tora* L.
	Atriplex semibaccata R.Br.	*Crotalaria spectabilis* Roth.
	Atriplex spongiosa F.V.M.	*Pueraria lobata* (Wild.) Ohwi.
		Vicia sativa L.

Source: Compiled from Black, 1971.

Table 18.10 Dry Matter Production of Selected Forage Species in Different Countries

Species	Country	Growth duration (days)	Growth rate ($g\ m^{-2}\ day^{-1}$)	Dry matter ($t\ ha^{-1}$)
Digitaria decumbens Stent.	Cuba, 23°N	365	10.8	39.4
Digitaria decumbens Stent.	Australia, 27°N	365	6.6	24.4
Lolium perenne L.	England, 52°N	365	7.9	29.0
Lolium perenne L.	New Zealand, 40°S	365	7.3	26.6
Medicago sativa L.	United States, 38°S	250	13.0	32.5
Pennesetum purpureum Schumach.	El Salvador, 14°N	365	23.2	85.2
Panicum maximum Jacq.	Puerto Rico, 18°N	365	13.4	48.8
Panicum maximum Jacq.	Nigeria, 7°N	328	7.1	23.4
Panicum clandestinum Hochst.	Australia, 27°S	365	8.2	30.0
Stylonsanthes guyanensis (Aubl.) Scv.	Ghana, 7°N	365	5.8	21.1
Trifolium pratense L.	New Zealand, 40°S	365	7.2	26.4

Sources: Compiled from Eagles and Wilson, 1982; Rodrigues and Rodrigues, 1987.

different photosynthetic production capacities due to differences in plant anatomy, plant physiology, and plant biochemistry (Black, 1971). Genotypic variation for growth rate has also been found among forage species (Wilson, 1975, 1982). Table 18.9 shows forage grasses and legumes with high and low photosynthetic capacity. Table 18.10 shows the dry matter production of some forage grasses and legumes in different countries.

IV. GRASS-LEGUME MIXTURES

Many improved pasture systems consist of grass-legume mixtures rather than a monoculture. The grasses are expected to provide the bulk of the energy to cattle because of their larger dry matter production. The role of legumes in such mixtures is to supply N to the grasses and improve the overall nutritional content of the forage, particularly that of protein, phosphorus, and calcium (Sanchez, 1976).

Legumes grown in combination with cool-season grasses can provide symbiotic N to associated grass (Farnham and George, 1993; Gettle et al., 1996) and improve total yields (George, 1984). Legumes have also been successfully

Table 18.11 Soil Physical Properties in Different Pastures in a Medium-texture Oxisol of Carimagua, Colombia

Pastures	% of soil aggregates > 0.5 mm	Water sorptivity ($cm/sec^{1/2}$)
Native savanna	9.6 a	0.20 a
Improved grass	25.5 b	0.40 b
Grass-legume pasture	31.9 b	0.39 b

Means in a column not followed by the same letter differ at 5% probability level.
Source: Rao et al., 1992.

grown with warm-season grasses in the southern United States to extend the growing season (Evers, 1985). Brown and Byrd (1990) reported that yield of an alfalfa-bermudagrass [*Medicago sativa* L.–*Cynodon dactylon* (L.) Pers.] mixture was similar to bermudagrass fertilized with 100–300 kg N ha^{-1} in Georgia. Posler et al. (1993) observed greater yields in Kansas with five of six native, legume-switchgrass mixtures compared with unfertilized switchgrass.

Marten (1989) suggested that warm-season grasses may be incompatible with cool-season legumes because of differences in growth habit, relative maturity, harvest schedules, and poor persistence. Taylor and Jones (1983) observed that alfalfa and red clover were not compatible with switchgrass grown in Kentucky, regardless of cutting management. In this same study, however, bigflower vetch (*Vicia grandiflora* Scop.) persisted well with switchgrass, with the mixture containing 20% legume and yielding 9.4 Mg ha^{-1}.

Grass-legume mixtures improved soil aggregate size and water infiltration (Table 18.11). This means better environment for grasses as well as legumes and higher dry matter yield. The addition of legumes with grasses also improved biological activities (Table 18.12). As compared to native savanna, the biological activity of the improved pastures was markedly higher in terms of the population of nitrifying bacteria and mycorrhizal spores and percent of mycorrhizal root infection (Table 18.12). The activity of earthworms increased threefold in the improved pastures. This increased biological activity is beneficial to soil properties such as mineralization, humification, texture, porosity, water infiltration, and retention. Soil characteristics are both a determinant and the consequence of earthworm activities, since these macroorganisms greatly influence the functioning of the soil system. They build and maintain the soil structure and take an active part in energy and nutrient cycling through the selective activation of both mineralization and humification processes (Lavelle, 1988).

Table 18.12 Comparison of Soil Biological Activity in Different Pastures

Soil biological characteristic	Savanna	Improved grass	Grass + legume
Nitrifying bacteria (number/g soil)	3.9×10^6	2.7×10^8	—
VA mycorrhizal fungi spore population (spores/100 g soil)	50	190	275
Root infection with VA mycorrhizae (%)	31	54	58
Earthworm activity (casts/m^2)	0.9	2.1	3.1

Source: Rao et al., 1992.

The contribution of legume residues to soil organic matter quality and turnover together with improved soil fertility, soil structure, and biological activity were associated with a 1.7 t ha^{-1} yield increase in a rice crop following 10–year-old grass + legume plots that did not require any N fertilizer when compared with rice following a grass-alone pasture of the same age (Rao et al., 1992). These yield differences indicate that the technique of ^{13}C natural abundance in soil organic matter may prove to be a valuable aid in predicting the likely beneficial effects of a forage legume for subsequent production of pastures and/or field crops. Further, a grass-legume mixture improves distribution of crop production through time, less susceptibility to disease or lodging, and better balanced food for livestock (Cruz and Sinoquet, 1994; Thomas and Sumberg, 1995).

Botanically stable and productive grass-legume forage mixtures are often difficult to maintain because of the high degree of competition among their components (Jones et al., 1988). Grasses and legumes may compete for solar radiation, water, and nutrients when grown in mixture. Competition for solar radiation is often considered the most critical of these (Donald, 1961). Although legumes may have deeper root systems than grasses, they also exhibit lower water use efficiency (Chamblee, 1972; Snaydon, 1978; Haynes, 1980). The fibrous nature and low cation-exchange capacity (CEC) of grass roots give grasses an advantage over legumes in extracting monovalent cations from the soil (Haynes, 1980).

The roots of legumes generally have a CEC about twice that of grass roots. This may be a partial explanation for the poor ability of the legumes to compete with grasses for P, K, and S. Application of P, K, or S is generally likely to favor legumes at the expense of the grass components (Haynes, 1980). The difference in root systems of grasses and legumes is one reason

Table 18.13 Some Examples of Successful Grass-Legume Mixtures in Different Countries

Country	Grass species	Legume species
Australia	*Brachiaria mutica*	*Centrosema pubescens*
	Setaria sphacelata	*Desmodium intortum*
	Sorghum almun	*Medicago sativa*
Columbia	*Melinis minutiflora*	*Stylosanthes guyanensis*
Kenya	*Pennisetum clandestinum*	*Desmodium uncinatum*
Peru	*Panicum maximum*	*Pueraria phaseoloides*
United States	*Dactylis glomerata*	*Medicago sativa*
	Bromus inermis	*Medicago sativa*
	Phleum pratense	*Medicago sativa*
	Hemarthria altissima	*Aeschynomene americana*
Brazil	*Panicum maximum*	*Centrosema pubescens*
	Panicum maximum	*Pueraria phaseoloides*
	Brachiaria decumbens	*Centrosema pubsecens*
New Zealand	*Lolium perenne*	*Trifolium repens*

Sources: Compiled from Janick et al., 1969; Sanchez, 1976; Hutton, 1978; Kornelius et al., 1978; Haynes, 1980; Sollenberger et al., 1987; Casler, 1988.

for nutrient competition in a grass-legume mixture. Grass roots have more root hairs than legumes, and the effective root surface area for nutrient absorption is 3–8 times higher in grasses than in legumes (Barrow, 1975; Evans, 1977). Grasses have fibrous root systems and are generally concentrated in the top soil layers, whereas legumes have taproot systems and penetrate deeper into the soil. The immobile nutrients (e.g., P) are mostly concentrated in the surface soil layers, and it is likely that direct competition for phosphate ions occurs in the top soil with grasses having a physical advantage over legumes. If the soil is deficient in P, legumes are likely to undergo considerable P stress in mixture with grasses (Kendall and Stringer, 1985). The optimum ratio of tall fescue to white clover in the sward is about 2:1 (Leffel and Gibson, 1982).

The compatibility of grass and legume species is related to their growth habits and similar adaptation to the specific climatic, soil moisture, and soil fertility regime (Sanchez, 1976). Selection of proper grass and legume species is important for successful growth of grass-legume mixtures under given environmental conditions. Table 18.13 shows some examples of successful grass-legume mixtures in Colombia, Brazil, Peru, Kenya, Australia, the United States, and New Zealand. Other practices which may help to favor the legume include (Janick et al., 1969): 1) fairly close grazing (or mowing); 2) rotational grazing to provide a recovery period; 3) mowing and weed control of grass patches that become undergrazed; 4) making hay of growth that exceeds the capacity of livestock to consume; 5) using fertilizers rich in phosphate and potassium (legumes provide much of their own nitrogen), and 6) suppressing grass competition by physical or chemical means during the establishment of legumes sod-seeded into grass.

V. NUTRITIVE VALUE

The nutritive value of a forage is characterized by its chemical composition, digestibility, and the nature of the digested products (Mott, 1982). The most important criteria used to evaluate forage nutritive values are dry matter digestibility (DMD) and forage intake. Digestibility can be defined as the difference in value between the feed eaten and material voided by the animals, expressed as percentage of feed eaten (Crowder and Chheda, 1982). The dry matter digestibility can be calculated as:

$$DMD = \frac{DM \text{ eaten} - DM \text{ in feces}}{DM \text{ eaten} \times 100}$$

The voluntary intake (VI) is defined as the amount eaten in relation to the size of the animal.

Soil fertility level, season, growth stage, species, genotypes within species, and management practices affect the nutritive value of forages (Reid et al., 1970). The physiological processes of nutrient absorption, photosynthesis, and translocation do not proceed at similar rates throughout the growth cycle. Nitrogen, phosphorous, and potassium especially are absorbed most actively in the earlier stages of plant growth, while carbon assimilation and transpiration reach a maximum much later. As a consequence of the differential rates involved, the protein, phosphorus, and potassium content of most forages tends to decrease as the plant matures (Trumble, 1952). This is particularly marked in the case of grasses. In mixed pastures, grasses attain maximum nutritive value prior to the appearance of inflorescences, whereas legumes have their highest feeding value at or after this stage (Trumble, 1952).

The nutritive value of pasture species, even at similar stages of growth, varies widely in both DMD and VI. Minson and McLeod (1970) reviewed comparisons of over 1000 tropical and temperate grass samples and showed that tropical grasses were an average of 13% units lower in DMD. Most samples of temperate grasses had digestibilities above 65%, but few tropical grass samples were in that category. They suggested that lower DMD values of tropical grasses may be due in part to higher growing temperatures. But the data of Reid et al. (1973) support the view that for selected or improved species of tropical grasses (e.g., *Brachiaria*, *Chloris*, *Setaria*, and *Panicum*) DMD values are comparable to those of similarly managed temperate grasses. A detailed discussion of nutritive value of forage species is given by Crowder and Chheda (1982) and Van Soest (1982).

VI. NITROGEN FIXATION

The data on nitrogen fixation by forage legumes in grass-legume mixed stands are scarce. Most of the data available are for pure stands. This lack of information is unfortunate because, in practice, the forage legumes are often planted with grasses or become mixed with grasses over time. According to Larue and Patterson (1981), nitrogen fixation by alfalfa, white clover, red clover, and Korean lespedesa averages 212, 128, 154, and 193 kg N ha^{-1} annually, respectively. Carvalho (1986) compiled some data on N fixation by tropical pasture legumes and N transferred to associated grasses (Table 18.14).

Early studies of the Rhizobium-legume association revealed that there were many kinds of nodule bacteria and that various leguminous plants had their preferences (Fred et al., 1932). Table 18.15 shows the *Rhizobium* sp. which can form nodules with some important forage legumes. Leguminous plants mutually susceptible to nodulation by the same strains of rhizobia constitute a cross-inoculation group. The rhizobia capable of nodulating plants within a group are considered a species (Burton, 1972).

According to Burton (1972), the ability of rhizobia to induce nodulation is called infectiveness; N-fixing capacity indicates effectiveness. While nodulation is a prerequisite of growth enhancement, it does not assure it. Effects attributable to rhizobia are termed strain variation; those attributable to the plants are referred to as host specificity (Burton, 1972).

Nodulation of legumes requires inoculation of desired rhizobia into leguminous seeds or into soil. Normally, legume seeds are inoculated because of the ease and convenience of this method of implanting rhizobia in the rhizosphere. The number of rhizobia required for effective nodulation varies with seeds, form of inoculum, and environmental conditions. In Austria, inocula that provide 100 viable rhizobia per seed are considered satisfactory under normal conditions (Vincent, 1968). In New Zealand, inoculants supply approximately 1000 rhizobia per seed, whereas quality inoculants in the United States provide approximately 5000 rhizobia per alfalfa seed when used 2–3 months after treatment (Burton, 1972). The inoculants available for inoculation of legume seeds are in powder form, liquid or broth culture, or an oil-dried rhizobial preparation absorbed in pulverized vermiculite. The seeds can be inoculated either by sprinkle, slurry, or waterless methods.

Table 18.14 Estimated N Fixation by Tropical Pasture Legumes and Transfer to Associated Grasses

Species	N fixed[a] (kg ha^{-1} $year^{-1}$)	N transferred[b] to grasses (%)	Location	Reference
Centrosema mucunoides	90	20	Queensland, Australia (average of 4 yr)	Miller and Vanderlist, 1977
Centrosema pubescens	100	—	São Paulo, Brazil (average of 3 yr)	Mattos and Werner, 1979
Desmodium intortum	238	39	Hawaii, USA (average of 2 yr)	Whitney and Green, 1969
	172	20	Queensland, Australia (average of 4 yr)	Miller and Vanderlist, 1977
	100–140	17	Queensland, Australia (average of 5 yr)	Johansen and Kerridge, 1979
Galactia straita	122	—	São Paulo, Brazil (average of 3 yr)	Mattos and Werner, 1979
Lotononis bainesii	66	13	Queensland, Australia (average of 4 yr)	Jones et al., 1967
	51–74	12–15	Queensland, Australia (average of 5 yr)	Johansen and Kerridge, 1979
Macroptilium atropurpureum	135	39	Queensland, Australia (average of 4 yr)	Miller and Vanderlist, 1977
	100–140	12–15	Queensland, Australia (average of 5 yr)	Johansen and Kerridge, 1979
	85	—	São Paulo, Brazil (average of 3 yr)	Mattos and Werner, 1979
Neonotonia wightii	133	19	Queensland, Australia (average of 4 yr)	Miller and Vanderlist, 1977
	40	—	São Paulo, Brazil (average of 2 yr)	Paulino et al., 1983
Pueraria phaseoloides	99	14	Queensland, Australia (average of 4 yr)	Miller and Vanderlist, 1977
Stylosanthes guianensis	135	10	Queensland, Australia (average of 4 yr)	Miller and Vanderlist, 1977
	43	—	São Paulo, Brazil (average of 3 yr)	Mattos and Werner, 1979

[a] N fixed (NF) = $N_L + N_G^+ - N_G^-$, where N_L = N in legumes, N_G^+ = N in grass associated, and N_G^- = N in grass plot without legume.
[b] N transferred (NT) = $N_G^+ - N_G^-$.

Table 18.15 *Rhizobium* Species Associated with Nitrogen Fixation in Selected Forage Legumes

Legume species	*Rhizobium* species
Alfalfa	*Rhizobium meliloti*
Clover	*Rhizobium trifolii*
Crotalaria	*Rhizobium japonicum*
Kudzu	*Rhizobium japonicum*
Lathyrus	*Rhizobium leguminosarum*
Lespedeza	*Rhizobium japonicum*
Sweet clover	*Rhizobium meliloti*
Vetch	*Rhizobium leguminosarum*

VII. NUTRIENT REQUIREMENTS

Nutrient requirements of forage species vary with yield level, species, soil, and climatic conditions. In many parts of the world, forages are grown on land which is not suitable for grain crops. These soils are usually steep, erodible, infertile, or droughty. Forage grasses are able to thrive on such soils and respond favorably to good management practices. Fertilization of pastures offers an opportunity for improving forage production in many parts of the world.

Several metabolic disorders in cattle have been related to mineral imbalances in the forages they consume (Noller and Rhykerd, 1974; Reid and Jung, 1974). Most agronomists and plant breeders have concentrated on maximizing yield and persistence but have invested little effort in improving mineral concentrations of forages with respect to animal requirements.

The use of management practices and cultivars that maximize yield has contributed to mineral imbalances for animals because mineral requirements of plants and animals differ (Reid and Jung, 1974). Diet supplementation, choice of species, alternation of fertilizer application, and breeding for specific mineral concentrations are potential methods of correcting imbalances in cattle diets (Hill and Jung, 1975).

Grass tetany, a conditioned Mg deficiency, is a serious nutritional problem in various parts of the United States (Rendig and Grunes, 1979; Kubota et al., 1980). Incidence of tetany is generally reported to be low or absent in areas where grasses have 0.2% or more Mg (Kubota et al., 1980).

Important forage legumes such as alfalfa and clover can produce better yields where nutrients like P, K, Ca, and Mg are present in adequate amounts. Similarly, all the important forage grasses require an ample supply of N for high production.

High soil acidity is a known limitation to pasture development in tropical as well as temperate regions around the world (Andrew and Hegarthy, 1969; Recheigl et al., 1988). In acid soils, the major limiting factors include low pH, toxic concentrations of aluminum and/or manganese, and deficiencies of calcium, magnesium, phosphorus, and molybdenum. Liming is considered to be an essential practice in improving pasture production in acid soils. Many soils in areas of permanent grassland in England and Wales are acidic and require liming to achieve optimum production (Jarvis, 1984). Current advisory recommendations are that grassland soils should be limed to pH 6 as measured in a 1:2.5 suspension of soil in water (Jarvis, 1984).

Soil testing, along with plant analysis, helps in formulating a sound fertilization program for pasture species. In order to use soil testing as a criterion for fertilization, field calibration data are required for immobile nutrients and forage species in different agroclimatic regions and grass-legume mixtures.

A. Nutrient Concentration and Uptake

The basic principle of plant analysis is that the chemical composition of the plant or its parts reflects the adequacy of the nutrient supply in relation to growth requirements (Martin and Matocha, 1973), but nutrient supply to a growing plant is influenced by several environmental and plant factors which influence chemical composition. Therefore, although the values of nutrient concentration and nutrient removal presented in this section may provide some guidelines, they cannot necessarily be used for all agroclimatic conditions. Table 18.16 shows adequate levels of nutrients for two legumes and two grasses, and Table 18.17 shows concentrations of P, K, Ca, Mg, and S for tropical pasture species. Similarly, Table 18.18 shows dry matter production and nutrient removal by selected forage species under Brazilian conditions, and shows that N and K are elements which are removed by forage species in large amounts. This means that these two nutrients should be given special attention in pasture management. In pasture testing, it is extremely important that plant samples be taken at the proper growth stage for chemical analysis. According to Martin and Matocha (1973), alfalfa and clover plant samples should be taken at or near the early bloom stage or at the time when regrowth sprouts begin to appear. This may be true for other forage species, too.

In many acid soils, aluminum is known to inhibit root growth and mineral uptake by roots. Inter- and intraspecific differences in plant growth and mineral composition in reference to Al have been well documented in the literature (Foy, 1984). Table 18.19 shows the percent inhibition of essential elements in various species of legumes and grasses due to the presence of 100 μM Al in the growth medium. In all these forage species, Al inhibited the uptake of essential nutrients. Table 18.20 highlights the effects of different levels of Al

Table 18.16 Adequate Nutrient Concentrations in Two Legume and Two Grass Species

Species, growth stage, and plant part	g kg^{-1}						mg kg^{-1}					
	N	P	K	Ca	Mg	S	Fe	Mn	Z	Cu	B	Mo
Alfalfa, early flowering, whole tops	35–50	2.5–4.0	20–35	10–20	2.5–5.0	2.5–4.0	45–60	25–30	15–40	5–15	25–60	> 0.2
White clover, preflowering, whole tops	48–55	3.5–4.0	11–20	> 11	1.8–2.2	2.0–3.0	50–65	25–30	16–19	6–7	25–30	0.15–0.20
Perennial ryegrass, preflowering, leaf	25–40	2.0–4.0	15–30	2.5–6.0	2.0–3.5	2.0–4.0	50–60	50–300	15–50	5–12	5–15	0.3–0.4
Bermudagrass, 4–5 weeks, whole tops between clippings	25–30	2.6–3.2	18–21	—	—	1.5–2.0	—	—	—	—	—	—

Sources: Compiled from Martin and Matocha, 1973; Smith, 1986.

Table 18.17 Macronutrient Concentrations of Some Selected Tropical Pasture Species Under Brazilian Conditions

		g kg^{-1}				
Species		P	K	Ca	Mg	S
Grasses	*Panicum maximum*	1.7	11.5	6.0	2.0	1.5
	Hyparrhenia rufa	1.6	10.6	3.4	2.2	1.4
	Andropogon gayanus	1.0	9.5	2.3	1.3	1.3
	Brachiaria brizantha	0.9	8.2	3.7	2.4	1.2
	Brachiaria dictyoneura	1.3	9.8	2.5	2.2	1.2
	Brachiaria decumbens	0.8	8.3	3.7	2.1	1.2
	Brachiaria humidicola	0.8	7.4	2.2	1.6	1.1
Legumes	*Stylosanthes humilis*	2.7	6.0	20.0	2.5	1.4
	Stylosanthes guianensis	1.6	8.2	8.5	3.0	1.4
	Stylosanthes capitata	1.2	11.3	9.7	2.2	1.2
	Stylosanthes macrocephala	1.0	9.3	7.8	2.0	1.4
	Pueraria phaseoloides	2.2	12.2	10.4	2.0	1.7
	Centrosema pubescens	1.8	14.0	9.8	2.4	1.6
	Centrosema macrocarpum	1.6	12.4	7.2	2.2	1.6
	Zornia latifolia	1.2	11.6	8.2	2.0	1.4
	Desmodium ovalifolium	1.0	10.3	7.4	2.1	1.2

Source: Veiga and Falesi, 1986.

Table 18.18 Dry Matter Production and Corresponding Macronutrients Removed by Selected Forage Species

		kg ha^{-1}					
Species	Dry matter (t ha^{-1})	N	P	K	Ca	Mg	S
Alfalfa	15	335	30	207	40	—	—
Red clover	10	160	20	108	24	—	—
Napiergrass	25	302	64	504	96	63	75
Bermudagrass	25	570	63	308	—	—	—
Pangolagrass	24	299	47	358	109	67	45
Paragrass	24	307	43	383	115	79	—
Colonial grass	23	288	44	363	149	99	45

Source: Malavolta et al., 1986.

Table 18.19 Nutrient Uptake Inhibition in Various Legumes and Grasses by Presence of 100 μm Al in the Growth Medium

	Percent inhibition of uptake (PI)[a]					
	Legumes			Grasses		
Element	Alfalfa	Birdsfoot trefoil	Red clover	Orchard grass	Tall fescue	Timothy
P	98	86	61	88	70	58
K	95	72	–29	73	42	19
Mg	95	84	62	97	88	87
Cu	94	81	96	93	62	58
Fe	96	88	85	92	70	30
Mn	97	87	63	94	68	73
Zn	94	72	51	91	52	50

[a] $PI = \left[\frac{\text{uptake at } 0\mu\text{M Al} - \text{uptake at } 100\mu\text{M Al}}{\text{uptake at } 0\mu\text{M Al}}\right] \times 100$

Sources: Baligar et al., 1988; Baligar and Smedley, 1989.

Table 18.20 Uptake (U) and Percent Inhibition (PI) of Nutrients in Two Red Clover Cultivars at 0, 50, and 100 μM Al Levels

	Altaswede[a]			Kenstar[a]		
	U[b]	PI		U	PI	
Element	0 μM Al	50 μM Al	100 μM Al	0 μM Al	50 μM Al	100 μM Al
P	6.0	70	92	7.0	44	88
S	5.6	61	91	6.5	51	90
K	58.4	51	87	50.3	32	81
Ca	62.5	61	97	84.1	63	96
Mg	8.8	65	94	9.4	56	93
Mn	177.4	60	95	171.7	53	93
Fe	199.0	50	91	261.2	54	92

[a] Altaswede is a mammoth or single-cut Canadian cultivar; Kenstar is a medium- or double-cut U.S. cultivar.

[b] U = mg 10 plants^{-1} for P, S, K, Ca, Mg, and μg 10 plants^{-1} for Mn, Fe.

Source: Baligar et al., 1987.

in the growth medium on the uptake and inhibition of uptake of various elements in two morphologically different red clover cultivars. In both red clover cultivars, increasing the Al levels in the growth medium inhibited the various essential nutrients. Such reductions in uptake are also due to reduction in dry matter accumulation of the shoot. Plant demand for any given nutrient is a function of the plant's internal ionic concentrations and rate of growth. Aluminum is known to interfere with the uptake, transport, and use of several essential elements (Foy, 1984).

VIII. SUMMARY

Forages play an important role in world food resources by supplying feed to livestock and consequently meat and milk for human consumption. A grassland ecosystem is an excellent example of a renewable resource, and, if properly managed, the system may be productive over a very long time. A first principle of modern pasture establishment is the association of grass with legume in any pasture to improve the quality of pasture forages. Grass-legume associations are used in many parts of the world because a greater total herbage yield may be obtained by growing a grass and a legume in association rather than individual swards, where no fertilizer N is applied. The use of legumes in pasture may also result in increased N content, greater digestibility, and a higher well-balanced mineral content of herbage, all of which are of importance in animal nutrition. Grass-legume forage mixtures also have other advantages compared to pure grasses or pure legumes, including erosion and weed control and increased stand longevity. Due to the competition of grasses with legumes in a mixture, some management is needed to maintain this mixture successfully. It is necessary to select a good combination of grasses and legumes appropriate to the climatic and soil environment, fertilizer treatments, and grazing management.

In Brazil, National Rice and Bean Research Center of EMBRAPA developed a technology of improving degraded pastures on rainforest cleared lands by planting upland rice in association with *Brachiaria* brizanta. Approximately 5 kg of *Brachiaria* seeds are mixed with fertilizers and placed 8–10 cm deep, with upland rice about 3–5 cm at the top. The grass germination is delayed, and in this way rice escapes from competition with grass. When rice is mature, it is harvested along with *Brachiaria.* The grass regenerates within about 60–70 days. In this way the farmers can get extra income from upland rice and at the same time establish pastures for animal production. A detailed description of this technology is given by Kluthcouski et al. (1991).

REFERENCES

Ahlgren, G. H. 1956. Forage Crops, 2nd edition. McGraw-Hill, New York.

Alexander, C. W. and D. E. McCloud. 1962. CO_2 uptake (net photosynthesis) as influenced by light intensity of isolated bermudagrass leaves contrasted to that of swards under various clipping regimes. Crop Sci. 2: 132–135.

Andrew, C. S. 1978. Mineral characterization of tropical forage legumes, pp. 93–111. *In*: C. S. Andrew and E. J. Kamrath (eds.). Mineral Nutrition of Legumes in Tropical and Subtropical Soils. CSIRO, Melbourne, Australia.

Andrew, C. S. and M. P. Hegarthy. 1969. Comparative responses to manganese excess of eight tropical and four temperate pasture legume species. Aust. J. Agric. Res. 20: 687–696.

Baligar, V. C. and M. D. Smedley. 1989. Responses of forage grasses to aluminum in solution culture. J. Plant Nutr. 12: 783–796.

Baligar, V. C., R. J. Wright, T. B. Kinraide, C. D. Foy, and J. H. Elgin, Jr. 1987. Aluminum effects on growth, mineral uptake, and efficiency ratios in red clover cultivars. Agron. J. 79: 1038–1044.

Baligar, V. C., R. J. Wright, N. K. Fageria, and C. D. Foy. 1988. Differential response of forage legumes to aluminum. J. Plant Nutr. 11: 549–561.

Barrow, N. J. 1975. The response to phosphate of two annual pasture species. I. Effect of the soil's ability to absorb phosphate on comparative phosphate requirement. Aust. J. Agric. Res. 26: 137–143.

Bennett, O. L., V. C. Baligar, R. J. Wright, and H. D. Perry. 1986. Selection and evaluation of forage plant genotypes for low pH and high aluminum tolerance, pp. 176–183. *In*: F. M. Borba and J. M. Abreu (eds.) European Grassland. Paper presented at European Grassland Federation, 11th General Meeting, Troia, Portugal.

Black, C. C. 1971. Ecological implications of plants into groups with distinct photosynthetic production capacities. Adv. Ecol. Res. 7: 87–114.

Brougham, R. W. 1960. The relationship between the critical leaf area, total chlorophyll content, and maximum growth-rate of some pasture and crop plants. Ann. Bot. 24: 463–474.

Brown, R. H. and G. T. Byrd. 1990. Yield and botanical composition of alfalfa-bermudagrass mixtures. Agron. J. 82: 1074–1079.

Bula, R. J. 1982. Climatic factors in forage production, pp. 372–2. *In*: M. E. Heath, D. S. Metcalfe, and R. F. Barnes (eds.). Forages: The Science of Grassland Agriculture, 3rd edition. Iowa State University Press, Ames.

Burton, J. C. 1972. Nodulation and symbiotic nitrogen fixation, pp. 229–246. *In*: C. H. Hanson (ed.). Alfalfa Science and Technology. Monogr. 15, Am. Soc. Agron., Madison, Wisconsin.

Carvalho, M. M. 1986. Fixacao biologica como fonte de nitrogenio para pastagens, pp. 125–143. *In*: H. B. Mattos, J. C. Werner, T. Yamada, and E. Malavolta (eds.). Calagem e Adubacao de Pastagens. Associacao Brasileira Para Pesquisa de Potassa e Fosfato, Piracicaba, Brazil.

Casler, M. D. 1988. Performance of orchardgrass, smooth bromegrass, and ryegrass in binary mixtures with alfalfa. Agron. J. 80: 509–514.

Chamblee, D. S. 1972. Relationships with other species in a mixture. *In*: C. H. Hanson (ed.). Alfalfa Science and Technology. Agronomy 15: 211–228.

CIAT (Centro Internacional de Agricultura Tropical). 1978. Beef program. *In*: Annual Report 1978. Cali, Colombia. pp. B1–B174.

CIAT (Centro Internacional de Agricultura Tropical). 1980. Tropical pasture programs. *In*: Annual Report 1979. Cali, Colombia. 187 p.

Cochrane, T. T., L. G. Sanchez, J. A. Porras, L. G. Azevedo, and L. C. Garver. 1985. Land in tropical America. Centro de Pesquisa Agropecuaria dos Cerrados (CPAC), Emprisa Brasileira de Pesquisa Agropecuaria (EMBRAPA and CIAT), Cali, Colombia.

Crowder, L. V. and H. R. Chheda. 1982. Tropical Grassland Husbandry. Longman, London.

Cruz, M. C. P. 1994. Effect of liming on forage grass production. Pesq. Agropec. Brasileira, Brasilia 29: 1303–1312.

Cruz, P. A. and H. Sinoquet. 1994. Competition for light and nitrogen during a regrowth cycle in a tropical forage mixture. Field Crops Res. 36: 21–30.

Donald, C. M. 1961. Competition for light in crops and pastures. Symp. Soc. Exp. Biol. 15: 282–313.

Eagles, C. F. and D. Wilson. 1982. Photosynthetic efficiency and plant productivity, pp. 213–247. *In*: M. Recheigl, Jr. (ed.). Handbook of Agricultural Productivity. CRC Press, Boca Raton, Florida.

Evans, P. S. 1977. Comparative root morphology of some pasture grasses and clovers. N.Z. J. Agric. Res. 20: 331–335.

Evers, G. W. 1985. Forage and nitrogen contributions of arrowleaf and subterranean clovers overseeded on bermudagrass and bahiagrass. Agron. J. 77: 960–963.

Farnham, D. E. and J. R. George. 1993. Dinitrogen fixation and nitrogen transfer among red clover cultivars. Can. J. Plant Sci. 73: 1047–1054.

Foy, C. D. 1984. Physiological effects of hydrogen, aluminum and manganese toxicities in acid soil, pp. 57–97. *In*: F. Adams (ed.). Soil Acidity and Liming, 2nd edition. Monogr. 12, Am. Soc. Agron., Madison, Wisconsin.

Fred, E. B., L. L. Baldwin, and E. McCoy. 1932. Root nodule bacteria and leguminous plants. Studies in Adv. Science, No. 5, University of Wisconsin Press, Madison.

George, J. R. 1984. Grass sward improvement by frost-seeding with legumes, pp. 265–269. *In*: Proc. Am. Forage Grassl. Conf., Houston, Texas. Am. Forage and Grassl. Council, Georgetown, Texas.

Gettle, R. M., J. R. George, K. M. Blanchet, D. R. Buxton, and K. J. Moore. 1996. Frost-seeding legumes into established switchgrass: Forage yield and botanical composition of the stratified canopy. Agron. J. 88: 555–560.

Haynes, R. J. 1980. Competitive aspects of the grass-legume association. Adv. Agron. 33: 227–261.

Heath, M. E. 1982. Grassland agriculture, pp. 13–20. *In*: M. E. Heath, D. S. Metcalfe, and R. F. Barnes (eds.). Forages: The Science of Grassland Agriculture, 3rd edition. Iowa State University Press, Ames.

Heinrichs, D. H. 1963. Creeping alfalfas. Adv. Agron. 15: 317–337.

Hill, R. R., Jr. and G. A. Jung. 1975. Genetic variability for chemical composition of alfalfa. I. Mineral elements. Crop Sci. 15: 652–657.

Hino, M., K. Fujita, and H. Shutto. 1987. A laboratory experiment on the role of grass for infiltration and runoff processes. J. Hydrol. 90: 303–325.

Hsiao, T. C. 1973. Plant responses to water stress. Annu. Rev. Plant Physiol. 24: 519–570.

Humphreys, L. R. 1966. Subtropical plant growth. 2. Effects of variation in leaf area index in the field. Queensland J. Agric. Anim. Sci. 23: 337–358.

Hutton, E. M. 1978. Problems and successes of legume-grass pastures, especially in tropical Latin America, pp. 81–93. *In*: P. A. Sanchez and L. E. Tergas (eds.). Pasture Production in Acid Soils of the Tropics. CIAT, Cali, Colombia. 1969.

Janick, J., R. W. Schery, F. W. Woods, and V. W. Ruttan. Plant science: an introduction to world crops, 2nd ed. Freeman, San Francisco.

Jarvis, S. C. 1984. The response of two genotypes of white clover to addition of lime to an acid permanent grassland soil. J. Sci. Food Agric. 35: 1149–1158.

Johansen, C. and P. C. Kerridge. 1979. Nitrogen fixation and transfer in tropical legume-grass swards in southeastern Queensland. Tropical Grassland, St. Lucia 13: 165–170.

Jones, C. A. 1985. C_4 Grasses and Cereals: Growth, Development, and Stress Response. Wiley, New York.

Jones, R. J., D. J. Griffiths, and R. B. Waite. 1967. The contribution of some tropical legumes to pasture yield of dry matter and nitrogen at Sanford, southern Queensland. Aust. J. Exp. Agric. Anim. Husb. 7: 57–65.

Jones, T. A., L. T. Carlson, and D. R. Buxton. 1988. Reed canarygrass binary mixture with alfalfa and birdsfoot trefoil in comparison to monocultures. Agron. J. 80: 49–55.

Kendall, W. A. and W. C. Stringer. 1985. Physiological aspects of clover, pp. 111–159. *In*: N. L. Taylor (ed.). Clover Science and Technology. Monogr. 25, Am. Soc. Agron., Madison, Wisconsin.

Kluthcouski, J., A. R. Pacheco, S. M. Teixeira, and E. T. Oliveira. 1991. Renovation of pasture in cerrado with upland rice. EMBRAPA-CNPAF Document 33, Goiania-Goias, Brazil.

Kornelius, E., M. G. Saueressig, and W. J. Goedert. 1978. Pastures establishment and management in the cerrado, pp. 147–166. *In*: P. A. Sanchez and L. S. Tergas (eds.). Pasture Production in Acid Soils of the Tropics. CIAT, Cali, Colombia.

Kubota, J., G. H. Oberly, and E. A. Naphan. 1980. Magnesium in grasses of three selected regions in the United States and its relation to grass tetany. Agron. J. 72: 907–914.

Larue, T. A. and T. G. Patterson. 1981. How much nitrogen do legumes fix? Adv. Agron. 34: 15–38.

Lavelle, P. 1988. Earthworm activities and the soil system. Biol. Fert. Soils 6: 237–251.

Leffel, R. C. and P. B. Gibson. 1982. White clover, pp. 167–176. *In*: M. E. Heath, D. S. Metcalfe, and R. F. Barnes (eds.). Forages: The Science of Grassland Agriculture, 3rd edition. Iowa State University Press, Ames.

Maas, E. V. 1986. Salt tolerance of plants. Appl. Agric. Res. 1: 12–26.

Malavolta, E., T. H. Liem, and A. C. P. A. Primavesi. 1986. Exigeneias nutriconais das plantas forrageiras, pp. 31–76. *In*: H. B. Mattos, J. C. Werner, T. Yamada, and E. Malavolta (eds.). Calagem e Adubacao de Pastagens. Associacao Brasileira para pesquisa da Potassa e do Fosfato, Piracicaba, São Paulo.

Marten, G. C. 1989. Summary of the trilateral workshop on persistence of forage legumes, pp. 569–572. *In*: G. C. Marten et al. (eds.) Persistence of Forage Legumes. Proc. Trilateral Workshop, Honolulu, Hawaii, July 1988. ASA, CSSA, and SSSA, Madison, Wisconsin.

Martin, W. E. and J. E. Matocha. 1973. Plant analysis as an aid in the fertilization of forage crops, pp. 393–426. *In*: L. M. Walsh and J. D. Beaton (eds.). Soil Testing and Plant Analysis. Soil Sci. Soc. Am., Madison, Wisconsin.

Mattos, H. B. and J. C. Werner. 1979. Efeito do nitrogenio mineral e de leguminosae sobre a producao do capim coloniao (Panicum maximum Jacq.). Bol. Ind. Anim. Nova Odessa 36: 147–156.

Metcalfe, D. S. 1982. The botany of grasses and legumes, pp. 80–97. *In*: M. E. Heath, D. S. Metcalfe, and R. F. Barnes (eds.). Forages: The Science of Grassland Agriculture, 3rd edition. Iowa State University Press, Ames.

Miles, J. W. and S. L. Lapointe. 1992. Regional germplasm evaluation: A portfolio of germplasm options for the major ecosystems of tropical America, pp. 9–28. *In*: Pasture for the Tropical Lowlands. Centro Internacional de Agricultura Tropical (CIAT), Cali, Colombia.

Miller, C. P. and J. T. Vanderlist. 1977. Yield, nitrogen uptake, and live-weight gains from irrigated grass-legume pasture on a Queensland tropical highland. Aust. J. Exp. Agric. Husb. 17: 949–960.

Minson, D. J. and M. N. McLeod. 1970. The digestibility of temperate and tropical grasses. Proc. 11th Int. Grassland Congr., Surfers Paradise, Australia.

Mott, G. O. 1982. Evaluating forage production, pp. 126–135. *In*: M. E. Heath, D. S. Metcalfe, and R. F. Barnes (eds.). Forages: The Science of Grassland Agriculture, 3rd edition. Iowa State University Press, Ames.

Newman, P. R. and L. E. Moser. 1988. Grass seedling emergence, morphology, and establishment as affected by planting depth. Agron. J. 80: 383–387.

Noller, C. H. and C. L. Rhykerd. 1974. Relationship of nitrogen fertilizer and chemical composition of forage to animal health and performance, pp. 363–394. *In*: D. A. Mays (ed.). Forage Fertilization. Am. Soc. Agron., Madison, Wisconsin.

Paulino, V. T., P. G. Cunha, H. B. Mattos, and G. Bufarah. 1983. Estimativa do potencial de forneicimento de nitrogenio para a dieta animal atraves de leguminosas forrageiras. Zootecnica, Nova Odessa 21: 135–150.

Posler, G. L., A. W. Lenssen, and G. L. Fine. 1993. Forage yield, quality, compatibility, and persistence of warm-season grass-legume mixtures. Agron. J. 85: 554–560.

Purseglove, J. W. 1985. Tropical Crops: Monocotyledons. Longman, London.

Rao, I. M., M. A. Ayarza, R. J. Thomas, M. J. Fisher, J. I. Sanz, J. M. Spain, and C. E. Lascano. 1992. Soil-plant factors and processes affecting productivity in Ley farming, pp. 145–175. *In*: CIAT (ed.) Pastures for the Tropical Lowlands: CIAT's Contribution. Cali, Colombia.

Recheigl, J. E., K. L. Edmisten, D. D. Wolf, and R. B. Reneau, Jr. 1988. Response of alfalfa grown on acid soil to different chemical amendments. Agron J. 80: 515–518.

Reid, R. L. and G. A. Jung. 1974. Effects of elements other than nitrogen on the nutritive value of forage, pp. 395–435. *In*: D. A. Mays (ed.). Forage Fertilization. Am. Soc. Agron., Madison, Wisconsin.

Reid, R. L., A. J. Post, and G. A. Jung. 1970. Mineral composition of forages. West Virginia Agric. Exp. Stn. Bull. 589T.

Reid, R. L., A. J. Post, F. J. Olsen, and J. S. Mugerwa. 1973. Studies on the nutritional quality of grasses and legumes in Uganda. I. Application of in vitro digestibility techniques to species and stage of growth effects. Trop. Agric. (Trinidad) 54: 119–123.

Rendig, V. V. and D. L. Grunes. 1979. Grass Tetany. Spec. Publ. 35, Am. Soc. Agron., Madison, Wisconsin.

Rodrigues, L. R. A. and T. J. D. Rodrigues. 1987. Ecofisologia de plantas forrageiras, pp. 203–230. *In*: P. R. C. Castro, S. O. Ferreira, and T. Yamada (eds.). Ecofisologia da Producao Agricola. Associacao Brasileira para pesquisa da Potassa e do Fosfato, Piracicaba, São Paulo.

Rotar, P. P. and A. E. Kretschmer, Jr. 1985. Tropical and subtropical forages, pp. 154–165. *In*: M. E.Heath and D. S. Metcalfe (eds.). Forages; The Science of Grassland Agriculture, 4th edition. Iowa State University Press, Ames.

Sanchez, P. A. 1976. Soil management for tropical pasture production, pp. 533–605. *In*: Properties and Management of Soils in the Tropics. Wiley, New York.

Smith, F. W. 1986. Pasture species, pp. 100–119. *In*: D. J. Reuter and J. B. Robinson (eds.). Plant Analysis: An Interpretation Manual. Inkata Press, Melbourne.

Snaydon, R. W. 1978. Genetic changes in pasture population, pp. 253–269. *In*: J. R. Wilson (ed.). Plant Relations in Pastures. Brisbane, Australia. CSIRO, East Melbourne, Australia.

Sollenberger, L. E., J. E. Moore, K. H. Quesenberry, and P. T. Beede. 1987. Relationships between canopy botanical composition and diet selection in aeschynomene-limpograss pastures. Agron. J. 79: 1049–1054.

Taylor, T. H. and L. T. Jones, Jr. 1983. Compatibility of switchgrass with three sod-seeded legumes, p. 15. *In*: Progress Report, Kentucky Agric. Expt. Stn., University Kentucky, Lexington.

Thomas, D. and J. E. Sumberg. 1995. A review of the evaluation and use of tropical forage legumes in sub-Saharan Africa. Agriculture, Ecosystems and Environment 54: 151–163.

Trewartha, G. T., A. H. Robinson, and E. H. Hammond. 1967. Elements of Geography, 5th edition. McGraw-Hill, New York.

Trumble, H. C. 1952. Grassland agronomy in Australia. Adv. Agron. 4: 1–65.

Van Soest, P. J. 1982. Composition and nutritive value of forages, pp. 53–63. *In*: M. E. Heath, D. J. Metcalfe, and R. F. Barnes (eds.). Forages: The Science of Grassland Agriculture, 3rd edition. Iowa State University Press, Ames.

Veiga, J. B. and I. C. Falesi. 1986. Recomendacao e pratica de adubacao de pastagens cultivados na amazonia brasileira, pp. 257–282. *In*: H. B. Mattos, J. C. Werner, T. Yamada, and E. Malavolta (eds.). Calagem e Adubacao de Pastagens. Assoc. Brasileira para pesquisa de Potassa e do Fosfato, Piracicaba, São Paulo.

Vincent, J. M. 1968. Basic considerations affecting the practice of legume seed inoculation, pp. 195–158. *In*: Testkrift til Hans Lauritz Jensen. Publ. Gadgaard Nielsens Bogtrykkeri Lemvig, Denmark.

Whitney, A. S. and R. E. Green. 1969. Legume contribution to yields and compositions of *Desmodium* sp. Pangola grass mixtures Agron. J. 61: 741–746.

Wilson, D. 1975. Variation in leaf respiration in relation to growth and photosynthesis of *Lolium*. Ann. Appl. Biol. 80: 323–338.

Wilson, D. 1982. Response to selection for dark respiration rate of mature leaves in *Lolium perenne* and its effects on growth of young plants and simulated swards. Ann. Bot. 49: 303–312.

Zeigler, R. S., S. Pandey, J. Miles, L. M. Gourley, and S. Sarkarung. 1995. Advances in the selection and breeding of acid-tolerant plants: Rice, maize, sorghum and tropical forages, pp. 391–406. *In*: R. A. Date, M. J. Grundon, G. E. Rayment, and M. E. Probert (eds.) Plant-Soil Interactions at Low pH: Principles and Management. Kluwer Academic Publishers, Dordrecht, The Netherlands.

Index